Dav

Fire Department Incident Safety Officer

REVISED THIRD EDITION

JONES & BARTLETT
LEARNING

Jones & Bartlett Learning
World Headquarters
25 Mall Road
Burlington, MA 01803
978-443-5000
info@jblearning.com
www.jblearning.com
www.psglearning.com

Jones & Bartlett Learning books and products are available through most bookstores and online booksellers. To contact the Jones & Bartlett Learning Public Safety Group directly, call 800-832-0034, fax 978-443-8000, or visit our website, www.psglearning.com.

Substantial discounts on bulk quantities of Jones & Bartlett Learning publications are available to corporations, professional associations, and other qualified organizations. For details and specific discount information, contact the special sales department at Jones & Bartlett Learning via the above contact information or send an email to specialsales@jblearning.com.

20510-7

Production Credits
General Manager and Executive Publisher: Kimberly Brophy
VP, Product Development: Christine Emerton
Senior Managing Editor: Donna Gridley
VP, Sales, Public Safety Group: Phil Charland
Executive Editor: Bill Larkin
Senior Development Editor: Janet Morris
Editorial Assistant: Alexander Belloli
Project Specialist: Meghan McDonagh
Senior Marketing Manager: Brian Rooney

Product Fulfillment Manager: Wendy Kilborn
Composition: S4Carlisle Publishing Services
Cover Design: Kristin E. Parker
Rights & Media Research Assistant: Robert Boder
Media Development Editor: Shannon Sheehan
Cover Image: Courtesy of David Dodson
Printing and Binding: Lakeside Book Company
Cover Printing: Lakeside Book Company
Unless otherwise noted, all interior photographs courtesy of David Dodson.

To order this product, use ISBN: 978-1-284-21655-4

The Library of Congress has cataloged the first printing as follows:
Dodson, David W.
  Fire department incident safety officer / David W. Dodson. -- Third edition.
    pages cm
  Includes index.
  ISBN 978-1-284-04195-8
1. Fire extinction--Safety measures. 2. Fire fighters. 3. Safety engineers. I. Title.
  TH9182.D63 2015
  628.9'250289--dc23
                    2015017515

6048
Printed in the United States of America
28 27 26 25 24   10 9 8 7 6

Dedication

This book is for the individual who works while others sleep; whose body is coated in soot and sweat; who pledges to help those imperiled; who resolves to improve with each experience; who leverages knowledge to gain an edge; who strives for intellectual growth; who values team effort and honors the sacrifices of those who have gone before; and who is selflessly willing to risk death for the triumph of life.

These collective attributes will ultimately make a difference in the fire service. Therefore, I dedicate this book to that individual who tirelessly protects us all: *the firefighter*.

Brief Contents

Contents

About the Author

Dave Dodson

Chief Dodson is a 41-year fire service veteran, with 25 of those years as a responder. He began his fire service career with the U.S. Air Force, after which he spent almost 7 years as a fire officer and training/safety officer for the Parker Fire District in Parker, Colorado (now South Metro Fire Rescue Authority). He became the first career training officer for Loveland Fire and Rescue (CO), serving in many functions, including engine officer, hazmat technician, duty safety officer, and emergency manager for the city. He was a shift battalion chief for the Eagle River Fire District in Colorado before starting his current company, Response Solutions, which is dedicated to teaching firefighter safety and practical incident handling. He has presented more than 900 classes to 80,000 firefighters across the United States and Canada.

Chief Dodson authored the first two editions of this book as well as Fire Engineering's three-volume DVD training series, *The Art of Reading Smoke*. In 2014, he teamed with Battalion Chief John Mittendorf (Los Angeles Fire) to coauthor *The Art of Reading Buildings* (PennWell Corporation).

Chief Dodson has served as Chairman of the NFPA 1521 Task Group (Fire Department Safety Officer) and currently serves as a member of NFPA's Fire Service Occupational Safety and Health Technical Committee.

Chief Dodson continues to assist the National Institute for Occupational Safety and Health (NIOSH) as a subject matter expert for the review of Line of Duty Death technical investigative reports. Likewise, he has assisted several law firms as a topic expert in matters concerning firefighter safety. He is also a past president of the Fire Department Safety Officers Association. In 1997, the International Society of Fire Service Instructors honored Chief Dodson as the George D. Post Fire Instructor of the Year. In 2017, he was honored as the recipient of the first-ever "Lifetime Achievement Award" by the Fire Department Safety Officers Association.

Acknowledgments

Writing a book such as this can never be a solo project, and I wish I had the memory cells to acknowledge everyone who planted a seed, offered insights and suggestions, or made a willing contribution. I will probably forget someone, but I want to list and offer a colossal thank-you to those who opened a door, provided mentoring, or challenged me to raise the bar.

The Fire Department Safety Officers Association (FDSOA) adopted the first and second editions of this book as the primary study reference for their National Fire Service Professional Qualifications Board Certification for the Incident Safety Officer—Fire Suppression position. Thank you to the elected FDSOA board members, membership, and certification candidates for your continued support. Also a special thanks to Rich Marinucci and Linda Stone, whose efforts have helped reenergize the FDSOA. Your commitment, support, and friendship are deeply appreciated.

To those who have left distinct fingerprints on this manuscript, I owe more than a thank you:

- The late David Ross (Chief of Safety, Toronto Fire Services, Canada)
- Chief David C. Comstock (Western Reserve Joint Fire District, OH)
- The late Chief Alan Brunacini (Phoenix Fire)
- Chief Billy Goldfeder (Loveland-Simms, Ohio)
- Chief Bobby Halton (editor, *Fire Engineering* magazine)
- Battalion Chief John Mittendorf (retired, Los Angeles Fire)
- Chief Sam Phillips (retired, Clallam County Fire District 2, WA)
- Deputy Chief Bill Shouldis (retired, Philadelphia Fire, PA)

Thanks also to those outstanding fire officers and associates who continually support my efforts and are always willing to answer any of my silly questions or help me with an issue or project: Brian Kazmierzak, Dan Madrzykowski, Steve Kerber, Mike West, Mike Gagliano, Phil Jose, Bruce Clark, Ric Jorge, Rick Davis, Murrey Loflin, and Greg Ward.

Fire Chief Forest Reeder (Tinley Park, IL) became a co-conspirator for many of the updates in this book—we spent many hours brain-storming at pubs all over North America. In addition to writing the chapter-opening case studies, Forest assisted with NFPA JPR interpretations and NIOSH technical report research. Thanks, Brother!

The project team at Jones & Bartlett Learning helped take this text to a new level! I am grateful for the efforts of Kim Brophy, Janet Maker, Janet Morris, and the design team for creating an attractive and user-friendly resource. Thank you also goes to those individuals and fire departments that helped with photographic contributions: Keith Muratori, Kim Fitzsimmons, Corey Kneedler, James Grimmett, Justin King, Loveland Fire & Rescue Authority (CO), Fort Lauderdale Fire (FL), New Orleans Fire (LA), Frisco Fire (TX), Clallam County Fire District 2 (WA), and the Superstition Fire & Medical District (AZ). Additional thanks to the contributors and review team.

My personal support system helps me navigate travel and business needs and keeps me driving forward. Thanks to Becky Stafford, Donna Davis, Carol Lanza, and Karen Boor. Finally, and most importantly, I would like to thank my loving family. My daughter, Kelsie, and son, Dan, are intelligent and energetic young adults who make my life a joy. My best friend and the mother of our children, LaRae, is an inspiration with her unwavering love and ever-positive spirit.

Contributors

Anthony Kastros
Battalion Chief (ret.)
Sacramento Metro Fire District
Sacramento, California

Stephen Raynis
Chief of the Bureau of Training (ret.)
Fire Department, City of New York
Brooklyn, New York

Shadd A. Whitehead
Fire Chief (ret.)
Livonia Fire and Rescue
Livonia, Michigan

Reviewers

Patrick Beckley
Fire Training Officer 2
Ohio Fire Academy
Reynoldsburg, Ohio

Rick Best
Eastern Advocate Manager, National Fallen Fire-
fighters Foundation
Pataskala, Ohio

David L. Bullins
Department Chair, Fire Protection Technology and
Emergency Management
Guilford Technical Community College
Jamestown, North Carolina

Craig Dart, B.A. CMM III
Division Chief, Quality Management and
Accreditation
Toronto Fire Services
Toronto, Ontario

Jennifer Henson
Director, Fire Protection Technology and Fire-
fighter Certification
Catawba Valley Community College
Hickory, North Carolina

Matthew Hull
Lieutenant/Training Officer
Athens Fire Department
Athens, Ohio

Richard A. Marinucci
Executive Director
Fire Department Safety Officers Association
Ann Arbor, Michigan

Gary D. Peaton
Industrial Program Coordinator
South Carolina Fire Academy
Columbia, South Carolina

Michael Petroff
Past Chair of the Board
Fire Department Safety Officers Association
Ann Arbor, Michigan

Sam Phillips, MS, EFO
Fire Chief (ret.)
Clallam County Fire District No. 2
Port Angeles, Washington

William Shouldis
Deputy Chief
Philadelphia Fire Department (ret.)
Philadelphia, Pennsylvania

Todd Van Roo
Incident Safety Officer/Instructor
Milwaukee Fire Department/Milwaukee Area
Technical College
Milwaukee, Wisconsin

Richard Vober
Captain
Akron Fire Department Training Academy
Akron, Ohio

Al Wickline
Adjunct Instructor, Fire Science
Community College of Allegheny County (Boyce
Campus)
Monroeville, Pennsylvania

James M. Zeeb
Captain/Paramedic
Kansas City Kansas Fire Department
Kansas City, Kansas

Preface

The Journey to *Making a Difference*

The purpose of this third edition remains the same as it was in the first two editions: to make a difference! With all the efforts and improvements invested in protective equipment, incident management systems, technology, training, and standards, we still have not met our consensus goal of reducing firefighter injuries and deaths by half, although we are finally starting to see some progress in certain areas.

In the first edition, I predicted that a well-trained and experienced ISO could reduce at least half the anticipated deaths and injuries. I further expressed that a dedicated and trained ISO is the quickest, cheapest, and simplest solution to improve on-scene firefighter safety. I was probably over-optimistic, but I remain convinced that a trained and experienced ISO can make a difference. In this book, therefore, I am revising the phrase "trained and experienced" to read "trained, experienced, and persuasive." We need to step up the effort if we are to realize our goal.

Since the first edition (1998), I have had the pleasure of watching the role of the ISO mature. Of particular note, the Fire Department Safety Officers Association (FDSOA) realized the goal of creating and implementing an accredited National Incident Safety Officer Certification. The FDSOA should be very proud of this program, and those who have invested the energy to become certified should be equally proud. I have seen hundreds of fire departments implement comprehensive ISO programs, and I have received hundreds (if not thousands) of emails describing how an ISO recognized and prevented a sure injury or death. It is interesting that we as a profession don't track statistics on firefighter injury or death "saves" by an ISO. Perhaps it's too speculative, too subjective, or it could be that we are humbled by just doing our job. Regardless, the ISO does—and can continue to—make a big difference.

With the thousands of trained ISOs on the job, why have overall death and injury statistics not been reduced by half? The answers are unclear, but I may have some of them. Recent research tells us that residential and commercial structure fires have changed: They have never been as hot, as fast, or as explosive as they are now. The building that we fight fires in is continually changing. Lightweight-constructed buildings will sucker punch the firefighter who approaches them with the same comfort as in the 1980s and 1990s. Further, we are responding to more incidents and to a greater variety of incidents. Finally, I believe that some of the basic curriculum we use to train and certify firefighters is deeply rooted in the past and has not evolved to reflect the changes that society has brought to the typical building fire. If the fires have changed, the building has changed, and the number and types of incidents have changed—yet our training *hasn't* changed—then it stands to reason that the ISO must change if we are to meet our death and injury reduction goals.

To be more persuasive (and realistic) as a safety officer requires more front-loading. So, for this edition, we have expanded the "front-loading" section to include up-to-date technical data that the fire service has gleaned from the outstanding fire research conducted by the National Institute of Standards, Underwriters Laboratories, and several fire departments. Additionally, we updated and expanded each chapter to reflect current best practices and evolving trends and concerns.

Some may appreciate that the ISO Action Model that Terry Vavra and I developed in the early 1990s is being retired. In its place, we offer a function/outcome model that more closely represents the expectations we should have of an ISO.

In some places, you'll notice that the ISO abbreviation for the incident safety officer has been altered to the simple SO or SOF. NFPA is taking steps to delete the "I" from ISO in favor of the the NIMS-compliant SO for Safety Officer. That is already reflective in NFPA 1561 - 2020 edition. In future editions of 1521, the same change will be made. National Incident Management Teams use the abbreviation SOF for the Safety Officer.

How to Use This Book

The book has been written in a stair-step, knowledge- and skill-building manner. First-time readers are encouraged to read chapters in order, as they build upon each other. As a desk reference for repeat readers, the book is divided into three sections:

- **Section 1** offers an introduction to the incident safety officer (ISO) role as a way to start preparing for the assignment. General safety concepts, guiding documents, and effective system components are discussed.
- **Section 2** is perhaps the most important one, providing in-depth "front-loading" information to help the ISO gain knowledge and skills. Of particular note, an entire chapter is devoted to reading smoke and another to reading buildings. These two areas—more than any—can help the ISO predict hostile events that can trap firefighters.

- **Section 3** applies the information in Section 2 to actual incidents. A basic approach to ISO duties is offered, as well as specific considerations for unique incidents, such as those involving hazardous materials or the wildland–urban interface. Section 3 ends with an overview of the often-forgotten ISO postincident responsibilities and includes a new chapter on the ISO's role at fire training activities (and planned, nonemergency events).

New to This Edition

Previous editions of this book have triggered an avalanche of appreciated criticism, corrections, and ideas that have been incorporated into this edition. Additionally, I have had a 15-year opportunity to meet with hundreds of colleagues while traveling and teaching ISO and Reading Smoke classes all over the country. For this revised third edition, I have gathered suggestions from hundreds of working safety officers, incident commanders, and suppression fire officers to improve this text and broaden its application. Each chapter has been critically updated and expanded. Other highlights include:

- A new chapter dedicated to the use of ISOs at training activities and planned, nonemergency events
- Content and formatting changes to help align to the new NFPA 1521, *Standard for Fire Department Safety Officer Professional Qualifications* (2020 edition), which is written using job performance requirement (JPR) language
- A revision and update for firefighter rehabilitation strategies
- Updated material that reflects a more complete understanding of the issues and concerns associated with newer and alternative building construction methods, including the "green" construction trend and renewable/alternative electrical power sources
- Specific ISO concerns associated with fire suppression staffing, deployment, and NFPA 1710/1720 criteria
- Results from recent fire behavior research projects, conducted by the National Institute of Standards and Technology and Underwriters Laboratories, with discussion of how this information can be used to help the ISO better protect firefighters
- New mnemonic tools to help readers remember key functions and to assist with certification study and assessments
- Attention to National Incident Management System (NIMS) compliance and processes for the expansion of the ISO role at major fires and disaster incidents, including an overview of safety officer responsibilities for incident management teams at NIMS Type 1 and Type 2 incidents
- New information on structural fire, on-scene contamination reduction processes to help with cancer prevention.

Suggestions Encouraged

As author, I take full responsibility for any misinterpretations of the suggestions and ideas that so many have graciously offered. I encourage you to be critical in your use of this edition—envision its application, apply the knowledge and skills, and provide suggestions and comments so that the fourth edition will be an even better tool for preventing firefighter injuries and deaths. *Make it safe out there!*

—David W. Dodson, Thornton, Colorado, Winter 2019/2020
Email: davedodson@q.com

Preparing the Incident Safety Officer

Flames: © Ken LaBelle NRIFirePhotos.com; Paper: © silver-john/ShutterStock, Inc.; Officers: © Crystal Craig/Dreamstime.com

This text is divided into three sections. Sections 2 and 3 are "street" oriented, with a strong focus on practical knowledge and skill-building. This first section focuses on safety concepts and the guiding standards behind the role of incident safety officer (ISO). One might understand why a fledgling fire officer would want to skip this section, but doing so would be a mistake. Section 1 provides a core foundation for anyone who wishes to make a difference as an ISO. The goal of any ISO is to prevent injuries and death, and the path to prevention starts with an understanding of the ISO's role. The first chapter, *The Safety Officer Role*, examines the ISO role through an exploration of the history of safety officers, a description of their responsibilities, and an overview of how firefighter deaths and injuries are occurring at

incidents. The chapter *Safety Concepts* introduces us to the deep roots and concepts of any safety program, while the chapter *Guiding Laws, Regulations, and Standards* focuses on initiatives and publications specific to the fire service that have been developed to address injury and death issues.

Finally, the chapter *Designing an Incident Safety Officer System* recognizes that the fire service is quite diverse in the way resources are used (e.g., large or small; volunteer, combination, or paid; rural, suburban, or urban) and offers suggestions on how to develop an effective ISO program. Your study investment in Section 1 will pay dividends not only while on the fireground but throughout your career as you tackle firefighter safety issues as an ISO or in any other position with safety accountability.

Voices of Experience

I completed the Fire Department Safety Officers Association's ISO program, along with the Reading Smoke program that was held in Grand Rapids, Michigan, after taking the reins as training coordinator for our department. The assignment of Incident Safety Officer, as is typical with most departments, was filled on an as-needed basis by the incident commander (IC) and was driven by incident needs. My experience over several years as the lead trainer for our department had helped me begin to see things differently in terms of on-scene operational processes, both strategically and tactically. I was developing an awareness of better

practices as I attended the ISO and Reading Smoke classes that challenged the way I had previously viewed how we worked in the hazard zone.

The ensuing changes in our department became contentious, as prior beliefs passed on from credible and skilled senior firefighters and officers were being challenged by science and new knowledge of the enemy. Once I embraced my new awareness and began to dig deeper into new techniques, it was as if I had put on a new pair of glasses and could now see things that I wasn't skilled enough to see before. This became painfully evident during an incident that forever changed how I viewed the role of Safety Officer.

The incident occurred on an average spring weekday, in a typical suburban family residence in a subdivision. The tones dropped for a smoke investigation at an occupied single-family residence shortly after 1530 hours on a Wednesday. I, like everyone on duty, quickly ran through the location in my head, analyzing the type of construction for homes found in this early 1990s subdivision. This neighborhood, like many others in our community, comprises well-kept ranch and two-story colonial dwellings, ranging in size from 1,600 to 2,800 square feet (150 to 260 m²). I was just returning from the training grounds where I was prepping for extrication training that would be taking place the following day. I listened as the first-arriving engine made the scene and gave a quick size-up: "Engine 3 on the scene of a two-story residence, nothing showing, and with people standing in the driveway. We'll be investigating. This will be Stamford Command." It was the same initial transmission that is heard every day in our department. Shortly after this message, command followed with, "We have smoke inside the residence. Checking for the source"—another routine radio transmission. As additional apparatus arrived on scene, the next transmission is what caused me to point my vehicle in that direction: "Command to dispatch, we have heavy smoke now developing on the first and second floors." Command requested the next-due engine to stretch a supply to the first Engine 3, as the IC directed his crew to deploy a 1¾-inch preconnect interior and search for the fire. I could tell by the amount of radio traffic and heightened voices that conditions were rapidly changing. Still, there was no confirmation as to where the seat of the fire was located. The battalion chief arrived on scene, was updated by the IC, and assumed command.

As I turned into the subdivision, the fire's location became obvious. The smoke plume acted as a big arrow pointing downward to its origin. I parked well back from the apparatus-lined street. I donned my gear and began to walk up the block, passing the crowd of neighbors who had gathered. I began to size up the scene as I approached. Moderate-volume, nonpressurized smoke could be seen exiting multiple first- and second-story windows that had been manually opened on the Alpha and Delta sides. What was even more concerning was the visible nonpressurized smoke coming from the gable vents. I approached the IC and listened while he communicated with interior crews that were operating on both the first and second floors, still searching for the origin. The interior three-man crew on the first floor was reporting rapidly decreasing visibility, but little heat. What I had seen and heard thus far pointed toward a fire that was compartmentalized or in a concealed space—a fire that by now was probably well advanced. It was easy to discern that crews operating in the interior, as well as command, were becoming more anxious with each passing minute.

I was assigned as Safety Officer and asked if a 360 had been completed and if utilities were secured. I was informed that the initial 360 was completed and that utilities had not been secured. I advised that I would perform another 360 and get a visual on utilities and handle the gas meter if not secured. As I left to begin my 360, the IC directed the first-floor interior crew, armed with a 1¾-inch line, to check the basement. Within seconds the interior crew radioed that they had located the basement stairwell and upon opening the door were met by heat and dark smoke. Their next transmission relayed that they were in heavy heat and zero-visibility conditions as they descended into the basement. I reached the Bravo side and secured the gas meter. I then radioed command that a crew needed to secure the electric meter located on the Bravo–Charlie corner. When I turned the next corner to the Charlie side, my eyes grew big, seeing heavy black turbulent smoke pushing out from under the rear patio deck

Voices of Experience (continued)

below the kitchen window. I immediately radioed command with priority traffic that there was heavy fire in the basement, directly beneath the kitchen area. This information was relayed back to the interior crew that had just made its way down the basement stairs.

I met back up with command after completing my 360 and realized that the IC had poor situational awareness up to this point and that he wasn't amending his initial action plan to meet the rapidly changing conditions. My concern now was the crew operating in the most dangerous place they could be—the basement. After restating the conditions that I had just witnessed on the Charlie side, it was evident that the IC had what I call "focal lock" and couldn't see the obviously changing conditions. The next transmission from the interior basement crew was that they could not find the room of origin and that heavy heat and zero visibility were hampering their search. They requested that a thermal imaging camera (TIC) be brought to the basement. I recommended to command that he pull out the interior basement crew because conditions were rapidly deteriorating and there was still no water on the seat of the fire, stating that we should change to a defensive strategy based on a now well-advanced basement fire. His response: "I'm going to give them just a few more minutes." That's when I knew that he had completely lost his baseline for what was occurring.

I had been on scene for approximately 10 minutes and felt like we were no longer in control of the situation. As I was working my way back to evaluate the Charlie side again, I noticed the hose line running into the interior front door was caught on some landscape and the crewmember on the front porch was having difficulty feeding hose through the front door. While assisting in untangling the hose line, I got a sudden feeling that something bad was about to happen as the smoke that was pushing out the front door suddenly became much darker and increased in volume. I radioed to command with priority traffic that I believed we now had fire on the first floor and to order an emergency evacuation of all interior crews. As the order to evacuate was given, my eyes were glued to the front door to count helmets as they exited. I began to hear the crew communicating to each other as they worked their way from the top of the basement stairs to the common hallway, following the hose line toward the front door. Just as the crew made the hallway, only feet from their exit, I heard a large crunch and thud come from within, immediately followed by a huge push of dense smoke and heat out the front door and bay windows. The kitchen had just collapsed into the basement. Within seconds the assigned basement crew safely scrambled out the front door, collapsing on the front lawn as they looked back to see flames now lapping out the top of the front door. I have never been more thankful to see crews on the outside of a structure.

After a short defensive operation, the fire was brought under control and the visible evidence inside the smoking shell of the collapse area made me realize just how close we had come to potentially the worst day in our department's history. This fire and the corresponding scene management represented a defining moment for me. Wearing my "new glasses," I was able to see things that day in a totally different light, and I will never again enter a scene without those glasses on.

The good part of the story, beyond the fact that everyone went home, was the changes that took place after that incident. We revisited and revised many outdated operating guidelines and created new guidelines on TIC use with crews assigned to interior operations. TICs became mandatory on every frontline piece of apparatus, as did in-depth training on TIC use and limitations. Most importantly, the incident prompted the building and teaching of "situational awareness" skills in every training opportunity, no matter how simple or complex that training may be, continually working to move our firefighters away from being task-driven linear thinkers to being multidimensional thinkers who can analyze and predict outcomes. The ISO training proved to be only the beginning for me, and it was certainly the catalyst toward understanding and seeing an incident in a much different way.

Shadd A. Whitehead
Fire Chief (ret.), Livonia Fire and Rescue
Livonia, Michigan

The Safety Officer Role

Knowledge Objectives

Upon completion of this chapter, you should be able to:

- Discuss the history of the fire department safety officer. (pp 7–8)
- List the National Fire Protection Association (NFPA) standards that pertain to the incident safety officer. (p 9)
- Cite current trends in firefighter injuries and fatalities. (pp 8–11)
- Describe the relationship between empirical and image factors and the need for an incident safety officer. (pp 9–12)
- Define the roles of an incident safety officer at planned and unplanned events. (**NFPA 5.2.1** , p 12)

Skills Objectives

There are no ISO skills objectives for this chapter.

You Are the Incident Safety Officer

Your department has received notice from the city risk management division of a recent audit of fire department operations, after a significant fire ground injury determined that a gap exists in fire ground occupational health and safety practices. A review of standard operating procedures found that no clear designation of or means of accounting for an incident safety officer (ISO) exists. Past practice has delegated these functions to the incident commander and company officers. The Chief of Department has asked you to begin the process of developing a response-capable ISO position to be available on a 24-hour basis.

1. What standards and established fire service practices can you use in researching and identifying both requirements for the position and job responsibilities of the ISO during emergency incidents?
2. What trends can be identified nationally in fire service injuries and line-of-duty deaths, and how can this information be used to justify the responsibilities that you assign to the ISO position?

Introduction: Defining the Title

The title "safety officer" is used daily in fire departments around the country. Often, this title is used to refer to the individual in charge of a department's entire safety and health program. More often, the title is associated with a fire officer who reports to the incident commander (IC) and is delegated the safety officer task at an incident **FIGURE 1-1**. In some departments, the safety officer is actually an Occupational Safety and Health Administration (OSHA) compliance officer. In other departments, the training officer is the default safety officer; that is, the training officer's responsibility to ensure safe practices during training activities is extended to department routines and incident activities. Over the past decade, fire departments have discovered that the title safety officer is a bit too generic. In 1991, the Fire Department Safety Officers Association (FDSOA) developed and offered a class titled "Preparing the Fireground Safety Officer." As part of the class, the FDSOA suggested that the safety officer title and responsibilities be divided. A project team audited this class and went on to help the National Fire Academy (NFA) develop courses that split the safety officer role into two titles (and separate curricula): <u>health and safety officer (HSO)</u> and <u>incident safety officer (ISO)</u>. The two courses were then developed into field courses and taught across the United States. The titles were divided into HSO and ISO in recognition of certain realities within the fire service. Small fire departments usually had one person who performed both the HSO and ISO roles. Those departments soon realized that one individual could not possibly be present at every incident where an ISO would be desirable and at the same time fill the role of HSO. In large departments, it was clear that the management of a significant occupational health and safety program required knowledge and skills that were markedly different from those required of an ISO.

In the mid-1990s, the National Fire Protection Association (NFPA) task group assigned to update NFPA 1521, *Fire Department Safety Officer*, took the concepts developed by the FDSOA and the NFA and rewrote the standard to reflect the

HSO/ISO division. The standard provided a consensus definition for each title, as follows:

- A *health and safety officer (HSO)* is the individual assigned and authorized by the fire chief as the manager of the health and safety program.

FIGURE 1-1 The title of safety officer can mean many things, but it is most often associated with a delegated responsibility at incidents.

© Keith Muratori. Used with permission.

- An *incident safety officer (ISO)* is a member of the command staff responsible for monitoring and assessing safety hazards or unsafe situations and for developing measures to ensure personnel safety.

Thus, a division of the safety officer role was written into the NFPA 1521 standard.

Interestingly, the National Incident Management System (NIMS) still uses the generic title of safety officer for NIMS compliance texts and documents. NIMS was developed through Homeland Security Presidential Directive 5 (HSPD-5) to create and mandate a consistent nationwide approach to prepare for, respond to, and recover from domestic incidents regardless of cause, size, or complexity. On its website, NIMS offers a glossary of terms relating to incident command structures, defining safety officer as, "A member of Command Staff responsible for monitoring and assessing safety hazards or unsafe situations and developing measures for ensuring personal safety. The Safety Officer may have Assistants."

The HSO and ISO titles will be used in this text, although there are "NIMS compliance" discussions suggesting that the ISO title revert to the NIMS-preferred safety officer (SOF). The change is already under way. In the 2020 edition of NFPA 1561, *Standard on Emergency Service Incident Management System and Command Safety*, the ISO title has returned to just safety officer (with the corresponding abbreviation SO). (As a side note, the ISO abbreviation is also used in the fire service to denote the Insurance Services Office, an agency that rates fire department capabilities to help insurance companies set policy rates for individual communities.)

For the sake of clarity, let's take a look at some of the functions of HSOs and ISOs. As seen in **FIGURE 1-2**, the HSO focuses on health and safety administration, whereas the ISO focuses on scene-specific operations. It is clear that some overlap occurs—by design. Overlapping functions provides consistency and communication in the different roles. However, before we dive too deep into ISO functions, let's take a quick journey through the history of today's ISO.

History of the Safety Officer Role

Safety officers, in one form or another, have been present in the American work force for a long time. Some of the first safety officers came out of the fire service. In the late 1800s and early 1900s, "wall watchers" stood at corners of buildings and watched the walls for signs of bowing or sagging during a working fire. Speaking trumpets were used to shout out warnings and give orders **FIGURE 1-3**. This practice followed the catastrophic collapse of New York's Jennings Building on April 25, 1854.[1] In this tragedy, 20 firefighters were buried following a partial, then significant collapse of the building at 231 Broadway. In Colorado Springs, in 1898, a decision was made by on-scene officers to withdraw firefighters from a railroad car fire containing black powder. Thirty minutes later,

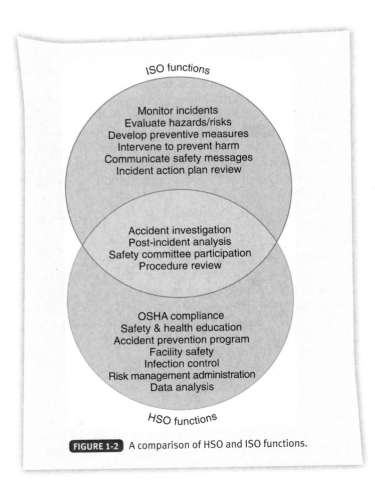

ISO functions

Monitor incidents
Evaluate hazards/risks
Develop preventive measures
Intervene to prevent harm
Communicate safety messages
Incident action plan review

Accident investigation
Post-incident analysis
Safety committee participation
Procedure review

OSHA compliance
Safety & health education
Accident prevention program
Facility safety
Infection control
Risk management administration
Data analysis

HSO functions

FIGURE 1-2 A comparison of HSO and ISO functions.

FIGURE 1-3 The first fire service "safety officer" used speaking trumpets to shout out warnings and give orders.

United States Library of Congress Currier & Ives Collection

the car detonated, causing a wind-fed fire that destroyed many buildings, including the famous Antlers Hotel. These are only a few examples of the safety officer role in its formative days. In some respects, the fire service was viewed as progressive in the "appointment" of safety officers as part of risk management.

As America became industrialized in the late 1800s, laborers suffered difficult working conditions that resulted in numerous injuries (and deaths), leading to employee strikes and riots (particularly in the steel-making industry). The need for an industrial safety professional was beginning to emerge. Signs of this need were felt in World War I when soldiers became mechanized, but it wasn't until World War II that the safety officer role was formalized. World War II, the first fully industrialized war, was unique in that it brought significant injury and death in *support* operations as well as in combat. The military started to look at why people were getting hurt outside of combat, and they appointed safety officers to effect immediate improvements while they developed safer procedures.

Even as World War II continued, factories and other manufacturing industries began looking at the safety of their workers, prompted in part by the inclusion of a significant number of women in the workforce. Some of this introspection came at the request of the insurance industry, while other safety measures came at the request of organized labor. Before long, safety inspections, posters, briefings, and other measures were commonplace in the manufacturing environment. In 1970, Congress passed the Williams-Steiger Act, which created the Occupational Safety and Health Administration (OSHA). President Nixon signed the act into law that December. The law gives equal rights and responsibility to employers and employees with respect to safe working conditions. Today you can find even small businesses with a dedicated safety manager or OSHA compliance officer. In large corporations, an industrial hygienist is tasked with administering risk management and employee safety programs. Safety in corporate America is so prevalent that many careers have been spawned in the safety arena. Colleges, universities, and even vocational schools offer degree and certificate programs in safety and safety-related programs.

Fire Department Safety Officer Trends

Even though some fire departments have been using safety officers for almost a century, the fire service as a whole was slow to catch on to the concepts of safety and risk management in all phases of fire department operations and administration—at least to the degree established by OSHA. The modern roots of risk management and a dedicated safety officer in today's fire service lie in the development and 1987 adoption of NFPA 1500, *Standard on Fire Department Occupational Safety and Health Program*. The late 1980s and early 1990s found fire departments trying to integrate the safety officer (risk manager) role into the department culture. It's not uncommon to hear individuals describe their appointment to the safety officer position with a story such as the following:

> Our fire chief went to a national conference and heard some guy talk about the need for a department safety officer. NFPA 1500 says you've got to have one. The best person for the job is your training officer. Next thing I know, I'm it!

Unfortunately, there was little fire service training material to tell the newly appointed safety officer what to do. Most training/safety officers received a personal copy of NFPA 1500, read it, and found that the standard was nothing more than a fire service twist on commonplace practices in the industrial world. Granted, some NFPA 1500 issues were controversial (staffing, equipment design, and the like), but the basic premise to develop and administer an active health and safety program was found to be its guiding purpose.

The original Safety Officer Standard, NFPA 1501, predates NFPA 1500 by ten years. That 1977 document had just a few requirements for the duties, responsibilities, and qualifications for the evolving position. NFPA 1501 has since been changed to NFPA 1521 in an effort to standardize NFPA numbering. Both 1500 and 1521 are updated on a regular revision cycle under the guidance of NFPA's Technical Committee on Fire Service Occupational Safety and Health.

Prior to the development of these standards, some safety officer trends were well under way in the fire service. In the 1970s, the FIRESCOPE (Firefighting Resources of Southern California Organized for Potential Emergencies) program was developed and used for multiagency incidents on the West Coast. A safety officer was listed as a command staff position to help the IC with delegated safety duties. In the late 1970s, Chief Alan Brunacini of the Phoenix Fire Department began teaching a "Fire Ground Command" seminar across the country. In this seminar, it was recommended that a safety officer, or safety sector, be established to provide a higher level of expertise and undivided attention to fireground safety. This sector was designed to report directly to the fire ground commander, as well as to advise and consult with other sector officers. In 1983, the International Fire Service Training Association published *Incident Command System*, a manual in which a safety officer position was integral to the command staff; a checklist and organizational chart were included[2] **FIGURE 1-4**. In other examples, fire departments in large cities, such as the Fire Department, City of New York (FDNY), were creating safety divisions and shift-assigned safety officers to provide injury investigation and incident safety duties.

The National Interagency Incident Management System (NIIMS), used by the National Wildfire Coordinating Group (NWCG), recognizes the safety officer as directly reporting to the IC. NIIMS is a direct descendant of the FIRESCOPE program. Note that the "single I" NIMS used by the Department of Homeland Security is not to be confused with the "double I" NIIMS used by the NWCG. In the NWCG NIIMS, the safety officer can be classified as one of the following:

- Type 1 Safety Officer (SOF1)
- Type 2 Safety Officer (SOF2)
- Line Safety Officer (SOFR)

An SOF1 is qualified to deploy nationwide as part of a national incident management team (IMT). An SOF2 is usually qualified at the state or local level to function at wildland and interface fires or other disasters. Interestingly, the SOF1 and the SOF2 must meet the same criteria for qualification.[3] The line

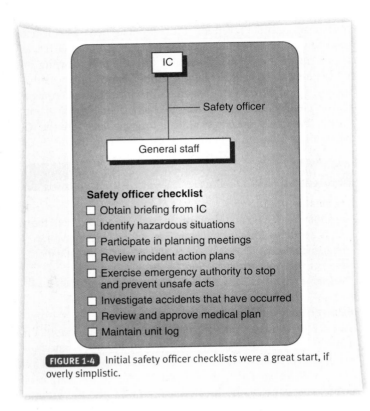

Safety officer checklist
- ☐ Obtain briefing from IC
- ☐ Identify hazardous situations
- ☐ Participate in planning meetings
- ☐ Review incident action plans
- ☐ Exercise emergency authority to stop and prevent unsafe acts
- ☐ Investigate accidents that have occurred
- ☐ Review and approve medical plan
- ☐ Maintain unit log

FIGURE 1-4 Initial safety officer checklists were a great start, if overly simplistic.

safety officer (SOFR) can be the safety officer assigned to initial attack operations, the safety officer assigned to incidents that have not escalated to a Type 1 or 2 incident, and the field assistant used by an SOF1 or SOF2. There are approximately 16 predesignated Type 1 IMTs and 32 Type 2 IMTs staged around the United States, responding to incidents on a rotational basis as needed. The National Interagency Coordination Center (NICC) is responsible for assigning the Type 1 and Type 2 IMTs.

In 2004, President George W. Bush signed Homeland Security Presidential Directive 5, *Management of Domestic Incidents*, which mandates the use of NIMS (single I) as part of the National Response Plan (NRP) administered through the Department of Homeland Security. The NRP has recently had its name changed to the National Response Framework (NRF).

The evolution of the safety officer role continues today. For example, the most recent editions of NFPA standards have relocated the duties and responsibilities of the HSO and ISO from the 1521 standard and placed them in NFPA 1500 and 1561 standards, respectively. NFPA 1521 is now written as a professional qualifications document using the job performance requirement (JPR) format to outline specific knowledge and skill objectives for the HSO and ISO. As such, the NFPA standards that currently pertain to safety officers are as follows:

- NFPA 1500: HSO requirements, duties, and responsibilities
- NFPA 1561: SO requirements, duties, and responsibilities
- NFPA 1521: HSO qualifications
- NFPA 1521: ISO qualifications for fire department ISOs
- NFPA 1026: SO qualifications for NIMS safety officers

The Need for an Incident Safety Officer

The role of a fire department ISO is based on a simple premise: We in the fire service have not done a good job of taking care of our own people. As Alan Brunacini, retired fire chief of the Phoenix Fire Department so aptly commented:

> For 200 years we've been providing a service at the expense of those providing the service.

Chief Brunacini is right on target. Thanks to him and many other fire service leaders and fire equipment manufacturers, there have been significant improvements in firefighting equipment, standards, and procedures, all with the intent of making the firefighting profession safer. Concurrently, the United States has seen a decline in the number of structure fires. One would think that the combination of better equipment and fewer fires would lead to fewer firefighter injuries and deaths. Looking at aggregate numbers (all fatalities and injuries), you'll see that progress is being made. Dig deeper, though, and you'll notice that the rate of fatalities and injuries associated with the fireground is not dramatically improving. More than ever, the fire service needs to step up its effort to prepare and use effective ISOs more often. Proof of this imperative can be found in the empirical (statistical) and image factors regarding firefighter fatalities and injuries.

■ Empirical Fatality and Injury Factors

An ISO who wants to truly make a difference embraces a value that lessons can be learned from each and every firefighter injury and fatality. Understanding statistical trends is part of that value. Various entities collect firefighter injury and fatality data and report them, typically on an annual basis or through investigative efforts. These agencies include the following:

- National Fire Protection Association (NFPA)
- U.S. Fire Administration (USFA)
- National Institute for Occupational Safety and Health (NIOSH)
- International Association of Fire Fighters (IAFF)

The NFPA collects U.S. firefighter injury and fatality statistics associated with line-of-duty activities, including incident response, training, station, and apparatus activities, and other nonemergency activities, while performing official department-related tasks. The USFA collects similar data but classifies line-of-duty deaths (LODDs) differently. The USFA includes what are known as "hometown hero" LODDs—that is, firefighters who have suffered a fatal heart attack *within 24 hours* of an incident response or physically stressful duty-related activity. (The NFPA omits the majority of these LODDs when collecting data.) NIOSH conducts LODD examinations of specific incidents to assist fire departments and the fire service in developing processes and procedures to prevent further LODD or injury occurrences. The IAFF collects and reports on injury and fatality data from its members, which may include those outside the United States.

Knowing which data are collected is a starting point, but what do the numbers say? Let's start with LODDs.

Firefighter Fatalities

Rather than reproducing cited numbers, graphs, and pie charts, the end of this chapter provides a list of references and additional resources where you can find real-time statistics. However, let's highlight some trends that the statistics suggest. We will draw from the USFA data, bearing in mind that prior to the hometown hero divergence (December 2003), the USFA used the NFPA data.[4] Since 1977, an average of 108 firefighters die in the line of duty each year (1997–2017). This average does not include the 343 FDNY firefighters who were killed in the collapse of the World Trade Center Towers on September 11, 2001. Most firefighters will never forget this act of terrorism and the sacrifices made. It is, however, considered a statistical anomaly. Some may argue that using data all the way back to 1977 does not reflect recent preventive efforts developed to improve firefighter safety. We have made progress: From 2004 to 2017, the yearly average for LODDs drops to 101. Some may argue that this number is acceptable considering the risks that a community expects firefighters to take; however, many fire officers believe that no fatalities are acceptable. More than 100 firefighter deaths per year suggest that more preventive efforts are necessary, even though the number is *slowly* trending downward.

Awareness Tip

Trends in Firefighter LODDs
- Overall number: Slowly declining—still averages over 100 per year
- Cardiac related: Declining
- Apparatus related: Declining
- Fire ground related: Increasing
- Leading cause: Overexertion

It is well documented that most LODDs are stress (cardiac) related; however, it is important to note that the percentage of cardiac-related LODDs has started to decline, thanks to many heart-healthy firefighter initiatives. Likewise, the percentage of deaths associated with apparatus incidents is trending downward. Now for the bad news: The percentage of firefighter LODDs due to noncardiac, on-scene causes (e.g., thermal insult, trauma, asphyxiation) is trending upward. The myriad reasons for this problem are being studied, although some fire service leaders believe that it boils down to the following dynamics:
- Society has changed the building (lightweight construction).
- Society has changed the fire (fuels with rapid, high-heat-release rates).
- The fire service has been slow to adapt to the societal changes.

These societal influences are being researched and solutions are being developed, but change is a slow process for most established organizations (not just the fire service). A quick, low-cost solution to the statistical trend of LODDs due to noncardiac events at fires includes the more frequent use of a trained, persuasive ISO. This solution isn't just the opinion of a lone author—NIOSH has repeatedly cited the need for an ISO to help prevent LODDs. The NIOSH "Fire Fighter Fatality Investigation and Prevention Program" has been established (and funded by Congress since 1998) to examine LODDs as a preventive measure to help the fire service. Following an LODD investigation, NIOSH publishes a summary report that includes key recommendations. A reoccurring recommendation cited in dozens of reports asks that fire departments ensure that a separate Safety Officer, independent from the IC, be appointed at each structure fire.

Safety Tip

Separation of ISO and IC
Ensuring that a separate safety officer, independent from the IC, is appointed at each structure fire is a key preventive recommendation cited in dozens of NIOSH LODD investigation reports.

Firefighter Injuries

The NFPA reports annually on firefighter injuries based on a survey sent to fire departments. Returned surveys are then extrapolated to arrive at an estimated injury experience. This method is considered to be accurate within a 5% margin of error.[5] The estimates do break out injuries at the fire ground versus nonfire incidents and other activities. In regard to trends, the estimates show that the combined number of injuries from all activities is slowly declining (a high of 103,340 in 1981 to a new low of 58,835 in 2017). This decline is good news, although 170 injuries per day—or roughly 7 injuries in the time it takes to read this chapter—can be quite alarming. A deeper look at injury statistics reveals an even more alarming trend. The number of injuries on the fireground has been declining—but not as fast as the decline in the number of fires. In fact, when viewed as the number of fire ground injuries per 1,000 fires, the number has been riding a roller-coaster, going up and down between 22 and 25 injuries per 1,000 fires over the past 30 years (*and* over the past 10 years). Given the improvements in personal protective equipment (PPE), equipment design, training, and attention to firefighter wellness, one could view this trend as a failure that demands immediate attention.

Awareness Tip

Trends in Firefighter Injuries
- Overall number: Slow and steady decline to an estimated 58,835 per year
- Nonincident related: Declining
- Apparatus related: Declining
- Fire ground related: Wavering
- Leading type: Strain, sprain, or muscle pain

The most common injuries for all fire service activities are strains, sprains, and muscle pain, followed by cuts, lacerations, and bruises. Considering only fireground injuries (and those that are considered moderate to severe), strains and sprains are still the most common injury type, followed by thermal burns.

Juxtapose firefighter death and injury numbers to the national trend of fewer fires and one can hypothesize that the effort to reduce firefighter LODDs and injuries may not be as effective as it could be. The appointment of an ISO, more often than not, is a prudent measure. Other programs, such as firefighter wellness and incident management systems, do reduce injury and death potential over time, but the use of an ISO can start to reduce these threats *today*.

How many firefighter injuries and deaths have been prevented by the action of a safety officer? We don't know for sure. One Missouri fire chief presented this question when arguing against the cost of formalizing a safety officer program for his department during an open forum on safety issues. No one present at the discussion could answer the question with hard data, although many felt that ISOs did, in their own experience, change a situation that "could have" led to an injury. In one case, an ISO at an Illinois strip mall fire called for the evacuation of a building, and the IC concurred. The firefighters present withdrew and, after protesting the pullout, witnessed the roof slowly collapse into the building. This incident came just a few short years after a firefighter died in a similar roof collapse in a neighboring department. The point is simple: We don't keep good data on what could have been, and at times our memories are too short.

Workers' Compensation

When studying the empirical effects of firefighter death and injury, it would be negligent to skip the effects of work-related injuries on the firefighters' worker compensation insurance coverage. Workers' compensation is statutory for each state, and each state's version has its own intricacies; however, rates are set by the National Council on Compensation Insurance (NCCI). Each state may also adjust rates for firefighters based on experience in that state; this is called an *experience modifier*. To determine a workers' compensation rate for a given department in a state, a formula of NCCI rate × payroll × experience modifier is used. The experience modifier is typically based on a 3-year loss experience.

A few points can be made about workers' compensation insurance. First, and most important, worker compensation programs are not free; they are costly and the cost is based on history: the number of claims and the cost of the claims. Second, a fire department cannot always shop around for a good rate, as individuals do for automobile insurance. If a firefighter is injured on the job, the ramifications may be felt for many years. The more serious the injury is, the longer its effects are felt. Further, this loss affects all employers with employees in the firefighter class. It is easy to imagine the effect of such injuries on the long-term financial status of a fire department.

Clearly, the ever-increasing cost of health care and the constant struggle of balancing operating costs with available tax or other income sources create additional effects when combined with workers' compensation fees. Even one moderate or severe firefighter injury can affect fire department budget-balancing for several years.

Creating and funding a total ISO response system may seem expensive, but the costs of *not* funding such a program can be extraordinary if a fire department experiences a firefighter injury or fatality, especially if the loss could have been prevented by a thoughtful and articulate ISO.

■ Injury and Fatality Image Factors

The image study (how the general public views us and how we perceive ourselves) of firefighter injury and fatality deals with less tangible issues than quantitative data. Generally speaking, the community image of a firefighter is one of a protector or hero FIGURE 1-5 . Likewise, the firefighting profession is often included on many "most dangerous" top 10 lists. Historically, firefighters have worn these images with a certain pride or swagger. More recently, the image fails to realistically depict the effect that a firefighter injury or death has on the *humans* involved.

A firefighter injury requiring hospital care or extended time off creates stress in the workplace. Small career departments have to devote time and effort to find a replacement, while small volunteer departments have to get by without the injured individual. Large departments have to shuffle people to fill-in for the injured firefighter. A less obvious consequence, perhaps, is the firefighter work slowdown following the loss of a fellow firefighter (whether through injury or death). Most fire officers have seen this reaction: Concern, introspection, and even trepidation fill the firehouse following a firefighter casualty event. The more serious the event is, the more pronounced and stressful these reactions can be. If an investigation follows a serious casualty event, the workplace stress can be multiplied to include finger-pointing and taking of sides.

Following a firefighter LODD, members of the affected fire department are likely to be pursued by local media journalists. Social media and blogs are likely to post video clips or images that can lead to critical comments, erroneous or unconfirmed information, and rumors. In worse cases, labor and management concerns might take the form of multiple investigations

FIGURE 1-5 The community image of a firefighter is one of a protector or hero.
© Kim Fitzsimmons. Used with permission.

Awareness Tip

An Emotional Reality
Concern, introspection, stress, and even trepidation fill the firehouse following a significant firefighter casualty event. Stress affects firefighter safety!

and/or attempts to minimize liability or place blame. Results of LODD investigations that attract media attention can quickly change the public's perception of the hero and protector, raising questions of accountability and the need for discipline. In some cases, career and volunteer officers have been demoted, suspended, and terminated as an outcome of the LODD. In almost all cases, an LODD causes tremendous department, community, and individual stress for those with close ties to the loss.

At the personal level, a firefighter injury can have a damaging effect on the involved families. Stress permeates the injured firefighter's family life. Many firefighters know of a peer who has resigned, been divorced, or developed substance abuse issues following a comrade's injury or death. Firefighters who have survived an incident in which another firefighter was killed are considered at risk of experiencing post-traumatic stress disorder (PTSD), a mental health disorder that can develop in those who have experienced a terrifying ordeal that involved physical harm or threat of harm (more on PTSD in the chapter titled *Post-incident Responsibilities and Mishap Investigations*). Alarmingly, the rate of firefighter suicides are increasing, likely due to the repeated exposure to human trauma, LODDs, and job stress.

Taken collectively, the empirical and image factors paint a compelling picture that more effort is needed to prevent firefighter injuries and deaths. This is not to say that the fire service isn't making an effort. The chapter *Guiding Laws, Regulations, and Standards* will outline several outstanding initiatives and efforts that are under way. The point is that *more* can be done, and one effort—the appointment of a trained and articulate ISO—can have an immediate influence that can help prevent injuries and death.

Incident Safety Officer Responsibilities

Simply put, and to use a popular fire service phrase, the responsibility of an ISO is to make sure "everyone goes home." Most adults assigned a responsibility want to know what their tasks are and the expected outcomes. "Everyone goes home" may be the expectation, but this expression does not define the tasks required to achieve it. In many ways, this entire text addresses the responsibilities of the ISO, as well as the required tasks, education and training, and specific tips that can help meet the expectation that everyone goes home. As a starting point, let's consider a more formal description of the ISO's responsibilities.[6] The ISO is a member of the command staff and is responsible for the following:

- Monitoring incident conditions and activities
- Evaluating hazards and unsafe conditions
- Developing measures that promote safe incident handling
- Intervening when an immediate or potential threat exists
- Communicating urgent and advisory safety messages that help prevent injuries or deaths

Breaking down this description of formal responsibilities can better define what tasks an ISO needs to perform to make sure everyone goes home. Let's start with the action verbs *monitor, evaluate, develop, intervene,* and *communicate*:

- *Monitor.* Actively survey (recon) the incident environment (e.g., buildings, smoke and fire, utilities) and watch the incident activities (e.g., civilians' actions, firefighters' actions).
- *Evaluate.* Assess the environment and activities and judge whether a hazard exists (or is being created) that can cause harm. The hazard judgment must be further assessed relative to a prudent risk-versus-gain mentality.
- *Develop.* Design and create preventive measures (proactive actions for forecasted hazards) that will minimize the chance of harm and promote safe incident handling.
- *Intervene.* Take deliberate actions to prevent harm from imminent and potential hazards.
- *Communicate.* Deliver urgent and advisory messages using multiple communication methods (e.g., face to face, radio, warning signs, written safety briefings).

In many ways, the ISO is the hazard "MEDIC" for an incident because this individual must <u>m</u>onitor the environment and activities, <u>e</u>valuate hazard potentials, <u>d</u>evelop preventive measures, <u>i</u>ntervene when a threat exists, and <u>c</u>ommunicate urgent and advisory safety messages. Some may say that the MEDIC mnemonic is simple or corny, but it can serve as a reminder for ISO responsibilities.

Regarding responsibilities, it is important to mention that the appointment of an ISO does not absolve other firefighters or fire officers from the responsibility to have situational awareness and to act in a safe manner. Ultimately, though, the IC is responsible for the safety of all those working an incident. The IC is encouraged to delegate the safety *focus* to an ISO, but he or she owns the responsibility. Ideally, *everyone* operating at an incident should have safety responsibilities commensurate with their assignment, as follows:

- *Incident commander.* Ultimately responsible for the safety of all members operating at an incident. Serves as the "safety officer" when no one has been delegated the task.

Getting the Job Done

ISO Responsibilities

Thinking of the ISO as the "hazard MEDIC" for an incident can remind you of the ISO's incident responsibilities:

M = Monitor the incident environment and activities
E = Evaluate hazard potentials
D = Develop preventive measures
I = Intervene when a threat exists
C = Communicate urgent and advisory safety messages

- *Incident safety officer.* Delegated the "hazard MEDIC" responsibility.
- *Company officers and group or division supervisors.* Responsible for accomplishing tactical directives in a safe manner and ensuring that the firefighters assigned to them operate as a team using appropriate PPE.
- *Individual firefighters.* Responsible for using appropriate PPE, performing directed tasks as trained, and maintaining team discipline.

All fire service members operating at an incident should be aware of their immediate environment and communicate observations of hazards that can injure unsuspecting teammates **FIGURE 1-6**. Likewise, everyone operating at an incident should have the responsibility to take immediate protective actions to prevent an imminent threat from causing harm.

FIGURE 1-6 All fire service personnel handling an emergency have the responsibility to prevent an imminent threat from causing harm.
© Kim Fitzsimmons. Used with permission.

Wrap-Up

Chief Concepts

- In 1991, the Fire Department Safety Officers Association (FDSOA) developed a class titled "Preparing the Fireground Safety Officer." As part of the class, the FDSOA suggested the division of the safety officer title and responsibilities. A project team then worked with the National Fire Academy (NFA) to split the safety officer role into two titles: health and safety officer (HSO) and incident safety officer (ISO).
- In the mid-1990s, the NFPA task group assigned to update NFPA 1521, *Fire Department Safety Officer*, rewrote the standard to reflect the HSO/ISO division, providing definitions for the roles.
- NFPA 1561 has recently changed the ISO title and acronym to the NIMS-compliant "Safety Officer" with the acronym "SO."
- The title *safety officer* is most often associated with a fire officer who reports to the incident commander (IC) and is delegated the safety officer task at incidents.
- The Williams-Steiger Act of 1970 established the Occupational Safety and Health Administration (OSHA) and solidified the responsibility of employers to create safe working conditions.
- The establishment of the FIRESCOPE project in southern California and Alan Brunacini's "Fire Ground Command" efforts in Phoenix, AZ codified the safety officer role within fire departments.
- The NFPA standards that affect and help guide the ISO include NFPA 1561 (requirements, duties, and responsibilities) and NFPA 1521 (qualifications).

- The U.S. fire service still averages more than 100 line-of-duty deaths (LODDs) annually.
- The National Institute for Occupational Safety and Health (NIOSH) has made a reoccurring recommendation that fire departments ensure that a separate safety officer, independent from the IC, be appointed at each structure fire to prevent LODDs.
- Firefighter injury estimates cite approximately 58,835 per year and show a gradual decline. Unfortunately, fire ground-related injuries have not declined even though the number of fires has.
- Firefighter injuries and deaths seriously affect the workplace through stress. Workplace activities, investigations, public perceptions, and family lives are all affected by the stress of the event.
- The ISO is a "hazard MEDIC." Responsibilities include <u>m</u>onitoring the incident environment and activities, <u>e</u>valuating current and potential hazards from a risk-versus-gain perspective, <u>d</u>eveloping preventive measures for forecasted hazards, <u>i</u>ntervening when immediate or potential threats exist, and <u>c</u>ommunicating urgent and advisory safety messages.
- The IC is ultimately responsible for the safety of all responders working an incident. Nonetheless, all fire service members operating at an incident should have the responsibility to be aware of their immediate environment and communicate observations of hazards that can injure unsuspecting teammates.

Key Terms

<u>health and safety officer (HSO)</u> The individual assigned and authorized by the fire chief as the manager of the health and safety program.

<u>incident safety officer (ISO)</u> A member of the command staff responsible for monitoring and assessing safety hazards or unsafe situations and for developing measures to ensure personnel safety.

<u>National Incident Management System (NIMS)</u> An incident response system developed by the Department of Homeland Security.

<u>National Interagency Incident Management System (NIIMS)</u> An incident response system developed by the National Wildfire Coordinating Group.

<u>post-traumatic stress disorder (PTSD)</u> A mental health disorder that can develop in individuals who have experienced a terrifying ordeal that involved physical harm or the threat of harm.

Review Questions

1. What is the difference between an incident safety officer (ISO) and a health and safety officer (HSO)?
2. In general terms, explain the history of today's safety officer in the industrial world as well as in the fire service.
3. List and discuss the NFPA standards related to the ISO.
4. What was the significance of the Williams-Steiger Act?
5. Discuss current firefighter injury and death trends and the need for ISO response.
6. How is the firefighter image affected by injuries and deaths of fellow firefighters?
7. What is meant by "hazard MEDIC" as it relates to the ISO's responsibilities?
8. Describe the safety responsibilities that are commensurate with the following positions:
   - Incident commander
   - Incident safety officer
   - Company officer
   - Firefighter

References and Additional Resources

Fire Protection Publications. *Incident Command System.* Stillwater, OK: Oklahoma State University, 1983. pp. 19–20, 61.

National Interagency Incident Management System. *Type 1/Type 2 Safety Officer, Task Book; PMS 311-04.* Boise, ID: National Wildfire Coordinating Group, 2009.

NFPA 1500, *Standard on Fire Department Occupational Safety and Health Program.* Quincy, MA: National Fire Protection Association, 2018.

NFPA 1521, *Standard on Fire Department Safety Officer Professional Qualifications.* Quincy, MA: National Fire Protection Association, 2020.

NFPA 1561, *Standard on Emergency Service Incident Management System and Command Safety.* Quincy, MA: National Fire Protection Association, 2020.

NFPA Firefighter injury and death reports and data are available at www.nfpa.org/research/reports-and-statistics/the-fire-service.

NIOSH firefighter fatality reports are available at www.cdc.gov/niosh/fire.

U.S. Fire Administration firefighter injury and death reports are available at www.apps.usfa.fema.gov/firefighter-fatalities/fatalityData.

Endnotes

1. Lyons, Paul Robert. *Fire in America.* Boston, MA: NFPA Publications, 1976.
2. Fire Protection Publications. *Incident Command System.* Stillwater, OK: Oklahoma State University,1983. pp. 19–20, 61.
3. National Interagency Incident Management System. *Type 1/Type 2 Safety Officer, Task Book; PMS 311-04.* Boise, ID: National Wildfire Coordinating Group, 2009.
4. The author conducted online research of data posted on the USFA website: www.usfa.fema.gov/fireservice/firefighter_health_safety/firefighter-fatalities. Trends were developed through plotting of data from multiple reports.
5. Two NFPA reports were used to generate the injury trends: Karter, Michael J. *Patterns of Firefighter Fireground Injuries.* NFPA, 2013; and Karter, Michael J., and Joseph L. Molis. *Firefighter Injuries in the United States.* NFPA, 2017.
6. NFPA, NIMS, and OSHA have differing definitions and descriptions for the (incident) safety officer position. The formal description of ISO responsibilities offered here is a hybrid using elements from each.

Smoke: © Greg Henry/ShutterStock, Inc.

In 2012, a volunteer lieutenant was killed and two firefighters were injured when a bowstring roof collapsed at a theatre fire. The responding fire department arrived shortly after noon and observed heavy fire conditions on Side A of the unoccupied, 50' × 100' masonry block building. A brick façade partially masked the presence of an arch-shaped roof over the auditorium portion of the theatre. A police officer reported that a large volume of smoke was issuing from the rear of the structure. After an initial exterior attack, crews proceeded into the building to find and extinguish active fire. The roof collapsed approximately 30 minutes into fire department operations.

The National Institute of Occupational Safety and Health (NIOSH) investigated the incident as part of their Fire Fighter Fatality Investigation and Prevention Program. The investigative report listed ten contributing factors, including the following:

- Risk management principles not effectively used
- Fireground and suppression activities not coordinated
- Incident safety officer (ISO) role ineffective
- Bowstring truss roof construction not recognized
- Fire burned undetected within the roof void space for unknown period of time

The report noted that the responding fire department had a designated safety officer who coordinated the departmental medical and safety and health issues of their members. When available, the SO would be used as an ISO at emergency incidents. According to the report, the SO had never received any formal training on emergency incident expectations of the position. The SO was available at this incident but was occupied at times with fireground activities such as water supply and setting up other fire suppression hose lines.

1. Given the circumstances listed above, what warning signs were present that the roof could collapse?

2. What recommendations would you make to prevent a similar occurrence?

3. Which resource would provide a list of the duties and responsibilities of an incident safety officer?
   A. NFPA 1026
   B. NFPA 1500
   C. NFPA 1521
   D. NFPA 1561

4. Recent data in firefighter LODDs are slowly declining in each of the following areas *except*:
   A. Cardiac
   B. Fireground
   C. Non-incident
   D. Apparatus

Safety Concepts

Knowledge Objectives

Upon completion of this chapter, you should be able to:

- Describe the safety and health practices accepted by the risk-reduction industry. (NFPA 5.2.1 , pp 17–26)
- Identify management principles needed to promote safety in the response environment. (NFPA 5.2.1 , pp 17–22)
- List the three components of the safety triad for the operational environment. (p 17)
- Differentiate formal and informal processes as well as procedures and guidelines. (p 17)
- List the qualities of a well-written procedure or guideline. (p 19)
- Discuss the external influences on safety equipment design and purchase. (p 20)
- List and discuss the three factors that contribute to a person's ability to act safely. (p 22)
- Define risk management. (p 24)
- Identify and explain the five steps of classic risk management. (pp 24–26)
- Describe the hierarchy of controls used to reduce accidents and injuries. (NFPA 5.2.1 , pp 25–26)

Skills Objectives

There are no ISO skills objectives for this chapter.

As you begin to develop and review the organizational documents necessary to establish the position of incident safety officer (ISO) and define the ISO's fire ground responsibilities, you reflect on the recent traumatic firefighter injuries and other near misses that could have resulted in tragedy or life-changing events for your firefighters. The department's administrators understand the necessity of the ISO position as an advocate for crew safety and health, but you sense there is much apprehension regarding the appointment of an ISO.

1. What type of documents or organizational procedures does your department use to direct members on risk-taking expectations and decision making?
2. How would you define the current culture of your department regarding safety concepts and practices? Are members and officers held accountable for failing to observe established guidelines and procedures?

Introduction: Theory Versus Reality

Let's face it: Theory can be boring. Many fire officers would prefer to skip the theory, or bookwork, necessary to become an incident safety officer (ISO); instead, they crave the practical, challenging, and critical aspects of the assignment. However, to become an ISO who can *make a difference*, fire officers must build a foundation of understanding, and that means theory. Although most fire officers agree that effective ISOs need a healthy dose of common sense (a sense of reality), they must also be well grounded in recognized safety concepts (theory), which gives them *uncommon* sense. Uncommon sense can be described as the ability of the ISO to ask and ponder two questions: What is the worst that can happen here? What is the probability of it happening? To answer both questions, the ISO needs a thorough understanding of safety concepts.

In general terms, these concepts represent acceptable health and safety principles and practices that are common in risk-reduction industries. The concepts are the operational safety triad, the five-step risk management model, and risk/benefit thinking.

The Operational Safety Triad

Creating a safe operational environment is dependent on three components (the triad): procedures, equipment, and personnel FIGURE 2-1. To make the operational environment safer, these components should be constantly evaluated, updated, and coached with solid risk management concepts.

■ Procedures

The word "procedure" is used in a very generic form here to describe all sorts of formal (written) and informal processes that are in place in a fire department. At an incident, it is desirable that working crews apply a series of prescribed procedures or processes to achieve a safe and standard outcome.

A formal process is defined in writing and can take on many forms: standard operating procedures (SOPs), standard operating guidelines (SOGs), departmental directives, temporary memorandums, and the like. In some departments, formal processes are derived from standard evolutions or lesson plans. These evolutions and lesson plans can be drilled periodically, on a rotating basis, to ensure that a crew's response to a given situation is appropriate. Some departments adopt training manuals as their operating standard. A manual may offer choices; for instance, for a hose load, the chosen load can simply be circled in the manual. No matter what the source, the key component is that formal processes and evolutions are in writing. In taking this approach, the department achieves consistency in its operations.

Many departments around the country have adopted SOGs in lieu of SOPs, reasoning that a guideline is more flexible and, therefore, more usable by line officers and incident commanders (ICs). A recommendation to use this term was made to the Lisle-Woodridge Fire Protection District (Illinois) by its insurance carrier as part of a scheduled audit. Some fire departments recognize both procedures *and* guidelines. In this context, procedures are strict directives that must be followed with little or no flexibility, and guidelines are adaptable templates that allow flexibility in application.

An informal process is a process or operation that is part of a department's routine but that is not written. Because they are not written, informal processes are typically learned through new member training, on-the-job training, and day-to-day routine. Both formal and informal processes play an important role in the overall safety of a department.

For the remainder of this chapter, we will use the term SOP. The first step in developing a formal SOP is establishing an administrative process to create, edit, alter, or delete established processes. Once the process is in place, a general format for SOP appearance and indexing is necessary. As seen in FIGURE 2-2, the department has chosen to classify SOPs by topic. FIGURE 2-3 shows a typical SOP format.

Once topics have been defined, the writing of SOPs can begin. It makes sense to write the most important ones first, but which topics are the most important? The department can approach this question in one of two ways, and either can be effective. One way is to perform a needs assessment and flag the areas in which line firefighters and officers need guidance. The other way is to look at external influences, such as Occupational Safety and Health Administration (OSHA) regulations, Insurance Services Office rating schedules, NFPA

FIGURE 2-1 Procedures, equipment, and personnel form the safety triad for incident operations.

1. 1.1 Incident Command System
2. Emergency Ground Operations
   - 2.1 Rapid Intervention Company (RIC)
   - 2.2 Gas/odor investigation*
   - 2.3 Auto alarms*
   - 2.4 Train fires*
   - 2.5 Vehicle fire*
   - 2.6 Fires at postal facilities*
   - 2.7 Emergent driving procedure
   - 2.8 Kaneb pipeline response
   - 2.9 Volunteer and fire apparatus placement for motor vehicle accidents (MVAs)
   - 2.10 Operations involving Thompson Valley ambulance
   - 2.11 Minimal staffing for interior firefighting*
   - 2.12 Fire ground formation and activation of companies*
   - 2.13 Standard fire attack procedures/dwelling fires*
3. Alarm Levels/Dispatching
   - 3.1 City alarm level assignments
   - 3.2 Rural alarm level assignments
   - 3.3 Fire resource officer*
   - 3.4 Fire alarm panel operation and response policy
   - 3.5 Mutual/automatic aid agreement*
   - 3.6 Staffing considerations during adverse weather conditions
   - 3.7 Cancellation procedures for emergency medical service (EMS) and MVA incidents
4. Hazardous Materials
   - 4.1 Hazardous materials operations*
5. Emergency Medical Services
   - 5.1 Duties for non-EMS certified personnel
6. Aircraft Rescue and Firefighting (ARFF)
   - 6.1 ARFF standby policy
7. Technical Rescue and Special Operations
   - 7.1 Vehicle extrication
   - 7.2 Rope
   - 7.3 Trench*
   - 7.4 Collapse rescue*
   - 7.5 Confined space*
   - 7.6 Farm equipment and industrial rescue*
   - 7.7 Loveland dive rescue standard operating procedures
   - 7.8 Use of Civil Air Patrol

*These policies still need to be developed and/or approved.

FIGURE 2-2 A sample SOP index.

standards, and other requirements, and determine which areas affect the department most by *not* having a related SOP. Departments choosing the latter route find that items such as personal protective equipment (PPE), self-contained breathing apparatus (SCBA), equipment maintenance, and patient care get high priority in the writing effort. As a starting point, SOPs should exist for the following:

- Use of PPE and SCBA
- Care and maintenance of PPE and SCBA
- Risk/benefit principles
- Incident response (emergent) driving

- Highway and traffic safety at incidents
- Accident/injury procedures and reporting
- Incident scene accountability
- Firefighter trapped and/or lost Mayday procedures
- Abandon building emergency procedures
- Use of the incident command system (ICS)
- Effective incident rehabilitation for responders
- Infection and chemical exposure control and reporting
- Cancer prevention and contamination reduction

Purpose:

To establish policy and direction to all department members regarding minimal staffing and resource allocation for safe and aggressive interior structural firefighting.

Responsibility:

It is the responsibility of all officers and firefighters engaged in firefighting operations to adhere to this policy. The Incident Commander is accountable for procedure included within this policy.[1]

Procedure:

1. This policy is applicable to situations where the Incident Commander (IC) has made a tactical decision to initiate an *offensive fire attack*, by firefighters, inside the structure. Additionally, tactical firefighting assignments that expose firefighters to an atmosphere that is *immediately dangerous to life and health* (IDLH) dictate the application of this policy.[1]

2. Prior to initiating interior fire attack or exposure of firefighters to an IDLH atmosphere, a *minimum of four (4) firefighters shall assemble on scene.*[2] These four members shall utilize a "two-in, two-out" concept.

3. The "*two-in*" firefighters that enter the IDLH atmosphere shall remain as partners in close proximity to each other, generally fulfilling the operational role as the *FIRE ATTACK GROUP*. As a minimum, the "*two-in*" firefighters entering the IDLH atmosphere shall have full PPE, with SCBA and PASS device engaged, and have among them a two-way portable radio, forcible entry tool. and flashlight or lantern.

[1]An *IDLH atmosphere* can be defined as an atmosphere that would cause immediate health risks to a person who did not have *Personal Protective Equipment (PPE)* and/or *Self-Contained Breathing Apparatus (SCBA)*. This includes smoke, fire gases, oxygen deficient atmospheres, or hazardous materials environments. For Loveland Fire and Rescue application, an IDLH atmosphere can be further defined as an environment that is *suspected* to be IDLH, has been *confirmed* to be IDLH, or *may rapidly become* IDLH. The use of full protective equipment including an activated SCBA and an armed PASS device is mandatory for anyone working in or near an IDLH atmosphere.

[2]The firefighters must be SCBA qualified and capable of operating inside fire buildings without immediate supervision.

FIGURE 2-3 A sample SOP format.

What makes a well-written SOP? The answer is simple: It's a well-written SOP if firefighters follow it! Achieving good writing is easier said than done, but at its root, it starts with a clear outline and the use of simple language.

The outline can come from an officers' meeting, direction from the chief, or a sample from another department. Using the format in Figure 2-3, the author should address the reason (purpose) for the SOP, followed by the responsibility of each affected member in achieving the SOP. Some SOPs have responsibilities at different levels. For example, firefighters may have the responsibility to ensure their accountability name tags have been placed on the company's passport. The company officer, on the other hand, may have an oversight responsibility to make sure that all crew members are represented on the passport and to process the passport based on his or her company assignment. The department would have to make sure that a usable policy exists for accountability and that training is provided for system use. Other qualities of a good SOP include the following:

- Simple language
- Clear direction
- Tested technique
- Easy interpretation
- Applicability to many scenarios
- Specificity only in relation to critical or life-endangering points

The benefits of a clear, concise, and practiced SOP are numerous. An SOP can become a training outline, a tool to minimize liability, and certainly a tool to guide your members. Above all, a well-applied SOP improves departmental *safety*.

The ISO's role in procedures deals with application and review, something like a quality control officer's function. To be effective, the ISO needs to know which SOPs are being applied to a given situation and whether the SOP is accomplishing what is intended. When the SOP is not being used appropriately or at all, the ISO needs to interpret whether the actions of firefighters meet the intent of the SOP or whether injury potential exists because the SOP is not being followed. The practical application of SOPs puts the ISO in the *best* place to suggest changes to SOPs or even help create new ones for the department. The ISO who witnesses a failure to follow SOPs during an incident should make a notation and bring up the infraction during post-incident analysis or the next scheduled safety committee meeting. If the failure to follow an SOP presents a potential or imminent danger, the ISO must intervene.

Getting the Job Done

Formal and Informal Processes

Both formal processes (written) and informal processes (not written but customary) play a role in increasing the overall safety of a department.

Getting the Job Done

SOP Quality Assurance

The ISO's role in procedures deals with application and review. The ISO who witnesses a failure to follow SOPs during an incident should intervene if an imminent threat of harm exists. Otherwise, the infraction should be brought up in a constructive manner following the incident. The ISO is also in a unique position to spot performance- and safety-related trends and feed-forward those observations to help develop or adjust SOPs.

■ Equipment

In the past few years, the fire service has seen a veritable explosion in new equipment designed uniquely for improved safety. What works? What doesn't? What's a fad? What's here to stay? What's essential? How much does the equipment cost? Is it worth buying? How long will it last? Will it be outdated soon? Fire departments must reach sound, defensible answers to such questions before spending limited resources and chancing adverse outcomes on new equipment.

With so many questions and so much time spent answering the questions, too often fire department efforts to improve firefighter safety become focused on equipment, and there is a tendency to blame equipment following an accident. (We have all seen how much easier it is to blame equipment than to blame a person.) To some degree, this blame is predictable. Let's call it the Blame Game.

Blame Game 1:

Example 1: "Chief, it wasn't my fault—the darn [*insert name of equipment*] broke."

Example 2: "Chief, if I only had one of those new [*insert name of equipment*], this would have never happened."

Equipment helps, but it is arguably the least important factor in the operational triad of procedures, equipment, and personnel. Yet when building an understanding of the safety triad concept, we have to explore equipment and how it can improve a department's safety. The following factors can be used to evaluate equipment, its selection, and its use.

Department Mission

By looking at a fire department's scope of offered services, we can quickly determine whether it lacks the equipment necessary for safe operations. This is actually quite easy to accomplish. To start, department officers should get together and make a list of the types of incidents handled by their jurisdiction. This list is accompanied by a corresponding list of equipment necessary to *safely* handle the incidents (to the degree that the fire department is responsible). As an example, many departments faced an influx of service calls for the activation of residential carbon monoxide detectors, which are designed to activate with as little as 20 parts per million of carbon monoxide present in the air. Yet many fire and rescue agencies lacked calibrated instrumentation to confirm the presence of carbon monoxide in a home. From this national experience, many departments began carrying high-tech, multi-gas monitors to assist in the safe handling of this type of incident.

With the two lists in hand, officers must discuss the equipment possibilities and place a check mark next to the items that are *essential* to safe operation and a circle next to the *nice-to-have* items. They need to stay focused in this process. The officers then compare the list of required equipment to the equipment on hand. Items that need to be obtained can then be prioritized for evaluation, budgeting, and appropriation.

External Influences

When looking for equipment to make incident operations safer, officers need look only to the advertising pages of the many trade journals or scan through the dozens of safety supply catalogs sent to the firehouse. A better tack, however, is to look at *required* equipment. Although requirements vary from state to state, you can consult the following for help in determining what is required:

- *OSHA regulations.* Known as the <u>Code of Federal Regulations (CFR)</u>, these regulations often outline the equipment required for a given process to be accomplished. Currently, states covered under a state-sponsored OSHA plan may have more stringent equipment requirements for public agencies. Those without a plan do not require OSHA compliance from public agencies. For example, Colorado has no state plan; fire departments have no obligation to follow OSHA. However, the state of Washington has its own plan (Department of Labor and Industries), and compliance is mandatory for all public agencies. OSHA reform is constantly being debated at the federal level. Soon, all public agencies may fall under more restrictive federal OSHA regulations.
- *NFPA standards.* The vast majority of fire service equipment is tailored to meet or exceed NFPA standards. These consensus standards are designed to offer a minimum acceptable standard for equipment design, application, and maintenance.
- *National Institute for Occupational Safety and Health (NIOSH), American National Standards Institute (ANSI), Factory Mutual (FM) Approvals, and Underwriters Laboratories (UL).* Many equipment manufacturers use these agencies to show that their equipment meets or exceeds design and performance requirements.

Equipment Maintenance

As most firefighters know, equipment used for incident operations requires a dedicated care and maintenance program. Following an injury accident, much time is spent evaluating the performance of involved equipment. Often, the piece of equipment is found to be inappropriate for the application or not operationally sound.

Because many firefighters may use and maintain a piece of equipment, the complete documentation of repairs and maintenance is essential. Further, a complete set of guidelines should be developed or adopted for essential equipment. Rich Duffy, Director of Occupational Health and Safety for the International Association of Firefighters (IAFF), and Chuck Soros, retired Chief of Safety for Seattle, Washington, suggest considering seven items when writing equipment guidelines:[1]

1. Selection
2. Use
3. Cleaning and decontamination
4. Storage
5. Inspection
6. Repairs
7. Criteria for retirement

The Right Equipment

A quick look at firefighter injury and death statistics shows *what* equipment can make a difference. The following are equipment items that have made a difference in firefighter safety over the past few years. This list is designed to stimulate conversation in your department, in the hope of leading to wise equipment changes or purchases. Like any equipment, the following equipment is worthless if it is not used and maintained by trained firefighters.

- **Personal protective equipment (PPE)** FIGURE 2-4
  - Task-specific PPE ensembles
  - Accountability passports and electronic personnel tracking systems
  - Disposable EMS masks/gloves
  - Water-free hand disinfectant
  - Integrated PASS devices and heads-up displays for SCBAs
  - "High-visibility" materials/colors/vests
  - Nomex®/PBI/P84/Kevlar® materials
- **Apparatus** FIGURE 2-5
  - Enclosed cabs
  - Intercom/radio headsets
  - Three-point, oversized seat belts for all riding positions
  - Quick-deploy scene lighting

- Mobile data terminals (laptop computers and smartphones and tablets)
- Ergonomically friendly hose beds
- Vertical exhaust pipes
- Wide reflective trim and rear-collision avoidance striping
- Roll-up compartment doors and roll-out trays
- Global Positioning Systems (GPS)
- Automatic vehicle locators (AVLs)
- **Tools** FIGURE 2-6
  - Multi-gas detectors/monitors
  - Speed Shores
  - Rehabilitation kits
  - Command/accountability status boards
  - On-scene contamination reduction kits
  - Two-way radios for each firefighter on a crew
  - Thermal imaging cameras
- **Station equipment**
  - Exhaust removal systems
  - Aerobic and strength exercise devices
  - Dedicated disinfection systems/areas
  - Fire suppression sprinkler systems
  - Extractors for washing structure firefighter clothing
  - Open-air/forced-air protective gear storage systems

The effective ISO understands the relationship of equipment to safety. Remember, equipment is arguably the least important facet of the safety triad. In some cases, equipment designed to improve safety can actually lead to greater risk-taking. As an example, consider the structure fire suppression ensemble. The protective (insulative) quality of structural gear is given as a relative value known as the thermal protective performance (TPP) rating. A TPP rating is quite scientific, although in simple terms it is a measurement given to the durability of equipment when exposed to a flash fire event. Today's gear has such high insulative qualities (i.e., high TPP) that it could mask the sensation of heat and allow a firefighter to move into a dangerously hot environment. Unfortunately, many firefighters are trained to use the sensation of heat to help monitor their environment. If the PPE gear masks the

FIGURE 2-4 Firefighters have a wide array of task-specific PPE ensembles to choose from.

FIGURE 2-5 Safety features continue to evolve in the design and manufacturing of fire apparatus.

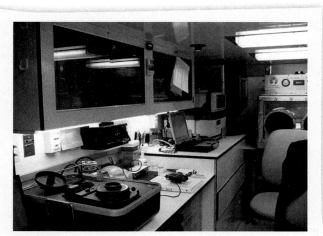

FIGURE 2-6 High-tech tools can help firefighters monitor conditions and improve their situational awareness.

sensation of heat, by the time a firefighter finally senses alarming heat levels, he or she is in a perilous situation. The effective ISO understands this.

■ Personnel

When discussing the effect of people on safety, many opinions, philosophies, and emotions must be considered. For the most part, firefighters won't throw their fellow firefighters under the bus when figuring out why an accident occurred. It is easy to blame the accident cause on an equipment deficiency or on a poor or nonexistent procedure. Once again, we can see this in our day-to-day station dialogue:

Blame Game 2:

Example 1: "Chief, if we only had a procedure that addressed [*insert issue*], this would have never happened. Example 2: "Chief, I'm really sorry Firefighter Jones got hurt, but I followed SOP Number 302 exactly."

It is more difficult to address the "people" component of the safety triad because of the opinions and emotions involved. Regardless, personnel are an essential factor in improving safety. There are three factors that must be addressed as part of the personnel leg of the safety triad: training, health, and attitude.

Safety Tip

The "People" Part of Safety
Three factors contribute to an individual's ability to act safely:
■ Acquired training and education
■ The person's physical and mental health
■ The person's general and current attitude
Addressing each factor is essential, although doing so can be difficult because of the opinions and emotions that may be involved.

Training

A successful safety program usually works in tandem with a successful training program. Conversely, an organization plagued by injuries or suffering from costly accidents usually has a deficiency in its training effort.

As it relates to safety, what makes a training program effective? First, some specific qualities should be present in the training:
■ Clear objectives
■ Applicability to incident handling
■ Established proficiency level
■ Identification of potential hazards
■ Definition of the acceptable risk to be taken
■ List of options, should something go wrong
■ Accountability to act as trained

Second, the training program must include the right subjects. Although arguments can be made for which training subjects or behaviors are most important for safe operations, a compelling list can be developed based on firefighter injury and death statistics. **FIGURE 2-7** lists training subjects that

---

Essential Training Subjects for Increased Incident Safety

Subject	Degree of Understanding
• Personal protective equipment	Mastery
• Accountability systems	Mastery
• Company formation and team continuity	Mastery
• Fire behavior and phenomena	Proficient
• Incident command systems	Proficient
• Apparatus driving	Proficient under stress
• Fitness and rehabilitation	Practitioner

FIGURE 2-7 Injury and death statistics help to suggest the essential training topics that lead to safer operations.

---

directly affect incident safety; if personnel are trained in these subjects and if they appropriately apply the training, incident operations will become safer. For each item in the list, there is an expectation of the depth of understanding and methods required.

Often, the terms training and education are used synonymously. In reality, they are different. Training is the process of learning and applying knowledge and skills. Education is the process of developing one's analytical ability using principles, concepts, and values. Simply stated, training deals with *how* to do something and education is the understanding of *why* you do something. The fire service rightfully focuses on training; we want firefighters to perform tasks in a safe and standard manner. Ideally, we should also invest in safety education (and many fire departments do). In many ways, it is safety *education* that helps to shape an individual's values and attitudes (covered later).

Health

The safety and well-being of firefighters are enhanced when they improve their overall health. Much has been written on the benefits of healthy firefighters, most of which centers on *physical* health. Yet stress or overexertion continues to lead in causes of firefighter line-of-duty deaths and is a significant contributor to injuries. To handle the inherent stress of firefighting, each firefighter's body must be accustomed to and capable of handling stress. Additionally, firefighters need to protect themselves from, and prevent the spread of, communicable diseases and infections. The following are some keys to improving physical health and therefore department safety:
■ Annual health screening for all firefighters and line officers
■ Vaccination and immunization offerings
■ A process to determine an individual's firefighting fitness
■ Access to a department-designated physician
■ Work hardening and mandatory ongoing fitness programs
■ Firefighter fueling (nutrition) education
■ Effective rehabilitation strategies, including hydration, active cooling, and refueling

Attention to physical health is indeed important, but mental health is also important to firefighter safety. Historically, the fire service has been slow to address mental health issues. Thankfully, the issue of responders' mental health is gaining much more attention. Starting with the formal and informal critical incident stress management (CISM) programs of the 1980s, the fire service has been shifting toward the inclusion of behavioral health professionals to help address atypical stress events, long-term occupational exposure to human suffering, post-traumatic stress disorders, and the alarming rate of firefighter suicides. The following actions are fundamental in supporting firefighter mental health:

- Training firefighters to recognize atypical incident stress signs and symptoms
- Creating professional and peer outreach options for suicide prevention
- Including firefighters' families in social, educational, and team-building events
- Accessing local and national resources to help in implementing a behavioral health program for department members

Attitude

Of all the people factors affecting safety, attitude is the hardest to address. Perhaps this is why firefighter attitudes receive the least amount of attention when it comes to safety efforts. As mentioned, firefighters generally take care of each other and are more likely to "blame" equipment and procedures as contributors to a mishap. Unfortunately, there seems to be a general societal trend of placing blame on others. This disturbing attitude trend can make its way into the fire house blame game.

Blame Game 3:

Example 1: "Chief, it's not my fault. How was I to know [insert someone's name] was going to [do whatever]?"
Example 2: "Chief, I was just doing what we've always done. I didn't think [insert the event] could ever happen."

Many factors affect the attitude of an individual, and attitudes are dynamic. Of the many factors affecting safety attitudes, the following few are especially prevalent in the fire service:
- The safety culture of a department
- The firefighter death or injury history of a given department
- The example set by chiefs, line officers, and veteran firefighters

The Safety Culture

The department's safety culture is made up of the ideas, skills, and customs that are passed from one "generation" to another. How does one see and, more importantly, measure a department's safety culture? To illustrate a safety culture, consider the following two firehouse conversations.

Sometown Fire Station 1:

Apparatus Operator: Hey, Cap', I left the ground ladders up on the drill tower. We'll use them to do rescue drills with Engine 2 after lunch.
Captain: Hope they don't blow over and hit someone.
Apparatus Operator: Nah, it's a nice day.

Captain: OK, but if something happens, I didn't see anything.

Anytown Fire Station 1:

Apparatus Operator: Hey, Cap', I left the ground ladders up on the drill tower. We'll use them to do rescue drills with Engine 2 after lunch.
Captain: That may not be a good idea. There's people coming and going and I'd hate to see one blow over and hit someone.
Apparatus Operator: Naw, it's a nice day.
Captain: I know, and I know it's a pain to put them all away just to put them back up, but I'm serious. We shouldn't leave an unattended ladder up. How about tying them off with webbing?
Apparatus Operator: Good idea Cap'. I'm on it. [*Walks away.*]

It is easy to see the two different attitudes toward safety. Ideally, your personnel are working for Anytown Fire Station 1. The culture of the department may be reflected in its daily conversations or in its actions. In a unique example, the Denver Fire Department experienced a significant accident in which two apparatus collided en route to a reported fire. The department ruled that the most significant factor leading to the accident was the department's attitude that condoned competition between companies to be the first one to get water on the fire.

Fire departments with a long and proud history of no line-of-duty deaths or significant injuries can fall into another trap: Simply stated, they believe such events cannot happen in their department. The conversation at this firehouse goes something like this:

Firefighter 1: Did you hear about the firefighter death in [*insert name of state*]?
Firefighter 2: Yeah, I saw the fire on [*insert social website*]. They're amateurs. That ain't gonna happen here.

Do you see the trap?

The Death and Injury History

The department's firefighter death or injury history is a factor in the attitude of its firefighters. A firefighter line-of-duty death often shocks a department's members into an attitude change. Dr. Morris Massey calls this a "significant emotional event" in his renowned video, *What You Are Is Where You Were When.* A traumatic death is capable of changing a person's value programming, often in the direction of a more healthy safety attitude. While some departments may dismiss a death as purely accidental, most seek to change the way they do business to ensure that the event never repeats itself. The death of Bret Tarver at the Southwest Supermarket fire in Phoenix, Arizona (March 14, 2001), triggered a sweeping change in procedures, training, and attitudes.[2] On June 18, 2007, nine firefighters died following a rapid fire spread and collapse of the Sofa Super Store in Charleston, South Carolina. The event and subsequent investigation reports triggered major changes for the Charleston Fire Department as well as many other fire departments around the country. As these examples demonstrate, a significant emotional event can trigger changes.

The Example Set Within the Department

The example (or lack of it) set by the line officers and veteran firefighters is very important. Legendary Notre Dame football coach Knute Rockne once said, "One man practicing sportsmanship is far better than a hundred teaching it." The same can be said about safety. Is safety merely being taught, or is it being practiced? One look at your own department can reveal whether the following safety indicators are being practiced:

- *Crews or company members are watching not only themselves, but also their team members.* Some examples are crews that make quick, head-to-toe checks of each other just prior to an interior firefighting entry; crew integrity that prevails at *all* incidents *all* the time; company officers who give brief safety reminders prior to tactical assignments; tool operators who voluntarily pass a tool to another operator when initial efforts are unsuccessful; firefighters who offer protective equipment reminders that are welcome and expected; rehab and SCBA attendants who are organized for quick recognition of fatigue and equipment problems.
- *Work areas are neat and organized.* A Pennsylvania safety officer once said that he could tell if a department had embraced safety from a simple tour of the apparatus bay and the firehouse lounge. Although the evidence is clear that a clean workplace is a safe workplace, it is best to look to the actions of individuals. Do firefighters routinely correct trip hazards while working on a project? Are swing-open compartment doors closed as soon as a tool is retrieved? Are doorways kept clear at the station as well as at the incident site? Do apparatus operators routinely point out obstacles to firefighters wearing masks? Is out-of-service equipment immediately flagged at the incident scene?
- *Drivers are calm, consistent, and attentive.* Safe drivers are usually the ones who follow a simple routine that begins with a confirmation of the incident location with the company officer. The driver then proceeds to the apparatus in such a way as to get a 360-degree (circle) check of the apparatus. The driver does start-up and seat belt checks and then a passenger check (is everyone ready?). After a go signal, he or she does a mirror check, looks up at the bay door, and visually scans the apron. Finally the vehicle moves. The driver stops before entering the roadway. Out on the road, the driver gives the sense of control with very few quick-jerk movements: Acceleration is smooth, braking is firm and straight, and cornering is like riding a rail. The driver's eyes are always moving and attentive. Face muscles are relaxed, and both hands are graceful in steering (and shifting).
- *Observations are openly shared.* ISOs see one of the most reassuring measures of instilled safety values when firefighting teams and company officers report hazards to *them.* Another positive indicator is when personnel are spending time looking *up* and looking *around.* Teams are pointing at walls, wires, and windows. Among the crews is heard, "Watch out for this . . ." or "Keep an eye on that . . ." The crews themselves will put up exclusionary barrier tape around firefighter hazards or collapse zones. The more you see and hear of these behaviors, the further advanced are the safety values of the firefighters.

Presumably, you can look at this list and assess where your department stands on the attitude scale. Remember, however, that attitude changes are slow and often emotional, and they require a lot of buy-in. Set goals for yourself. Be the example, and then work for small but steady changes in the department.

Five-Step Risk Management

Every day we take risks. Risk can simply be defined as the chance of damage, injury, or loss. Risk management is the process of minimizing the chance, degree, or probability of damage, injury, or loss. Risk managers in most industries use a five-step process called *classic risk management* **FIGURE 2-8**. An understanding of this process can help the ISO make a difference.

■ Step 1: Hazard Identification

Identifying hazards is the primary function of an ISO. In the fire service, we may view many operations as routine and not as hazardous, but they are dangerous nonetheless. A great example is smoke. A firefighter breathing through an SCBA does not see smoke as a hazard, whereas the unprotected civilian avoids the smoke to prevent coughing and tear-filled eyes. With today's plastics, one breath of dark smoke can cause dizziness, loss of sensation, and even unconsciousness. Benzene, a known carcinogen, can cause lung cancer with one exposure. Hydrogen cyanide is a more prevalent smoke by-product gas today and can linger long into overhaul operations. Yet how often do you see firefighters breathing smoke during overhaul?

For most ISOs, identifying hazards is a real-time process of monitoring the incident environment and activities. Previous training and education efforts help the ISO spot current and evolving hazards. This real-time hazard identification approach is desired and beneficial but is *not* how the risk management community identifies hazards. Those in the risk profession use loss history to identify

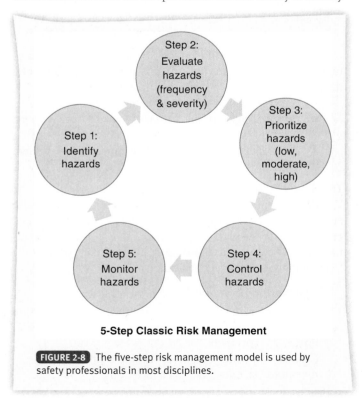

5-Step Classic Risk Management

FIGURE 2-8 The five-step risk management model is used by safety professionals in most disciplines.

hazards; namely, they research injury, death, and property damage data to spot trends and identify hazards. The historical approach for spotting hazards might be appropriate for a fire department HSO or for training ISOs, but it is much too reactive for a real-time incident. Once a hazard has been identified, it needs to be evaluated.

■ Step 2: Hazard Evaluation

Once a hazard has been identified, it has to be assigned relative importance. In this step, a value is established for a hazard in terms of frequency and severity. *Frequency* is the probability that an injurious event can happen, and it can best be described as low, moderate, or high based on the number of times that a particular hazard is present or the number of times injury results from the hazard. The same descriptions of low, moderate, or high can be applied to severity. *Severity* can be viewed as a harmful consequence or cost associated with injury or property damage from a given hazard. Using this approach, a risk matrix can be plotted, resulting in nine categories of risk **FIGURE 2-9**.

Once again, those in the loss-prevention profession use data to help establish the values of severity and frequency. The ISO at an incident uses training, education, and, frankly, intuition to make these judgments. The ISO rarely plots a matrix on scene but can mentally picture it. With this matrix, a value can be assigned to a given hazard. This value helps to determine the priority, or level of importance, of the hazard.

■ Step 3: Hazard Prioritization

Clearly, a hazard that ranks as high frequency and high severity is one we want to avoid or immediately correct at all cost. Conversely, a low-frequency, low-severity hazard typically warrants less attention. A good example is the classic division of fire ground strategies: offensive and defensive. A well-involved fire that has captured the attic space in a lightweight wood construction is a high-frequency, high-severity situation; that is, it will produce a devastating collapse (and potentially severe injury) in nearly every case. On the other hand, we do not spend much time worrying about the hazards associated with investigating smoke caused by overcooked popcorn in a microwave: Sure, there are hazards associated with the smoke, a potential for the kernels to smolder, and possible damage to the microwave, but all of them are easily corrected and rarely (frequency) lead to an injury or costly damage (severity). One method to simplify this matrix and priority system is to divide the matrix into three hazard classes—priorities 1, 2, and 3 **FIGURE 2-10**.

As a starting place, the ISO should address any hazard that falls in the priority 1 category. During some incidents, the ISO may never get an opportunity to address priority 3 items. If the incident is such that only priority 1 hazards get attention, then the ISO or IC may consider expanding the safety role to include assistant safety officers (ASOs) or change the incident action plan to better fit the hazards present.

■ Step 4: Hazard Control

Once a hazard has been prioritized, efforts can be made to minimize exposure to the hazard or to correct the hazard. The overall strategy of hazard control is called mitigation. Mitigation is accomplished using a hierarchy of controls to reduce the potential for accidents and injuries. Safety professionals have developed many variations of the "accident prevention hierarchy of controls." Some are developed to address workplace engineering while others are designed to address human behaviors. In the simplest form, a control hierarchy includes the following steps: (1) Design, (2) guard, and (3) warn. A more complex hierarchy would include the following:

1. *Eliminate* through design.
2. *Substitute*. For example, use a less dangerous chemical or material.
3. *Isolate*, usually through containment or enclosure.
4. *Adopt engineered controls*, including add-on devices such as fans or spark-arresting features.
5. *Apply administrative controls*. Examples include training, procedures, etc.
6. *Use PPE*. Bear in mind, this measure is a last resort.

Control hierarchies are well represented in the design of fire stations, apparatus, equipment, and hopefully training activities. They may not, however, be applicable to firefighters handling a real-time dynamic incident at a "who-knows-where" type of location involving a "who-knows-what" type of circumstance. The hierarchy concept, however, can be molded to the incident-handling aspect of the fire service.

The fire service mitigation hierarchy refers to a preferred order of hazard control strategies:

- Elimination
- Reduction
- Adaptation
- Transfer
- Avoidance

Hazard Evaluation Matrix

		Frequency		
		High	**Moderate**	**Low**
s e v e r i t y	**High**	High/High	High/Moderate	High/Low
	Moderate	Moderate/High	Moderate/Moderate	Moderate/Low
	Low	Low/High	Low/Moderate	Low/Low

FIGURE 2-9 Identified hazards should be judged as low, moderate, or high in terms of frequency and severity. This approach can be represented by a matrix.

Hazard Priorities

		Frequency		
		High	**Moderate**	**Low**
s e v e r i t y	**High**	High/High	High/Moderate	High/Low
	Moderate	Moderate/High	Moderate/Moderate	Moderate/Low
	Low	Low/High	Low/Moderate	Low/Low

☐ First Priority ▣ Second Priority ■ Last Priority

FIGURE 2-10 Once classified, hazards can be prioritized. This can help the ISO juggle multiple hazards.

For firefighting operations, hazard avoidance and transfer are not always possible. In most cases, the fire department was called to *eliminate* the hazard! Prior to elimination, though, hazard *adaptation* and *reduction* are the control methods most often employed at an incident. Hazard adaptation can be accomplished in many ways and in many forms. The actual action used for mitigation is called a <u>countermeasure</u>. For example, if the mitigation strategy of *adaptation* is used at a structure fire, the countermeasures would include a high gallon-per-minute flow rate, flow path management, sound tactics/procedures, PPE, and the like. An example of hazard *reduction* would be the strategy of fuel alteration in the path of an advancing wildfire using the countermeasures of dozer breaks, burn-outs, and wet lines. An example of hazard *avoidance* is the strategy of just letting something burn itself out. The countermeasures to achieve avoidance can include creating no-entry or exclusionary zones and marking them with barrier tape or posting sentries to make sure nobody enters the zoned space.

Getting the Job Done

Hazard Control Language

- *Mitigation:* The overall strategy used to control a hazard or hazards
- *Mitigation hierarchy of control:* The preferred order used for mitigation—elimination, reduction, adaptation, transfer, and avoidance
- *Countermeasures:* The specific actions used to accomplish mitigation (e.g., PPE, fire streams, zoning, barriers, ventilation, shoring, lock-out/tag-out)

■ Step 5: Hazard Monitoring

If the risk management approach is effective, the department should see a decline in injuries, accidents, and close calls over time. Changes in equipment, staffing, and procedures can create, alter, or eliminate hazards. Constant monitoring can catch the changes and lead to proactive hazard control. In one example, a city government experienced a notable increase in employee back injuries. The city risk manager hired a back-injury prevention specialist and in less than two years virtually eliminated back injuries. The program paid for itself in those two years by a reduction in workers' compensation claims. This, in turn, lowered the city's loss history, which lowered its annual premium. The savings were used to help fund employee benefits (health club memberships).

At an incident, the ISO is always monitoring hazards, even after hazard countermeasures are implemented. This is cyclic thinking—that is, the ability to revisit hazards and continually assess the operations and the environment to determine whether a hazard is truly being mitigated. Just as a fire is dynamic, so must ISOs be cyclic in their evaluation of risk. An excellent phrase that captures the essence of this last step was presented in the U.S. Fire Administration's publication *Risk Management Practices in the Fire Service*: "Risk management is a system, not a solution."[3]

Risk/Benefit Thinking

The five-step risk management model is a process for assessing and addressing hazards. One weakness of the five-step model is that there are no provisions for acceptable risk-taking. For this reason, fire service practitioners of the model often overlay the specifics of the second step and fourth step, hazard evaluation and hazard control, with a simple question: Are the risks being taken by firefighters (hazard exposure) worth the benefit that can be gained (such as saving another human)? We call this risk/benefit thinking. The hallmark of a good ISO—and any decision maker for that matter—is the ability to continually reassess risk versus benefit. We talk about this process in greater depth in the *Reading Risk* chapter.

Wrap-Up

Chief Concepts

- Acceptable safety and health practices begin with an understanding of the concepts of the operational safety triad and the use of a five-step classic risk management model.
- The management principles needed to promote safety in the response environment are defined in the *operational safety triad*. The triad includes three elements that need to be addressed: procedures, equipment, and personnel.
- Various formal and informal processes make up the procedure arm of the triad. Formal processes are defined in writing, while informal processes are practiced day to day but are not defined in writing. More specifically, the word "procedure" is used to describe strict processes that should be followed with little deviation. "Guideline" is used to describe processes that are outlined in a flexible template with room for deviation or exception.
- The best quality of a well-written procedure or guideline is that everyone adheres to it. Other qualities include clear direction, simple language, easy interpretation, and broad application except for life-endangering points.
- When making equipment purchasing decisions, several external influences need to be considered. They include the OSHA Code of Federal Regulations (CFR), compliance with NFPA standards, and product listing with Underwriters

Laboratories (UL), American National Standards Institute (ANSI), and other product testing agencies.

- The personnel leg of the safety triad includes programs that address member training, health, and attitude. Training topics can be prioritized using data assembled from firefighter injury and death incidents. Health issues include efforts to develop and implement programs that impact physical and mental needs. Attitude issues are typically harder to address but should include attention to the department's safety culture and values, injury and death history, and example set by personnel.

- Risk can be defined as the chance of injury or loss, whereas risk management is the process of minimizing the chance, degree, or probability of injury or loss. The most common form of risk management includes the use of a five-step process: Identify, evaluate, prioritize, control, and monitor hazards. The person managing this process often must apply a risk/benefit mentality to the second and fourth steps (evaluate/control).

- The hierarchy of controls used to reduce accidents and injuries is accomplished using mitigation strategies and countermeasures. An incident-handling hierarchy of mitigation includes hazard elimination, reduction, adaptation, transfer, and/or avoidance. These strategies are achieved through specific countermeasure actions, such as the use of PPE, high flow rates, flow path management, well-trained and applied procedures/tactics, zoning, and lock-out/tag-out.

Key Terms

assistant safety officer (ASO) A member of the fire department appointed by the incident commander to assist the ISO in the performance of the ISO functions at an incident scene.

Code of Federal Regulations (CFR) OSHA regulations that often outline the equipment required to accomplish a given process.

countermeasure An action used to effect hazard mitigation.

education The process of developing one's analytical ability using principles, concepts, and values.

formal process A process defined in writing. It can take on many forms: standard operating procedures, standard operating guidelines, departmental directives, temporary memorandums, and the like.

guideline An adaptable template that offers wide flexibility in application.

informal process A process or operation that is part of a department's routine but that is not written. Because such processes are not written, they are typically learned through new member training, on-the-job training, and day-to-day routine.

mitigation The overall strategy of hazard control.

mitigation hierarchy A preferred order of hazard control strategies: elimination, reduction, adaptation, transfer, and avoidance.

procedure A strict directive that must be followed with little or no flexibility.

risk The chance of damage, injury, or loss.

risk management The process of minimizing the chance, degree, or probability of damage, loss, or injury.

thermal protective performance (TPP) A value given to the protective (insulative) quality of structural firefighting PPE and equipment.

training The process of learning and applying knowledge and skills.

Review Questions

1. List the three elements that make up the operational safety triad.
2. Explain the difference between formal and informal processes.
3. Describe four qualities of a well-written procedure.
4. List and describe the external influences that can affect safety equipment design and purchase.
5. List and briefly describe the three factors that influence a person's ability to act safely.
6. Define risk management.
7. List and explain the five steps of classic risk management.
8. What is the difference between mitigation and countermeasure?
9. List an accident prevention control hierarchy that can be applied to fire department incident handling.

References and Additional Resources

Federal Emergency Management Agency. *Risk Management Practices in the Fire Service, FA-166*. Emmitsburg, MD: U.S. Fire Administration, 1996.

Kipp, Jonathan D., and Murrey E. Loflin. *Emergency Risk Management*. New York: Van Nostrand Reinhold, 1996.

NFPA 1250, *Recommended Practice in Fire and Emergency Service Organization Risk Management*. Quincy, MA: National Fire Protection Association, 2015.

Safeopedia provides articles, definitions, and more information on safety concepts and the accident prevention hierarchy. It is available at http://www.safeopedia.com.

Stephans, R. *System Safety for the 21st Century*. Hoboken, NJ: John Wiley and Sons, 2004.

U.S. Fire Administration. *Developing Effective Standard Operating Procedures for Fire and EMS Departments*. Washington DC: Federal Emergency Management Agency, 1999.

Endnotes

1. Duffy, Richard, and Chuck Soros. *The Safety Officer's Role*. Ashland, MA: Fire Department Safety Officers Association, 1994.
2. Final Report: Southwest Supermarket Fire, Incident #01-045301, Phoenix Fire Department. https://www.phoenix.gov/firesite/Documents/072604.pdf. Accessed April 29, 2015.
3. Federal Emergency Management Agency. *Risk Management Practices in the Fire Service, FA-166*. Emmitsburg, MD: U.S. Fire Administration, 1996.

Smoke: © Greg Henry/ShutterStock, Inc.

INCIDENT SAFETY OFFICER
in action

The death of a fellow firefighter conducting fire suppression activities usually gives us a chance to pause and reflect on our own mortality. Most of us want to know the circumstances of the event so that we can minimize the risk of a similar occurrence. In February 2005, a fire captain with eleven years of career service (three as captain) died after being trapped by the partial collapse of the roof of a vacant one-story wood frame dwelling in Texas. The house was abandoned and in obvious disrepair. Area residents referred to the building as a "crack house." The 1950s ranch-style house was small and had two additions built onto the back of it sometime after its initial construction. Arriving firefighters reported fire venting through the roof at the rear of the house. The captain and a firefighter entered the front of the house as part of the initial attack. Visibility was reported to be good in the front of the house but changed quickly as they advanced toward the rear. The crew had just started applying water to the burning ceiling area near the rear of the house when the building addition roof collapsed, trapping the captain under burning debris. The collapse pushed fire toward the front of the house, which also ignited trapped smoke and sent a fireball rolling toward the front entrance. The fireball engulfed other crews that had entered the house, causing burn injuries to them. Immediate efforts to rescue the trapped captain proved difficult; he was pronounced dead at the scene.

The subsequent investigation involved several representative agencies. One of the agencies was the National Institute of Occupational Safety and Health (NIOSH). In their report, several recommendations were made to help minimize the risk of a reoccurrence:

- Ensure that the incident commander (IC) continuously evaluates risk versus gain when determining whether the fire suppression operation will be offensive or defensive.
- Train firefighters to communicate interior conditions to the IC as soon as possible and then provide regular updates.
- Use thermal imaging cameras (TICs) during the initial size-up and search phases of a fire.
- Ensure firefighters open ceilings and overhead concealed spaces as hose lines advance.
- Ensure that team continuity is maintained during fire suppression operations.
- Consider using exit locators such as high-intensity floodlights or flashing strobe lights to guide lost or disoriented firefighters to the exit.
- Train firefighters on the actions to take while waiting to be rescued if they become trapped or disoriented inside a burning structure.
- Consider developing and implementing a system to identify and mark dangerous and/or abandoned structures to improve firefighter safety.

Each of the recommendations should be viewed as preventive, not as a judgment on what did or did not happen in Texas. Given that consideration, each suggestion can be tied to a safety concept discussed in this chapter.

1. With the limited information available in this case, what specific indicators could have been used to help make risk versus gain judgments?

2. For each recommendation, which operational safety triad elements are at play? Why?

3. What is typically the most difficult factor to address regarding the "personnel" leg of the safety triad?
   A. Attitude
   B. Training
   C. Education
   D. Health

4. The hazard control countermeasures listed in the NIOSH recommendations most closely align with which mitigation strategy?
   A. Elimination
   B. Reduction
   C. Adaptation
   D. Avoidance

Note: This case study was excerpted from NIOSH report F2005-09. The complete NIOSH report is available at www.cdc.gov/NIOSH/fire.

Guiding Laws, Regulations, and Standards

Flames: © Ken LaBelle NRIFirePhotos.com

Knowledge Objectives

Upon completion of this chapter, you should be able to:

- Explain the motivation for the development of guiding publications. (p 30)
- List the significant players and their roles in developing guiding publications. (pp 31–32)
- Define the differences between regulations, codes, laws, and guides. (pp 31–33)
- State the technical and regulatory areas pertinent to emergency response within the incident command system. (NFPA 5.2.1 , pp 33–34)
- List significant publications that can affect the incident safety officer. (pp 33–34)
- Identify applicable legislation, regulations, codes, and standards that identify levels of risk in fire department operations. (NFPA 5.2.2 , pp 33–34)

Skills Objectives

There are no ISO skills objectives for this chapter.

You Are the Incident Safety Officer

A mound of resource and research documents are piling up on the desk and it seems your computer will soon run out of storage space to contain all of this information. You've identified many national, state, and local agencies that use different requirements and language in defining their safety compliance standards. You must gain a better understanding of these publications and determine which ones "trump" others.

1. Which agencies carry what levels of authority relating to the ISO position and the overall safety program you are establishing? Do the terminology or requirements of these agencies conflict with one another?
2. How can you use these agencies' regulations and standards to make your work consistent with national or state best practices for the ISO?

Introduction: The Reasons Behind the Rules

Where did all these rules come from? Before we answer this question, the point needs to be made that the fire service has very few rules compared to most other developed professions, trades, and crafts. It's true. Historically, fire departments have enjoyed the role of the "good guys." Mostly, firefighters have a reputation of being charitable, reliable, flexible, positive, and willing to do whatever it takes. They work with good intent. They are willing to risk their lives for a stranger. Each of these attributes has helped to protect the fire service from overly restrictive rules—and persecution in the court of law when an outcome was tragic. In fact, there are several legal principles that have a long history of minimizing fire department and firefighter liability, namely the principles of sovereign immunity and discretionary function.[1] Historically, the fire department existed to protect the community, and the community protected the fire department. Fire officers and firefighters who were caught misbehaving or doing something inappropriate (misdemeanor) were often dealt with behind closed doors or outright dismissed. It was rare to find a media source that published disparaging stories about firefighters and departments. Starting in the 1960s, rules began to be created, and by the 1990s, fire departments were asking, "Where did all these rules come from?" So what happened? Answering the question could take a few chapters, but we can slim the discussion down to three influencing factors that changed the fire service to a seemingly regulated profession:

1. An alarming rate of fires and other tragic events have claimed the lives of firefighters. The 1970s and early 1980s brought a rate of firefighter line-of-duty deaths (LODDs) that fire service leaders found unacceptable. Subsequent investigations and reports uncovered many cases of incident mishandling, and the loss of firefighters' lives was deemed preventable.
2. Fire departments' responsibilities have expanded to include emergency medical services (EMS) and hazardous materials response. Both of these functions carry high liability and a certain standard of care.
3. There has been a societal shift to hold public services accountable (litigation and demand for monetary compensation) for injuries, deaths, and suffering that occurred from the perceived mishandling of incidents.

Though the fire service may not be as highly regulated as the airline, heavy manufacturing, or food and drug industries, there is a growing need for incident safety officers (ISOs) to understand the many regulations, standards, and procedures that address incident operations. For the remainder of this chapter, we use the phrase "guiding publications" to refer to regulations, codes, laws, standards, and procedures.

One last point before we proceed. ISOs have been, and will most likely continue to be, named in criminal and civil court proceedings. When they are, the guiding publications become tools that help lawyers, judges, and juries weigh the circumstances and evidence presented. The ISO is wise to view the same publications as tools to avoid litigation and, more importantly, to help keep firefighters safe. The majority of the fire-service-specific publications and rules are written as a result of a tragic event; therefore, the ISO can use them as a basis to prevent similar tragedies from occurring. It is important to understand that fire officers have participated in the development of many guiding publications because they felt it was likely that a similar tragic event may occur in the future—and that probability was unacceptable. The ISO who understands this basic premise is on the path to making a difference.

The purpose of a guiding publication is better understood when the ISO knows which organization issued the publication and what factors typically motivate that organization. In other words, who are the "players"? Knowing the role of the Occupational Safety and Health Administration versus the role of the National Institute for Occupational Safety and Health is an example. These organizations exist for unique purposes and endorse various publications tailored to these purposes. This chapter looks at the roles of the players and the differences in the official publications. It also explores some of the more applicable publications from an ISO perspective.

Note: The information presented here is applicable to situations in the United States. Our firefighter friends in Canada, Mexico, and across the ocean may not find it as applicable, although similar systems probably exist with different names and authorities.

The Players

Knowing who the players are can help the ISO understand the breadth of publications and their effect at incidents to help prevent injury and death. Simply stated, hundreds—if not thousands—of established groups are involved in creating guiding publications. Fire service personnel routinely participate in these groups to help make these documents a usable prevention tool or to ensure that they are practical for incident handling. In some cases, the firefighters represent their own department; in others, the firefighters represent the voice of a trade organization or association. Let's look at some of the players that have a direct effect on incident operations **FIGURE 3-1**.

■ National Fire Protection Association (NFPA)

The NFPA was established in 1896 to address a multitude of fire prevention and fire protection issues. The NFPA is recognized for developing consensus standards, guides, and codes for a whole realm of fire-related topics **FIGURE 3-2**. These standards, guides, and codes are developed through committees who are appointed based on the needs of the fire service, private interests, and other technical specialties. Over time, the NFPA has also become a data collection resource for many fire-related issues, such as firefighter injury and death statistics, information on civilian fire deaths, and national fire and rescue incident trends. The NFPA also offers educational materials, training services, and investigative assistance. It is important to note that NFPA standards are often used to help define what is "acceptable" for fire service equipment, procedures, and professional qualifications. Additionally, NFPA standards could be—and have been—viewed by the courts as *common practice* or the *standard of care* when considering legal questions.

■ Occupational Safety and Health Administration (OSHA)

OSHA is part of the U.S. Department of Labor and is tasked with the creation and enforcement of workplace law. OSHA uses the Code of Federal Regulations (CFR) as the body of laws to improve workplace safety. Not all laws included in the CFR are enforceable for the public sector. Individual states adopt OSHA-approved *state plans*. Fire officers should contact their state department of labor to ascertain whether their public entity is covered by a state plan. Regardless, OSHA carries a pretty big stick when it comes to workplace safety: Federal OSHA, and where state labor authorities exist, can write citations and fine fire departments for noncompliance. In cases where a "willful disregard" is found, the fines can be extreme. OSHA also provides a great resource in addressing workplace

AGENCY	ROLE
NFPA	Development of national minimum consensus standards, codes, and guides. Also collects data and reports trends on a wide range of fire-related topics.
OSHA	Develop and enforce the Code of Federal Regulations (CFRs) dealing with occupational safety and health.
NIOSH	Research, investigate, and recommend safe procedures, processes, and habits.
DHS	Develop and implement a national response plan
EPA	Issue and enforce regulations and provide training for issues regarding hazardous materials and processes.

FIGURE 3-1 The significant fire service "players" and their roles.

- *Standards.* A developed body of work that gives minimum consensus direction for procedures, programs, equipment performance, and professional training and qualifications. Standards are written using *mandatory* language.
- *Guides.* A group of publications that NFPA calls *Recommended Practices*, which are written in a language that offers suggestions and in some cases options. Historically, many recommended practices go on to become standards.
- *Codes.* A complete work designed to be adopted as law by an authority having jurisdiction to do so. NFPA's *Life Safety Code* is the best known.

FIGURE 3-2 NFPA guiding publications: overview of NFPA standards, guides, and codes.

safety issues through training programs, audits, and employee "right to know" literature and promotions.

■ National Institute for Occupational Safety and Health (NIOSH)

NIOSH is the safety and health research and educational arm of the federal government and is part of the Centers for Disease Control and Prevention (CDC) under the Department of Health and Human Services. In 1998, President Clinton directed—and congress funded—NIOSH to investigate all duty-related firefighter fatalities. This service is voluntary for the department that experienced the loss. NIOSH uses firefighter fatality investigations to help others prevent similar occurrences. These investigative reports can be found online through the CDC website (http://www.cdc.gov/niosh/fire). NIOSH has no enforcement responsibilities, but it can recommend the adjustment or creation of regulatory measures to OSHA. NIOSH has written several guides for specific hazards to help firefighters better prevent injuries and fatalities. Most are available for free download at the NIOSH website. A few notable examples include the following:

- *Preventing Injuries and Deaths of Fire Fighters due to Truss System Failures*

- *Preventing Deaths and Injuries of Fire Fighters Using Risk Management Principles at Structure Fires*
- *Preventing Deaths and Injuries to Fire Fighters During Live-Fire Training in Acquired Structures*

■ Department of Homeland Security (DHS)

Following the September 11, 2001, attack on the World Trade Center and the Pentagon, President George W. Bush authorized the creation of the DHS through the Homeland Security Act of 2002. The purpose of the act and the DHS's creation is to better prepare for, defend against, and respond to terrorist acts and other disasters within the United States. One of the first charges of the DHS was to develop a National Response Plan (NRP) to help manage catastrophic events that are beyond the capabilities of state and local agencies. The NRP title has since been changed to the National Response Framework (NRF). On February 28, 2003, President Bush issued the Homeland Security Presidential Directive (HSPD-5) that directed the DHS to develop and administer the National Incident Management System (NIMS) as part of the NRF. Federal grant money to fire departments is tied to their compliance with NIMS. Currently, the Federal Emergency Management Agency (FEMA) and the U.S. Fire Administration (USFA) fall under the cabinet-level DHS. Specifically, FEMA plays an important role in the response and recovery of natural and man-made disasters and is empowered to make judgements regarding cost-recovery for fire departments that have responded to the disaster. The USFA serves as an important resource for fire departments in that they collect data, develop safety-related training publications (and courses), and serve as the conduit to collect and approve various fire service–related federal grant requests.

■ Environmental Protection Agency (EPA)

The devastating results of hazardous materials (hazmat) release incidents (beginning generally in the 1970s) spurred the creation of the EPA to better prevent, respond to, and recover from hazmat incidents. The EPA has issued many regulations and offers support to fire departments for hazmat training. It also helps manage Superfund monies for cleanup and hazmat training. Even though some state fire agencies are not compelled to follow OSHA's CFR, they are required to follow EPA regulations.

■ National Institute of Standards and Technology (NIST)

Founded in 1901, NIST is a nonregulatory federal agency within the U.S. Department of Commerce. The mission of NIST is to promote innovation and industrial competitiveness by advancing measurement science, standards, and technology in ways that enhance security and improvement in our quality of life. How does this make NIST a player in the fire service? One arm of NIST, the Fire Research Division, is a tremendous resource center that conducts testing and collects a vast amount of information on fire- and building-related subjects. The reports generated by NIST serve as an excellent source of training and education tools to help ISOs understand fire behavior in buildings. In recent years, NIST has joined with other agencies (Underwriters Laboratories, fire departments, and fire-related interest groups) to research and develop science-based solutions to fire behavior in the modern structural environment, which poses unprecedented challenges due to high heat release rates.

While far from inclusive, the preceding players are responsible for most of the guiding publications we talk about next. Remember that many fire service–oriented groups also figure in to the "player" group (see the *Fire Marks* box).

Fire Marks

Other Players in the Creation of Guiding Publications
It is important to note that many players are involved in addressing fire service issues—especially programs and publications designed to prevent firefighter deaths and injuries. Many of these groups are nonprofit, yet are intimately involved in the processes to create, alter, and implement guiding publications. It is an impressive list:

- International Association of Fire Fighters (IAFF)
- International Association of Fire Chiefs (IAFC)
- National Volunteer Fire Council (NVFC)
- National Fallen Firefighters Foundation (NFFF)
- Fire Department Safety Officers Association (FDSOA)
- International Society of Fire Service Instructors (ISFSI)

Defining the Terminology Used by Guiding Publications

There are thousands of publications that may have an effect on fire and emergency service personnel. It is important to understand not only the differences among these publications and their applicability, but also that all of them have a common goal: safety. At the street level, fire service personnel often throw around terms such as "codes" and "standards" interchangeably. Doing so may cause confusion and certainly misrepresents the specific applicability of each publication. Let's look at some of the intricacies of these publications' terminology.[2]

- *Laws.* A <u>law</u> is an enforceable rule of conduct that helps protect a society. From a legal perspective, laws are divided into *statutory law* and *case law*. A <u>statutory law</u> deals with rules of conduct in civil and criminal matters. <u>Case law</u> refers to a precedent established over time through the judicial process. As a hypothetical example, an incident commander (IC) is charged with criminally negligent homicide (statutory law) following a firefighter fatality. In the court proceedings, the IC may be found not guilty because of a precedent set by "*United States v. Gaubert*, 486 U.S. 315, 111 S. Ct. 1267, 1991 (case law), *which ruled ....*"

  Although it may sound confusing, some jurisdictions use the term *code* for their statutory laws. Local research can help fire officers understand how their jurisdictions use the two terms.
- *Regulations.* A <u>regulation</u> typically outlines details and procedures that have the force of law issued by an executive government authority. The regulations included in OSHA's CFR and EPA regulations are examples.
- *Codes.* A <u>code</u> is a work of law established or adopted by a rule-making authority. Codes serve to regulate an approach, system, or topic for which they are written. The *Uniform Fire Code* and *Life Safety Code* are examples.

- *Standards.* NFPA standards are perhaps the most familiar to fire service personnel. The term <u>standard</u> can apply to any set of rules, procedures, or professional measurements that are established by an authority. To have the effect of law, a standard must be adopted by an authority with the legal responsibility to enact the standard as law (that is, promulgate it). For example, a city may enact a local ordinance that adopts a standard. In the case of NFPA standards, a formal consensus approach is used to develop the documents through representative technical committees, a standards council, and membership voting.
- *Guides.* A <u>guide</u> is a publication that offers procedures, directions, or standards of care as a reasonable means to address a condition or situation. Guides do not have the impact of a law, but they can be used in negligence cases to provide clarity on the general duty or standard of care. NIOSH has written several guides to address firefighter safety issues, and occasionally it issues an *alert*, which is another form of a guide. Alerts are issued in response to a disturbing trend of injuries or deaths by a specific cause (such as the failure of truss systems during structure fires) and typically describe case studies, technical research, and preventive measures. Textbooks (like this one) can also be considered guides and are often cited in investigations and legal proceedings to establish a standard of care.

NFPA 1500 CHAPTER ORGANIZATION (2018 EDITION)

Chapter 1: Administration
Chapter 2: Referenced publications
Chapter 3: Definitions
Chapter 4: Fire department administration
Chapter 5: Training, education, and professional development
Chapter 6: Fire apparatus, equipment, and drivers/operators
Chapter 7: Protective clothing and protective equipment
Chapter 8: Emergency operations
Chapter 9: Traffic incident management
Chapter 10: Facility safety
Chapter 11: Medical and physical requirements
Chapter 12: Behavioral health and wellness programs
Chapter 13: Occupational exposure to atypically stressful events
Chapter 14: Exposure to fireground toxic contaminants
Annex A: Explanatory material
Annex B: Monitoring compliance with a fire service occupational safety, health and wellness program
Annex C: Building hazard assessment
Annex D: Risk management plan factors
Annex E: Hazardous materials ppe information
Annex F: Sample facility inspector checklists
Annex G: Informational references

FIGURE 3-3 NFPA 1500 chapter organization.

Reproduced with permission from NFPA 1500, Standard on Fire Department Occupational Safety and Health Program; © 2018, National Fire Protection Association. This is not the complete and official position of the NFPA on the referenced subject, which is represented only by the standard in its entirety.

Publications That Affect the Incident Safety Officer

Once you understand the roles of the players and guiding publications in emergency services, you can look at individual publications that may have a direct impact on firefighter safety. To cite them all here would take volumes; it is a compelling list. Instead, presented here are some of the more important documents that apply to incident handling and firefighter safety. The effective ISO will spend time researching the depth of these publications beyond what is offered here.

■ NFPA 1500: *Standard on Fire Department Occupational Safety and Health Program*

First published in 1987, the NFPA 1500 standard has become the "mother ship" of fire department safety and health because it ties together many other NFPA standards by referring to them by their applications. NFPA standards for professional qualifications, protective equipment, tools, apparatus, incident management, and training are all cited in the document. NFPA 1500 continues to evolve through the efforts of the Technical Committee made up of a diverse group of stakeholders. The current edition is divided into 14 chapters and an extensive annex section with explanatory information, checklists, and examples **FIGURE 3-3**.

Within NFPA 1500, Chapter 8 is essential information for any current fire officer and especially those who may be assigned the position of ISO. Of particular importance is the section on risk management during emergency operations. The section outlines a standard for risk levels that are considered acceptable while performing incident activities. The ISO is responsible for

making sure the activities fall within these criteria (discussed in much greater depth in the *Reading Risk* chapter).

Of particular note to the ISO, NFPA 1500 addresses safety at the incident scene in Chapter 8, Emergency Operations. The chapter is broken into 11 areas:

1. Incident Management
2. Communications
3. Crew Resource Management (CRM)
4. Risk Management During Emergency Operations
5. Personnel Accountability During Emergency Operations
6. Members Operating at Emergency Incidents
7. Hazard Control Zones
8. Rapid Intervention for Rescue of Members
9. Rehabilitation During Emergency Operations
10. Scenes of Violence, Civil Unrest, or Terrorism
11. Post-Incident Analysis

■ NFPA 1521: *Standard for Fire Department Safety Officer Professional Qualifications*

In previous editions of NFPA 1521, the requirements, roles, responsibilities, functions, and authorities of an ISO and Health and Safety Officer (HSO) were outlined. It was "one-stop shopping." For many reasons, the 1521 document has shifted to that of a professional qualifications document, much like those for the firefighter and fire officer positions. The aforementioned requirements, responsibilities, functions, and authorities for the ISO have been moved to the NFPA 1561 standard (covered in the following NFPA 1561 section). The 1521 standard includes job performance requirements (JPRs) for the HSO and ISO. These JPRs outline a requirement by listing conditions and a desirable outcome followed by a list of requisite knowledge and skills objectives that support the topic. You may have noticed these JPRs and supporting knowledge and skills objectives cited at the start of the chapters in this text. ISO JPRs are included in Chapter 5 of NFPA 1521. The chapter is divided into seven areas:

- General statements

- General JPRs for all incidents
- JPRs for Fire Suppression Operations
- JPRs for Technical Search & Rescue Operations
- JPRs for HazMat Operations
- Accident Investigations and Review JPRs
- Post-Incident Analysis JPRs

One of the general statements is a requirement that the ISO meet the Fire Officer I JPRs that are specified in NFPA 1021, *Standard for Fire Officer Professional Qualifications*. The Fire Officer I requirement establishes a baseline and recognizes that the knowledge and skill set required of an ISO is much more analytical than those for a company officer. The 1521 standard also includes some useful annex material, including sample checklists and reports.

■ NFPA 1561: *Standard on Emergency Service Incident Management System and Command Safety*

In many ways, the real substance that establishes the ISO role is contained in the NFPA 1561 standard. While the entire standard defines and describes the essential elements of an incident management system (and those required by HSPD-5, *Management of Domestic Incidents*), Chapters 5 and 8 go into the specific requirements, functions, and authorities that pertain to the ISO. As mentioned previously, NFPA 1561 has already dropped the "I" from the ISO abbreviation in favor of the NIMS-preferred SO (safety officer). Some of the more important ISO elements are highlighted and paraphrased here:

- The fire department shall develop a policy to ensure that a separate SO responds automatically or is appointed to all working incidents.
- The IC shall appoint assistant safety officers (ASOs) when the size, scope, or technical complexity of an incident warrants doing so.
- Where utilized, the designated SO is a member of the command staff and reports directly to the IC.
- The SO shall have the authority of the IC to stop, alter, or suspend activities that present an imminent threat to firefighters. The SO must immediately inform the IC of any actions taken to correct the imminent threat.
- The SO has certain major responsibilities at all incidents (see the *Getting the Job Done* box).

Getting the Job Done

Major Responsibilities of the SO—NFPA 1561
- Participate in planning meetings.
- Identify hazardous situations associated with the incident.
- Review the incident action plan for safety implications.
- Exercise emergency authority to stop and prevent unsafe acts.
- Investigate accidents that occurred within the incident area.
- Assign assistants as necessary.
- Review and approve the medical plan.
- Maintain a unit log.

- The SO shall ensure that an incident rehabilitation area has been established.

OSHA Title 29 CFR

OSHA's primary focus area is in the private sector. Some regulations of the CFR, however, are intended to apply to the public sector as well. In some cases, the CFR specifically speaks to the rescue of employees engaged in certain activities (like confined space work). The series known as Title 29 CFR includes numerous subtitles that are specific to public sector members who engage in rescues and exposure to environments that are immediately dangerous to life and health (IDLH). The following entries of the Title 29 CFR may have some impact on the functions of the ISO:

- 29 CFR 1910.120, *Hazardous Waste Operations and Emergency Response Solutions*
- 29 CFR 1910.134, *Respiratory Protection*
- 29 CFR 1910.146, *Permit-Required Confined Spaces*
- 29 CFR 1910.147, *The Control of Hazardous Energy (Lockout/Tagout)*
- 29 CFR 1910.1030, *Blood-Borne Pathogens*
- 29 CFR 1910.1200, *Hazard Communication*
- 29 CFR 1910.1926, *Excavations, Trenching Operations*

The regulations of the CFR that pertain to emergency response to incidents and rescues include specific language instructing that responders use NIMS for the command and control functions of the incident. As it relates to the ISO, the CFR emphasizes the need to have a written site safety plan for operations involving hazmat, confined spaces, trenches, and hazardous energy emergency incidents. When the fire department is engaged in rescue activities involving these elements, the ISO should develop a site safety plan (in writing) and present a safety briefing to those working the incident. The development of safety plans and briefings is discussed in later chapters.

National Fallen Firefighter Foundation—16 Firefighter Life Safety Initiatives

The National Fallen Firefighters Foundation (NFFF) hosted a first-of-its-kind Firefighter Life Safety Summit on March 10–11, 2004, in Tampa, Florida. The summit, consisting of more than 200 fire and emergency service representatives from more than 100 organizations and departments nationwide, was convened to support the USFA's stated goal of reducing firefighter fatalities by 25% within 5 years and 50% within 10 years. After the event, the NFFF and the USFA released a report that details 16 initiatives and recommendations for drastically reducing firefighter fatalities and injuries (see the *Fire Marks* box).

It is clear that the goal of LODD reduction has not been made even though significant efforts have been made by many departments and organizations **TABLE 3-1**. The NFFF and USFA have hosted several follow-up summits to revisit the initiatives. The focus of these follow-ups has been to define some of the underlying human behaviors that are delaying the goal

TABLE 3-1	LODD Statistics, 1988–2018.

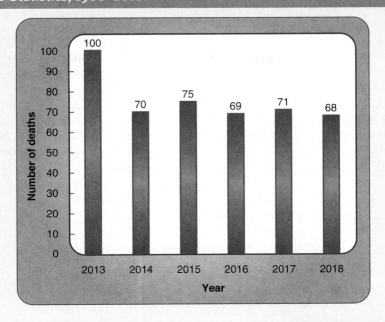

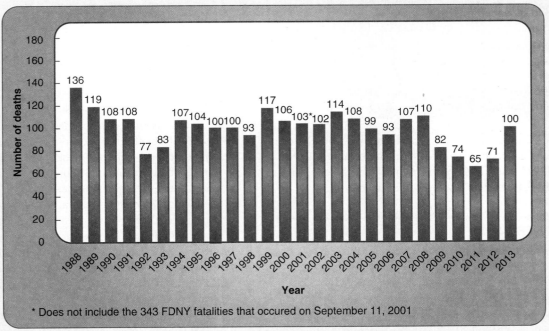

* Does not include the 343 FDNY fatalities that occured on September 11, 2001

Modified with permission from NFPA's report, "Firefighter Fatalities in the United States-2103" by Fahy, Rita F. © 2013, National Fire Protection Association.

Fire Marks

The NFFF 16 Firefighter Life Safety Initiatives

1. Define and advocate the need for a cultural change within the fire service relating to safety, incorporating leadership, management, supervision, accountability, and personal responsibility.
2. Enhance the personal and organizational accountability for health and safety throughout the fire service.

3. Focus greater attention on the integration of risk management with incident management at all levels, including strategic, tactical, and planning responsibilities.
4. All firefighters must be empowered to stop unsafe practices.
5. Develop and implement national standards for training, qualifications, and certification (including regular recer-

Continues

Fire Marks (*Continued*)

tification) that are equally applicable to all firefighters based on the duties they are expected to perform.

6. Develop and implement national medical and physical fitness standards that are equally applicable to all firefighters, based on the duties they are expected to perform.

7. Create a national research agenda and data collection system that relates to the initiatives.

8. Use available technology wherever it can produce higher levels of health and safety.

9. Thoroughly investigate all firefighter fatalities, injuries, and near misses.

10. Grant programs should support the implementation of safe practices and/or mandate safe practices as an eligibility requirement.

11. National standards for emergency response policies and procedures should be developed and championed.

12. National protocols for response to violent incidents should be developed and championed.

13. Firefighters and their families must have access to counseling and psychological support.

14. Public education must receive more resources and be championed as a critical fire and life safety program.

15. Advocacy must be strengthened for the enforcement of codes and the installation of home fire sprinklers.

16. Safety must be a primary consideration in the design of apparatus and equipment.

Wrap-Up

of a significant reduction of LODDs. The reports from these meetings are forthcoming. Until then, it's important for ISOs to know the 16 initiatives and help champion the effort in their own departments and sphere of influence.

Chief Concepts

- Starting in the late 1960s, the fire service has seen a dramatic increase in the number of laws, regulations, and standards (guiding rules and publications) that impact the fire service. The majority of these rules are a direct result of a tragic occurrence. In many cases, the tragedy resulted from incident mishandling and the loss of lives was deemed preventable.

- Guiding publications become tools that help lawyers, judges, and juries weigh the circumstances and evidence presented. The ISO is wise to view the same publications as tools to avoid litigation and, more importantly, help keep firefighters safe.

- Fire service personnel routinely participate in the development of guiding publications. Some represent their own department and others represent the voice of a trade organization or association.

- Some of the leading players in the development of guiding publications include the NFPA, OSHA, the EPA,

NIOSH, the DHS, NIST, and several trade associations such as the IAFC, IAFF, and FDSOA.

- Defining various regulations, codes, and standards is not always easy. Fire officers are encouraged to consult their local jurisdiction's legal administrators to help understand the definition and applicability of each. As a starting place, some general definitions can help:
  - *Laws* are rules of conduct that help protect a society. From a legal perspective, laws are divided into *statutory laws* and *case laws*.
  - *Regulations* outline procedures that have the force of law issued by an executive government authority.
  - *Codes* are bodies of work established or adopted by a law-making authority.
  - *Standards* are any set of rules, procedures, or professional measurements that are established by an authority. To have the effect of law, a standard must be promulgated by an authority with the legal standing to do so.
  - *Guides* are publications that offer procedures, directions, or standards of care to address a condition or situation. Guides do not have the impact of a law but can be used in cases of negligence to provide evidence of general duty or standard of care.

- NFPA 1500 is a mother-ship document that outlines the requirements for an occupational safety and health program for the fire service. Chapter 8 addresses emergency scene operations and identifies the levels of risk-taking that are considered acceptable.

- NFPA 1521 is now a professional qualifications standard containing job performance requirements (JPRs) for HSOs and ISOs. The standard also requires that ISOs meet the qualifications of a Fire Officer I as defined in NFPA 1021.
- NFPA 1561 is the incident management standard that establishes the requirement, responsibilities, and functions of the SO position. It requires that fire departments create a policy to ensure that a separate SO, independent of the incident commander (IC), responds automatically to working incidents. Further, the document requires that the SO be a command staff position that reports directly to the IC.
- According to several regulations of OSHA's CFR, an ISO must be appointed at certain emergency response events (hazmat, confined space, and technical rescue) and is responsible for the development of a formal safety plan and safety briefings for responders.
- The National Fallen Firefighters Foundation published 16 Firefighter Life Safety Initiatives in 2004 to help reduce the number of line-of-duty deaths (LODDs) by 50% in 10 years. Although the goal has not been met, all firefighters need to take ownership in their respective departments to keep the initiatives in focus and work toward the goal.

Key Terms

case law A precedent established over time through the judicial process.

code A work of law established or adopted by a rule-making authority. It serves to regulate an approach, system, or topic for which it is written.

guide A publication that offers procedures, directions, or standards of care as a reasonable means to address a condition or situation.

law An enforceable rule of conduct that helps protect a society. From a legal perspective, laws are divided into *statutory laws* and *case laws*.

regulation A rule, issued by an executive government authority, that outlines details and procedures that have the force of law.

standard A rule, procedure, or professional measurement established by an authority. To have the effect of law, it must be adopted by an authority with the legal responsibility to enact the standard as law (that is, promulgate it).

statutory law A rule of conduct in civil and criminal matters.

Review Questions

1. What has typically motivated the establishment of guiding publications?
2. How are OSHA and NIOSH different?
3. What is the significance of the U.S. Department of Homeland Security to the fire service?
4. Define regulations, codes, laws, and guides.
5. List the 11 topical areas in the NFPA 1500 chapter on emergency operations.
6. What does IDLH stand for?
7. What responsibility does the ISO have in the use of OSHA's Title 29 CFR?
8. Where can the ISO find the consensus acceptable risk management levels for incident operations?

References and Additional Resources

Callahan, Timothy. *Fire Service and the Law.* 2nd ed. Quincy, MA: National Fire Protection Association, 1987.

NFPA 1500, *Standard on Fire Department Occupational Safety and Health Program.* Quincy, MA: National Fire Protection Association, 2018.

NFPA 1521, *Standard on Fire Department Safety Officer Professional Qualifications.* Quincy, MA: National Fire Protection Association, 2020.

The National Institute for Occupational Safety and Health (NIOSH) website is available at http://www.cdc.gov/niosh.

The National Institute of Standards and Technology (NIST) website is available at http://www.nist.gov.

The Occupational Safety and Health Administration (OSHA) website is available at http://www.osha.gov.

Endnotes

1. Callahan, Timothy. *Fire Service and the Law.* 2nd ed. Quincy, MA: National Fire Protection Association, 1987.
2. Author's note: A special thanks to attorney and Fire Chief David Comstock, Jr., Western Reserve Joint Fire District, Poland, Ohio, for his assistance in the development and review of this material.

On September 25, 2001, Firefighter Trainee Bradley Golden died during a live-fire training exercise in Lairdsville, New York. Assistant Chief Alan G. Baird III, the designated instructor for the training exercise, was charged with second-degree manslaughter and later found guilty of criminally negligent homicide by an 11-person jury.

The case against Baird arose out of a live-fire rapid intervention exercise during which Golden and another firefighter posed as victims. Golden, who had been a volunteer firefighter for only a few weeks, had never had any formal training, and had never worn an SCBA in a live-fire environment. Despite this, Golden and another firefighter were placed in the second-floor front bedroom of a residential duplex and covered with debris to simulate entrapment. A burn barrel was ignited on the second floor to develop smoke, and shortly thereafter, Baird ignited a foam mattress on the first floor. Within minutes of the mattress ignition, flames rolled across the ceiling, up the stairway, and out the windows of the front bedroom where the simulated victims were. No hose lines had been stretched prior to the ignition of the fires. A total of three firefighters were trapped on the second floor with limited egress options. One firefighter self-extricated by prying open a boarded-up window and the others were removed from the building by crews that had been staged for the drill. Golden was found unresponsive and was transported to the hospital, where he was pronounced dead. The other two firefighters suffered burns and were airlifted to the hospital.

In his defense, Baird testified that he was not the officer in charge and that the incident commander and designated safety officer were aware of the drill objectives and knew what fires were to be set. Baird pointed out that he was not the highest-ranking officer on scene. Baird also testified that at the time of the exercise he was not aware of the NFPA standards that address live-fire training. Additionally, Baird testified that the training was being accomplished following procedures that they used at a previous live-fire training event several years earlier that was under the direction of a state fire inspector. Baird's counsel argued that Baird should not be held accountable for not knowing about the NFPA standards and that he should have never been charged because the tragedy was an "accident." Prosecutors cited NFPA 1403, *Standard for Live Fire Training* in their arguments and ultimately convinced the jury that Baird violated many nationally known standards.

1. What role do NFPA standards play in the defense and prosecution of fire officers in criminal matters or in the judgment of civil cases?

2. What accountability do the incident commander and safety officer have in similar situations?

3. The legal precedent established above in *People of the State of New York v. Alan G. Baird III* is an example of which kind of guiding publication?
   A. OSHA CFR
   B. Case law
   C. Statutory law
   D. Promulgated code

4. In addition to criminal and civil lawsuits, a fire department can be levied fines for noncompliance with recognized safety practices. Which agency has a responsibility to write citations to help enforce safe practices?
   A. NFPA
   B. NIOSH
   C. NIST
   D. OSHA

Designing an Incident Safety Officer System

Knowledge Objectives

Upon completion of this chapter, you should be able to:

- Discuss the reasoning for preplanning the response of an incident safety officer (ISO). (pp 40–42)
- List six examples of when an automatic ISO response should take place. (pp 43–44)
- List seven examples of when an incident commander should automatically delegate the safety responsibility to an ISO. (pp 44–45)
- List and discuss the advantages and disadvantages of using various methods to ensure that an ISO arrives on scene. (pp 45–47)
- Discuss issues relating to the ISO's authority, as defined by NFPA standards. (p 47)
- List several tools that will help the ISO be effective on scene. (pp 48–49)

Skills Objectives

There are no ISO skills objectives in this chapter.

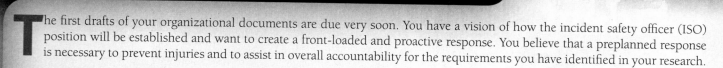

You Are the Incident Safety Officer

The first drafts of your organizational documents are due very soon. You have a vision of how the incident safety officer (ISO) position will be established and want to create a front-loaded and proactive response. You believe that a preplanned response is necessary to prevent injuries and to assist in overall accountability for the requirements you have identified in your research.

1. What minimum qualifications for persons assigned as an ISO have you identified in your research of the position in the governing documents?

2. In addition to the requirements established by the various standards and regulations, what skills and knowledge bases should a qualified ISO possess? Which of these areas are your individual strengths and weaknesses?

Introduction: Making an ISO System Work

The design and implementation of a fire department incident safety officer (ISO) program or system can make the difference in whether or not a program will be effective. When the first edition of this text came out, many fire departments took suggestions from it to help make their ISO system work. Since then, departments have discovered through trial and error ways to maximize the effectiveness of their ISO program. The options now available for designing an ISO system are numerous—and endless—involving such dynamics as fire department size, deployment strategy, and unique needs. Nevertheless, the design of the ISO program should address some major questions, such as:

- Who responds and fills the ISO role?
- What types of incidents necessitate the use of an ISO?
- What tools and training are necessary to maximize ISO effectiveness?

When designing a system, it is important to keep in mind the requirements for the appointment of an ISO from NFPA 1561. Key requirements are listed below. In this chapter, we explore the rationale for the NFPA requirements as well as suggestions on how to design an ISO system.

Incident Management System Requirements for a Safety Officer

NFPA 1561, *Standard on Emergency Service Organization Incident Management System and Command Safety*, outlines several key points pertaining to the use of safety officers within the incident management system, including the following:

- SOPs shall define criteria for the response or appointment of a safety officer. The safety officer shall be appointed as early in an incident as possible.
- A policy shall exist for the appointment of a safety officer and to ensure that a separate safety officer (independent from the incident commander) responds automatically to predesignated incidents.

- If a predesignated SO is not available, the incident commander shall appoint a qualified member to fill the SO role.
- An additional assistant safety officer(s) shall be appointed when the activities, size, or need of the incident warrants extra safety personnel.
- The safety officer shall be integrated within the incident management system as a command staff member.
- The SO has authority to stop, alter, or suspend activities that present an imminent threat of harm.
- The SO and ASO(s) shall be readily identifiable at the incident scene.

Reproduced with permission from NFPA 1561, *Standard on Emergency Service Organization Incident Management System and Command Safety*, © 2020. National Fire Protection Association. This is not the complete and official position of the NFPA on the referenced subject, which is represented only by the standard in its entirety.

Proactive ISO Response

Most incident commanders (ICs), fire chiefs, and firefighters would agree that ISOs are necessary and valuable at significant or complex incidents. However, the discovery that an incident is significant or complex typically comes after incident operations are under way and the incident management system has been set up. This leads to a situation in which the delegation of an ISO is *reactive*. For example, an emergency call is received and a programmed response is initiated. Once on scene, the incident command system is implemented and actions are taken to begin mitigation. If the incident is significant or overcomes the initial response, additional resources are requested. At about this time, the IC realizes that his or her responsibility for the safety of all responders requires more attention and that a separate ISO can help. In this case, the IC is reactive in the delegation of firefighter safety duties. For an IC to truly make a difference at an incident scene, the delegation of the safety function to an ISO needs to be *proactive*. To be proactive in the delegation and placement of an ISO, the fire department needs to *preplan* the ISO response.

Before we delve into the means of developing an effective ISO system (i.e., the *when*, *where*, and *how*), let's take a moment to more fully consider *why* departments need to preplan an ISO response.

■ *Why* Preplan the ISO Response?

A few ICs believe that any fire officer should be able to fill the ISO position, at any time, under any circumstance, at the will and demand of the IC; therefore, the agency really doesn't need to create an ISO system. Not only is this thinking flawed, it is dangerous. Just as ICs have various levels of knowledge and expertise, so do other fire officers. Likewise, the requirements to be a fire officer may change from department to department—a problem if the need for mutual aid arises. Further, the emphasis placed on safety may vary from one IC to another. Firefighter death and injury statistics suggest that more can be done to improve firefighter safety (as stated in the chapter *The Safety Officer Role*). Those same statistics show that the majority of deaths and injuries on the fire ground occur at residential structure fires. Logic (as opposed to traditional fire ground thinking) suggests that we send an ISO to all residential fires. However, we know that some ICs are thinking, "Residential fire, a couple engines, a dozen people—we can handle that. We don't need a safety officer." The statistics show otherwise. Simply stated:

> *The ISO is most effective when he or she arrives early in the incident.*

The National Institute for Occupational Safety and Health (NIOSH), through its line-of-duty death (LODD) investigation and prevention program, has repeated this recommendation in numerous cases. A few pragmatic graphs, based on a typical residential working fire, can further illustrate the need to have an ISO assigned early at fires.

Graph 1: Environmental Change

As applied to a residential structure fire, "environmental change" means fire propagation, building degradation, and smoke volatility. As shown in **FIGURE 4-1**, the rate of change is measured and rated over time. During a fire in a structure, there is actually a routine, or even a rapid, rate of change upon arrival of fire crews.

Granted, the environment encountered by fire personnel depends on response time. In most cases, the fire is intensifying upon arrival (that's probably why someone called 9-1-1). At arrival (zero on the time line), the fire is starting to develop significantly. With that comes smoke production as more contents (fuel) become heated. Equally, the building itself is being attacked and thus becoming structurally degraded—especially in lightweight structures. If flashover happens prior to fire department control efforts, the environmental change becomes ultra-rapid. Other events can cause ultra-rapid change: smoke explosions, backdrafts, partial collapse, the presence of accelerant fuels, and other phenomena. Once control efforts have begun, the rate of change can be stabilized and even reduced. It would seem to make sense to have an ISO on hand early to help evaluate hazards during these periods of rapid and ultra-rapid change.

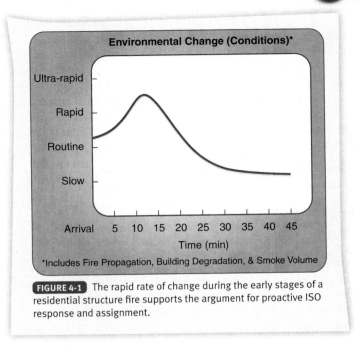

FIGURE 4-1 The rapid rate of change during the early stages of a residential structure fire supports the argument for proactive ISO response and assignment.

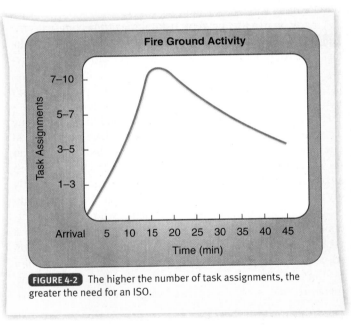

FIGURE 4-2 The higher the number of task assignments, the greater the need for an ISO.

Graph 2: Fire Ground Activity

At a scene with lots of fire ground activity, early ISO assignment is highly advantageous. Upon arrival at a typical residential fire, you can argue that no tasks are being performed **FIGURE 4-2**. Even though all responders are going through their personal size-ups and potential ICs are performing a lot of mental evaluations, no crews are performing physical, active tasks in the dangerous environment upon arrival. As on-scene time passes, the activities needed to control the incident increase: Perhaps one to three tasks are being accomplished based on the number of resources that arrive initially. Within 20 minutes, the IC may be orchestrating 7 to 10 simultaneous assignments, which may include search and rescue, exposure protection, flow path management, preparing attack and backup lines, rapid intervention planning, etc. For some departments, the simple task of water supply is a significant chore that incorporates two or three simultaneous operations. The point is simple: About 15 to 20 minutes into the incident, several tasks are going on all at once. A key to preventing injuries is enabling the IC to monitor the numerous simultaneous activities during the first 20 minutes after arrival. The good IC wants an ISO appointed early on in the incident. Wouldn't it be great if you, as the IC, knew that an ISO was predesignated and responding before five or six tasks were being performed?

Graph 3: Relative Danger to Firefighters

Practically speaking, the first-arriving firefighters at a working fire should perform some kind of risk/benefit analysis. This initial risk analysis comes with a degree of uncertainty—simply because the firefighters may not know the full extent of the dangers at hand upon arrival. Firefighters actually begin taking risks upon arrival—or at "zero time." At times, risks are taken only to determine the risk of the incident! Here is an actual tragic example: The Denver Fire Department was called to assist the police department on a welfare check. A firefighter was assigned to climb a ladder to check for a possibly unlocked upstairs window. He reached the top of the ladder and, without warning, was shot by the occupant of the house!

Early on in a fire incident, risk-taking can become extreme **FIGURE 4-3**. The perfect example of extreme risk-taking is the task of search and rescue. In a typical scenario, a first-arriving crew starts setting up for an aggressive interior attack; they're laying supply lines, pulling attack lines, donning SCBA masks—preparing for an offensive attack. Suddenly, the first-due officer comes from the back of the building after doing a "360" and reports that there is an immediate need for rescue. Three victims are hanging on the third floor balcony and they need to be evacuated at once. The crew drops the planned attack and begins taking risks to get to the people. It is early on in an incident, when risk taking is high. Volunteer, paid on-call, or pager-notified fire departments, who allow firefighters to respond directly to the incident scene, may face an even greater degree of risk-taking because firefighters may arrive on scene prior to apparatus and equipment. Once again, the point is simple: Risks are usually greater early on in an incident; therefore, that is when a safety officer is needed.

Overlapping the Graphs

Fire and rescue departments that wish to reduce firefighter injuries and death develop a system to get the ISO on scene or appointed early in an incident. The overlapped graphs in **FIGURE 4-4** clearly show that the first 20 minutes of an incident warrant close monitoring of firefighters and firefighting operations. The early appointment of an ISO gives the IC another set of eyes, another viewpoint, and another consultant. The logic illustrated by the graphs is sound. Although no accurate data suggest that firefighter deaths and injuries happen early at an incident, most fire officers would agree that the most dangerous time at an incident is the critical point when they are making aggressive efforts to rescue victims and stop a rapidly growing fire. It is also at this time that key decisions are made, and while the act of making decisions may not be dangerous in itself, the consequences of those decisions can be. The only way to ensure the early arrival of an ISO at an incident is to

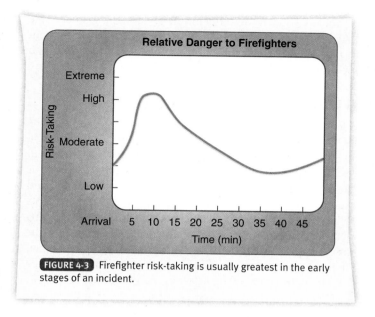

FIGURE 4-3 Firefighter risk-taking is usually greatest in the early stages of an incident.

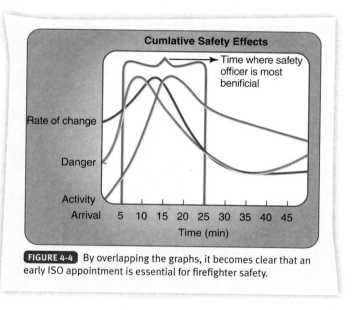

FIGURE 4-4 By overlapping the graphs, it becomes clear that an early ISO appointment is essential for firefighter safety.

predefine *when* the ISO is required to be appointed and design a system to make sure a trained ISO arrives on scene early.

■ *When* Does the ISO Respond?

If, as suggested, an ISO should be present on all working residential fires, common sense suggests that an ISO should also be planned for highly technical or complex incidents. Once again, the response of an ISO should be preplanned (proactive). It is not suggested that the IC's ability and authority to make decisions be taken away; however, firefighter injury statistics show that we need to have a dedicated ISO more often and sooner. If an IC can get an extra set of eyes on scene sooner, more hazards can be recognized, which helps to reduce death and injury occurrence. In defining *when* an ISO is required to respond, the individual fire and rescue department should develop standard operating procedures (SOPs) for an *automatic ISO response* as well as SOPs for, or circumstances that mandate, an *automatic ISO delegation* during the incident.

Automatic ISO Response

The goal is to have a predesignated, trained ISO programmed to respond to certain incidents that typically present a high risk to firefighters. Generally speaking, the types of incidents that should require an automatic ISO response are as follows:

Residential or Commercial Structure Fire

Upon receipt of a credible report of a fire in a residential or commercial fire, an ISO should be programmed into the response **FIGURE 4-5**. For some departments, dispatch personnel may need to add a level of 9-1-1 screening so that a "burnt toast" call is not classified as a structure fire. Some may argue that an automatic response would lead to having an ISO at every fire, something that a large department (cost) or small department (available staffing) may not find acceptable. Unfortunately, statistics do not justify this argument. Residential fires hurt and kill firefighters! Once we, as a fire service, force a significant decline in injuries, then maybe sending an ISO to every structure fire is not necessary. Until then, the merits of the requirement outweigh its cost. In fact, small or critically staffed fire agencies can make the argument that the fewer people who are available, the more important an ISO becomes; in essence, each person on scene is taking a greater risk than they would in an appropriately staffed response. Likewise, an argument and comparison can be made regarding the welcome trend of adding another engine or truck company to a structure fire assignment to serve as the rapid intervention crew (RIC). The fire service has embraced this concept. Should we also embrace the use of an automatic ISO response to help minimize the need for the RIC?

Commercial buildings present a profusion of hazards that require immediate understanding—again requiring an early ISO response. High-rise building fires also have unique hazards that need to be addressed early on in an incident.

Wildland–Urban Interface Fires

Fires in the wildland or at the wildland–urban interface (I-zone) can cause unique problems and present unusual choices for first responders. First-arriving company officers have to evaluate fuels, weather, topography, fire conditions, access, and the defensibility of threatened structures **FIGURE 4-6**. This combination leads to an environment well suited for the partnership of an IC and ISO early in the incident. In some jurisdictions, the time of year and local fuel conditions may indicate the need for an automatic ISO response. For example, when fire danger is rated high or extreme by the forest service or other agency, the local fire department may "bump up" its resource response to include the automatic response of an ISO to any fire report.

Specialty Team Incidents

Many fire and rescue departments have specific-function teams designed to handle the array of incidents outside the scope of standard engine and truck company functions **FIGURE 4-7**. Examples are hazardous materials (hazmat), dive rescue, heavy (urban) rescue, rope rescue, and wildland "hot shot" teams. Often, the activation of these teams takes a separate request or page, and an ISO should be included with this activation. In the case of technician-level hazmat and technical rescue incidents, the assignment of a dedicated ISO is mandated by OSHA CFRs (more on this in later chapters). Some of these specialty teams include a safety officer who is technically trained to interface with the needs of the particular team (NFPA 1006 and 472). Under the National Incident Management System (NIMS), these safety officers work at the tactical level and are subordinate to the primary ISO, who works at the command staff level. Therefore, the specialty team safety officers are actually assistant safety officers (ASOs) and should be titled ASO-Hazmat, ASO-Tech Rescue, or ASO-Line. The now famous urban search and rescue (USAR) teams around the nation have technically trained (NFPA 1006 and 472) safety officers who travel with the team. If any specialty team does not include its own ASO, a basic-trained ISO can function as a "common-sense" consultant to the team, support personnel, and IC prior to the appointment of the specially trained ASO.

Target Hazard Incidents

In nearly every city, local firefighters can name a building where they hope they never get a fire—"cause it'll be a killer." One advantage of computer-aided dispatch (CAD) systems is the

FIGURE 4-5 The report of an actual hostile fire should trigger an ISO response.

© Jones & Bartlett Learning. Photographed by Glen E. Ellman

FIGURE 4-6 Fires in the wildland or the wildland–urban interface (I-zone) should trigger an ISO response.

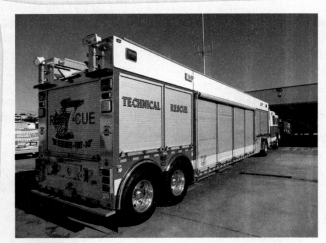

FIGURE 4-7 A designated ISO should respond whenever specialty teams are activated.

FIGURE 4-8 Weather conditions that are considered extreme in a given jurisdiction should trigger a proactive ISO response.

© Keith Muratori. Used with permission.

ability to flag locations and occupancies that present unique hazards to firefighters. If an incident is reported at one of these locations, an ISO can be preprogrammed into CAD to respond. Examples may include chemical and industrial plants, historical buildings, large warehouses, stadiums, underground structures, difficult- or limited-access occupancies, and the like.

Aircraft Incidents

The potential for high loss of life, mass fire, and the release of hazardous materials is prevalent in aircraft incidents. In Fresno, California, for example, a military contract Learjet crashlanded on a street in the middle of town and instantaneously started an aircraft, apartment, and commercial building fire, as well as a hazmat and collapse incident. At this type of incident, an IC would surely embrace the early arrival of a predesignated ISO. Departments with airfield aircraft rescue and firefighting (ARFF) responsibilities have specialized equipment and procedures to minimize the associated risks. An ALERT 3 (crash) can stretch these safety measures, especially if the airport has an ALERT 3 with an out-of-index aircraft. ("Indexing" of aircraft refers to its size and passenger load that are common to the given airport.)

Weather Extremes

Firefighting efforts in subfreezing temperatures or in high-humidity and high-temperature environments present additional hazards to firefighters **FIGURE 4-8**. Changes in strategies, tactics, and action plans may be required just because of the weather. If firefighters are unaccustomed to working in a given weather extreme, injury risk rises. As an example, for the Colorado Front Range area (Denver, Colorado Springs), temperatures on either side of 0°F (−17.8°C) and 100°F (−37.8°C) are rare but usually do happen a few days each year; therefore, firefighters may be forced to work in extremes to which they are not acclimated. In this case, accelerated rehabilitation and monitoring of firefighters is a solid argument for an automatic response by an ISO. Likewise, it makes sense to get an ISO on scene early to help monitor the risks associated with the added stress of excessive snow or rain/wind (such as from tropical

storms or approaching hurricanes). Granted, firefighters in parts of Minnesota or upstate New York may think it silly to send an ISO just because it is cold and the snow is a little deep, but on a hot, humid day, they might benefit from someone monitoring their reaction to the extreme weather. Our firefighter friends in Florida and Arizona are probably well versed in hot-weather firefighting, but throw them a −20°F (−28.9°C) environment and behold the stress! Acclimation is the key. If the weather becomes extreme in terms of *your* jurisdiction, then consider adding an ISO to the response.

Automatic ISO Delegation

"Plan for the worst" is a personal code that most responding fire officers live by. The odds are that fire officers have enough resources to handle most "routine" situations. However, they are occasionally caught short or, worse, think the situation has been handled—only to have "Murphy" show up. In these cases, it pays to have a predefined set of circumstances that lead to the automatic delegation of the ISO role where one hasn't been made. Having an SOP that tells the IC that he or she *shall* delegate the safety function to a qualified ISO under certain conditions is what we mean by automatic delegation. By having these situations predefined, the IC is reminded of the valuable assistance that an ISO can provide for escalating or multiple-risk situations. Let's look at some examples.

Working Incident

Generally speaking, a <u>working incident</u> can be defined as one at which the initial response assignment is 100% committed and more resources are needed. In such cases, the control effort is behind the power curve, which leads to greater risk-taking by personnel **FIGURE 4-9**. The IC's call for additional assistance becomes the indicator that an ISO is needed. Although language varies among jurisdictions, the sounding of a second alarm can serve as a similar impetus for an automatic ISO delegation.

A Span of Control in Excess of Three

Most incident management systems recognize an ideal span of control of five or fewer for emergency operations. However,

FIGURE 4-9 The first-due report of a "working incident" tells the IC to rapidly delegate a qualified ISO.

FIGURE 4-10 Mutual-aid incidents give rise to an array of safety concerns. An ISO can help.
© Keith D. Cullom/www.fire-image.com

those well versed in the use of an incident command system (ICS) choose to delegate sections or functions before the one-to-five mark. Once the span of control grows past three, the IC should include an ISO. If the ICS is handling an incident using groups and divisions, the addition of the fourth group or division should be the signal to delegate an ISO. Also, the IC's decision to delegate the existing groups or divisions into an operations section (with an ops section chief) could be the signal to appoint an ISO. Regardless of the type of ICS used, the department should include a "mark" that signifies the point at which an ISO is automatically assigned into the system.

Mutual-Aid Request

Even though many mutual-aid companies work well together **FIGURE 4-10**, the mutual-aid environment asks for firefighters with different cultures, tactics, training, and equipment to work together. If for no other reason, an ISO should be appointed at all mutual-aid incidents to monitor and ensure that the action plan is being carried out as intended by the visiting team. This makes sense for moral reasons (taking care of our own), if not for legal reasons (liability). When mutual-aid systems train and respond together on a regular basis (say, quarterly), the need for an automatic ISO delegation to all mutual-aid requests is not as great.

Firefighter Down, Missing, or Injured

A seemingly simple incident can get turned upside down and shaken loose when a firefighter emergency takes place. The well-prepared fire and rescue department has a firefighter emergency (Mayday) plan and a RIC in place to deal with such an emergency—although they hope never to use these measures. One main part of the plan should be the immediate appointment of an ISO (if not already assigned) to help the IC implement it. Once the Mayday is handled, it is the ISO who begins the accident documentation and investigation process—another reason to automatically delegate the safety function for a Mayday.

Incidents Requiring Victim Extrication

When responders discover that a lengthy, complicated, or difficult victim extrication is warranted, an ISO should be assigned. Granted, some fire departments are well versed in power tool use for vehicle mishap extrication and may find the ISO delegation to be overkill. Key triggers here are the words lengthy, complicated, or difficult, or incidents in which the extrication is from an entrapment mechanism that is unusual or unfamiliar to the responders.

Need for Hazardous Environment Monitoring

Incidents that require the monitoring of a hazardous environment (such as one in which carbon monoxide has been released) can benefit from the delegation of an ISO. Note that the ISO does not do the actual gas monitoring; rather, the ISO can help evaluate system requirements that support crew entry and task accomplishment (rehab, zoning, personal protective equipment [PPE] requirements, and the like).

IC Discretion

Incident management systems recognize that the IC is ultimately responsible for firefighter safety and the incident outcome. At any time and under any circumstance, the IC should have the discretion to delegate the safety function. For this reason, a trained ISO needs to be available for delegation. Likewise, and perhaps arguably, fire departments should give an on-duty ISO the discretionary option to respond to incidents that don't fall into an "automatic response" SOP. In these cases, an IC may question or take exception to the ISO showing up, but their mutual concern for safety should override personality differences and/or turf protection. A well-written policy can help avoid complications.

■ Where Does the ISO Come From?

Once a fire and rescue department decides to preplan its ISO response, a system to get an ISO on scene is imperative. Who is this person? Where does he or she come from? How is he or she alerted? There are probably as many answers to these questions as there are fire departments. Let's look at some typical arrangements.

Training Officer(s) on Call

Having the training officer on call to function as the ISO is a popular means of ensuring that an ISO is available on scene; however

Getting the Job Done

Preplanning the Use of the ISO

A fire department should have SOPs that address the mandatory response and use of an ISO according to NFPA 1561. SOPs are needed for automatic ISO responses, which include, at a minimum, the following:

- Reported structure fires
- Wildland–urban interface fires
- Specialty team activation
- Incidents at target hazard complexes
- Aircraft incidents
- Responses during extreme weather periods

SOPs are also needed for automatic ISO delegation incidents, which include, at a minimum, the following:

- Working incidents (all resources committed)
- Growing ICS span of control
- Mutual-aid incidents
- Firefighter Mayday situations
- Incidents requiring lengthy or difficult victim extrication efforts
- Incidents requiring hazardous environment monitoring
- IC discretion for any incident

it may not be the most effective means. In these systems, the officer can either monitor incident activity and self-dispatch or rely on a pager to be notified when an ISO is needed.

- *Advantages.* This system is popular because many departments believe that the training officer is a readily available fire officer who knows how the department operates. Further, the responsibilities of training and safety are embodied in one person—an attractive package to sell in the preparation of budgets, staffing reports, and other planning. Training officers often have the radios, vehicles, and other tools that make their transition into the ISO role quick. Individual training officers may like this option because it gives them a chance to get out of the daily routine and help maintain their incident skills.
- *Disadvantages.* Training officers are not always available around the clock, putting a burden on the department and the officer. In small- to medium-sized departments, the training officer is actually the Training/Safety/Infection Control/Research and Development/Special Projects/Recruitment/Wellness/ Quality Control/Accreditation/Do Everything the Chief Needs Officer for the department. Wearing all these hats usually comes at the expense of not wearing any one of them well. In these cases, the training officer is probably better suited to train a contingent of ISOs to be available for response.

Health and Safety Committee Members

In this system, the members of the department's health and safety committee are trained to serve as on-duty ISOs. Typically, an on-duty schedule is made up among the members and the on-duty person monitors the radios and self-dispatches

according to department policy. This system is often seen in small volunteer departments.

- *Advantages.* The members are familiar with safety issues and have a forum to communicate. A pool of people to draw from is defined and available. ISO training is easily accomplished. An ISO network is formed.
- *Disadvantages.* Career (paid) departments may incur overtime expenses due to committee training, meetings, need for members to be on call, and actual incident responses by members. Extra radios, pagers, and other ISO equipment must be either purchased for each member or passed from member to member according to the duty schedule. Many health and safety committees are represented by members from all ranks. Some members may not have officer status or the experience/knowledge commensurate with ISO functions (an ISO should be a Fire Officer I, at a minimum).

All Eligible Officers

Strong disciples of the ICS structure use this method to delegate an ISO. In essence, the IC appoints an eligible officer to the ISO position. The eligibility level is established by the department based on which level of officer (lieutenant, captain, battalion chief, and so on) it feels is trained and experienced enough to be an effective ISO. A typical example of this type of ISO system is the automatic assignment of a second battalion chief (BC) to the incident. The first BC is responsible for assuming command while the second BC fills the ISO role. In another example, some departments design the ISO system so that a later-arriving company officer (third or fourth due) automatically reports to the IC as the ISO (the officer's crew fills support tasks such as running the accountability system, initiating crew rehab, or other like tasks). In both examples, for the system to be effective, the department *must* provide specific ISO training to each and every officer who may fill the role.

- *Advantages.* The pool to draw from is as large as the department has companies or on-duty BCs. The officer filling the ISO role may have a crew to fill critical on-scene safety roles like rehab or running the crew accountability system. If an ISO is not needed by the IC, other assignments can be given or the officer is rotated back into staging as an available resource. In the case of BCs serving as the ISO, a certain level of credibility and respect might be assumed given the chief officer designation.
- *Disadvantages.* The department must ensure that all eligible officers are ISO trained. This obligation requires specialized training for what could be a large group of officers.

Dedicated ISO

Of all the systems, the use of a designated ISO is perhaps the most desirable. The system amounts to a 24/7 duty position with an ISO always available. In large metropolitan areas, multiple ISOs can be on duty to cover the potential for simultaneous significant events.

- *Advantages.* The ISO is responding for a specific assignment and is positioned to have focused, duty-related training, experience, and proficiency. The ISO can assist with other health and safety assignments, documentation, preplanning, or proactive safety training when not actively engaged at an incident. Departments with a strong safety attitude and commitment to firefighter safety and available staffing are more likely to have a dedicated ISO system.

- *Disadvantages.* The system requires commitment from the department, including funding additional positions. Departments with a significant incident injury rate, a significant volume of working incidents, or a high potential for firefighter injury are the ones that typically choose this method.

Some fire departments use the officer of the RIC to serve as the ISO due to the standby nature of the assignment. Using the RIC officer for the ISO is *not recommended* by this text mainly because the RIC officer's focus is to first and foremost prepare and preplan resources for firefighter rescue. ISO duties such as risk/benefit judgments, rehab evaluation, incident action plan review, resource allocations, and traffic observations can distract from Mayday preplanning efforts. If the RIC is activated for a real or potential rescue situation, who takes the ISO role when it is arguably most needed?

The concept of "embedded safety officers" is being used by some fire departments. The concept centers on an ideology that *all officers* working at an incident should act as safety officer for their assigned crew and function. Departments that use the embedded safety concept do not use a separate, dedicated ISO unless it is mandated by the incident type (i.e., technician-level hazmat incident). Departments with a strong safety culture that have trained *all* officers in the functions of an ISO might benefit from this ultraistic concept, although a point can be made that the accomplishment of tactical/task objectives requires linear thinking, whereas an effective, dedicated ISO is free to use cyclic thinking (more on this in Section 3 of this text). Further, NFPA standards (1500, 1561) and NIOSH LODD investigative reports strongly support the appointment of a dedicated ISO, separate from the IC, at working incidents. Given these arguments, the use of a dedicated ISO, reporting as a command staff member, is *recommended* by this text—even for those who are using, or transitioning to, an embedded safety officer approach.[1]

Preparing the Incident Safety Officer

■ Define the ISO Plan

The whole point of a fire and rescue department's ISO plan is to ensure that the ISO is used at significant incidents. To get the job done, the ISO needs the department to define the options (many of which are covered in this text) and commit them to writing (SOPs) for use by the line officers. Appendix A includes sample SOPs for the design of an ISO system.

When a department does not formalize a system for automated or rapid delegation of the ISO function, or chooses not to, the ISO, if ever appointed, faces a difficult task. One area that has not been discussed is that of *authority* in designing an ISO system. What kinds of authority should the ISO have? This is an area of some controversy in the fire service ranks.

Issue: Should the ISO have the authority to stop an unsafe act and correct it on the spot? Consider the following arguments:

- *Yes.* Those who agree that the ISO should be able to stop an unsafe act recognize that the whole purpose of the ISO is to make the incident safer. NFPA 1500, NFPA 1561, and numerous requirements of the OSHA Code of Federal Regulations (CFR) give the ISO the authority to stop, alter, or terminate activities if an imminent threat exists.
- *No.* Those who disagree believe that the ISO, by stopping or correcting any act on the fire ground, is actually countermanding the IC. Acts deemed unsafe by the ISO should be immediately reported to the IC for a decision to stop, alter, allow, or correct them.

Regardless of the arguments, an important point is that the ISO can do tremendous damage to an effective ISO program by exploiting the authority of the position. As with the other components of an effective ISO system, the authority issue needs to be addressed *prior* to the incident. Obviously, the intent of this text is to support the ISO's authority to stop, alter, or suspend operations that pose an imminent threat, and then report it to the IC.

There is an evolving trend that takes the ISO's emergency authority (stop, alter, and suspend activities for imminent hazards) and extends it to *all* responders operating at an incident. This trend was first recommended following the South Canyon fire in Colorado that claimed the lives of 14 firefighters on July 6, 1994 (also known as the Storm King Incident).[2] In fact, the National Fallen Firefighters Foundation supports the trend in its 16 Firefighter Life Safety Initiatives. Initiative 4 states, "All firefighters should be empowered to stop unsafe acts."

■ Train the ISO

As with any fire service discipline, a system for initial and ongoing training is essential. Many departments assume that the training/safety officer or any company officer knows what it takes to fill the ISO role. This assumption is dangerously flawed and is one of the reasons that NFPA 1561 requires ISOs to meet the requirements of Fire Officer I as written in NFPA 1021, *Standard for Fire Officer Professional Qualifications.* The fire service members of the NFPA committees responsible for standards 1561 and 1521 recognize that the ISO position requires a level of training and education that exceeds that of a company officer: It's a different knowledge and skill set. Section 2 of this text is dedicated to the key training areas that help develop an ISO and makes it clear that the knowledge and skills required of an ISO exceed those of the typical Fire Officer I. Chapter 5 in NFPA 1521 lists the professional qualification requirements for an ISO, and most, if not all, of these are addressed in this text. Combine the topics from this text with an ongoing continuing education schedule and the department has an ISO training program.

The Fire Department Safety Officers Association (FDSOA) can serve as a training resource for the ISO. The FDSOA website, www.FDSOA.org, offers network forums, products, and information on training opportunities to help the ISO.

Departmental drills, exercises, and other training sessions can serve as an opportunity to help keep ISOs proficient and can help prepare new ISOs by shadowing an assigned, qualified ISO.

■ Give the ISO Tools to Do the Job

The ISO program should consider the tools required for the ISO to get the job done **FIGURE 4-11**. The following list of items is created from experienced ISOs from around the country.

- *Radio/phone.* As explained in subsequent chapters, the effective ISO is constantly roving and watching. A radio is essential to maintaining contact with the IC as well as monitoring the working crews. Smartphones are valuable because they provide a secondary method to communicate with the command post and can be loaded with applications that contain useful information that the ISO may need to access. Further, they provide a quick, easy-to-use camera to capture images. A warning though: The portable, personal digital device age seems to have brought a "heads-down" distraction as users manipulate their device. The ISO role is very "heads-up" (remember the hazard MEDIC mnemonic?).

- *High-visibility identification.* Many departments use vests to help identify members filling various ICS positions. It is recommended that the safety officer–labeled vest be a unique color. Not only does this make the ISO easy to spot, but it serves as a reminder to all crews that their safety is important. It's interesting that many ISOs report a remarkable change in firefighter behavior when the firefighter discovers that the "safety vest" is nearby. This may seem trivial, but it works. The San Jose (California) Fire Department uses green vests to separate the ISO from other ICS position titles. The Superstition Fire and Medical District in

Arizona uses a green helmet for an assigned ISO. This text recommends green as a color choice for the SO identifiers because of its association with safety organizations (such as the National Safety Council or Fire Department Safety Officers Association). As with all vests, bright-colored, light-reflecting trim should be included to help with night time and low-light visibility. Wearing a self-contained breathing apparatus (SCBA) will obviously cover up an identifying vest. A few vendors offer NIMS title position sleeves that fit over the SCBA bottle. They are also easy to make. Likewise, green arm bands are easy to make and can help with the SCBA issue.

- *Personal protective equipment (PPE).* To tour the scene, the ISO may have to cross into "hot zones" or other areas that require PPE. If assigned to check out interior, immediately dangerous to life and health (IDLH), or high-risk areas, the ISO should request a partner from the IC and be processed through the crew accountability system just like any crew. Also, the ISO sets a good example by using appropriate PPE.

- *Clipboard file box.* An ISO can make a difference if he or she is prepared for things the IC expects. A metal clipboard file box can give a working surface for notes, sketches, checklists, and other documents. The file box can contain accident report forms, quick-reference sheets, required ICS forms, safety briefing forms and reminders, tablets, pencils, and other supplies. One idea worth considering is to attach a hard-plastic sheet to the back side of the clipboard and keep a grease pencil or dry-erase marker handy so quick signs can be

FIGURE 4-11 Essential ISO tools include position identification, radio, phone, box light, documentation kit, PPE, and zone-marking tape.

FIGURE 4-12 A hard-plastic sheet and grease pencil can be used to flash messages to crews in a hot zone.

made—like a *"Look up!"* message that can be flashed to crews in a hot zone **FIGURE 4-12**.

- *Zoning tape.* Yellow barrier tape is perhaps overused and, in many ways, ignored by firefighters (it's for public warning—not us!). It is even used in "creative" retail store displays—further overuse. The ISO should have access to red and white slashed or chevron tape to denote "no-entry" zones for firefighters. (Zone-marking schemes are further discussed later in the text.)
- *Box light.* Most firefighters carry a personal flashlight in their pocket. The ISO should carry a personal flashlight as well as a more powerful, flood-type box light to help spot hazards in low-light and night-time environments. A flood-type projection lens (as opposed to a narrow spot light) helps to illuminate more potential hazards in a single sweep and assists in the reading of smoke.
- *Miscellaneous.* Thermal imaging cameras, helmet-cams, gas detectors and monitors, whistles, electrical current detection instruments, multipurpose hand tools, lock-out/tag-out kit, and other pieces of equipment are used by many ISOs. These and other items may be necessary based on local needs and experiences.

Designing an ISO system may sound like an overwhelming task, but know that there are many departments with good systems in place today. Your department should be able to use this chapter to outline a workable system to ensure that a qualified and well-equipped ISO is available for significant incidents. Remember that the sooner the ISO position is filled, the better the chances are for a safe operation.

Getting the Job Done

What's in Your Pockets?

Numerous textbooks, trade magazine articles, social-media posts, and firehouse kitchen table discussions have centered on the topic of what firefighters should carry in their PPE ensemble pockets. Items such as extra gloves, personal flashlight, and some kind of personal multipurpose tool usually make the mandatory list. Veteran firefighters expand the list to include door-stops, marking chalk, and nylon webbing. Some take it to the extreme and list so many items that you wonder how they fit it all in the limited pocket space.

Specific to the ISO function, what pocket items do you think would be most essential? The ISO pocket item question is personal; therefore, no answer is likely to be judged wrong or right. However, we can share a list of items that practicing ISOs seem to use often (you will likely discover items that could be added here):

- *Grease pencil.* This item is used to mark doors or create messages on the back of a clipboard that can be flashed to crews. The grease pencil can also be used to mark a crack in a wall. The initial mark serves as a starting place to monitor the crack to see if it is getting wider or worse.
- *Personal flashlight.* The ISO should not rely on a pocket flashlight at night or for low-light areas; rather, a separate flood-type box light should be used for such applications. Still, a pocket flashlight should be carried. It should be of the narrow-beam variety, ideal for peering into holes or cracks and getting someone's attention. (Do not point your high-intensity, Xenon-gas, smoke-cutting, laser-focused flashlight into someone's eyes!)
- *Vice-grip pliers.* A commercial-grade pair of vice grips can be helpful to the ISO in numerous ways. They can be used to attach a warning sign or barrier tape to an object, hold a door or gate open (clamp the hinge), clamp and twist-break a seal tag on equipment, and so on. Although the ISO should avoid getting caught up in performing operational tasks, the vice grips can help flag or temporarily address a threat while conducting incident reconnaissance.
- *Contractor's flagging tape.* This is not the wide barrier tape used to mark zones. Flagging tape is the narrow vinyl variety that can easily fit into a pocket. Hardware stores typically carry a selection of bright colors. The tape can be used to flag hazard items or to create hang-down streamers that warn of an overhead danger. It can also be used to signify that an item or territory has already been checked.
- *Nylon webbing.* A 1-inch wide by 20-foot long piece of webbing can fit in most PPE pockets. Some ISOs also like to have a wide-opening carabineer to go with the webbing. The webbing can be tied off to hold doors and gates open, to secure loose or flapping materials, or even to loop and move a downed electrical wire (DANGER: as a life-saving measure only).

Wrap-Up

Chief Concepts

- To be successful, an ISO response system needs to be preplanned and captured in department standard operating procedures and policy. Department size, deployment strategy, and unique needs help mold the ISO program. At a minimum, the program should address the ISO program requirements that are outlined in NFPA 1561. Highlights of NFPA 1561 include the following:
  - Criteria for the response or appointment of a safety officer and assistant safety officers when the activities, size, or need warrants extra safety personnel.
  - A policy to ensure that a separate ISO (independent from the IC) responds automatically to predesignated incidents.
  - Provision that safety officers shall meet the requirements for a Fire Officer I (NFPA 1021).
  - Provision that the ISO shall be integrated within the ICS as a command staff member.
- Too often, ICs delegate the ISO function after an incident grows and becomes difficult or complex. Fire ground realities and many recommendations from line-of-duty death reports support the argument that a separate ISO is assigned in the early stages of an incident. Factors such as environmental rate of change (fire growth, structural degradation, and smoke volatility), number of task assignments, and responder risk-taking are at their worst in the early part of the incident. For that reason, departments should create a response system that ensures that a predesignated ISO responds and arrives in the early stages of incident handling.
- Types of incidents that warrant an automatic ISO response include all reported structure fires, wildland fires, incidents at target hazard sites, aircraft crashes, incidents requiring a specialty team, and incidents that occur during extreme weather conditions.
- In cases where no ISO was required to respond, the IC should have a written policy specifying when to delegate the safety function to a qualified ISO. Examples include situations where the initial response resources are 100% committed and more are needed (a *working incident*), the IC span of control exceeds three, mutual-aid incidents, and firefighter Maydays.
- There are advantages and disadvantages to various ISO response programs. The most desirable one is a dedicated 24/7 ISO position. Other ISO response programs include the use of a second battalion chief, coverage by training officers, or the use of safety and health committee members. It is *not* recommended that the officer assigned to a rapid intervention crew be used as an ISO because the focus of the two positions is different.

- An effective ISO response program addresses the authorities, training, and tools necessary to be successful. NFPA 1500, NFPA 1561, and various requirements of the CFR give an assigned ISO the authority of the IC to stop, alter, or suspend activities that present an imminent threat of harm. ISOs who exercise that authority must report the action to the IC immediately.
- NFPA 1521 and the contents of this text help outline the requirements that lead to a qualified ISO. Ongoing continuing education and participation as an ISO at drills and other department training events can help keep ISOs proficient.
- In terms of tools and equipment, the ISO should have a radio, smartphone, PPE, highly visible and unique position identifier (vest or SCBA sleeve), zoning tape, and box light.

Key Term

working incident An incident at which the initial response assignment will be 100% committed and more resources are needed.

Review Questions

1. Explain the reasons that the ISO role should be preplanned.
2. List four examples of when an automatic ISO response is beneficial.
3. List four examples of when automatic ISO delegation should take place.
4. List three methods to get an ISO on scene, and discuss the advantages and disadvantages of each.
5. Explain the authority given to the ISO by NFPA standards.
6. List four tools that can help the ISO be effective on scene.

References and Additional Resources

Fire Department Safety Officers Association (FDSOA) website contains network forums, products, and training opportunities and information on ISO certification. It is available at https://www.fdsoa.org.

National Fallen Firefighters Foundation. *Understanding and Implementing the 16 Firefighter Life Safety Initiatives*. Stillwater, OK: Fire Protection Publications, 2011.

NFPA 1021, *Standard for Fire Officer Professional Qualifications*. Quincy, MA: National Fire Protection Association, 2020.

NFPA 1500, *Standard on Fire Department Occupational Safety and Health Program*. Quincy, MA: National Fire Protection Association, 2018.

NFPA 1521, *Standard on Fire Department Safety Officer Professional Qualifications*. Quincy, MA: National Fire Protection Association, 2020.

NFPA 1561, *Standard on Emergency Service Incident Management System and Command Safety.* Quincy, MA: National Fire Protection Association, 2020.

Endnotes

1   The author believes that the goal of any safety officer is to eliminate the need for safety officers. To accomplish this, *everyone* operating at an incident scene needs to think and act like a dedicated ISO: Continually monitor and predict hazardous outcomes, stop unsafe acts, and embrace a risk/benefit approach to tactical/task accomplishment. Injury and fatality data suggests that we, the fire service, have much work to do before this goal is obtained. Until then, the use of a separate, dedicated ISO is warranted—even for those that are utilizing or transitioning to an embedded safety officer approach.

2   *Final Report of the Interagency Management Review Team, South Canyon Fire,* was issued June 1995 and is available at http://www.wildfirelessons.net. Members of a hotshot team were critical of their assignment and location, but supervisors with a "can-do" attitude decided to stay. The team was trapped in a blow-up event, and 14 firefighters died. The follow-up investigations and recommendations pointed to the need of changing safety cultures: Everyone has a responsibility for their own safety and those of their team.

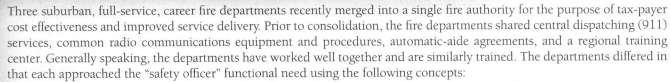
Three suburban, full-service, career fire departments recently merged into a single fire authority for the purpose of tax-payer cost effectiveness and improved service delivery. Prior to consolidation, the fire departments shared central dispatching (911) services, common radio communications equipment and procedures, automatic-aide agreements, and a regional training center. Generally speaking, the departments have worked well together and are similarly trained. The departments differed in that each approached the "safety officer" functional need using the following concepts:

- Fire department #1: A dedicated safety officer staffed 24/7 with ISO-trained captains who respond to incidents and report to a battalion chief who serves as the IC.
- Fire department #2: Staff chiefs (training, EMS, and fire prevention) who take turns being on duty to respond from work or home and to serve as the ISO for a battalion chief IC.
- Fire department #3: Use an "embedded safety officer" concept where all captains and chiefs are expected to act as a safety officer in the performance of their tactical assignments (e.g., division/group supervisor, ops section chief). The battalion chief serving as the IC rarely appoints a dedicated ISO and relies on embedded safety thinking by officers at each tactical level to correct hazards.

The new fire authority will serve a geographical area of 150 square miles, a population of approximately 150,000, and an anticipated annual call volume of 18,000 incidents. There will be two battalion chiefs for incident command/control and the management of a suppression force that includes twelve fire stations. The battalion chiefs wish to support each other at incidents using a "command advisor" concept where one is the IC and the other develops and troubleshoots the incident action plan and assists with resource management (as opposed to serving as an ISO). Each engine and truck company will be staffed with three personnel (captain, apparatus driver/firefighter, and paramedic/firefighter).

As part of the merger, a task group has been created to weigh the pros and cons of each safety officer system and recommend a uniform approach that includes SOPs, budget/staffing needs, and training requirements.

1. What are the advantages and disadvantages of each of the safety officer systems described?
2. What are the operational challenges that the merged authority will face as it relates to suppression force safety?
3. Regardless of the system used, what types of incidents mandate the assignment of a separate ISO?
   A. Working structure fires
   B. Wildland–urban interface fires
   C. Technician-level hazmat incidents
   D. Aircraft incidents

4. What is considered the minimum level of training required for anyone serving as an incident safety officer?
   A. NFPA 1021, Fire Officer I
   B. NFPA 1001, Firefighter II
   C. NIMS ICS, 100 level
   D. NIMS ICS, 400 level

Smoke: © Greg Henry/ShutterStock, Inc.

Front-Loading Your Knowledge and Skills

The first section of this book focused on introducing the role of a fire department incident safety officer (ISO) and presenting the basic concepts for using the ISO. In this section we offer specifics on how you can prepare yourself to address the ISO's functions. Working ISOs around the country agree that to be effective they must acquire additional knowledge and skills well past the Fire Officer I level. Your efforts to acquire the knowledge, skills, and attitude needed to be an effective ISO are what we will call "front-loading."

The goal of ISO professional development is to achieve "mastery." Generally speaking, mastery can be defined as the ability of an individual to achieve 90% of an objective 90% of the time. As it relates to the ISO—and incident handling—mastery may seem hard to gauge. Perhaps a better way to describe mastery for any incident responder is the ability to perform with a certain "unconscious competence." That is, you perform your functions with an analytical automation that is deeply connected with and appropriate for the incident. Some argue that mastery of the ISO function is not obtainable—a compelling argument, given the diversity of the emergency response world. The intent, however, is to encourage the *pursuit* of mastery—and, therefore, ever-increasing efficiency and effectiveness in filling the ISO role.

In this text, we use the word *efficiency* to mean doing things right and the word *effectiveness* to mean doing the right things. To become efficient and effective, the ISO must learn, *then* perform. Learning and performance are not the same. Learning is the acquisition of the knowledge, skills, and attitude needed to achieve mastery. Performance is the demonstration of acquired mastery in the incident environment. Coming full circle, learning is the "front-loading," and performance is the application of that learning. Here in Section 2 we look at the learning areas, and in Section 3 we look at the performance side.

Voices of Experience

On the afternoon of Sunday April 12, 2012, I responded to a fire in a 150-year-old, balloon-frame 20- by 50-foot (6- by 15-meter) three-story, wood-frame building. The importance of front-loading knowledge and skills couldn't have been more apparent than at this fire.

Voices of Experience (continued)

Fire originated in the cellar and extended into an alley on the B side that was about 3 feet (1 meter) wide. The exterior wall of the fire building was covered with asphalt shingles that ignited. Fire then extended from the exterior wall of the fire building into the second and third floors of the adjoining building on the B side.

The first-arriving engine arrived in 3 minutes and transmitted a working fire to the dispatcher. At first, there was some confusion by first-arriving units trying to determine the original fire building; fire was burning in two adjacent buildings. The wood-frame building was quickly identified as the original fire building. Following the initial knockdown of the main body of fire inside the original fire building, units began opening up the interior wall on the B side to locate and expose fire in the bays of the exterior wall. The interior stairs located on this B side wall created difficulties in opening up this area of concern; the fire continued to progress as these walls were opened up.

A member operating in the exposure on the B side transmitted an urgent message on his portable radio that the exterior wall of the fire building was showing signs of deterioration. The ISO notified the incident commander (IC) to evacuate the fire building immediately. During the evacuation, the B side wall of the fire building collapsed, causing the second and third floors to collapse onto the first floor.

Two members were trapped in the collapse, and the fire continued to expand rapidly. Maydays were transmitted, and the rapid intervention crew (RIC) was deployed. As units continued to fight the fire, the ISO along with the RIC entered the fire building to assist the trapped firefighters. The RIC quickly removed the trapped firefighters, who suffered serious but non-life-threatening injuries. The fire eventually extended to both exposures on the B and D sides, and a fourth alarm assignment was required to bring it under control.

This collapse occurred 40 minutes after the initial alarm and 25 minutes after the arrival of the ISO. According to the National Incident Management System, the IC is responsible to fill the role of the safety officer until that role is delegated to an ISO.

As the health and safety officer (HSO) for the Fire Department, City of New York (FDNY), I was on call that Sunday afternoon and was notified to respond on the transmission of a third alarm. I arrived at the scene shortly after the collapse. As the HSO, I operated at the command post with the IC and sectored the safety officer duties on the scene. One safety officer supervised the removal of the trapped firefighters; the other was deployed to the C side of the collapse to monitor for a secondary collapse.

For the ISO, education and experience work hand in hand. *Reading the building, reading the smoke, reading the risk, and reading hazardous energy* all needed to take place simultaneously and quickly. Conditions were changing rapidly.

The skill of reading buildings enhances the ability of the ISO to predict potential collapse. The ISO must understand wood-frame building construction and know that these buildings, especially the older, freestanding type, will present a serious collapse danger. Fire burning in the concealed wall spaces on the bearing walls of a wood structure is a serious collapse indicator. This information must be conveyed to the IC in a timely manner. In the scenario presented here, the experienced ISO used recognition-primed decision making (RPD) to make quick, effective decisions when faced with the complex situations of a fast-moving fire and collapse.

This was an intense fire that was burning for a long time, more than 20 minutes. This fact should initiate communications between the ISO, operating members, and chief and company officers that a change in tactics may be necessary. Indicators of collapse must be communicated to all units on the scene. The ISO reacted quickly to the report of the deterioration of the exterior wall, but the evacuation of the fire building by operating forces took a considerable amount of time. Firefighting forces need to understand the urgency when the order to evacuate is given by the IC.

As the ISOs were addressing the collapse and the removal of the trapped firefighters, I continued to *read the building*, assigning another safety officer to monitor the fire building for a secondary collapse. I also *read the smoke*, assisting the IC to determine the extent of the expanding fire and the effectiveness of the fire streams. The potential danger of overhead electrical lines in front of the fire buildings was a concern; therefore, *reading hazardous energy* was also required. This danger needed to be communicated continuously to all members throughout the operation to enhance situational awareness and safety. At the same time, *reading the risk* was extremely important; firefighters were trapped, the RIC firefighters were performing extrication, and firefighters were operating the hose lines to protect the trapped members. As firefighters, we are expected to put ourselves in harm's way to save a life, especially one of our own. In this scenario, the risks taken were considered acceptable.

Finally, as the trapped firefighters were removed, treated by EMS, and transported to a hospital, *reading the firefighters* became a priority. Operating at this fire and removing the trapped firefighters was arduous work. It was also emotionally taxing for everyone involved. A rehabilitation group was established for members to provide rest and recovery. Operating firefighters were demobilized; they doffed their PPE, rehydrated, and were monitored by EMS personnel.

A postincident analysis was conducted following this incident to assess and analyze the operation. It was used to determine what occurred at this incident and why it happened. We focused on resources, procedures, equipment, and improving operational effectiveness. It generated lessons learned and lessons reinforced on how we can do things better the next time. This review included the operations of the HSO and ISO.

Stephen Raynis
Chief of the Bureau of Training (ret.)
Fire Department, City of New York
Brooklyn, NY

CHAPTER

5

Reading Buildings

Knowledge Objectives

Upon completion of this chapter, you should be able to:

- Describe the relationship of loads and load imposition in a building. (p 58)
- List the three types of forces created when loads are imposed on materials. (p 58)
- Describe the effect that fire has on building materials, loads, and forces. (NFPA 5.3.3 , pp 58–61)
- Define columns, beams, and connections. (pp 61–63)
- List and define the influences used for building construction classifications. (p 64)
- Define and list several types of hybrid buildings. (p 67)
- State the structural collapse hazards of common building construction classifications. (NFPA 5.3.3 , pp 65–69, 71–73)
- List, in order, the five-step analytical approach to predicting building collapse. (p 70)
- Describe the structural collapse indicators present at an incident scene. (NFPA 5.3.3 , pp 68, 71–73)
- Describe the additional structural collapse indicators present after a collapse has occurred. (NFPA 5.3.3 , p 74)
- Identify the structural collapse issues that should be communicated to the rapid intervention crew by the incident safety officer. (NFPA 5.3.2 , p 74)

Skills Objectives

Upon completion of this chapter, you should be able to:

- Identify common building construction types. (NFPA 5.3.3 , pp 64–67)
- Identify, analyze, and make judgments regarding collapse potential at building fires. (NFPA 5.3.3 , pp 70–74)
- Adjust an incident action plan based on incident evaluation to improve member safety. (NFPA 5.3.3 , p 74)

One of the areas of knowledge that you have identified as a requirement of the incident safety officer (ISO) position is an up-to-date understanding of building material, construction, and the forces that hold buildings up and allow them to fail. Arriving at the scene of a structure fire, you begin to triage the building by assessing its type, size, use, and condition, and the location and estimated impact of fire on the structural components.

1. What information does this assessment of the building provide for the ISO and the incident commander relating to the safety of the crews operating and the overall incident action plan?
2. In your response area, which types and eras of building construction are the most prevalent?

Introduction: The Collapse Warning

Incident safety officers (ISOs) must be able to give an incident commander (IC) explicit detail and their judgment regarding the collapse potential of a given building being attacked by fire. Likewise, the ISO needs to communicate building construction considerations and observations to a designated rapid intervention crew (RIC) so that contingencies can be developed. The ISO must draw from a significant knowledge base in order to make collapse judgments and communicate them effectively. Fire officers who draw only from experience to predict building collapse are fooling themselves into a situation in which a building will collapse without warning. Francis Brannigan, noted author of *Building Construction for the Fire Service*, says it best: "Relying on experience alone is not sufficient. Firefighters must be aware of the theories and principles involved [in building construction]."[1] To say that a building collapses without warning is a flawed statement. The warning for structural collapse is found in the ISO's ability to understand building construction and the effects of fire on the building. Granted, once collapse occurs, unpredictability increases because structural elements are no longer in a physical position to carry load as they were designed.

This chapter is not a substitute for an in-depth study of the building construction texts that focus on the subject; rather, the chapter takes some essential building construction topics and thought processes and shows you how to evaluate and communicate judgment on collapse potential. We will start with some fundamental terms and concepts that help you understand how buildings are built **FIGURE 5-1**. These concepts will assist us as we describe the various methods used to classify a building, which is the starting point for predicting collapse at your next fire.

Essential Building Construction Concepts

Predicting building collapse is dependent on the application of essential building construction concepts. ISOs use this knowledge base to help them "read" a building. Combined with smoke and fire observations, ISOs must apply skill to predict and communicate collapse potential and establish collapse

Fire Marks

The Ol' Professor

For the past 50 years, the name Francis L. Brannigan has been synonymous with building construction in the fire service. Although not a builder, the "Ol' Professor" was passionate about teaching firefighters the hazards and traps that come with fires in buildings. First published in 1971, his book, *Building Construction for the Fire Service*, has become a must-read for any firefighter. The book has gone through three editions, each becoming important study references for promotional tests and for those seeking to better understand buildings. Brannigan was writing the fourth edition when he died in his sleep in January of 2006. The fourth edition was published posthumously and has since been updated to a sixth edition with the help of Glenn Corbett.

Over the years, Mr. Brannigan has coined many powerful—and lifesaving—phrases. These bits of advice have remained part of the teachings of many fire instructors. The following quotes are taken from various editions of his book and from his teachings around the country:

- "The building is the enemy—know your enemy."
- "Beware the truss!"
- "The bottom chord of a truss is under tension—it's like you hanging on a rope. If the rope gets cut, you will fall. So it is with a truss."
- "The slightest indication of column failure should cause the building to be cleared immediately."
- "From an engineering point of view, [lightweight] buildings are made to be disposable. . . . We don't make disposable firefighters!"
- "Don't make light of a partial collapse. . . . A partial collapse is very important to at least two groups—those under it and those on top of it!"
- In response to those who claim a building collapsed without warning during a fire: "The warning is the brain—in your ability to understand buildings and anticipate how they will react to a fire."

The Brannigan student is better prepared to identify and analyze buildings that are being attacked by fire and to make judgments regarding collapse potential. The fire service is forever indebted to the Ol' Professor—a true *Fire Mark* in our history!

FIGURE 5-1 Understanding how a building is built and classified is the first step in predicting how it will fail.

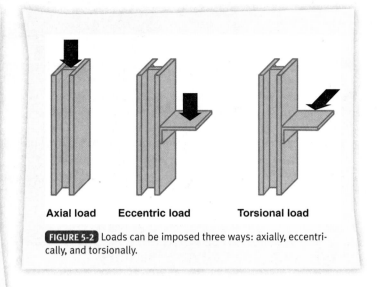

Axial load   Eccentric load   Torsional load

FIGURE 5-2 Loads can be imposed three ways: axially, eccentrically, and torsionally.

zones. The main topics include loads, characteristics of materials, and structural elements.

■ Imposition and Resistance of Loads

Because buildings are constructed to provide a protected space to shield occupants from the outside environment, the building must be built to resist the forces of wind, snow, rain, and gravity. Additionally, the intended use of the building can add a tremendous amount of weight, placing more stress on the building's ability to resist gravity. In building terms, these stressful elements create *loads*. Loads are nothing more than static and dynamic weights that are applied to buildings. More specifically, loads are divided into *dead loads* and *live loads*. A dead load refers to the weight of the building itself and anything permanently attached to it. A live load refers to any force or weight, other than the building itself, that a building must carry or absorb. In simple terms, a live load is anything other than a dead load. The engineering world further defines loads based on the manner in which the load is delivered—concentrated, distributed, static, suspended, or impact.

Loads are then *imposed* on building materials. This imposition causes stress and strain on the materials—called *force*. Forces must be delivered to the earth for a building to be structurally sound. The basis of all building construction techniques is to carry a load to earth. The load itself must be described or classified (such as a "distributed live load"). The load is then applied to another component in some fashion; this is called the *imposition* of the load. Loads can be imposed three ways **FIGURE 5-2**:

- *Axially*. An axial load is imposed through the centroid of another object.
- *Eccentrically*. An eccentric load is imposed off-center to another object.
- *Torsionally*. A torsional load is imposed in a manner that causes another object to twist.

The material that has been imposed must resist the load, which creates a *force* within the receiving material. The three types of forces are compression, tension, and shear **FIGURE 5-3**.

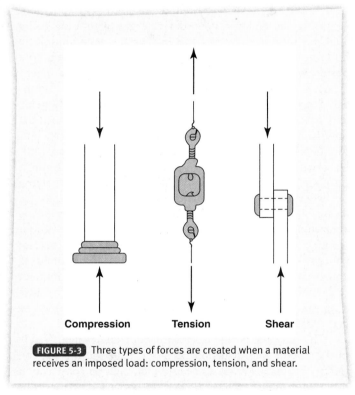

Compression   Tension   Shear

FIGURE 5-3 Three types of forces are created when a material receives an imposed load: compression, tension, and shear.

- *Compression*. A force that causes a material to be crushed or flattened axially through the material.
- *Tension*. A force that causes a material to be stretched or pulled apart in line with the material.
- *Shear*. A force that causes a material to be torn in opposite directions perpendicular or diagonal to the material.

We can exercise the use of the above concepts using **FIGURE 5-4**, which represents the structure of an all-steel building. The floor above is a *distributed dead load* that is *imposed axially* to the vertical columns *creating a compressive force*. When wind (or an earthquake) creates a *live, torsional impact load* on the wall, the columns will want to twist. The "X" bracing you see (called a *moment frame*) is designed to limit the twisting of the columns. Likewise, an applied wind will create a *shear*

heat resistance, and heat resistance is time. From an ISO's point of view, the more mass a material has in a given surface area, the more time (or heat) is required before the material starts to degrade.

Before we look at the specific characteristics of materials, we must introduce another concept from the engineering community. When a material attempts to resist forces, a reaction is created that is attempting to twist, smash, or pull the material's molecules apart. In other words, the forces are trying to deform the material. Some materials can deform yet remain strong, while others will actually break apart as they deform. Thus, material is classified as being <u>brittle</u> or <u>ductile</u> based on its reaction to imposed loads and resistance to forces:

- *Brittle.* Materials that will fracture or fail as they are deformed or stressed past their design limits.
- *Ductile.* Materials that will bend, deflect, or stretch—yet retain some strength—as a force is resisted.

Simply stated, a brittle material breaks before it bends, and a ductile material bends before it breaks. Masonry, tile, and cast iron are brittle, whereas wood, plastics, and most alloyed metals are ductile. With an understanding of material characteristics, we can now discuss individual materials.

In the past, the fire service looked at the characteristics of four basic material types: wood, steel, concrete, and masonry. These materials can be found together or separately, and each reacts to fire in a different way TABLE 5-1 . Today, advanced material technologies have found their way into structural elements. Buildings are being assembled using plastic, graphite, wood derivatives, and other composites. In this section, we cover the four basic building materials as well as some of the new composites.

Wood

Wood is a ductile material that has marginal resistance to forces compared to its weight, but it does the job for most residential and small commercial buildings. We also know that wood burns and in doing so gives away its mass. The more mass a section of wood has, the more material must burn away before its strength is lost. For the purposes of understanding the manner in which wood burns, it can be divided into two basic types—native wood and engineered wood.

Native wood is wood that has been cut from a tree. Wood that has been rough sawn from a tree will develop a surface char when exposed to flame, slowing its burn rate. Native wood that is smooth finished (planed) has a tendency to alligator-check (i.e., take on an alligator skin-like pattern) and burn more quickly. Investigators can use these unique burn patterns to "read" the type of wood involved and thereby estimate the burn rate.

<u>Engineered wood</u> can react differently when exposed to heat from a fire. Engineered wood includes a host of products that consist of many pieces of native wood (chips, veneers, and sawdust) glued together to make a sheet, a long <u>beam</u> (trees grow only so tall), or a strong <u>column</u>. The glues that bind engineered wood products require only heat to break down. In some cases, the heated smoke from a distal (distant) fire can cause engineered wood products to degrade.

FIGURE 5-4 Can you envision the loads, imposition of loads, and forces created in these structural features?

force on all the bolted connections. Clearly, the connections must resist that force, which brings us to the characteristics of materials.

■ Characteristics of Building Materials

When forces are created through the imposition of loads, a receiving material must resist and transfer the force. A material's response to applied force depends on its load-bearing characteristics. These characteristics include:

- Type of material (e.g., wood, steel, concrete)
- Shape of the material (e.g., round, square, rectangle)
- Orientation or plane of the material (e.g., vertical, horizontal)
- Mass of the material (e.g., surface-to-mass ratio, density, thickness)

Additionally, the fire service looks at how these materials react during a fire and how their ability to resist a load changes during fire conditions. Of importance is the dual concept of mass and fire resistance. The *mass* (or surface-to-mass ratio) of a material directly affects its fire resistance. In essence, mass is

TABLE 5-1	Performance of Common Building Materials Under Stress and Fire			
Material	**Compression**	**Tension**	**Shear**	**Fire Exposure**
Brick	Good	Poor	Poor	Fractures, spalls, crumbles
Masonry block	Good	Poor	Poor	Fractures, spalls
Concrete	Good	Poor	Poor	Spalls
Reinforced concrete	Good	Fair	Fair	Spalls
Stone	Good	Poor	Fair	Fractures, spalls
Wood	Good across grain Marginal with grain	Marginal	Poor	Burns, loss of material
Engineered wood	Good	Fair	Fair	Burns, rapid loss of materials, heat can degrade glue, causing rapid failure
Structural steel	Good	Good	Good	Softens, elongates, loses strength
Cast iron*	Good	Fair	Fair	Easily fractures when heated

*Some cast iron may be ornamental in nature and not part of the structure or load bearing.

Steel

Steel is a mixture of carbon, iron ore, and alloys that are heated and rolled/extruded into structural shapes to form elements for a building. Steel is a ductile material that has excellent tensile, shear, and compressive strength. For this reason, steel is a popular choice for girders, lintels, cantilevered beams, and columns. Additionally, steel has high factory control; that is, it's easy to change its shape, increase its strength, and manipulate it during fabrication. It is important to note that the shape of steel is engineered for a very specific application. For example, a steel *column* used for compression forces is best shaped as a square or circle and is typically oriented vertically with the load imposed on top. A steel *beam* will be shaped as an "I" when viewed from its end and is typically oriented horizontally with the imposed load on the top or bottom of the "I" stem. Any change to the shape or orientation of the imposed load will have a negative implication on material integrity.

In a fire, steel loses strength as temperatures increase; the specific range of temperatures at which it loses strength depends on how the steel was manufactured. Cold-drawn steel, such as cables, bolts, rebar, and lightweight fasteners, loses 55% of its strength at 800°F (427°C). Extruded structural steel used for beams and columns loses 50% of its strength at 1,100°F (593°C). Structural steel also elongates or expands as temperatures rise. At 1,100°F (593°C), a 100-foot (30-m) long beam can elongate 10 inches (25 cm). Now, imagine what that could do to a building. If a steel beam is fixed at two ends, it tries to expand and likely deforms, buckles, and collapses. If the beam sits in a pocket of a masonry wall, it stretches outward and places a shear force on the wall, which was designed only for a compressive force. The expansion of the beam can knock down the entire wall. Steel softens, elongates, and sags when heated, leading to collapse.

Concrete

Concrete is a mixture of Portland cement, sand, gravel (or other aggregate), and water that cures into a solid mass. Concrete has excellent compressive strength when cured. The curing process creates a chemical reaction that bonds the mixture into a solid mass to achieve strength. The final strength of concrete depends on the ratio of these materials—especially the ratio of

Safety Tip

Cool Steel

Yes, steel is pretty darn cool because it is so strong and is amazingly versatile: During production you can shape it, size it, and add many other strengtheners to make it fit almost any application. Unfortunately, steel is not so cool when exposed to the heat of a fire. Steel softens, elongates, and sags when heated to 800°F (427°C) for cold-drawn (cables, rebar, bolts) or 1,100°F (593°C) for extruded (columns, beams). These traits cause a loss of shape and load orientation. Collapse can happen quickly, depending on the load the steel is carrying. *Cooling structural steel with fire streams is just as important as attacking the fire.*

water to Portland cement. Because concrete has poor tensile and shear strength, steel is often added as reinforcement. Steel can be added to concrete in many ways. Concrete can be poured over steel rebar, which becomes part of the cured concrete mass. Cables can be placed through the plane of concrete and be tensioned, compressing the concrete to give it the required strength. Cables can be pre-tensioned (at a factory) or post-tensioned (at the job site). Precast concrete consists of slabs of concrete (with embedded steel) that are poured at a factory and then shipped to a job site. Precast slabs are "tilted up" to form load-bearing walls; hence the phrase "tilt-up" construction.

All concrete contains some moisture and continues to absorb and wick moisture (humidity) as it ages. When heated, this moisture content expands, causing the concrete to crack or spall. Spalling refers to the crumbling and loss of concrete material when exposed to heat. Spalling can take away the critical mass of the concrete—that is, the mass used for strength. Reinforcing steel that becomes exposed to a fire can transmit heat within the concrete, causing catastrophic spalling and failure of the structural element. Unlike steel, concrete is a heat sink and tends to slowly absorb and retain heat rather than conduct it. This heat is not easily reduced. Concrete can stay hot long after the fire is out, causing additional thermal stress to firefighters performing overhaul.

Masonry

Masonry is a common term that refers to brick, concrete block, and stone. Masonry is used to form load-bearing walls because of its compressive strength, but it can also be used to build a veneer wall. A veneer wall supports only its own weight and is commonly used as a decorative finish. Masonry units (blocks, bricks, and stone) are held together using mortar. Mortar mixes are varied but usually contain a mixture of lime, Portland cement, water, and sand. These mixes have little to no tensile or shear strength; they rely on compressive forces to stay bonded. In fact, a masonry wall relies on axially imposed compression forces to keep it strong. If the wood roof on a masonry wall burns away, the wall is no longer "sound" because it lost the load that was strengthening it. Any lateral impact load on the wall (like wind or fire streams) can cause collapse.

Brick, concrete block, and stone have excellent fire-resistive qualities when taken individually. Many masonry walls are typically still standing after a fire has ravaged the interior of the building. Unfortunately, the mortar used to bond the masonry is subject to spalling, age deterioration, and washout. During a fire, masonry blocks (or bricks) can absorb more heat than the mortar used to bond them, creating different heat stresses that can crack the binding mortar. Whether from age, water, or fire, the loss of bond causes a masonry wall to be very unstable.

Composites

New material technologies have posed interesting challenges for the firefighting community. The term "composite" can be used for many things but in this case refers to a combination of the four basic materials, as well as various plastics, adhesives, and assembly materials. Of particular interest are the many engineered wood products that are widely used for structural elements.

Lightweight wooden I-beams (the slang term is "I-joists") are a combination of two engineered wood products: laminated veneer lumber and oriented strand board. Laminated veneer lumber (LVL) is created by gluing and pressing together sheet veneers of native wood (with the same grain direction) to form a piece of lumber. LVL is typically used to make beams and columns. Oriented strand board (OSB) is sheeting created with wood slivers (the grains are oriented in multiple directions) and an emulsified glue. LVL is used for the top and bottom chord (flanges) of the I-beam and OSB is used for the web (stem) between the chords **FIGURE 5-5**. Although the wooden I-beam is structurally strong (stronger than a comparable solid wood joist), it fails quickly when heated. Actually, no fire contact is required; ambient heating (radiant from the flame or convection from smoke) causes the failure of the binding glue and the potential for a quick collapse.

■ Structural Elements

Buildings are an assembly of structural elements designed to transfer loads to the earth. Structural elements can be defined as the primary load-bearing columns, beams, and connections used to erect a building. (A wall is nothing more than a long continuous column.) Each of these elements facilitates load transfer through its own unique characteristics, although they share a common goal—to resist gravity and deliver loads to

FIGURE 5-5 An engineered wooden I-beam uses LVL for the top and bottom chords (flanges) and OSB for the web (stem). Heat alone can cause the adhesives in the materials to degrade, leading to a rapid collapse.

earth through a foundation or footers. To predict collapse, the ISO must constantly evaluate structural elements to determine if they can still transfer the load as designed. If an element can no longer perform as designed, gravity takes over and pushes the element to earth.

Columns

A column is any structural element that transmits a compressive force axially through its center. Columns can be formed as a wall or as a post, and they typically support beams and other columns **FIGURE 5-6**. Columns are typically viewed as the vertical supports of a building, even though they can be diagonal or even horizontal. The guiding principle is that a column is totally in axial compression. **FIGURE 5-7** shows a horizontal column used to keep two buildings from falling into the alleyway. Notice that the shape of the horizontal column is square as opposed to a horizontal I-beam. Squares and circles are preferred shapes for columns because the load can be transmitted more equally around a center of axis with that shape (recall *axial loads*).

The load-bearing wall created by stacking material such as brick or concrete masonry units (CMUs) (also known as concrete blocks or cinder blocks) is a wall column. Wall columns can also be frame-built using spaced sticks of wood or channel steel. These assemblies are called stud walls, and the individual vertical columns are called *studs*.

Beams

A structural element that transfers loads perpendicularly to its imposed load is called a beam. Obviously, something must support the beam, and it is usually a column or another beam. It stands to reason that beams are used to support a covering for a space (floor or roof). In doing so, it is subjected to the imposed load of itself and of anything placed on it (floor or roof dead loads *and* live loads such as people or furnishings). Loads placed on a beam create opposing forces as the beam tries to deflect: The top of the beam is subjected to a compressive force while the bottom of the beam is subjected to

FIGURE 5-6 A cast iron post column used to support a roof beam.

FIGURE 5-7 The horizontal steel supports between these buildings are columns, meaning they are resisting compression.

tension **FIGURE 5-8**. The distance between the top of the beam and the bottom of the beam dictates the amount of load the beam can carry or the distance the beam can span. If you want to strengthen a beam (to carry more load) or increase the distance that a beam spans, then you must increase the distance between the top and bottom of the beam. The term "I-beam" reflects the shape of certain beams, viewed from either end. The top of the "I" is known as the *top chord*; the bottom of the "I" is the *bottom chord*. Some builders call the chords flanges. The material between the chords is known as the *web* or *stem* of the beam. Webs can be closed (a solid stem) or open (utility openings, cutouts, or open spaces between the web members of a truss).

There are numerous types of beams, although the principal method of load transfer remains the same. Some types of beams are as follows:

- *Simple beam.* A beam that is supported at the two points near its ends
- Continuous beam. A beam that is supported in three or more places
- Cantilever beam. A beam that is supported at only one end (or a beam that extends well past a support in such a way that the unsupported overhang places the top of the beam in tension and the bottom in compression)
- Lintel. A beam that spans an opening in a load-bearing masonry wall, such as over a garage door opening (often called a "header" in street slang)

- Girder. A beam that carries other beams
- *Joist.* A wood framing member used to support floors or roof sheeting
- Truss. A series of triangles used to form an open-web structural element to act as a beam (in many ways, a "fake" beam because it uses geometric shapes, lightweight materials, and assembly components to transfer loads just like a beam)
- *Purlin.* A series of beams placed perpendicularly to other trusses or beams to help support roof decking

Beams are essential for the creation of floors and roofs. Those with solid wood or steel beams in the floors and roof are often called *conventional* construction, whereas those with open-webbed beams are called *truss* construction **FIGURE 5-9**. A typical peaked roof in conventional construction is achieved by using a series of solid wood joists that are diagonally supported at one end by a ridge beam (the peak, sometimes called a ridge *board*) and the other by a top plate (beam) on a stud wall. The open underneath part of the peaked roof can remain open (creating a vaulted ceiling) or can be closed off using ceiling joists. In truss construction, the same peaked roof is achieved by building a large triangle with reinforcing web triangles. The sloped portions of the truss are the top chords, and

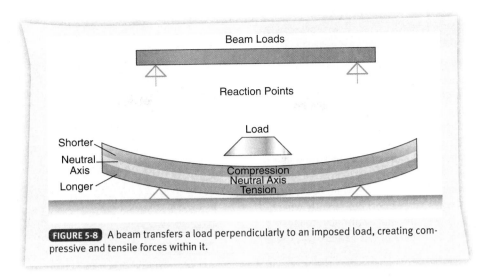

FIGURE 5-8 A beam transfers a load perpendicularly to an imposed load, creating compressive and tensile forces within it.

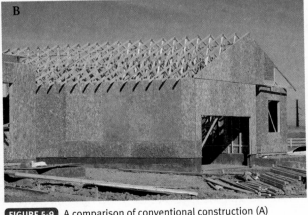

FIGURE 5-9 A comparison of conventional construction (A) versus truss construction (B).

- *Parallel chord truss.* In a parallel chord truss, the top and bottom chords run in the same plane (for a flat roof or floor). Parallel chord trusses can be engineered so that just the top chord is attached to the support (column/other beam) or traditionally resting on the bottom chord. The top chord of a parallel truss is loaded in compression and the bottom cord is in tension. Parallel chord trusses can be manufactured using wood, metals, or a combination of wood and metal.
- *Arched truss.* In an arched truss, the top chord is arched and the bottom chord is straight. The arched truss can be created using a rigid frame of chords and web member (called a rigid-arch) or using a steel-tied bottom chord and web (a true bowstring). Interestingly, many firefighters call any arch-shaped roof a bowstring. The reality is that there are many ways to achieve an arched roof (see the *Fire Marks* box).

Connections

As mentioned previously, beams and columns (inclusive of walls) must be assembled in some fashion to effectively transfer loads. Connections provide the transfer. A <u>connection</u> can be defined as a structural element used to attach other structural elements to one another. Often, the connection is the "weak link" in structural failure during fires. As can be imagined, the connection point is often a small, low-mass material that lacks the capacity to absorb much heat, thereby failing more quickly than an element with more mass, such as a beam or wall.

Connections are loaded in shear force for the most part. There are three general types of connections: pinned, rigid, and gravity. *Pinned* connections use nuts and bolts, screws, nails, rivets, and similar devices to transfer load through attached elements. In a *rigid* connection, the elements are bonded together to form a solid union. Embedded rebar in reinforced concrete, beaded welds, and adhesives are all considered rigid connections. *Gravity* connections are just that: The load from an element is held in place by gravity alone. An example is a beam sitting in a masonry wall pocket (sometimes called a "let").

Together, structural elements defy gravity and make a building sound. A series of post columns and beams used to hold up a building are often referred to as *skeletal frame* or *post*

the flat base is the bottom chord (the bottom chord can include pitches for a vaulted ceiling, making it a scissor-truss).

Trusses can be built in many configurations using a multitude of materials and shapes. Examples include the following:
- *Triangular truss.* This type is the most common truss used to form a peaked roof.

Fire Marks

"That Arched Roof Is a Bowstring"—NOT!

The original bowstring truss roof is a derivative of the bowstring truss bridge developed by inventor Squire Whipple in the mid-1800s.[2] Bowstring truss roofs are notoriously dangerous for firefighters, because they collapse fairly quickly when attacked by fire. Many firefighters have been killed by collapsing bowstrings. This unfortunate history has led many firefighters to label all arched roofs as bowstring, which is a safe and protective approach. More accurately, arched roofs can be created numerous ways and perform differently when exposed to fires. Granted, they all will eventually collapse given enough fire and time. Let's look at the various ways to build an arched roof:

- *Simple arch.* In a simple arch roof, the beams are shaped like an arch and the ends simply rest on walls. The beams are typically a heavy timber glulam or solid steel. A simple arch is not a trussed roof; there is no bottom chord. Older geometric domes are like a 3-D simple arch; the arched roof beams must be restrained somehow and most are done by concrete piers in the earth.
- *Tied arch.* The tied arch has an arched top beam and a tie rod or steel wire bottom restraint to help hold the ends of the top beam in compression (like an archer's bow). Technically, it's not a truss, since there are no web members.
- *Arched truss.* A true *bowstring truss* is like a tied arch but has diagonal web members (forming triangular connections) along the length of the top chord and bottom tie chord to help keep the bottom chord tensioned. Another type of arched truss, known as a *rigid- or rib-arch truss*, uses a rigid frame of chords and web members, usually heavy timbers with steel-plate connects and bolts (it does not use tensioned steel rods or ties). Wooden rigid-arch trusses that began to sag or crack were often repaired using a steel tie rod to help keep the bottom chord tensioned, making it a bowstring truss.
- *Lamella arch.* An arched roof that uses a weave of octagon, triangle, or diamond-patterned roof beams to help form the arch is known as a lamella or Summerbell roof **FIGURE 5-10**. Again, this is not a trussed roof. If you look at it from underneath, it looks like a concaved honeycomb pattern of short wood or steel beams that seem to be weaved together. Newer domed roofs use the lamella principle but are done with reinforced lightweight concrete panels that look like an egg crate when finished.

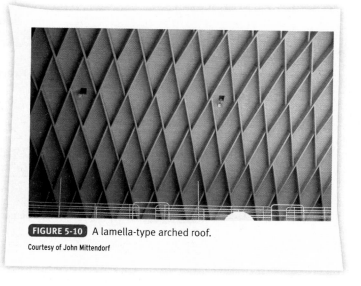

FIGURE 5-10 A lamella-type arched roof.
Courtesy of John Mittendorf

and beam. Beams resting on exterior walls are simply called *wall-bearing* buildings. *Center-core* buildings are those with internal load-bearing wall columns and an attached post-and-beam frame. The exterior walls don't bear any weight other than their own (more on this method later).

As you can see, there are many ways to combine materials and structural elements to form a building that will carry loads. Over time, basic construction classifications have been developed to help us understand more specifically how a building is assembled.

Construction Classifications

The general classifications of construction methods can help the fire officer understand the materials, methods, and components present in a structure. There are many different ways to classify the construction of a building. Codes define building construction classifications one way, contractors another. Even the fire service classifies construction styles differently within the profession. As the ISO, your ability to articulate the various construction classifications is not nearly as important as your ability to look at a building and understand how the building is assembled and what materials were used. For example, when you arrive at a reported fire in a building and begin size-up, you must "read the building." Reading the building begins with a general classification of the building you are dealing with. There are four construction influences that help the fire officer classify the building:

1. The general *type* of construction
2. The *era* or historical time period in which the building was constructed
3. The intended *use* of the building
4. The *size* of the building

■ Categorizing by Building *Type*

Over time, five broad categories of building construction types have been developed that help marry occupancy needs with the materials and methods needed for code compliance. The "five types," also known as the NFPA 220 system, are:

- Type I, Fire Resistive
- Type II, Noncombustible
- Type III, Ordinary
- Type IV, Heavy Timber
- Type V, Wood Frame

The five types can give the ISO a starting place to understand the basic materials used for structural elements. Let's look at each and outline its basic definition and some associated historical fire/collapse issues.

Type I: Fire-Resistive Construction

In a Type I (fire-resistive) construction, the structural elements are of an approved noncombustible or limited combustible material with sufficient fire-resistive ratings to withstand the effects of fire and prevent its spread from story to story. Concrete-encased steel, monolithic-poured cement, and post-and-beam steel with spray-on fire protection coatings are typical of Type I. Examples of Type I buildings are high-rises, megamalls, large stadiums and arenas, large parking garages, and larger hospitals. Due to their size, most Type I buildings rely on protective systems to rapidly detect and extinguish fires. If these systems do not contain the fire (or they don't exist), a difficult firefight is required. Fire can spread from floor to floor on high-rises as windows break and the windows on the next floor up fail and allow the fire to jump. Fire can also make vertical runs through utility conduits, elevator shafts, and unsealed or compromised utility poke-throughs. Regardless of how the fire spreads, firefighters are relying on the fire-resistive properties to protect the structure from collapsing. The collapse of fire-resistive structures can be massive, as we saw in the World Trade Center collapse in New York City on September 11, 2001.

Type II: Noncombustible Construction

In Type II (noncombustible) construction, structural elements do not qualify for Type I construction and are of an approved noncombustible or limited-combustible material with sufficient fire-resistive rating to withstand the effects of fire and prevent its spread from story to story. More often than not, Type II buildings are steel. Modern warehouses, small arenas, and newer churches and schools are built as noncombustible. Because the steel is not required to have significant fire-resistive coatings, Type II buildings are susceptible to steel deformation and resulting collapse. Fire spread in Type II buildings is influenced by the contents. Although the structure itself does not "burn," rapid collapse is possible by means of the burning contents' heat release, stressing the steel.

Suburban strip malls, with CMU load-bearing walls and steel truss roof structures, can be classified as Type II. Fires can spread from store to store through wall openings and shared ceiling and roof-support spaces. The roof structure is often of lightweight steel trusses that fail rapidly. More often than not, the fire-resistive device used to protect the roof structure is a dropped-in ceiling. Missing ceiling tiles, damaged drywall, and utility penetrations can render the steel unprotected. These buildings may have combustible attachments, such as facades and signs, as well as significant content fire loading.

Type III: Ordinary Construction

By definition, ordinary construction includes buildings in which the load-bearing walls are noncombustible (masonry) and the roof and floor assemblies are wood. Typically, this building is made of load-bearing brick or CMUs with wood roofs and floor beams and joists. Ordinary construction is prevalent in most downtown or "main street" areas of established towns and villages. Firefighters have long called ordinary construction "taxpayers" because the buildings are laid out with shops and/or restaurants on the first floor and apartments above, allowing landlords to maximize income to help pay property taxes. Newer Type III buildings include strip malls with CMU walls and wood truss roofs **FIGURE 5-11**.

The primary fire and collapse concerns with ordinary construction are the many void spaces in which fire can spread undetected. While the outside load-bearing walls are masonry, the balance of the building's structure is wood: floors and roof joists/trusses, interior columns, stairs, and so on. Common hallways, utility spaces, and attics can also communicate fire rapidly. Masonry walls hold heat inside, making for difficult firefighting.

Type IV: Heavy Timber Construction

Type IV (heavy timber) buildings can be defined as those that have block or brick exterior load-bearing walls and interior structural elements of a substantial dimension (greater than 8 inches [20 cm] in thickness and width). As the name suggests, heavy timber buildings are stout, making them ideal for warehouses, manufacturing buildings, and some churches. In many ways, a Type IV building is like a Type III, but larger-dimension lumber is used and void spaces are minimized **FIGURE 5-12**. Some firefighters call a Type IV building "mill construction," which is fine, although a true mill-constructed building is a much more stout, collapse-resistant building that may or may not have block walls, and it is constructed without hidden voids. A new Type IV building is hard to find; the cost of large-dimension lumber and/or laminated wood beams makes this type of construction rare, though it may be found in newer expensive resorts or churches.

Fire spread in a heavy timber building can be fast due to wide-open areas and content exposure. The exposed timbers contribute to the fire load (heat release). Because of the mass and large quantity of exposed structural wood, fires burn a long time. If the building housed machinery at one time,

FIGURE 5-11 Newer strip malls are typically Type III construction, using CMU load-bearing walls with lightweight wood trusses for the roof. The decorative brick, stucco finish, and slate tiles are all veneers.

oil-soaked floors add heat to the fire and accelerate collapse. Once floors and roofs start to sag, heavy timber beams may release from the walls. The release is actually designed on purpose to help protect the load-bearing wall. It is accomplished by making a "fire cut" on the beam, and the beam is gravity-fitted into a pocket within the exterior load-bearing masonry wall **FIGURE 5-13**. As the floor sags, it loses its contact point with the wall and simply slides out of its pocket without damage to the wall. However, because a masonry wall requires

compressive weight from floors and roofs to make it sound, once the weight of floors and roofs is lost, the wall becomes an unstable cantilevered beam not designed for lateral loads like wind and fire streams. Type III buildings from the early 1900s may also have fire cut beams.

Type V: Wood-Frame Construction

Wood frame is perhaps the most common construction type; homes, small businesses, and even chain hotels are built primarily of wood. Some call wood frame "stick-built." Some wood-frame buildings may appear more like Type III ordinary construction because of the appearance of brick walls. Remember, brickwork may be a simple veneer to add aesthetics. The primary concern with wood-frame buildings is just what the term *wood frame* suggests—they're made from a combustible material. Gypsum wall board (drywall or sheetrock) is applied to the interior portions of the wooden structural elements to protect them from fire. Once finished, wood-frame buildings typically have many rooms that can help compartmentalize content fires. Fire and heat that penetrate or degrade the protective drywall wall will then attack the wooden elements (floors, attic spaces, and walls), creating a collapse threat, especially in newer buildings. Often, the only warning that fire has penetrated these spaces is the issuance of smoke from crawl space vents, gable-end vents, and eaves.

FIGURE 5-12 An example of Type IV construction. The lighter perpendicular beams are called *purlins* and are used to help support roof decking.

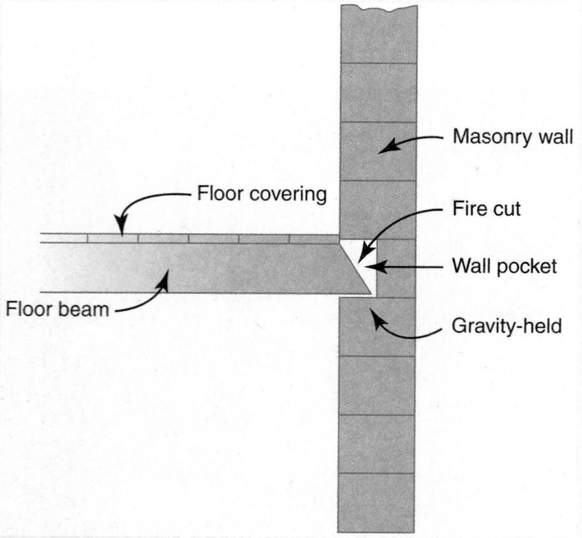

FIGURE 5-13 Wood and heavy timber beams were often "fire cut" so that a fire-damaged, sagging floor would slide out of a gravity wall pocket to preserve the wall.

Floor covering — Masonry wall — Fire cut — Wall pocket — Floor beam — Gravity-held

Awareness Tip

Drywall Calcination

Gypsum wall board (also called drywall or sheetrock) is often used to protect a Type V wood-frame building from fire. When installed, the drywall isn't really dry. The sheets of gypsum are manufactured with a "water of crystallization" process, meaning they are chemically hydrated. When exposed to the heat of a fire, the water is sacrificed (called calcination), thus protecting the wood framing. Once the hydrate is gone, the remaining gypsum dust and fillers (usually paper) become fragile and may bow, crumble, and ultimately fall. The time it takes to wick away all the hydrate depends on the drywall thickness and the amount of heat that is being absorbed. Most tests give a horizontal, 5/8-inch-thick drywall assembly a 30-minute rating. From a firefighter's view, this rating should be viewed with a critical eye: It came from a test lab with a controlled furnace that produces the heat. Today's fire loads release a tremendous amount of heat in a short time, causing more rapid calcination. More realistic tests conducted by Underwriters Laboratories (UL) show that drywall delays collapse of an engineered wooden I-beam floor assembly only 20 minutes more than if the assembly is not protected! That gives firefighters a dangerously small window of opportunity for interior firefighting operations. *Source:* "Changing Severity of Home Fires Workshop," sponsored by the U.S. Fire Administration. December 2012. Available at: http://www.usfa.fema.gov/downloads/pdf/publications/severity_home_fires_workshop.pdf.

Other Construction Types (Hybrid Buildings)

Frankly, and with no disrespect to the NFPA 220 five types, there are buildings that, from a firefighter's perspective, aren't covered by the definitions of Types I through V. Although no official definition exists, we can define a <u>hybrid building</u> as one that is a mix of multiple NFPA 220 construction types or one that doesn't fit into any of the five construction types.

The advent of material technologies has produced construction types that are impressive from an engineering perspective. The fire service has very little research information on the stability of hybrid buildings during *actual* fires. One thought can be certain: Expect rapid collapse as a result of the low-mass, high-surface-to-mass exposure of structural elements. The list of hybrid buildings that are being built currently can fill an entire book. You are encouraged to conduct some research and also be curious about buildings being built in your jurisdiction. Go look! Let's briefly examine two common hybrid construction types that present interesting challenges to firefighters.

Insulated concrete forming (ICF) buildings use expanded polystyrene (EPS) to form a concrete mold for walls. Once the concrete cures, the EPS form is left in place as insulation. There are three types of ICF: grid block, flat panel, and post and beam. Grid-block ICF is similar to a stack of Styrofoam blocks. After the blocks are stacked, small cavities within the blocks form a waffle-like grid that is then filled with free-flowing concrete (minimal gravel). Flat-panel ICF uses two parallel panels with a gap between them to form the concrete mold. Concrete is poured into the gap, creating a solid wall. Post-and-beam ICF is like a reinforced concrete post-and-beam building, except that EPS molds are left in place to help insulate the columns and beams. Types of walls used to fill the spaces between the posts and beams can be myriad, including brick veneer, wood- or steel-framed panels, and composite sheeting. Of concern for firefighters is the grid-block ICF. The load-bearing walls are mostly EPS! Granted, there is concrete in the wall, but the EPS is supporting the waffle grid. Fire will rapidly deteriorate the EPS, and the concrete will then simply crumble and fall.

Structural insulated panel (SIP) buildings are those in which the load-bearing walls and roof are made from panels of OSB and EPS. The panels are simply two sheets of OSB glued to both sides of an EPS sheet that is typically 6 to 12 inches (15 to 30 cm) thick **FIGURE 5-14**. These panels are assembled like the house of playing cards you may have built as a child, only they are glued together using a system of splines and adhesives. Long wood screws typically secure the roof to the walls. Once the structures are finished, they may resemble a typical wood-frame building, but look for extended window and door jambs that indicate the wall is thicker than that of typical wood or masonry buildings. Heat alone can cause the interior EPS to contract (melt), leaving large combustible voids between the sheets of OSB. Fire can also enter the wall space through utility cuts and conduits. We do not have much fire experience with buildings constructed with SIP walls. Simple analysis should lead a firefighter to expect rapid structural failure of these buildings during fires.

FIGURE 5-14 Stacked SIP wall panels that are awaiting assembly. The SIP building is held together using adhesive splines and wood screws.

■ Categorizing by Building *Era*

Categorizing a building by one of the five types or as a hybrid can lead to some disastrous misunderstandings at building fires. For example, a Type III building built originally in 1930 is much more resistive to a fire-caused collapse than a Type III building built in 2010. For this reason, ISOs must look at the time period in which the primary structure was built. We use the term *era* to describe various time periods, although some use the term *vintage* or *heritage*. For our purposes, we shy away from the term "age" as a descriptor mostly because of the effects of *aging*. A building's age deals more with issues of maintenance, deterioration, alterations/additions, and renovations.

Various methods and materials used to construct buildings have evolved over time and have been influenced by some significant events. Although no official date line separates various eras, some generalizations can be made that have impacted the building of structures in the United States. The following time periods represent significant changes in the history of building construction:

- The founders' era: 1700s to WWI
- The industrial era: WWI to WWII
- The legacy era: WWII to roughly 1980
- The lightweight era: 1980s to present

An examination of each of these eras can yield some clues regarding the materials, methods, and codes used to construct buildings and, therefore, their effect on fire resistance and collapse potential.

Founders' Era

From the 1700s up to World War I, the principles of building a sound building improved gradually—mostly because it took our forefathers time to transition from the use of hand tools to cut trees and build buildings to the creation of foundries, wood mills, and stone quarries that supplied more substantial building materials and tools.

From a firefighter's perspective, the founders' era buildings generally have lots of combustible voids, alterations made as utilities improved, open/narrow stairways and hallways, lack of modern fire-stopping, minimal fire code requirements, and issues of aging (see the *Safety Tip* box). Most collapse threats are usually self-announcing as wood floors and roofs begin to sag. Most floor-to-wall connections use gravity. Major fires present the threat of catastrophic wall collapse for stacked-brick, load-bearing walls, also called unreinforced masonry (URM).

Industrial Era

WWI was considered the first mechanized war—a direct outcome from the developing use of steel (as opposed to wrought iron and cast iron). Steel allowed humankind to develop all manner of tools, appliances, and equipment that were larger, stronger, and more durable. For firefighters, the industrial era brought much larger buildings and the associated risks that come with unprotected steel, which softens and elongates when heated by fire. For commercial buildings, the primary threat of a major collapse shifted from walls (which during the founders' era failed later in a fire) to roofs (which with the advent of steel failed earlier in a fire). The use of balloon framing for wood buildings was prevalent in the industrial era (although it began in the late 1850s). In balloon framing, continuous wood studs run from the foundation to the roof, and floors are placed on a shelf, called a ribbon board, that hangs on the interior surface of the studs. Fires that enter the stud voids can run straight to the attic and through the spaces between floor and roof joists.

Even though fire codes were beginning to mature during this time, firefighters can expect minimal fire-stopping, open hallways and stairways, and issues of aging with the industrial-era building. Sagging is still a good indicator of impending collapse for most original industrial-era wood buildings.

Legacy Era

If WWI was the first mechanized war, WWII was the first industrialized one. Maybe we should call it the first engineered war. Nations involved in the war efforts found themselves rationing all manner of food, raw materials, and other resources to help build more tanks, planes, ships, weapons—and transport huge quantities of them to all points of the earth. Arguably, the United States won the war because of its innovative engineering methods to minimize material use and maximize durability for, well, everything! The engineering innovations of the war can be found in the way legacy buildings are constructed: better building/fire code requirements, durable strength,

---

Safety Tip

What Do the Stars Say?

Many buildings from the founders' and industrial eras have what appear to be decorative stars on the exterior masonry walls. In most cases, these are not decorations—they are holding the building together! The effects of aging take their toll on buildings: Wood shrinks and cracks, soft bricks erode as does the binding mortar, gravity wall pockets degrade, and unprotected wrought iron and steel rust. The result is sagging beams, walls, and roofs that require corrective measures to keep a building from falling. The decorative stars (you can also find diamonds, "S" shapes, or simple square and round metal plates) are called spreaders, and they are used to distribute force over more bricks or blocks as part of an unseen corrective measure that exists inside the building **FIGURE 5-15**.

The two most common corrective measures that use external wall spreaders are *tie rods* and *joist/rafter tie plates*. A tie rod is used to correct or prevent the sagging of walls or floors. A steel rod runs through the building to an opposite load-bearing wall and is tensioned using a turnbuckle or bolts threaded onto each end of the rod. Joist/rafter tie plates indicate the use of wood screws or L-shaped hooks to secure the end of a beam into the masonry wall. It is likely that the wood beam sits in a gravity wall pocket and, as the wood shrank or sagged, it was dangerously close to falling out of the wall. Likewise, tie plates are often added to older buildings as an earthquake collapse-prevention measure.

From a firefighter's perspective, the use of tie rods presents the more dangerous collapse potential. Often, the tie rod is unprotected and runs the entire distance of a building to the opposite wall. When heated during a fire, the tie rod elongates, negating its purpose and accelerating the likelihood of a large collapse. Joist/rafter tie plates are more likely to help firefighters, as the wood screws are mostly shielded from an interior fire and they help the beam stay in place. A warning though: Tie plates indicate that the building has aging issues.

A visual clue that can help you discriminate between tie rods and anchor plates is the number and spacing of the spreaders. Although not absolute, where multiple spreaders are arranged in a uniform spacing fashion that aligns with floors (or roof attachment), suspect joist/anchor tie plates. Where spreaders are misaligned, are few in number, or don't necessarily line up with a floor, think tie rod. Better yet, don't wait for a fire incident; try to arrange a preplan visit to examine the building's interior!

---

improved detection and suppression systems, and more reliable utility systems. Although debatable, many believe that, overall, the legacy building is a friend to firefighters. It is very predictable for fire spread and there are many things that work in our favor regarding collapse. Still, there are fire spread and collapse issues that come with the legacy era building that include the following:

- Balloon framing was all but supplanted by platform framing in wood buildings. Platform framing is

- The legacy era saw a gradual transformation away from conventional construction. Trusses were starting to be used more frequently.

Lightweight Era

Unlike the previous war events used to separate construction eras, the division between the legacy and lightweight eras is harder to define. The use of lightweight trusses (e.g., small-dimension lumber, gusset plates) for roof construction can be traced to the 1960s, although their use was not widespread. If a division can be made, it would have to be the 1980s due to an evolving trend to transform *prescriptive* building codes into *performance-based* codes. A prescriptive code tells the user *how* something is to be accomplished, whereas a performance-based code tells the user *what* something needs to do when it's done. As a simple example, if a builder wanted to build a wood floor for a given use and space, the prescriptive code would require a certain size and type of wood for joists, a certain nail size and nailing pattern, and a certain type of decking to create the floor. A performance-based code, in contrast, would state that the floor must hold such-and-such amount of weight and resist a certain amount of fire for a certain amount of time when it's completed. The latter approach allowed architects, engineers, and builders to choose the materials and methods they used to achieve the performance requirements, spawning innovation, material conservation (profits), and the use of composites and synthetics.

The lightweight building may be stronger than its predecessors for carrying live and dead loads—under normal conditions. Unfortunately, the new era's buildings have structural elements that are of low mass, or better stated, of a high surface-to-mass ratio. Couple the lightweight nature of the structural elements with an ever-increasing heat-release rate of modern building contents and the stage is set for disaster. Granted, the use of automatic fire sprinkler systems is required in more types of occupancies than ever before, but the majority of residential and smaller commercial buildings lack that feature.

The evidence is rolling in and the conclusion is frightening for firefighters: The lightweight-era building collapses very quickly in fire conditions.

■ Categorizing by Building *Use*

In a simpler time, firefighters could classify buildings as residential or commercial (and many still do for the sake of rapid on-scene radio reports). Today, the intended use of a building is tied to code requirements and the manner in which the building must be built to host the people, equipment, and processes that will take place within. In code speak, the *use* of a building is called *occupancy type*. Here, we will stick with occupancy *use* to avoid confusing this category with the five *types* of buildings listed earlier, but keep in mind that established codes tie together construction type and occupancy type and are very specific and much more detailed than what is presented here. For firefighters and the ISO, some easy groupings can be made for most building uses:

- Single-family dwellings
- Multiple-family dwellings
- Offices/hotels

FIGURE 5-15 The visible stars are spreaders used to secure steel tie rods that extend through the building. The tension on the tie rods helps to prevent or correct floor and wall sagging.

achieved when a single-story wall is built and the next floor is built upon the tops of the studs, creating fire-stopping to help minimize fire spread.

- Native lumber was smooth finished (as opposed to rough-sawn) and was of nominal dimension (a 2- by 4-inch plank was no longer a full 2 inches by 4 inches), resulting in more rapid burning and mass loss. Recall that rough-sawn lumber exposed to flame will initially form a char crust that slows burning. Smooth-planed wood readily checks (alligatoring) and burns more quickly.
- Steel and concrete commercial buildings constructed using center-core or post-and-beam methods began using curtain walls. A curtain wall is a non-load-bearing wall that supports only itself and is used only to keep weather out. Curtain walls can collapse quickly during fires and endanger outside crews, although the building itself may remain upright.
- The legacy era brought bigger interior spaces for most buildings. The added space can lead to higher fire loads, more storage (increased weight), and easy-to-alter floor plans (changing the building's use, legally or otherwise).
- The use of drywall (gypsum wallboard) replaced lath and plaster as an interior wall finish and fire spread deterrent. Likewise, plywood replaced wood slats for floor and roof decking.

- Commercial retail
- Manufacturing/warehouses
- Schools/hospitals
- Public assemblies (stadiums/arenas/theaters/churches)
- Mixed/miscellaneous use

The purpose of categorizing a building by its apparent use is to forecast the potential fire spread and collapse concerns that may lie within. Fire and collapse issues for each of the occupancy groups discussed here are numerous, yet somehow obvious. Fire spread issues are closely related to the contents that can be found in a given building, and the building use can clue the ISO into the potential. For residential buildings, the ISO can look at the building arrangement to understand the fire load: single or multiple family; location of bedrooms versus living areas; garages attached or separate; and so on. For commercial buildings, the signage provides excellent clues to the fire load **FIGURE 5-16**. Likewise, the arrangement and quantity of the live load (as suggested by the occupancy use) may help a fire officer envision collapse potential.

■ Categorizing by Building *Size*

The size of a building affects a fire suppression operation in many ways. Staffing needs, hose lay distances, ladder reach, rate of water flow (gallons per minute), and search times are only a few examples. The ISO needs to be mindful of these effects and the hazards they create for firefighters. More specifically, the ISO views building size as a multiplier for risk-taking: Increase the measurement of any element of a building's size and you've exponentially increased the risks that might be taken. These size elements can be defined as follows:

- The footprint, or single-floor, square footage of a building (width, length/depth)
- The interior arrangement of walls and the volume of space for any one room
- The number of floors above ground and basement levels below
- The distance that must be traveled to reach a fire or potential victims

FIGURE 5-16 Commercial signage can provide clues to the building use and associated fire load and collapse potentials.

Each of these size elements is used to help the ISO develop a judgment regarding the obstacles, conditions, and hazards that might be faced by firefighters in a given building.

You should have discovered the trap that is created when you classify a building by just the classic five types. The ISO who classifies a building using the *type/era/use/size* approach is more likely to forecast the fire spread and collapse issues that may be present and to communicate them more clearly to the IC and the designated RIC. However, recall that classifying a building is only the first step in predicting collapse. To truly make a difference, the ISO uses a mental process that takes the building classification step to another level.

Predicting Collapse

With a keen understanding of the preceding topics, the ISO can begin to analyze buildings during structure fires to establish whether they will "behave" during firefighting operations in and around the involved structures. Obviously, the ISO must perform this analysis and make a judgment about collapse potential *before* the collapse occurs. A simple approach is presented here to provide a thought process for predicting collapse. As with most incident dynamics, nothing is absolute about predicting collapse. There is no perfect formula. For the ISO, however, the effort to analyze and predict collapse is essential to firefighter safety. The approach explained in this section can be applied to other structures, such as bridges, cranes, mine shafts, and wells, or any engineered "thing" attacked by fire. Building analysis during any incident should be cyclic—that is, performed on a regular basis as conditions change and time elapses.

To predict collapse, the ISO uses a classic "identify–analyze–decide" method. We can take that three-part method and apply it using a five-step process:

- *Step 1.* Classify the building's construction using the *type/era/use/size* approach.
- *Step 2.* Determine structural involvement (read the smoke and flames).
- *Step 3.* Visualize and trace loads.
- *Step 4.* Evaluate time.
- *Step 5.* Predict and communicate collapse potential (foundation for zoning).

Steps 1 and 2 in our process are when the identification takes place. Steps 3 and 4 are analytical, and step 5 is when the decision (regarding collapse) is made and communicated. Let's explore some specifics of each step and show how the process can work.

■ Step 1: Classifying the Building's Construction

Using the *type/era/use/size* approach helps the ISO identify the strengths and weaknesses associated with a given building and interpret how the materials and arrangement of structural elements might be impacted by fire and heat.

■ Step 2: Determining Structural Involvement

Determining whether a fire is a contents or structure fire is imperative **FIGURE 5-17**. Too often, fire departments use the term "structure fire" to dispatch crews. This may serve as a

FIGURE 5-17 Once load-bearing structural elements are attacked by fire, collapse may come quickly.

warning, but many incidents are merely content fires. Fire service leaders have suggested that crews should be dispatched to a "fire in a building." Only upon arrival and size-up is a fire classified as structural. For the sake of this text, "structure fire" means that the load-bearing components of a building are being attacked by fire or heat.

Fire in concealed spaces, content fires in unfinished basements, attic fires, a heated exposed beam or truss, and the like are all examples of fires that have become "structural." The point is obvious: Once the structure is involved, the attention to possible collapse should be immediate. From the outside of a building, how would you know if the "hidden" structural elements are being attacked by fire or heat? The observation of fire coming from structural spaces should be obvious. Less obvious is what the smoke is doing (discussed separately in the next chapter). Without jumping too far ahead, know that dark gray or black smoke venting under pressure from structural seams, ridge/rim boards, eaves, and attic vents is a significant indicator that heat and likely fire are present in that space. Also, unfinished wood that is being rapidly heated emits a brownish smoke—a collapse warning sign in lightweight wood construction.

Step 3: Visualizing and Tracing Loads

This step is more an art than a science—and is analytical. The ISO visually scans the building, tracing any load to the ground. In doing so, the ISO determines whether any structural element is carrying something it should not and whether key elements are being attacked by fire and/or heat. In visualizing all this information, the ISO mentally "undresses" a building. This step helps the ISO define the weak link, which typically precipitates the collapse. Historical weak links are listed here:

- *Connections.* Structural failure is often the result of a connection failure. Usually, the connection (bolt, pin, weld, screw, gusset plate, or the like) is made of a material with lower mass, or fire resistance, than the assemblies it connects. In some cases, the connection relies on gravity and an axial load to hold it in place; a shift or lateral load may cause the connection to fail.

- *Overloading.* Excessive live and dead loads place stress on building components, which are vulnerable to early collapse when attacked by fire. In some cases, the building may not have been designed and built to handle excessive loads. Two feet of wet snow may have been considered when a flat roof building was built, but the addition of a second roof structure, storage, and/or rooftop signs—plus the two feet of snow—can overload the roof. Attic spaces are seldom designed to be used as storage areas, but the open space between trusses often lures the occupant into storing old files, parts, furniture, and other things. Indications that an occupant is a hoarder should always evoke a concern of overloading.

- *Occupancy conversion.* Many buildings are used for occupancies they were never built to support. For example, a lawyer moves his or her practice into a remodeled home: Walls are removed, increasing floor and roof spans to accommodate law books, legal files, heavy furniture, and office equipment. This massive weight addition leaves little room for fire tolerance and collapses quickly. Other examples include buildings constructed for a simple retail operation and converted to a manufacturing, storage, or medical use, which can bring excessive weight, higher fire loads, and alterations to the structure that were never meant in the original construction.

- *Trusses.* Trusses are nothing more than "fake beams." Made of extremely light materials, trusses' geometric shape (triangles) and open space are used to replace a solid beam. This is a deadly combination when fire and/or heat are introduced. Other questions arise when dealing with trusses, each indicating a weak link: Does the truss span a significant distance? What materials are being used to piece together the truss? Gusset plates typically hold together trusses. Gusset plates pop out quickly during rapid heating and fall off completely when a small portion of the wood's surface burns away—after as little as 5 minutes of burn time. Trusses are used in roof, floor, and mezzanine assemblies.

- *Void spaces.* In Type III ordinary construction, voids are numerous. Some voids may pass through masonry walls, causing fire spread from one store to the next in a row of buildings. The obvious collapse danger with void spaces is that the fire may be undetected with the simultaneous destruction of structural elements. Floor and roof truss spaces in lightweight wood construction create large combustible voids (usually packed with utility runs). Any signs of heat or fire from these spaces should trigger an immediate collapse warning.

- *Stairs.* First-arriving firefighting crews rely on internal stairways to gain access for rescue and fire attack. Previously, firefighters have found stairways to be durable and a bit stronger than other interior components. This is a dangerous assumption in newer wood-frame buildings. Stairs are now being built off-site and simply hung in place using light metal strapping. Additionally, stairs are being made of lightweight engineered wood products and recycled wood pieces that fail quickly when heated. Remember, press-glued wood chip products can fail from the heat of the thermal column or heat in the by-products of combustion (smoke)—no flame is required!

- *Large, open-span interior spaces.* Creating a large, unobstructed open space within a building interior requires the use of an engineered truss roof or floor assembly. Although all trusses are considered a weak link (see above), combining the truss with a large, unobstructed span creates an increased collapse risk. Clearly, the scale or severity of the collapse will be much greater with clear open spans.
- *Parapet walls.* A parapet wall is the extension of a wall past the top of the roof. Parapets are used to hide unsightly roof equipment and heating, ventilation, and air-conditioning (HVAC) units, and they give a building a "finished" look. Typically made of masonry materials, these walls are freestanding with little stability. Collapse may be caused by the sagging of the roof structure, which has the tendency to lift the parapet. Business owners hang signs, utility connections, and other loads on the parapet. During a fire, the steel cables and bolts holding these loads weaken and sag, placing additional eccentric load on the parapet, which accelerates collapse. Some of the newer, lightweight-era buildings have parapets made entirely of foam plastic. They will not support even the weight of a firefighter on a ladder before breaking.
- *Facades.* Facades are nonstructural facings or false fronts used for decoration or paneling or to form a porch or protective overhang. Being nonstructural, these facades are typically hung on a building. Under fire conditions, the connections fail quickly and the facade simply falls off the building. Many strip malls have a cantilevered facade over the store fronts to provide shade and rain protection and to hang lighted signage. Small fires can easily run through the void space of the facade, resulting in a rapid collapse.

■ Step 4: Evaluating Time

Some structural elements fail as soon as fire (heat) reaches the material. Other materials absorb incredible heat for a long duration before they become susceptible to collapse. Time as a factor should be brought into the collapse equation, although the time it takes for gravity to overcome the structure during a fire is not predictable. A number of variables determine the amount of time a material can resist gravity and heat before failure. Factors that can accelerate the potential collapse time include:

- Low material mass, or high surface-to-mass ratio
- An imposed overload
- Higher British thermal unit (Btu) development (fire load)
- Alterations (undesigned loading)
- Age deterioration or the lack of care and maintenance of the structure

- Firefighting impact loads (fire stream force, accumulated water, forcible entry and ventilation efforts, weight of firefighting teams)
- Breakdown or loss of fire-resistive barriers

Although no formula can predict collapse time, a few truisms can be applied regarding time (see the *Safety Tip* box).

Safety Tip

Time Truisms for Predicting Collapse
- The lighter the structural elements, the faster the structure comes down.
- The heavier the imposed load, the faster the structure comes down.
- Wet (cooled) steel buys time.
- Gravity and time are constant; resistance is not.
- The time opportunity window for interior operations in lightweight buildings has been reduced.
- There is no safe time window for interior operations when a building is under construction, being renovated, or being disassembled. This is a definite "no-go" situation, and you should always establish a defensive attack.
- Brown or dark smoke coming from lightweight engineered wood products means that time is up.

The National Institute for Occupational Safety and Health (NIOSH) has released an alert (Publication No. 2005-1432) to help prevent injuries and the deaths of firefighters due to truss system failures. One of the recommendations contained in the alert states:

> Ensure that firefighters performing firefighting operations under or above trusses are evacuated as soon as it is determined that the trusses are exposed to fire—not according to a time limit.

The Federal Emergency Management Agency (FEMA) funded a grant through which UL partnered with the Chicago Fire Department, International Association of Fire Chiefs, and Michigan State University to conduct full-scale test burns that compared lightweight engineered wood structures to legacy wood structures. The highly scientific studies compared numerous performance aspects as well as collapse times. The floor assemblies tested were 14 feet by 17 feet (4 m by 7 m), had typical flooring, and included the live load of two firefighter mannequins. Various combinations of conventional and lightweight engineered wood floors were tested using a consistent flame burner furnace from below. The time to collapse results shown in **TABLE 5-2** are sobering.

The UL University has posted an online training program for firefighters that details more of the report, which is titled "Structural Collapse: The Hidden Dangers of Residential Fires." This report should be mandatory reading for ISOs (see the additional resources at the end of this chapter).

The message is clear: Do not rely on some "20-minute rule" or other such nonsense in trying to predict collapse times. Many factors dictate the time it takes to trigger a collapse, and

TABLE 5-2	Collapse Times for Various Floor Assemblies	
Structural Elements	**Ceiling (Protecting the Underside of the Elements)**	**Firefighter Breach (Collapse) Time (Minutes:Seconds)**
2- × 10-inch solid wood joist	None	18:35
2- × 10-inch solid wood joist	Lath and plaster	> 79 minutes
2- × 10-inch solid wood joist	Gypsum wallboard	44:40
Engineered wooden I-beam	None	6:00
Engineered wooden I-beam	Gypsum wallboard	26:43
Wood and metal hybrid truss	None	5:30

Source: USFA/UL as presented at the Changing Severity of Home Fires Workshop, December 2012.

we should always predict the early collapse of trusses and engineered lightweight wood assemblies to ensure crews are safely outside the structure before the collapse occurs.

■ Step 5: Predicting and Communicating the Collapse Potential

Predicting collapse is the "decision" part of the identify–analyze–decide model. The hope is that the previous four steps have caused you to anticipate collapse potential before obvious visual signs are present. However, in some situations, certain observations point to obvious collapse potential; let's call these "late signs." Although waiting for visual signs that a building will collapse is dangerous, especially in newer buildings, the following observations make collapse likelihood obvious:

- Deterioration and cracks of mortar joints and masonry
- Signs of building repair, including reinforcing cables, spreaders, and bracing
- Bulges and bowing of walls
- Sagging floors and roofs
- Abandoned buildings with missing roof, wall, or floor segments
- Buildings under construction, renovation, or demolition (see the *Safety Tip* box)
- Large volumes of fire impinging on structural components and spaces
- Multiple fires in the same building or damage from previous fires
- Settling noises
- Doors out of plumb or jammed
- Water flowing out of the building that doesn't match water going in (in regard to quantity)

Most firefighters can equate a sagging roof to a collapse potential. Remember, though, the ISO needs to identify, analyze, and decide the collapse potential before it becomes obvious. This decision should be communicated to command and include the establishment of a collapse zone. A collapse zone consists of the areas that are exposed to trauma, debris, and/or thrust should a building or part of a building collapse. A collapse zone is a more specific form of a no-entry zone for anyone—including firefighters. Most fire service texts suggest that a collapse zone be at least 1½ times the height of the structure that is anticipated to fall. Although this may be appropriate for wood structures, it may not be adequate for unreinforced masonry block/brick walls (URM). There are many documented collapses where the falling masonry wall propelled blocks three times the height of the wall!

Safety Tip

Strategy for Buildings Under Construction
Buildings are especially unsafe during construction, demolition, renovation, or restoration. Add a hostile fire and the word unsafe becomes disastrous. For this reason, ICs should *default to a defensive strategy* for any fire in a building that is under construction, demolition, or structural alteration. An ISO who arrives to find an offensive attack in a building under construction should immediately consult with the IC and strongly urge a defensive strategy. This seemingly inflexible directive is not without merit. It is well supported in logic and, unfortunately, historical losses.

Buildings need to meet fire and life safety codes only once they are completed. During construction (or any structural alteration), many of the protective features and fire-resistive components are incomplete. Additionally, stacked construction material may overload other structural components. This is not to say that contractors are using unsafe practices, but rather that exposed structural elements, incomplete assemblies, and material stacks all contribute to a rapid collapse if a fire develops. History has proved this out: Every year it seems that a large and costly fire, with rapid collapse, occurs in a building or complex under construction. Hence, firefighters call a fire in a building under construction a "loser."

One of the reasons for this loss, even in concrete buildings, is the use of false work during construction. False work can be defined as temporary shoring, bracing, or formwork used to support incomplete structural elements during building construction **FIGURE 5-18**. Often, false work uses 2- by 4-inch and 4- by 4-inch wood columns and beams to create a shoring framework. A raker is also used as false work. Rakers are diagonal braces that serve primarily as a column but must absorb some beam forces as well. Fires involving false work are hot (essentially a wood crib fire), spread rapidly, and lead to a sudden, general collapse. A building under construction is insured and can be replaced—a firefighter's life cannot. Default defensive!

FIGURE 5-18 False work is temporary shoring used to support unfinished structural elements. Here, the false work is shoring concrete floors that are still curing.

When it is not possible or desirable to honor the 1½ distance rule, some have chosen to flank the building corners so that fire streams can still reach. The thinking is that the building corners are inherently stronger and that a diagonal outward collapse is not likely. These assumptions are not absolute—corners have collapsed diagonally, though it is rare. Regardless, when flanking a collapse zone, firefighters should use spotters and have rapid-withdrawal signals and routes preestablished and communicated prior to the approach.

Typically, a building experiences a partial or general collapse. In a <u>partial collapse</u>, the building can accept the failure of a single component and still retain some strength (like a curtain wall collapse). A <u>general collapse</u> is the complete failure of the building to resist gravity. Following a partial collapse, additional collapse indicators are present that can lead to secondary partial or general collapse in most, but not all, cases. In the case of an outward curtain wall collapse, the weight of the wall is removed from the primary structural elements and may actually reduce secondary collapse potential, assuming the primary structural elements are not damaged by the loss. In most partial collapse cases, though, the weight of the felled portion is picked up by other elements. A secondary collapse becomes a real threat either by the stresses it places on those elements or because the fire

continues to attack those elements. For example, the collapse of a floor or roof may impose weight on interior partition walls that were not designed for the load. These become a "loaded gun," meaning that the slightest movement could cause an explosive release and secondary collapse. Likewise, a "V"-type roof or floor collapse can put lateral forces on load-bearing walls that are designed for primarily an axial, compressive load. In this case, the lateral force can cause the bottom of the wall to kick out and lead to a secondary, general collapse.

Only a few of the indicators that can suggest a secondary collapse have been discussed here. This text will not endeavor to include them all, but many resources are available and ISOs are encouraged to seek this information as part of their professional development.

Last, we must address communication of the ISO's collapse potential decision. There are three communication options the ISO can use. The choice of which to use is based on the urgency of the message (time before collapse), described as follows:

- <u>Emergency abandonment</u>. An emergency abandonment is a strict order for all crews to *immediately escape* from a building interior or roof, leaving hose lines and tools that can impede rapid retreat behind. Once in a safe area outside of collapse zones, a personal accountability report should be initiated (more on this in other chapters). Obviously, the ISO should use the emergency abandonment order when the collapse potential is judged to be imminent and crews are within the collapse zone. In most cases, the emergency abandonment order is initiated by the ISO via radio and followed up with the activation of air horns on apparatus.

- <u>Precautionary withdrawal</u>. A precautionary withdrawal is a directive for crews to exit a building interior or roof in an orderly manner, bringing hoses and tools along. The decision to make a precautionary withdrawal is typically made after the ISO shares the collapse potential decision with the IC. The order to crews then comes from the IC.

- *Planning awareness*. As the ISO forms his or her judgment regarding collapse potential, an opportunity exists to share this information with the IC and the RIC. By doing so, the IC can proactively adjust incident action plans. The RIC also benefits from the ISO-shared information in that it can develop rescue contingencies, make tool choices, and preplan access/egress options. The ISO gathers and evaluates many building-related factors that are useful for the RIC crew: exit/egress options, construction material types and arrangements, utility hazards, reach distances, potential loads, and fire spread predictions. Each of these factors can help the RIC plan for a collapse or other firefighter emergency event.

An ISO who truly wants to make a difference will be a lifelong student of building construction, constantly seeking out opportunities to conduct pre-incident building surveys to improve his or her ability to read buildings.

Chief Concepts

- ISOs must be able to give an IC and the RIC officer a judgment regarding the collapse potential of a given building being attacked by fire. To do this, the ISO must draw from a significant knowledge base and not rely solely on "experience."

- Loads are static and dynamic weights applied to a building and are divided into dead loads (the weight of the building) and live loads (all other weights). These loads are imposed on other materials in order to be delivered to the earth below a building. The three ways loads are imposed are axially (through the center), eccentrically (off-center), and torsionally (through twisting).

- Imposed loads create a resistive force in the materials they are acting on. The forces can be either compression (crushing), tension (pulling), or shear (tearing).

- A given material's composition, orientation, shape, and mass all factor into its ability to resist forces. The reaction of a material exposed to forces will cause it to deform. If the material can deform and still retain strength, it is called *ductile*. If the material breaks as it deforms, it is called *brittle*.

- Fire is an obvious destructive force, and materials succumb to fire in various ways. Wood contributes to the fire and loses mass. Steel doesn't really burn but will soften and elongate. Concrete and masonry products will crack and spall but will not contribute to the fire. Spalling is the crumbling of concrete and masonry due to the expansion of internal moisture when heated.

- Newer, lightweight composites are being created as a building material. Two of these materials are especially prevalent in wood construction: laminated veneer lumber (LVL) and oriented strand board (OSB). Both are engineered wood products that use native wood chips and shavings bonded together with an adhesive. These products burn rapidly but can also degrade just from the heat of smoke or radiant energy.

- The primary load-bearing portions of a building are called structural elements. The elements include columns, beams, and connections. Columns are those elements that are resisting compressive forces. Beams transfer loads perpendicularly, creating opposing compression and torsional forces within the beam. Beams are essential supports for floors and roofs and use solid members (conventional construction) or trusses (truss construction). A connection can be gravity, pinned, or rigid.

- Building materials and construction methods have changed over time, leading to many construction classifications. Influences for these classifications include:
  1. The NFPA 220 five types
  2. Hybrid buildings
  3. The era or vintage of the construction
  4. The intended occupancy use of the building
  5. The size of the building

- The NFPA 220 system includes Type I, Fire Resistive; Type II, Noncombustible; Type III, Ordinary (masonry walls with wood floors and roof); Type IV, Heavy Timber; and Type V, Wood Frame.

- A hybrid building is one that combines one or more of the five building types in a single building or that doesn't readily fit any of the five types. Notable hybrid examples include insulated concrete form (ICF) and structural insulated panel (SIP) buildings.

- Common eras that have influenced building construction include the founders' era (prior to WWI), industrial era (WWI to WWII), legacy era (WWII to roughly 1980), and lightweight era (1980s to present). Each era has unique features that influence fire spread and collapse potentials.

- Era and age of a building are different. Era refers to the time period that influenced the building's materials, methods, and features. Age refers to issues of aging, including sagging, rot, shrinkage, wear and tear, and lack of maintenance. Spreaders are seemingly decorative metal stars or other shapes that indicate that a building has hidden issues of deterioration. Spreaders can anchor the ends of unprotected tie rods that pass completely through a building or hold metal anchors that are securing wooden beams into gravity wall pockets.

- The occupancy use that was intended when a building was built influences fire spread and collapse potentials. The ISO can look at the building's arrangement (for residential dwellings) or the signage (commercial buildings) to glean clues as to the fire and live loads present within.

- The size of a building impacts firefighting operations, including issues related to hose lengths, rate of water flow, ladder reach, and search times. Size elements that need to be considered include the total square footage of a given floor, number of floors above and below grade, the arrangement and interior volume of individual rooms, and the access routes and distances that must be traveled.

- The ISO uses a five-step process that incorporates the identify–analyze–decide method of decision making:
  1. Identify the building classification using a *type/era/use/size* approach.
  2. Identify the fire and heat (smoke) locations, and determine whether structural elements are being attacked.
  3. Analyze the transference of loads that pass through the building and envision weak links.
  4. Analyze the passage of time and its impact on materials.
  5. Determine collapse potential and communicate warnings and collapse zones.

- Typical weak links (collapse indicators) at structure fires include connection points, truss construction, overloaded elements, clear/open large floor and roof spans, occupancy switches (alterations), facades, parapets, lightweight stairs, and fires in void spaces.
- Late indicators of collapse include sagging floors and roofs, cracks in masonry, settling noises, bulging or leaning walls, signs of construction or alteration, and large-volume fires attacking structural elements.
- The time that a building will resist collapse during a fire is not absolute and is influenced by many factors. As a rule, though, lightweight/low-mass materials fail very quickly. Buildings undergoing construction, demolition, or structural alteration are considered "losers," with no time left before failure. They require defensive strategies.
- Collapse potential should be communicated using one of three options that are based on urgency: emergency abandonment, precautionary withdrawal, and planning awareness. Emergency abandonment is an order given by the ISO when the threat of collapse is imminent. A precautionary withdrawal is typically ordered by the IC after consultation with the ISO. Planning awareness is a proactive sharing of collapse concerns between the ISO, IC, and RIC officer.
- The ISO must communicate established collapse zones, which are exclusionary or "no-entry" zones for everyone, including firefighters. The 1½ rule is a starting place for creating zone distances, although unreinforced masonry (URM) walls can collapse outward over three times the height of the wall.
- A partial collapse is an event in which the building can accept the failure of a single component and still retain some strength. A general collapse is the failure of the entire building to resist gravity. The biggest threat created by a partial collapse is a secondary partial or general collapse.
- RICs benefit from shared ISO observations pertaining to exit/egress options, construction material types and arrangements, utility hazards, reach distances, potential loads, and fire spread predictions.

Key Terms

axial load A load that is imposed through the centroid of another object.

balloon framing A construction method in which continuous wood studs run from the foundation to the roof, and floors are placed on a shelf—called a ribbon board—that hangs on the interior surface of the studs.

beam A structural element that transfers loads perpendicularly to the imposed load.

brittle Description for a material that will fracture or fail as it is deformed or stressed past its design limits.

cantilever beam A beam supported at only one end, or a beam that extends well past a support in such a way that the unsupported overhang places the top of the beam in tension and the bottom in compression.

collapse zone The area that is exposed to trauma, debris, and/or thrust should a building or part of a building collapse. It is a more specific form of a no-entry zone.

column A structural element that transmits a compressive force axially through its center.

compression A force that causes a material to be crushed or flattened axially through the material.

connection A structural element used to attach other structural elements to one another.

continuous beam A beam that is supported in three or more places.

curtain wall A non-load-bearing wall that supports only itself and is used just to keep weather out.

dead load The weight of the building itself and anything permanently attached to it.

ductile Description for a material that will bend, deflect, or stretch as a force is resisted, yet retain some strength.

eccentric load A load that is imposed off-center to another object.

emergency abandonment A strict order for all crews to *immediately escape* from a building interior or roof, leaving hose lines and tools that can impede rapid retreat behind.

engineered wood A host of products that consist of many pieces of native wood (chips, veneers, and sawdust) glued together to make a sheet, a long beam, or a strong column.

false work Temporary shoring, bracing, or formwork used to support incomplete structural elements during building construction.

general collapse The complete failure of a building to resist gravity.

girder A beam that carries other beams.

hybrid building A building that is a mix of multiple NFPA 220 types or that does not fit into any of the five types.

lintel A beam that spans an opening in a load-bearing masonry wall, such as over a garage door opening (often called a "header" in street slang).

live load Any force or weight, other than the building itself, that a building must carry or absorb.

partial collapse An event in which the building can accept the failure of a single component and still retain some strength (such as a curtain wall collapse).

platform framing A construction method in which a single-story wall is built and the next floor is built on the tops of the wall studs, creating vertical fire-stopping to help minimize fire spread.

precautionary withdrawal A directive for crews to exit a building interior or roof in an orderly manner, bringing hoses and tools along.

raker A diagonal brace that serves primarily as a column but must absorb some beam forces as well.

shear A force that causes a material to be torn in opposite directions perpendicular or diagonal to the material.

spalling The crumbling and loss of concrete material when exposed to heat.

spreader A seemingly decorative star or other metal plate used to distribute force over more bricks or blocks as part of an unseen corrective measure that exists inside a building.

structural element The primary load-bearing column, beam, or connection used to erect a building.

tension A force that causes a material to be stretched or pulled apart in line with the material.

torsional load A load that is imposed in a manner that causes another object to twist.

truss A series of triangles used to form an open-web structural element to act as a beam (in many ways, a "fake" beam because it uses geometric shapes, lightweight materials, and assembly components to transfer loads just like a beam).

veneer wall A decorative wall finish that supports only its own weight.

Review Questions

1. What are three ways loads are imposed on materials?
2. List the three types of forces created when loads are imposed on materials.
3. What is the definition of a beam?
4. Explain the effects of fire on steel structural elements.
5. How does a masonry wall achieve strength?
6. List and define the five common types of building construction using the NFPA 220 system.
7. What is a hybrid building? List several types.
8. List, in order, the five-step analytical approach to predicting building collapse.
9. List several factors that decrease the time that it takes for a structural element to fail under fire conditions.
10. Define the difference between an emergency abandonment, precautionary withdrawal, and planning awareness.
11. What observations and conditions should be communicated to a RIC officer by the ISO?

References and Additional Resources

Brannigan, Francis. *Building Construction for the Fire Service.* 5th ed. Quincy, MA: National Fire Protection Association and Jones & Bartlett Learning, 2015.

Building Construction for Fire Suppression Forces—Wood and Ordinary. Emmitsburg, MD: National Fire Academy, National Emergency Training Center, 1986.

Dunn, Vincent. *Collapse of Burning Buildings.* 2nd ed. Tulsa, OK: Fire Engineering Books and Video, PennWell Publishing Company, 2010.

Frechette, Leon A. *Build Smarter with Alternative Materials.* Carlsbad, CA: Craftsman Book Company, 1999.

Mittendorf, John, and David Dodson. *The Art of Reading Buildings.* Tulsa, OK: Fire Engineering Books and Video, PennWell Publishing Company, 2015.

Modern Construction Considerations for Company Operations. DVD Training Program and Instructors Guide, ISFSI, Pleasant View, TN, 2010. Available through www.ISFSI.org.

Morley, Michael. *Building with Structural Insulated Panels (SIPs).* Newtown, CT: Taunton Press, 2000.

NIOSH Alert, *Preventing Injuries and Deaths of Fire Fighters due to Truss System Failures.* Publication No. 2005-1432. Cincinnati, OH: NIOSH Publications, 2005.

Report on Structural Stability of Engineered Lumber in Fire Conditions. Project Number 07CA42520. Northbrook, IL: Underwriters Laboratories, 2008.

Structural Collapse: The Hidden Dangers of Residential Fires. Online Firefighter Training Program. Northbrook, IL: Underwriters Laboratories, 2008. www.fireengineering.com/content/dam/fe/online-articles/documents/FEU/Dalton.pdf. Accessed May 19, 2015.

U.S. Fire Administration. Changing Severity of Home Fires Workshop. December 2012. www.usfa.fema.gov/downloads/pdf/publications/severity_home_fires_workshop.pdf. Accessed May 19, 2015.

U.S. Fire Administration. *High Rise Office Building Fire, One Meridian Plaza, Philadelphia, PA.* Technical Report Series. Washington, DC: 1991.

Endnotes

1   Brannigan, Francis. *Building Construction for the Fire Service.* 3rd ed. Quincy, MA: National Fire Protection Association, 1992.

2   Guise, David. *Abstracts and Chronology of American Truss Bridge Patents.* Society for Industrial Archeology, 2009. http://www.sia-web.org/publications/other-publications/. Accessed April 26, 2015.

"Reading" a building is a skill that is developed through book study, pre-incident building visitations, and actual incident experience. Most fire officers would agree that building visitations are a great opportunity to improve the skill set—especially if the building is under construction. **FIGURE 5-19** shows a "podium-built" hybrid apartment project in various stages of construction. Using the photo as a reference, answer the following questions:

FIGURE 5-19

Smoke: © Greg Henry/ShutterStock, Inc.

1. What are the construction types visible in the picture and how would you classify this building once it is finished (type/era/use/size)?

2. Assuming the building is finished and occupied, what are the fire spread and collapse issues associated with a fire that originates in a third-floor condominium?

3. The wooden materials stacked in the foreground are examples of what structural element?

   A. Post columns

   B. Parallel cord trusses

   C. Lightweight rakers

   D. Engineered wooden I-beams

4. Which of the following hazards should be prioritized during the phase of construction pictured?

   A. Difficult apparatus access and construction equipment obstacles

   B. Lack of fire hydrants and potential damage to hose lines

   C. Incomplete stairway access and minimal options for interior attack lines

   D. Quick fire spread and a high potential for a rapid, general collapse

Reading Smoke

Flames: © Ken LaBelle NRIFirePhotos.com

Knowledge Objectives

Upon completion of this chapter, you should be able to:

- Define smoke. (pp 80–81)
- List the four attributes of smoke. (p 80)
- Identify the fire behavior conditions (phases) that lead to flashover or other hostile events. (NFPA 5.3.4 , pp 82–84)
- Define flow path and fuel load characteristics that affect the potential for flashover or other hostile fire events. (NFPA 5.3.4 , pp 83–86)
- Describe how fire behavior affects firefighting efforts. (NFPA 5.3.4 , pp 84–86)
- Analyze the factors presented at an incident related to fire behavior and fuel load/type to develop a proper incident action plan. (NFPA 5.3.4. , pp 80–90)
- Adjust an incident action plan based on assessment of fire behavior and fuel load characteristics. (NFPA 5.3.4 , p 90)
- Identify common products of combustion and pyrolysis that lead to the need for contamination reduction and cancer prevention processes for exposed firefighters. (NFPA 5.2.15 , pp 80–82, 87)

Skills Objectives

Upon completion of this chapter, you should be able to:

- Read smoke conditions of volume, velocity, density, and color to predict future hostile fire events. (NFPA 5.3.4 , pp 86–90)

Smoke is a predictor of future events. The incident safety officer (ISO) is in a unique position on the fire ground to assess smoke conditions and evaluate their hazards. As you respond to one of your first incidents as the designated ISO, you find a bedroom fire on the alpha-bravo corner of the second floor of a modern lightweight two-story home. Smoke is issuing from the bedroom and you also notice brown smoke under pressure pushing from the eaves and soffit area below the roof.

1. What factors should be considered when reading smoke coming from a structure? Do the conditions you described warrant action by the ISO?

2. In your experience, have you assessed smoke as a factor to determine your actions or used your reading smoke skills to anticipate a hazardous event? How were your observations communicated?

Introduction: A "Lost" Art Resurrected

The current popular trend of "reading smoke" was triggered by incident safety officer (ISO) academies that were developed by the late David Ross (then Health and Safety Chief, Toronto Fire Services) and the author of this text for the Fire Department Safety Officer's Association in the 1990s. Though the concept of reading smoke is not new, prior to the ISO academies' effort, few had tried to define a process to assist ISOs in predicting fire behavior and to help them gauge firefighting progress. The practice of reading smoke has been around for many decades; fire officers handling America's fire epidemic in the 1970s became quite proficient at the skill. Unfortunately, these sound tacticians felt that the ability to read smoke was based on experience and intuitiveness and could not be taught in a classroom environment. However, a few fire officers tried to define the concepts of reading smoke, namely Battalion Chief John Mittendorf (Los Angeles City Fire) and Deputy Chief Vincent Dunn (Fire Department, City of New York).

The author of this text took the teachings of Mittendorf and Dunn and developed a reading smoke teaching model that was presented around the country and then in text form for the first edition of this book back in 1998. Since then the fire service has acknowledged that changes have led to more severe content and structure fires—and more flashover entrapments of firefighters. As a result, the skill of reading smoke for predicting fire behavior became an essential firefighter safety "must have" but not a cure-all. The fire service needed to better understand the dynamics of the modern interior firefighting environment, leading to significant grant studies conducted by the National Institute of Standards and Technology (NIST) and Underwriters Laboratories (UL), among others. These studies have concluded that the modern fire behavior environment is hotter, develops tremendous energy (heat release rate), and contains smoke that is extraordinarily explosive. The studies are now evolving to include the development of tactical solutions that can help firefighters better manage these environments.

Reading smoke is a developed skill that helps you predict fire behavior. This chapter presents the reading smoke process and offers suggestions on how an ISO can apply the process in a way to predict hostile fire events (like flashover) and monitor fire attack effectiveness so that he or she can prevent harm and help the incident commander (IC) with incident action plan (IAP) issues.

Smoke Defined

Reading smoke is not difficult, although for most fire officers, it takes an effort to break the "heavy smoke or light smoke" mentality that has come out of rapid size-up radio reports. Smoke leaving a structure has four key attributes: *volume*, *velocity* (*pressure*), *density*, and *color*. A comparative analysis of these attributes can help the ISO determine the size and location of the fire, the effectiveness of fire streams, and the potential for a hostile fire event, such as a flashover. Before we can look at the meaning of each attribute, we must understand the science underlying what is seen in smoke.

In a simpler time, smoke was viewed as a by-product of incomplete combustion, specifically particulates (solids) that were suspended in a thermal column. In today's world, that oversimplification is dangerous: Smoke is so much more than particulates from incomplete combustion. Smoke is better defined as the products of incomplete combustion and pyrolitic decomposition that include an aggregate of particles, aerosols, and fire gases that are toxic, flammable, and volatile[1] **FIGURE 6-1**.

Breaking down this definition, we find that smoke at a building fire is being developed by two sources:

- *Incomplete combustion.* Materials contributing smoke through incomplete combustion are actually on fire, but the rapid oxidation hasn't fully reduced the fuel.
- *Pyrolitic decomposition.* Materials contributing smoke through pyrolitic decomposition aren't necessarily burning but are being chemically degraded by heat. Pyrolysis (the shorter and more common term for pyrolitic decomposition) is the chemical breakdown of compounds into other substances by heat alone. Pyrolysis often leads to combustion.

Most of the smoke at a typical building fire is *not* coming from burning materials, but from materials breaking down without burning (pyrolysis)![2] Heat energy from materials that are burning is radiated and convected throughout the building and is transferred to objects that are cooler (remember from high school science class that heat seeks cold, and cold doesn't seek heat). These cooler objects serve as a heat sink. A heat sink will continue

FIGURE 6-1 Smoke is a toxic and explosive aggregate of particulates, aerosols, and gases from incomplete combustion and pyrolysis.
© Keith Muratori. Used with permission.

to absorb heat based on its composition and surface-to-mass ratio. At some point, the composition of the material will begin to degrade, and smoke is produced. The term *off-gassing* is often used for pyrolysis. It is not, however, accurate, so we will refer to it as slang. More accurately, surfaces that are smoking—not burning—are *off-aggregating*. That is, they are releasing a mix of particulates and aerosols as well as gases. This leads us to a further breakdown of the smoke definition, the *aggregate*:

- *Particulates.* The solids (particulates) suspended in smoke are "high surface-to-mass" materials, meaning they have lots of surface area to collect heat and no mass to "sink" heat. Clearly, this leads to more volatility for ignition. Soot and ash are the more prevalent solids in smoke. Previous editions of this text classified soot and ash as being similar, an oversimplification that should be corrected. Soot is carbon (officially "black carbon"), and carbon can support flaming. Ash is the trace metals and minerals (depleted salts) that can no longer support flame. This is not to say that ash is less dangerous than carbon. Ash can carry tremendous volumes of heat and cause other materials to ignite. As they relate to reading smoke, carbon will add a flat (dry) black color to smoke, whereas ash will add a dirty white color.

Other solids in smoke include dust, fibers, and pulp. All of these particulates take up space and give smoke its density, texture, and color. For most interior fires, the volumetric composition of smoke is mostly particulate matter. When smoke is confined in a compartment (e.g., room, hallway, attic), the particulates displace air, which can cause a fire to become ventilation-limited (some use the term air-limited).

- *Aerosols.* An aerosol can be defined as a suspended or propelled liquid. At a structure fire, the liquids in smoke include primarily moisture and hydrocarbons (oil and tar), although there are also acids, aldehydes, and ketones, to name a few. Water vapor is a by-product of combustion (flaming) and pyrolysis (materials steaming off saturated moisture). In the early stages of heating, a material will first release its moisture (steam), which is a very pure, dissipating white vapor. As that material breaks down further, the vapor becomes more of a true, nondissipating white (usually indicating ammonia and other chemicals). We can thank plastics for adding an incredible amount of hydrocarbons to smoke. Flaming plastics release hydrocarbons, as do plastics that are decomposing due to heat. These hydrocarbons give smoke a satin (wet) black color. It is important to note that some very common hydrocarbons in smoke can self-ignite as low as 450°F (232°C), but they often do not ignite because the particulates have made the smoke too rich to burn. This makes smoke at a typical house fire amazingly explosive.

- *Gases.* There are hundreds of chemically named fire gases in smoke, although most are transient and trace. Some gases of significant quantity that affect fire behavior are carbon monoxide (CO), hydrogen cyanide (HCN), benzene (C_6H_6), acrolein (C_3H_4O), and hydrogen sulfide (H_2S). These gases are singled out because, due to the flammable nature of each, they ultimately influence fire behavior. What we haven't mentioned is the toxicity of each. They are extraordinarily dangerous to health and can enter the human body through various methods, including inhalation and skin absorption (see the *Safety Tip* box).

Safety Tip

A Toxic Cocktail
Although the intent of this chapter is to discuss fire behavior influences, we must state how important it is to always wear self-contained breathing apparatus (SCBA) and personal protective equipment (PPE) for fire suppression *and* overhaul activities due to the toxicity of smoke. Carbon monoxide, hydrogen cyanide, benzene, acrolein, acid gases, phosgene, formaldehyde, polynuclear aromatic hydrocarbons, polyvinyl chloride, phenol, nitrogen oxides, and ammonia are only some of the toxic chemicals that are routinely found in today's residential structure fire smoke. This toxic smoke cocktail can lead to short-term and chronic health issues, including an increased risk of cancer. The use of SCBA and PPE cannot be overstated—don't breathe smoke!

TABLE 6-1	Properties of Gases Typically Found in Smoke			
Gas	Flashpoint	Self-Ignition Temperature	Flammable Range in Air	Notes
Carbon monoxide (CO)	See notes	1,128°F (609°C)	12–74%	CO is considered a gas only and, therefore, doesn't have flashpoint. The flammable range of CO is 12–74% only at the ignition temperature. The flammable range narrows as temperatures decrease. Below 300°F (149°C), CO is not likely to ignite.
Hydrogen cyanide (HCN)	0°F (–18°C)	1,000°F (538°C)	5–40%	HCN is produced when high temperatures break down nitrogen-containing products (most synthetics). HCN is quite flammable and is considered extremely toxic.
Benzene (C_6H_6)	12°F (–11°C)	928°F (498°C)	1–8%	Most plastics release benzene while burning or pyrolyizing. Benzene is also a common product from the burning of fuel oils.
Acrolein (C_3H_4O)	–15°F (–26°C)	450°F (232°C)	3–31%	Acrolein is a by-product from the incomplete combustion of wood, wood products, and other cellulosic materials. "Poly-plastics" can also render acrolein.

Source: http://www.cdc.gov/niosh/npg/npg.html.

TABLE 6-1 lists the properties of some of these gases. While reviewing these properties, you should note that smoke is ignitable as low as 450°F (232°C) and has a collective flammable range of 1% to 74% in air!

Defining smoke and discussing its composition will help you understand why the process of reading smoke works. The bottom line: The modern structure fire smoke environment is extremely flammable (explosive, even) and ultimately dictates fire behavior. Let's now look at predicting fire behavior within a building.

Predicting Fire Behavior

The ISO who focuses on the fire (flaming) to determine fire behavior is being set up for a sucker punch. To predict fire behavior in a building, the ISO needs to understand some realities that govern fire outcome. This understanding begins with the notion that "open flaming" is actually a good thing: The products of combustion are minimized because the burning process is more complete. In fact, the complete combustion of common materials renders heat, light, carbon dioxide, and water vapor (none of which can burn). As previously stated, within a building, the heat from flaming (exothermic energy) is absorbed in other materials that are not burning (contents and the walls/ceiling). In today's plastic-rich world, many materials lack material mass to absorb heat and therefore breakdown quickly when exposed to just a small amount of exothermic energy. At that point, smoke becomes a flammable concern, and two conditions must come together to trigger its ignition within a building.

■ Smoke Ignition Triggers

Within a box (a room), the off-gassed smoke from so many low-mass materials displaces air, leading to what is termed an *underventilated* or *ventilation-limited* condition. Some fire behaviorists use the terms *air-limited* or *ventilation-controlled* to say the same thing. Ventilation-limited fires do not allow the open flaming to complete a reaction with pure air, which in turn leads to increasing volumes of carbon monoxide as well as the aforementioned smoke products. Bright, open flaming is reduced although the fire continues to spread and release heat. Smoke is now "looking" to complete what was started. Two triggers may cause accumulated smoke to ignite: the right temperature and the right mixture.

Smoke gases that are below their ignition temperature (but above flashpoint) need only a proper air mix and a sudden spark or flame to complete their ignition (called a piloted ignition). Distal to the actual fire, a simple glowing ember or failing lightbulb can spark an ignition where an appropriate air mix exists. Once smoke gases reach ignition temperature, they don't need a spark or flame to ignite—only the right mixture in air. The ignition of smoke that has pressurized a room (or "box") likely results in an expansive surge. The ignition of accumulated smoke also changes the basic fire spread dynamics; instead of flame spread across surfaces of contents, the fire spreads with the smoke flow. The ISO who watches what the smoke is doing makes better decisions than the one focused on flaming, because the smoke tells you how intense the fire is about to become as opposed to how bad it currently is FIGURE 6-2. John Mittendorf, noted author, instructor, and retired battalion chief from Los Angeles Fire Department, said it best: "Smoke is the fire talking to you—it's telling you the future."

Watching smoke flow can help the ISO understand what is about to happen, if the ISO understands the phases that a building fire typically follows.

Compartmentalized Fire Growth Phases

Most basic fire schools have long taught that a building fire goes through four phases or stages: ignition, growth, fully developed, and decay (see the *Fire Marks* box). In actuality, the four-phase growth model is for a fuel-controlled fire that has plenty of air. The sheer quantity of smoke being developed from the pyrolysis of low-mass fuels has rendered the four-phase teaching model outdated for interior building fires.

FIGURE 6-2 The ISO who reads the smoke (versus the flames) can predict how severe a fire can become.

© Keith Muratori. Used with permission.

Compartmentalized fires are influenced by several factors that can lead to a six-phased growth model (ventilation-controlled model). Factors include:

- The size of individual compartments.
- The type, quantity, and continuity of the fire load (contents and combustible finishes).
- The presence or lack of smoke and fire control systems.
- Available <u>flow paths</u> through adjoining compartments and the exterior of the building. Flow paths are the avenues that heat, smoke, flames, and combustion air follow. Flow paths can be divided into the "air track" (intake combustion air moving toward the fire) and "exhaust" (heat, smoke, and flames moving away from the fire). Flow paths are typically influenced by the interior geometry of the building (room volume, interior doors, stairways, hallways, and the like) as well as openings to the exterior environment (windows and doors).
- The size and status of exterior openings (closed, partially or fully opened).

Although the influences are many, the modern, synthetic-based fire load produces overwhelming quantities of smoke that can exceed available flow paths, leading to the

Fire Marks

Busting the Fire Growth Myths

Curriculums used to teach fire behavior to new firefighters have slowly evolved over the past three decades but are still considered fairly elementary. For example, the teaching of the early 1970s included fire growth models that showed a three-phase progression: incipient, growth, and smoldering. More advanced training of that era divided the growth phase into *steady-state* and *clear-burn*. Still more advanced was the division of the last phase into *hot-smoldering* (setup for a backdraft) and *cold-smoldering* (decay of heat).

In the 1980s a revision was made to core fire behavior lesson plans and included the four-phase growth model: ignition, growth, fully developed, and decay. The growth phase and fully developed phase were bridged by a flashover event. The 1990s brought shared stories of hotter fires (due to plastics), unusual fire behavior, and classic fire attack tactics rendered useless. Although not mythical, these anecdotal experiences led to some very unscientific explanations and the development of well-meaning solutions that were argued, debated, or otherwise refuted by others within the fire service.

Enter the fire behavior "rocket scientists." In the late 1990s (and continuing today), NIST and UL secured federal grant monies to research modern fire behavior. Why? Simply answered, firefighters were being caught in flashovers at an alarming rate (resulting in close calls and line-of-duty deaths), and many fire officers have experienced fire behavior events that they've called unusual or extreme. Classic tactical ventilation methods were not effective and interior hand lines were proving inadequate for certain fires in relatively small buildings. Scientific research and data were needed, hence the NIST and UL involvement.

In a 2005 conversation with two leading fire behaviorists (Daniel Madrzykowski from NIST and Stephen Kerber from UL), a question was posed: Should we (fire service instructors) change our classic four-phase fire growth curriculum? Both experts replied with a resounding *yes*. In their teachings, Madrzykowski and Kerber explain that the four-phase model is for a "fuel-controlled" fire event that has plenty of air. The new interior (or compartmentalized) fire can transition through six phases, prompting what is now known as the *ventilation-controlled fire growth model*.

The scientific studies continue and several explanations and solutions have already been documented—saving the lives of firefighters. We owe a huge thank-you to the modern-day fire behavior myth busters!

potential of a ventilation-controlled event. The six phases of a ventilation-controlled fire are illustrated in **FIGURE 6-3** and further described below.

1. *Ignition (incipient) phase.* The ignition phase includes the event (or events) that brings together heat, fuel, and oxygen to start the self-sustaining process of combustion.
2. *Initial growth phase.* The word "initial" has been added here for reasons that you'll see in the fourth phase. The initial

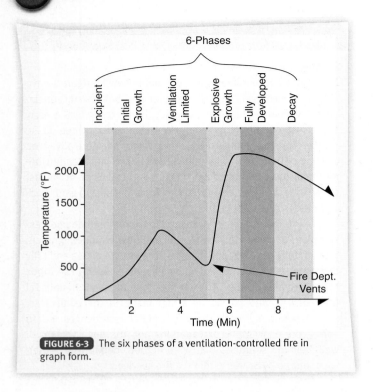

FIGURE 6-3 The six phases of a ventilation-controlled fire in graph form.

Unfortunately, the short time lag it takes for an air track flow path to reform back to the fire is near the same time that it takes for firefighters to open the door and start advancing through floor-to-ceiling smoke to find the fire. The incredible insulation that modern PPE affords can keep the firefighters from feeling heat, yet the smoke surrounding them is hot enough to ignite. When air reaches the ventilation-limited fire, the explosive growth occurs and the ensuing flame wave overtakes the unsuspecting firefighters.

The explosive growth phase can include some detrimental effects on firefighting efforts. First, firefighters advancing toward the seat of the fire in zero-visibility conditions can be totally engulfed in flames. These flames can originate from the door they entered and move toward the fire seat; the firefighters may not see the situation coming. Second, the heat release rate will be substantial as the dense smoke ignites, easily surpassing the cooling capabilities of common hand lines. A proactive solution to prevent the explosive growth phase includes *protective cooling*, which is the application of water to ceilings and walls to help quench and cool moving smoke.

5. *Fully developed phase.* The explosive growth event typically leads to total flame involvement of the interior flow paths.
6. *Decay phase.* In the decay phase, available fuels become consumed and the fire begins to wane. Some call this the *fuel-limited phase.*

Some may interpret the explosive growth phase as nothing more than the old "hot smoldering" backdraft event. Taking this oversimplified approach may not do justice to the science of modern fire behavior (see the "Backdraft" section that follows). The primary difference is that an explosive growth event can be delayed as a volume shift takes place after opening a door or window; in this situation, a huge volume of smoke is allowed to exhaust and an air track redevelops as the volume of smoke allows. This volume shift delay can fool unsuspecting firefighters: "It didn't ignite, so we must be good to go." The firefighters enter the smoke environment only to have "the right mixture and right temperature" collide, engulfing the crew. The explosive growth phase of a compartmentalized fire is a very real hostile fire event that is likely in today's building fires. Let's discuss the other hostile fire events that can occur.

■ Hostile Fire Events

Many believe that any uncontrolled fire in a building is a hostile fire event. This philosophy is right on; however, in this text, we use the phrase hostile fire events to describe fire behavior phenomena that can suddenly harm firefighters: flashover, backdraft, smoke explosion, and flame-over. The aforementioned explosive growth phase is actually a combination of flashover and flame-over. During interior firefighting operations, committed firefighters may have restricted visibility and reduced ability to see the "big picture" of fire behavior. Similarly, the IC is busy with communications, status reports, tactical worksheets, and strategic planning at a fixed command post. The ISO is typically in a unique position to roam and watch smoke leaving the fire

growth phase has also been labeled the fuel-controlled phase. In this phase, fire growth is controlled by the proximity of burning fuels to other burnable fuels. Air is abundant and the fire grows in an upward and outward fashion as flames touch other materials. Room visibility is still good at lower levels, and smoke from the flames begins to accumulate at the ceiling. Exothermic radiant heat waves are released and start heating up other fuels in the area of the fire. Pyrolysis begins and smoke production becomes abundant. The convection smoke column can also cause fire spread in the initial growth phase.

3. *Ventilation-limited phase.* This is a new phase not typically included in previous fire behavior teaching models. The ventilation-limited phase is a compartmentalized fire condition where open flaming decreases because smoke production displaces and limits available combustion air, although heating continues to produce smoke. The condition is considered dangerously explosive because heat and volatilized fuels are readily available—only air is missing. Introduction of air will cause the fire to rapidly transition into an explosive growth phase.

4. *Explosive growth phase.* This phase is a rapid fire growth phenomenon that occurs when combustion air is reintroduced into a ventilation-controlled fire, leading to smoke flame-over and room flashovers. Some have called this the flashover phase or delayed flashover phase. It may be better to call it explosive growth because a flashover is typically a single-room event. In an explosive growth phase, the room of origin will likely flashover but the fire growth may include ignition of all the smoke in the building (a true flame-over). In too many cases, fire department efforts to reach the fire (such as the opening of more doors) will cause explosive growth.

building. Given this flexibility, the ISO must watch smoke for warning signs that a hostile fire event is looming. Historically, hostile fire events have been defined and argued by scientists, scholars, and fire experts. A review of fire phenomena definitions in available texts and elsewhere shows that the debate continues. Regardless, some middle ground can be found to help us better define and predict hostile fire events so that smoke can be read. We'll start with a brief description of some phenomena terms and events followed by some warning signs TABLE 6-2 . Again, these descriptions are arguable, but they provide one way of understanding that helps the ISO read smoke.

Ghosting

Ghosting is a hostile event warning sign and is characterized as the intermittent ignition of small pockets of smoke, usually seen as fingers of flame that dance through the upper smoke layer. Previous teachings have labeled this same phenomenon as "rollover." Ghosting is not necessarily a hostile fire event but serves as a warning sign of impending flashover or flame-over.

Flame-Over

Flame-over is a hostile fire event that includes the ignition and sustained burning of the overhead smoke layer within a room and/or hallway. Lower contents don't necessarily ignite, but when they do, they will ignite quickly. Where plenty of air exists, a flame-over typically originates at the seat of the fire and travels along the heat flow paths. The flame-over travel is usually opposite in a ventilation-limited phase fire that finally gets air: The flames start near the ventilation opening and burn back to the fire seat where they can trigger a flashover of the room (ignition of all surfaces and contents).

Smoke Explosion

Several fire service teachings categorize smoke explosions and backdrafts as the same thing. Doing so is perhaps an oversimplification that can lead to undesirable outcomes (firefighters get hurt). As defined here, a smoke explosion is a hostile fire event that occurs when a spark or flame is introduced into a pocket of smoke that is below ignition temperature but above some aggregate flashpoint. The result is a split-second ignition (and rapid expansion) of that pocket with no sustained burning. The resulting expansive surge can blow out windows, collapse building elements, or knock over firefighters. Smoke explosions typically occur in a trapped-smoke area away from the fire, such as a dead-end hallway, the top of a stairwell, an uninvolved room above the fire, or a void space.

TABLE 6-2	Hostile Fire Events	
Event	**Warning Signs**	**Notes**
Flashover	• Turbulent smoke flow that has filled a compartment • Ghosting • Vent-point ignition (exterior autoignition of the exhaust flow path) • Rapid change in smoke volume and velocity (getting worse in seconds)	Flashover is an event triggered by radiant heat reflected within a room or space. All surfaces reach their ignition temperature at virtually the same time due to rapid heat buildup in the space. If air is present, the room becomes fully involved. If air is not present, the flashover is delayed until air is introduced (ventilation-limited phase).
Explosive growth phase	• Dense smoke appears to have totally filled a building, floor to ceiling • Slow but steady smoke flowing from closed doors or windows • Smoke that rapidly speeds up when an exterior door is opened • The development of an air track below the smoke when an opening is made • Exhaust flow paths that intermittently puff or try to suck air (open doors and windows) • Smoke flame-over upon the breaking of windows or opening of doors (late sign)	Explosive growth is a phase that occurs when air is introduced to a ventilation-limited fire. It can include smoke flame-over in flow paths and flashover of individual rooms that are heat saturated.
Backdraft	• Yellowish-gray smoke emitting from cracks and seams • Bowing, black-stained windows • Closed pressurized box with signs of extreme heat • Puffing from the cracks and seams of a closed box	Backdraft occurs when oxygen is introduced into a closed, pressurized space where fire products are above their ignition temperature. Note: Sucking or puffing of smoke from an *open* door or window is a warning for explosive growth. Puffing of the cracks and seams of a *closed* pressurized space is a backdraft warning.
Smoke explosion	• Smoke that is being trapped in a separate space above the fire • Signs of a growing fire • Signs of smoke starting to pressurize	A smoke explosion occurs when a spark or flame is introduced into trapped smoke that is below its ignition temperature but above its flashpoint. A proper air mix is necessary. Carbon monoxide (CO) is ignitable (with a spark or flame) around 300°F (149°C) but has a small flammable range in air mix. As trapped CO heats up, its flammable range widens, making it easier to ignite with a spark or flame as it gets hotter. Smoke explosions typically happen in spaces away from a growing fire.
Flame-over	• Increase in smoke speed • Ghosting • Laminar flow of smoke that is becoming turbulent • Smoke flowing from hallways and stairways faster than a firefighter can move	Flame-over occurs when smoke reaches sustaining temperatures that are above the fire point of prevalent gases. The gases can suddenly ignite when touched by an additional spark or flame. Fire spread changes from flame contact across content surfaces to fire spread through the smoke. This marks a significant change in fire spread behavior.

Flashover

Flashover is a sudden hostile fire event that occurs when all the surfaces and contents of a space reach their ignition temperature nearly simultaneously, resulting in full-room fire involvement. In most cases, a flashover occurs because the room itself (walls, ceiling) can no longer absorb heat and begins to reflect radiant energy back into itself. The super-heated upper gas layer expands downward and rapidly heats floor-level fuels to their ignition temperature. As mentioned before, a room may have been heated to the point of flashover, but the ignition can't occur because no combustion air is available—the setup for an explosive growth event.

Backdraft

A backdraft is an explosive event that occurs when air is suddenly reintroduced into a closed space that is filled with pressurized, ignition-temperature, and oxygen-deprived products of combustion and pyrolysis. In many ways, explosive growth and backdraft are similar; both are the result of air introduction into an ignition-temperature smoke environment. They differ in the rate of air introduction and the resulting force and speed of the ignition. The explosive growth event (flame-over and delayed flashover) is triggered by a volume shift between smoke and air that occurs over a 10- to 90-second time span, whereas the backdraft is instantaneous upon the introduction of air to the oxygen-depleted environment. Upon ignition, the explosive growth event will spread the fire at a rate that is consistent with the exhaust flow path (smoke speed). The backdraft event will ignite more like a detonation (burning above the speed of sound) and will likely include a shock wave.

Firefighters are usually taught *reactive* warning signs for hostile fire events. For example, most firefighters list sudden heat buildup (that forces them to the floor) as a warning sign of flashover. Given the insulation provided by today's structural PPE ensemble, the sensation of heat is a dangerously late warning sign. Low-ignition-temperature gases/aerosols are already ignitable when the firefighter feels heat. For this reason, firefighters—and the ISO—must learn about the *proactive* warning signs listed in Table 6-2.

An obvious starting place for the skill of reading smoke is the knowledge of hostile event warning signs. In the next section, we will get much more specific and show you how the skill can work.

The Art of Reading Smoke

Reading smoke is a skill set that combines the ability to identify smoke characteristics, analyze factors that are influencing the smoke, and make judgments regarding the location, phase, and spread potential of a fire in a building, as well as the likelihood of a hostile fire event **FIGURE 6-4**. Those judgments become the basis for an ISO to evaluate threat potentials relating to fire behavior or fire load and to take actions to prevent harm, or to communicate IAP issues to the IC.

Identifying smoke characteristics goes beyond the simple classifications of light or heavy. Smoke has four distinctive characteristics or attributes that combine to tell a story: volume, velocity, density, and color (VVDC). Let's consider how each of the attributes contributes to understanding fire behavior.

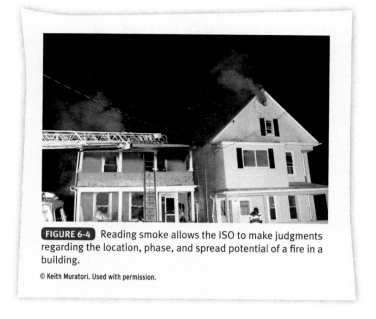

FIGURE 6-4 Reading smoke allows the ISO to make judgments regarding the location, phase, and spread potential of a fire in a building.
© Keith Muratori. Used with permission.

■ Volume

Smoke volume by itself tells very little about a fire, but it sets the stage for understanding the amount of fuels that are off-gassing in a given space. A hot, well-ventilated, and clean-burning fire emits very little visible smoke (low smoke volume), yet a hot fire in a ventilation-limited environment will develop a tremendous volume of smoke. In the latter environment, synthetic, high surface-to-mass fuels develop large volumes of smoke even though little flame is present. Given these considerations, the volume of smoke *can* create an impression of the fire relative to the size of the space it fills. For example, a small fast-food restaurant can be totally filled with smoke from a small fire. Conversely, it would take a significant fire event to fill the local big-box store with smoke. Once a container is full of smoke, pressure builds if adequate ventilation is not available; therefore, smoke volume becomes one of the generators of smoke velocity.

■ Velocity

Smoke velocity is an indicator of pressure that has built up in the building. When smoke moves through a building, it has certain speed, direction, and flow characteristics that are created by pressure variants. In context, we don't really see smoke pressure, but we do see the smoke movement or *velocity*. From a practical fire behavior point of view, two forces cause smoke to pressurize in a building: convection heat or smoke volume. When smoke is leaving the building, its velocity is caused by *heat* if it rises quickly and then slows gradually. Heat-pushed smoke warrants attention. Smoke caused by restricted *volume* will immediately slow down and balance with outside airflow (and often sinks). Smoke that is only volume-pushed is notable but of lesser concern.

Generally speaking, the faster smoke moves, the more heat it has. In addition to the speed of smoke, the ISO needs to look at its flow characteristic. The flow of smoke can be *turbulent* (other descriptions may include "agitated," "boiling," or "angry") or *laminar* (smooth, calm, or following a straight line).

Turbulent smoke is caused by serious heat; the molecules of the smoke are bouncing around, colliding, and expanding in a violent fashion. The smoke coming off of the flame tips of an uncontrolled fire is almost always turbulent, but it smoothes out as it mixes with cooler air and flows across surfaces that sink (absorb) heat. Smoke that is being developed by pyrolysis begins laminar. When the fire compartment can no longer absorb heat, the reflecting radiant energy causes all the smoke to become turbulent; therefore, turbulent smoke that has filled a compartment is a warning sign of impending flashover **FIGURE 6-5**. If the box is still absorbing heat, the heat of the smoke is subsequently absorbed, leaving a more stable and smooth flow characteristic—laminar flow. The most important smoke observation is whether its velocity is turbulent or laminar. Turbulent smoke is ready to ignite and indicates a flashover environment that may be delayed by improper air mix (ventilation-limited). Laminar smoke leaving a building doesn't mean *safe*; it only means that the box is still absorbing heat.

Comparing the velocity of smoke at different openings of the building can help the fire officer determine the location of the fire: Faster smoke is closer to the fire seat. Remember, however, that the smoke velocity you see outside the building is ultimately determined by the size and restrictiveness of the exhaust opening. Smoke follows the path of least resistance and loses velocity as the distance from the fire increases. To find the location of fire by comparing velocities, you must compare only like-resistive openings (doors to doors, cracks to cracks, and so on). More specifically, you should compare cracks in a wall to other cracks in walls. Do not compare the speed of smoke leaving a crack in window glass to the cracks in a wall; the resistance to smoke flow is different.

■ Density

While velocity can help you understand much about a fire (how hot it is and where it is), density tells you how bad things are going to be. The density of smoke refers to its thickness.

FIGURE 6-5 Turbulent smoke that has filled a compartment is a warning sign of impending flashover.

© Keith Muratori. Used with permission.

Safety Tip

Turbulent Versus Laminar Smoke Flow
The most important smoke observation is whether its velocity is turbulent or laminar. Turbulent smoke velocity that is thin and dark indicates that a well-ventilated fire is nearby. *Turbulent smoke velocity that has volume-filled a compartment indicates that a flashover is imminent.* Firefighters should not crawl into a box (compartment) filled with turbulent smoke, and an ISO should intervene if they are doing so. Laminar (or smooth) flow means that the interior of the space near the fire is still absorbing heat or that the smoke you see has traveled some distance from the fire. Generally speaking, turbulent smoke is always fast. Laminar smoke can be fast or slow; a faster laminar flow indicates the presence of more heat.

Since smoke is *fuel*—containing airborne solids, aerosols, and gases that are capable of further burning—the thickness of the smoke tells you how *much* fuel is laden in the smoke. In essence, the thicker the smoke, the more spectacular the flashover or fire spread will be. Smoke thickness also sets up fuel continuity. On a practical basis, thick smoke spreads a fire event (like flame-over and flashover) farther than less dense smoke. Even though we use turbulent smoke as a flashover warning sign, thick, laminar-flowing smoke can ignite because of the continuity of the fuel to the flaming source.

One other point regarding smoke density: Thick, black smoke in a compartment reduces the chance of life sustainability due to smoke toxicology. A few breaths of thick, black smoke renders a victim unconscious and causes death in minutes. Further, the firefighter crawling through zero-visibility smoke is actually crawling through ignitable and contaminating fuel. These fuels are cancer-causing and must be removed from PPE surfaces to minimize spread and reduce the risk of skin/lung exposure once gear is doffed. Smoke density is thus an indicator that gross decontamination processes must be implemented on-scene.

■ Color

Most fire service curriculums teach us that smoke color indicates the "type" of material that is burning. This is misleading. Granted, a burning rubber car tire produces a thick, deep-black smoke, but that same tire can smoke white in the early stages of pyrolysis. That is true for almost all materials—they start off by smoking a whitish color, and as applied heat is increased, they turn gray, then black. Likewise, most materials give off black smoke from flame tips.

In typical residential and commercial fires, it is rare that a single fuel source is emitting smoke; the smoke seen leaving a building is a mix of colors. Further, we know that most of the smoke coming out of a building was produced by pyrolysis; therefore, smoke color tells the stage of heating and points to the location of the fire in a building. Nearly all solid materials emit a white "smoke" when first heated. This white smoke is mostly moisture but can contain other gases and aerosols such as ammonia and phenols. As a material dries out and heats up, the color of the smoke changes from white. Smoke produced by natural, unfinished materials (like wood) changes to tan or brown, whereas that produced by synthetics and painted or stained surfaces changes to gray. As materials are further heated, the smoke will eventually be all black. When flames touch surfaces that are not burning, the surfaces begin to off-gas black smoke almost immediately. Therefore, the more black the smoke you see, the hotter the smoke. Black smoke that is high velocity and very thin (low density) is flame-pushed. Interpreted from outside a building, thin, black smoke means that open (and ventilated) flaming is nearby.

Smoke color can also tell you the distance to a fire. As smoke leaves an ignited fuel, it heats up other materials and the moisture from those objects can cause black smoke to turn gray or even white over distance. As smoke travels, any carbon and hydrocarbon content from the smoke is deposited along surfaces and objects, eventually lightening the smoke color. So the question is whether white smoke is a result of early-stage heating or of late-stage heating smoke that has traveled some distance. The answer is to look at the velocity. Fast-moving, dirty white smoke indicates that the smoke you see has traveled some distance, but a hot fire exists.

One more important note about smoke color—namely, brown smoke: Unfinished wood gives off a distinctive brown smoke as it approaches mid- to late-stage heating (just prior to turning black or flaming). In many cases, the only unfinished wood in a structure is what makes up the wall studs, floor joists, and roof rafters and trusses. Brown smoke from structural spaces indicates that significant heat is present **FIGURE 6-6**. Using our knowledge of lightweight wooden building construction, the issuance of brown smoke from gable-end vents, eaves, and floor seams is a warning sign of impending collapse. Remember that engineered wood products like oriented strand board (OSB) and laminated veneer lumber (LVL) lose strength when heated. The glues of these products break down with heat and do not necessarily need flames to come apart. Brown smoke from structural spaces containing glued

trusses, OSB, or LVL can indicate that critical strength has been already lost. The ISO who witnesses tan/brown smoke from the structural area(s) of lightweight wood buildings should immediately confer with the IC to communicate the collapse warning and address the IAP risk/benefit issues that might be affected by it.

By knowing the meaning of smoke VVDC and processing the attributes observed, the ISO can paint a mental picture of the fire. Compare smoke velocity and color from various openings to locate the fire. Faster and/or darker smoke is closer to the fire seat, whereas slower and/or lighter smoke is farther away. Typically you see distinct differences in velocity and colors from various openings. When the smoke appears uniform—that is, it is the same color and velocity from multiple openings—you should start thinking that the fire is in a concealed space (or deep-seated). In these cases, the smoke has traveled some distance or has been pressure-forced through closed doors or seams (walls and concealed spaces), which "neutralizes" color and velocity prior to exiting the building. Upon seeing smoke that is the same color and velocity being pushed from multiple building seams, the ISO should inform the IC that the fire may have extended to concealed spaces.

Black Fire

Black fire is a slang term that fire officers have used for years. Although not scientific, it is a good phrase to describe smoke that is high-volume, has turbulent velocity, is ultra-dense, and is deep black (in other words, smoke VVDC has gotten as bad as it can be). Black fire is a sure sign of impending autoignition and flashover. In actuality, the term "black fire" is accurate—the smoke itself is doing all the destruction that flames would cause: charring, heat damage to steel, content destruction, concrete spalling, and victim death. Black fire can reach temperatures of over 1,000°F (538°C). The ISO should report black fire conditions, and no firefighter should be in or near compartments emitting black fire. The solution to black fire is the same as that for flames: Vent and cool!

■ Factors That Influence the Reading of Smoke

Weather, thermal balance, container size, and firefighting efforts can change the appearance of smoke. Let's look at how each of these factors can influence smoke.

Weather

It is important to know how weather affects smoke. Once smoke leaves a building, the outside weather, including temperature, humidity, and wind, can influence its appearance. Virtually every element of smoke is heavier than air, yet it rises due to heat. If the outside air is cooler than the smoke, the smoke should rise until the outside air cools it. It stands to reason that cold air temperatures cool smoke faster and cause it to stall and/or fall. A hot day finds the smoke rising much farther because cooling is prolonged. Humidity in the air increases resistance to smoke movement by the cohesive qualities of water vapor. Low humidity allows smoke to dissipate more easily. Humid climates keep the smoke plume tight.

FIGURE 6-6 Brown smoke from structural spaces indicates that unfinished wood is being heated and is decomposing—a warning sign for collapse in lightweight wood buildings!

© Keith Muratori. Used with permission.

When outside air temperatures are well below freezing, hot, dark smoke leaving the building turns white almost instantly. This phenomenon is the result of moisture or humidity condensing.

Practically speaking, the effects of wind on smoke should be obvious and should be taken into account when viewing the smoke leaving a building. Wind can even keep smoke from leaving an opening. Wind can rapidly thin and dissipate smoke, making it difficult to fully view its velocity and density. In a well-ventilated building, wind can speed up smoke velocity and give a false read on heat or location, although it should fan flaming. Firefighters engaged in an interior fire attack downwind of a wind-fed fire are in danger of being overrun by the fire! If the weather is cold and humid, the smoke should sink rather quickly and remain dense; if the smoke doesn't behave that way (instead, it is rising straight up and not cooling quickly), you should expect an exceptionally hot fire. Inversely, on a hot, humid day, if instead of finding the smoke climbing straight up into the atmosphere, you see that the smoke is coming out and lying down, then you know that the smoke has been cooled, either by distance (a deep-seated fire) or by a sprinkler system. If the building is sprinkler-protected, then the presence of low-lying smoke can indicate that the fire is not being controlled by the system.

Thermal Balance

Most buildings do not allow fires to maintain thermal balance for long. Simply stated, thermal balance is the notion that the heat of combustion will collect at the ceiling and cool air will flow to the fire below it. The separation of the two is known as the boundary layer or neutral plane. If the box can ventilate the heat and allow air for combustion, it can keep thermal balance until the box surfaces can no longer absorb heat. Likewise, ceilings, windows, doors, and inadequate airflow disrupt thermal balance. If a fire cannot draft and draw air, it soon drafts and draws smoke, thereby choking itself, leading to more incomplete combustion and creating denser smoke. As explained earlier, the thicker the smoke is, the more explosive it becomes. Viewed from outside a building, a fire out of thermal balance shows signs that air is being "sucked" through the smoke (an air track flow path). Sucking, puffing, and "breathing" signs indicate that a fire is out of thermal balance. From the ISO's perspective, signs of air being sucked into a building indicate that the fire is intense, yet struggling for proper airflow. A sudden inflow of air can cause the fire to take off—trapping firefighters.

Container Size

All smoke observations must be analyzed in proportion to the building. For example, smoke that is low volume, slow velocity, very thin, and light colored may indicate a small fire, but only if the building (or "box") is small. The same smoke attributes from several openings of a big-box store or large warehouse can indicate a large, dangerous fire. Historically, firefighters have been killed at fires that were reported as "light smoke showing." Remember, the size of the building is an important indicator of the significance of the smoke leaving. Light, thin smoke showing from more than one opening of a very large building is a significant observation.

Firefighting Efforts

The ISO is usually in the best position to tell whether firefighting efforts are succeeding. By watching the smoke outside a building, the ISO can determine the effectiveness of the firefight. All four attributes of smoke should change in a positive, continuous manner if fire stream and ventilation efforts are appropriate. Smoke volume should rise (as steam displaces smoke). Contents that were flaming should start "smoking white" as they are cooled, adding more smoke to the mix. Smoke velocity initially surges as steam expands but should gradually slow as heat is reduced in the building. Turbulent smoke velocity should change to laminar flow. Smoke density should thin. Obviously, the smoke color should eventually turn to pure, clean white. If all four attributes are not changing quickly (as described here), the ISO should judge the firefighting efforts as insufficient and share the observations with the IC.

Forced-ventilation tactics should cause an increase in smoke velocity at the designated exhaust flow path exit. When ventilation fans are being used, the ISO should watch the smoke for signs that the fans are helping, rather than hurting, the fire suppression efforts. If smoke becomes darker and thicker, the fan is actually making conditions worse. The use of positive-pressure ventilation (PPV) tactics for fire attack has gained in popularity. Known more appropriately as positive-pressure attack (PPA), this tactic should be performed in a strict and disciplined manner. Occasionally, the use of PPA can lead to disastrous results if not used properly. ISOs operating at scenes where fire departments use PPA should be well versed in PPA tactics. A full discussion of this training is outside the scope of this text, but several good sources on the topic exist. Still, a few smoke observation warnings are shared here so that an ISO can make an appropriate intervention to shut down or alter the PPA tactic (see the *Safety Tip* box).

Safety Tip

Warnings to Shut Down PPA Tactics

Positive-pressure attack (PPA) is a specific tactical maneuver that uses a high-CFM (cubic feet per minute) ventilation fan to achieve a controlled air and exhaust flow path to improve smoke conditions and firefighter safety. The ISO should monitor smoke conditions and firefighter actions and make an intervention to stop or alter PPA tactics in the following situations:

- Fire behavior and smoke flow do *not* increase at the exhaust opening after a full minute of air introduction.
- Smoke VVDC leaving other portions of the building (not the exhaust opening) intensify, getting darker, thicker, and/or turbulent.
- Fire is discovered behind advancing firefighters.
- Firefighters entering the building (after the fan is introduced) report zero visibility and high heat while advancing to the fire. Firefighters should cool their environment and retreat until conditions improve with alternative ventilation and cooling tactics.

These warnings are brief and conservative and do *not* replace an in-depth study of the PPA tactic.

The Three-Step Process for Reading Smoke

The ISO may view the principles of reading smoke as complicated or time-consuming. However, once you capture the basics and start practicing them, your ability to read smoke improves exponentially and you can read smoke in seconds. The principles can be incorporated into a process for rapid application. By following three simple steps, the ISO can refine and rapidly apply the principles.

- *Step 1.* View the smoke VVDC; then compare the differences in the attributes from each opening from which smoke is emitting. This exercise should give you a strong understanding of the fire size, location, and spread potential, and it allows you to capture any warning signs of hostile fire events.
- *Step 2.* Analyze the contributing factors to determine whether they are affecting VVDC. Remember, the size of the box can change the meaning of VVDC. Likewise, weather and firefighting efforts can change each attribute. This step should refine and/or confirm your read on fire behavior and firefighting effectiveness.
- *Step 3.* Determine the rate of change of each attribute. As you watch smoke, each attribute may be changing before your eyes. If the rate of attribute deterioration can be measured in seconds, it is likely that firefighters will be trapped or injured by fire spread. In other words, the firefighters are at risk. As a rule, the ISO should consider an intervention if the smoke is getting faster/thicker/darker in seconds while interior operations are underway.

These three steps should give ISOs a good understanding of the fire and help them predict what the fire will do next. If a hostile fire event appears imminent, immediate action must be initiated to prevent firefighter injury or death. As firefighting efforts progress during the incident, the ISO should look for positive changes in the smoke conditions. If positive changes are not taking place, the ISO should confer with the IC and share observations regarding fire behavior and potential fire loads that serve to adjust the IAP.

TABLE 6-3 outlines some of the shortcuts presented in this chapter. Remember, however, that shortcuts are just that— quick ways of summarizing what you see. Nothing is absolute about reading smoke. Smoke, just like fire, is dynamic and is influenced by many variables.

Simply reading literature on how to read smoke does not make one a smoke-reader. To convert this text and other instructions on reading smoke into a useable, on-scene skill takes practice. Practice comes from watching real smoke coming from actual fire incidents, as opposed to simulated smoke animations and theatrical smoke at burn training facilities. ISOs are encouraged to practice the skill by the use of raw, fire-ground video footage from various sources. A simple online search for the terms "reading smoke," "flashover," and "house fires" will help you gain access to hundreds of practice sources.

TABLE 6-3	Reading Smoke Shortcuts
What You See	**What It Can Mean**
Turbulent smoke that fills a box	Warning sign of impending flashover
Thick, black, fast smoke	Close to the seat of the fire, super-hot smoke capable of instant ignition; may be a ventilation-limited fire that needs air
Thin, black, fast smoke	Flame-pushed smoke; fire nearby that is well ventilated
Dirty white smoke with velocity	Heat-pushed smoke that has traveled a distance or has had the carbon/hydrocarbon filtered (like smoke through a crack)
Same color (white/gray) of smoke and same velocity from multiple openings	Deep-seated fire, possibly located well within a building or in combustible voids and concealed spaces
Low-volume white smoke from more than one location of a large building	Working fire deep within
Brown smoke	Unfinished wood reaching late heating (can support flame)—usually a sign that a contents fire is transitioning into a structure fire; *when coming from structural spaces of lightweight wood structures, a warning sign of collapse!*
Yellowish-gray smoke from cracks or seams	Sulfur compounds: a warning sign of impending backdraft

Wrap-Up

Chief Concepts

- The fire service has acknowledged that societal changes (the way we build buildings and the types/volume of contents we put in them) have led to more severe content and structure fires—and more flashover entrapments of firefighters. As a result, the skill of reading smoke in predicting fire behavior has become an essential firefighter safety "must have" but not a cure-all.

- Smoke is the products of incomplete combustion and pyrolysis that include an aggregate of particulates, aerosols, and gases.

- Most of the smoke at a building fire is being developed by pyrolysis. The quantity is exponentially greater than in the past. The sheer quantity and makeup of smoke has rendered the four-phase (fuel-controlled) fire behavior model obsolete for building fires. A six-phase ventilation-controlled model better reflects modern compartmentalized fire behavior. The phases, in order, are ignition, initial growth, ventilation-limited, explosive growth, fully developed, and decay.

- The explosive growth phase is often triggered by firefighters opening a door to make entry for fire attack. As the firefighters advance in zero-visibility smoke, a volume shift is taking place that can allow air to be reintroduced to the heated environment—and suddenly ignite the smoke (flame-over and ensuing flashover). An explosive growth event can render interior firefighting efforts useless due to the extreme heat-release rate of smoke ignition. The best solution is to begin protective cooling of ceilings and walls upon entry.

- Flow paths are the avenues that convection heat and combustion air travel. The combustion airflow path is often called the air track, and the convection current is called the exhaust or heat path. Factors that affect flow paths (and compartmentalized fire growth) are compartment size, interior geometry (hallways, stairs, doors), continuity of the fire load, and size and status of exterior openings.

- The ignition of smoke is triggered by two influences that meet: right temperature (starting as low as 450°F [232°C]) and right mixture (a flammable range that can be as wide as 1–74% in fuel to air).

- Hostile fire events are fire phenomena that can suddenly affect firefighters. They include smoke explosion, flame-over, flashover, explosive growth (a combination of flame-over and flashover), and backdraft. A smoke explosion is a spark-induced, momentary ignition of a smoke pocket distal to the actual fire; a flame-over is the ignition of the upper heat layer and exhaust flow path; a flashover is the more-or-less simultaneous ignition of all surfaces and contents in a room due to radiant heating; and a backdraft is the sudden detonation of oxygen-deficient smoke products within a closed, pressurized space upon the reintroduction of air.

- Hostile fire events have several warning signs, although a few can help differentiate which event is likely. They include:
  - *Smoke explosion.* Increasing pressure in trapped smoke away from a growing fire (e.g., room above, dead-end halls/stairways)
  - *Flame-over.* Ghosting, smoke flow paths that are moving faster than firefighters can crawl
  - *Flashover.* Turbulent smoke that has filled a room, ghosting, exterior vent-point ignition
  - *Explosive growth (flame-over and flashover).* Puffing and sucking of smoke at open doors and windows, ghosting
  - *Backdraft.* Closed pressurized box with signs of extreme heat and puffing of smoke at cracks and seams, yellowish/gray smoke

- Reading smoke can help the ISO understand fire behavior and fuel load/types to develop interventions or communicate IAP suggestions to an IC. Smoke has four attributes: volume, velocity, density, and color (VVDC). VVDC observations can be influenced by the compartment, size of flow path, outside weather, and firefighting efforts. It is the ISO's interpretation of VVDC and their influences that helps to predict future hostile fire events.

- Volume of smoke tells little about a fire but provides an impression regarding the potential size of an event. Velocity is very telling. Turbulent velocity indicates serious heat (flames nearby), or, when a box is full of turbulent smoke, it indicates that a flashover looms. Laminar velocity means the box is absorbing heat. Density is indicative of the severity of hostile events. Color of smoke rarely tells of the type of material burning; rather, it tells the level of heating that is present: Black is hot; white is cool. Color can be filtered by distance or resistance, so rely on velocity for the true heat story.

- As a general rule, smoke that is getting faster (and turbulent), thicker, and darker is a sign that things are getting worse. If they are getting worse within seconds, the ISO should make an intervention or inform the IC of the need to adjust the IAP. Black fire is a slang term used to describe smoke VVDC that has gotten as bad as it can be—a sure sign of impending flashover or explosive growth. No firefighters should be in black fire; they should immediately cool and withdraw if the condition develops while crews are interior.

- Smoke byproducts are extremely toxic and its residue on PPE can be absorbed into the wearer's skin and lungs. Firefighters exposed to dense smoke must use gross decontamination measures on-scene to reduce cancer threats.

Key Terms

backdraft An explosive event that occurs when air is suddenly reintroduced into a closed space that is filled with pressurized, ignition-temperature, and oxygen-deprived products of combustion and pyrolysis.

black fire A slang term for smoke that is high-volume, has turbulent velocity, is ultra-dense, and is deep black; a sign of impending autoignition and flashover.

explosive growth phase A rapid fire growth phenomenon that occurs when combustion air is reintroduced into a ventilation-controlled fire, leading to smoke flame-over and room flashovers.

flame-over A hostile fire event that includes the ignition and sustained burning of the overhead smoke layer within a room and/or hallway.

flashover A sudden hostile fire event that occurs when all the surfaces and contents of a space reach their ignition temperature nearly simultaneously, resulting in full-room fire involvement.

flow path An avenue that heat, smoke, flames, and combustion air follow.

ghosting A hostile fire event warning sign that is characterized as the intermittent ignition of small pockets of smoke; usually seen as fingers of flame that dance through the upper smoke layer.

hostile fire event A fire behavior phenomenon that can suddenly harm firefighters; events include explosive growth phase, flashover, backdraft, smoke explosion, and flame-over.

pyrolysis Also referred to as *pyrolitic decomposition*, the chemical breakdown of compounds into other substances by heat alone.

smoke The products of incomplete combustion and pyrolysis; it includes an aggregate of particles, aerosols, and fire gases that are toxic, flammable, and volatile.

smoke explosion A hostile fire event that occurs when a spark or flame is introduced into a pocket of smoke that is below ignition temperature but above some aggregate flashpoint. The result is a split-second ignition (and rapid expansion) of that pocket with no sustained burning.

Review Questions

1. What is smoke?
2. What are the six phases of a ventilation-controlled compartmentalized fire?
3. List common hostile fire events and their associated warning signs.
4. What are the four attributes of smoke?
5. How do the four smoke attributes contribute to the understanding of fire behavior within a building?
6. What is meant by the term "black fire"?
7. Explain the influencing factors that impact smoke VVDC.
8. List the three steps involved in reading smoke.

References and Additional Resources

Dodson, David W. *The Art of Reading Smoke: Practice Sessions*. Training DVD. Tulsa, OK: Fire Engineering Books and Video, PennWell, 2009.

Dodson, David W. *The Art of Reading Smoke: Volume 3*. Training DVD. Tulsa, OK: Fire Engineering Books and Video, PennWell, 2012.

Gagliano, Mike, Casey Phillips, Phil Jose, and Steve Bernocco. The Breath from Hell. *Fire Engineering*, March 2006.

Grimwood, Paul. The FireTactics website offers great articles and links to fire behavior studies and is available at http://www.firetactics.com.

National Fire Protection Association. *NFPA Fire Protection Handbook*. 19th ed. Section 3, Chapters 9 and 10; Section 8. Quincy, MA: NFPA, 2003.

National Institute of Standards and Technology. Numerous technical reports on fire behavior and fire dynamics at firefighter fatality incidents can be accessed through the NIST website, available at http://fire.nist.gov. Good technical information can be gleaned from the following incidents (search these key words): Cherry Road; Cook County office building; The Station nightclub fire; Sofa Super Store fire; wind-fed fire studies; use of positive-pressure attack on structure fires.

Underwriters Laboratories. The online training resource "Impact of Ventilation on Fire Behavior in Legacy and Contemporary Residential Construction" is available at http://content.learnshare.com/courses/73/306714/player.html.

U.S. Fire Administration. The 2012 USFA report "Changing Severity of Home Fires Workshop" is available at http://www.usfa.fema.gov/downloads/pdf/publications/severity_home_fires_workshop.pdf.

Wallace, Deborah. *In the Mouth of the Dragon: Toxic Fires in the Age of Plastics*. Garden City Park, NY: Avery Publishing Group, 1990.

Endnotes

1. National Fire Protection Association. *Fire Protection Handbook*. 19th ed. Vol. II, Sec. 8. Quincy, MA: NFPA, 2003. The NFPA definition of smoke is the "total affluent from burning or heating of a material." The author assembled the more complete definition presented in the text based on this NFPA definition.

2. This claim is constructed from years of live-fire training in acquired structures, video from dozens of NIST/UL test fires, teachings from noted fire protection engineer Daniel Madrzykowski (of UL), and hundreds of experienced fire incidents by the author.

INCIDENT SAFETY OFFICER
in action

The evidence is compelling. The building fires you respond to today and the ones you'll respond to tomorrow are producing some of the most explosive, toxic, and volatile smoke that we, as a fire service, have ever experienced. Thanks to the work of the National Institute of Standards and Technology (NIST), the Bureau of Alcohol, Tobacco, and Firearms (ATF), Underwriters Laboratories (UL), and numerous product safety test laboratories, we are beginning to document the "new" building fire environment. Many have already claimed that the fire is the least of your problems—it's the smoke coming from low-mass synthetics that are being heated without burning (pyrolysis). According to some research, smoke production is 500 times that of the "wood-based" fires we saw a few decades ago.

The list of chemical products found in the smoke of a "typical" residential fire will challenge the imagination of most chemists: hydrogen cyanide, polyvinyl chloride, hydrogen sulfide, hydrogen chloride, polynuclear aromatic hydrocarbons, formaldehyde, phosgene, benzene, nitrogen dioxide, ammonia, phenol, acrolein, and halogen acids. Further, the chemical decomposition process of pyrolysis is engaging the mechanisms of chain scission, chain stripping, and cross-linking (think rocket-science-like chemistry). It gets worse. We used to measure the typical compartment (room) fire time-temperature curve using a 10-minute cycle. Recent research tells us that the cycle is closer to 3 minutes.

Now we'll get personal. A review of firefighter news items, investigative reports, and incident data indicates that:

- Firefighter cancer rates are escalating.
- Firefighters that get just a few breaths of zero-visibility smoke either die or face significant medical issues that can become chronic.
- Firefighters are being treated for cyanide poisoning from just a "little" smoke exposure outside of burning buildings.
- The rate of firefighter flashover deaths is going up in an environment where the number of fires is going down.
- Otherwise healthy firefighters are experiencing deadly or debilitating heart attacks.

The bottom line: Today's building fire smoke is truly the smoke from hell. Noted Fire Chief, author, and firefighter safety advocate Billy Goldfeder offers this view: "The smoke today is not your Daddy's smoke."

1. How does zero-visibility smoke impact victim survivability and the risk-based decision making of the IC and ISO?

2. What changes need to take place in firefighter training to help account for the changes we're experiencing in fire behavior?

3. What actions can the ISO take to help protect firefighters from the harmful health effects of the modern smoke environment?

   A. Enforce the use of SCBA and PPE for all fire suppression and overhaul activities.

   B. Consult with the IC to reduce the exposure time of firefighters in smoky environments.

   C. Remind firefighters to attack fire and smoke from established flow paths.

   D. Encourage the use of positive-pressure mechanical fans.

4. The smoke warning sign of impending flashover is:

   A. yellowish smoke from a closed, pressurized compartment

   B. thin, black, fast laminar smoke

   C. turbulent smoke that has filled a compartment

   D. smoke that is the same color, same velocity from multiple openings

Note: This case study was inspired by the article, "The Breath from Hell," written by Seattle Fire Officers Mike Gagliano, Casey Phillips, Phil Jose, and Steve Bernocco. The article appeared in the March 2006 issue of *Fire Engineering.*

Reading Risk

Knowledge Objectives

Upon completion of this chapter, you should be able to:

- Describe the differences between *dangerous* and *risky*. (p 95)
- List the risk management principles outlined in NFPA standards. (p 96)
- Describe the risk management principle that would cause a threat to be imminent to firefighter survival. (NFPA 5.2.4 , p 96)
- Define the risk management criteria that are relative to an incident action plan. (NFPA 5.2.2 , pp 97–100)
- Define situational awareness. (p 98)
- List the factors that reduce situational awareness. (pp 98–99)
- List the three levels of survivability that can be assigned to a given space at an incident. (p 100)
- Describe the three-step process to read risk at an incident. (pp 99–100)

Skills Objectives

Upon completion of this chapter, you should be able to:

- Demonstrate the ability to evaluate hazards and assign the relative degree of risk to the hazard. (NFPA 5.2.4 , pp 97–100)

Firefighters recognize and assume great risk in many situations on a daily basis. Risk must be balanced against potential rewards. At a recent fire, a crew operating with very little reconnaissance information and little likelihood of rescuing occupants ignores possible hazards and attempts to search the floor above the fire without a hose line. You need to address this at the incident critique.

1. What is your department's policy on risk management during fire ground operations? If you do not have a written policy, how are you expected to assess risk taken at an incident?

2. Have there been incidents within your department at which you felt crews or individual members took actions that unnecessarily placed them at risk? How was this handled after the event?

Introduction: Here Comes the Risk-Taking Judge

NFPA 1561 states that the "Safety Officer shall monitor conditions, activities, and operations to determine if they fall within the criteria [of acceptable risk-taking] as defined in the department's risk management plan."[1] Simply stated, the incident safety officer (ISO) is the designated risk manager for incident operations and must offer judgment regarding the risk-taking of responders. How does the ISO monitor risk-taking to determine whether it falls within acceptable parameters? What are acceptable parameters? To answer these questions, the ISO needs to know what his or her department has preestablished as acceptable risk criteria. Where no criteria exist, the ISO needs to follow nationally accepted risk standards (that is, the way the courts and public opinion will view it). There has been much written and suggested related to risk management, loss control, and risk/benefit thinking. Yet, even with the many resources available, many in the fire service still have trouble figuring out what is acceptable or unacceptable risk-taking at incidents. Making a judgment regarding what is at risk—people or property—is perhaps the most important step in determining a risk profile. While easy to say, this is much harder to do. Many firefighters stand by the notion that all incidents are "people" events until proven otherwise. Likewise, some firefighters are willing to concede that an environment is incompatible with life and that, therefore, the only thing to save is property. Historically, the fire service has a poor history of changing risk-taking based on the people/property issue. The National Institute for Occupational Safety and Health (NIOSH) has pointed this out in numerous firefighter fatality investigations. The reoccurring recommendation from these reports states that ICs need to continually evaluate risk vs. gain and adjust the action plan accordingly.

Some of the most glaring evidence that we need to improve risk management is in the firefighter fatality trends presented in the *Guiding Laws, Regulations, and Standards* chapter. As mentioned, the National Fallen Firefighters Foundation recognized this need and has held several summit meetings to address these trends. The 45 fire service organizations represented at those meetings have voted to unanimously endorse the 16 Firefighter Life Safety Initiatives. One of the initiatives states:

Focus greater attention on the integration of risk management with incident management at all levels, including strategic, tactical, and planning responsibilities.

At an incident, the ISO must read the risks being taken and offer judgment to the incident commander (IC) regarding their acceptability. This chapter looks at firefighter risk-taking and offers thoughts to help the ISO read risk and offer sound judgments regarding its acceptability.

Firefighter Risk-Taking

Historically, firefighters have endeared themselves to the communities they serve through selfless acts of bravery and sacrifice. Underlying the public's positive regard for firefighters is the common belief that the firefighting profession is dangerous. Ask any firefighter if firefighting is dangerous, and the response may be a humble "sometimes" or a resounding "yes!" Now consider a different perspective: "Firefighting isn't dangerous; it's merely risky." This perspective came from fire service educator and former fire chief Dave Daniels. The point is eloquent and should be adopted as an "attitude" by the ISO. Look at it this way: From a community perspective, firefighters are expected to work in inherently dangerous environments. From a fire service perspective, the risks of many specific dangers are well known. Further, we learn, train, and equip ourselves to understand the dangers and take steps to avoid, control, or eliminate the dangers. When we view things this way, we make *choices* about the dangers we face. That is risk-taking **FIGURE 7-1**.

Firefighters are action oriented and results oriented, both of which are positive attributes. However, this action and results orientation can be the precursor for injuries and deaths because many firefighters are *arbitrarily* aggressive; their action approach to get results exposes them to dangers that they may not appreciate or acknowledge. Since the causes of most firefighter fatalities are well known—along with the situations that cause them—it seems reasonable that we can make decisions to avoid or control those situations. Instead of being *arbitrarily* aggressive, we should be *intellectually* aggressive. Being intellectually aggressive requires recognizing predictable dangers and taking steps to reduce risk-taking by our fellow responders. You can acquire and maintain this risk perspective by front loading an active knowledge of two things: the fire service consensus risk values and the values found at the local level.

FIGURE 7-1 The essence of risk-taking centers on *choices* regarding the dangers we face.

© Keith Muratori. Used with permission.

■ Consensus Fire Service Risk Values

NFPA

NFPA standards specifically address risk management principles related to the handling of emergency operations and appear in NFPA 1500, 1561, and 1521. These consensus standards clearly spell out acceptable risk-taking principles that should be incorporated into the incident action plan (IAP) for emergency responses. From these principles, decisions can be made regarding whether risk-taking is acceptable or unacceptable.

Risk Management During Emergency Operations (NFPA 1500, 8.4.2.1, 2018 Edition)

The concept of risk management shall be utilized during incident operations based on the following principles:

1. Activities that present a significant risk to the safety of members shall be limited to situations in which there is a potential to save endangered lives.
2. Activities that are routinely employed to protect property shall be recognized as inherent risks to the safety of members, and actions shall be taken to reduce or avoid those risks.
3. No risk to the safety of members shall be acceptable when there is no possibility to save lives or property.
4. In situations where the risk to fire department members is excessive, activities shall be limited to defensive operations.

Reproduced with permission from NFPA 1500, *Standard on Fire Department Occupational Safety and Health Program*; © 2018, National Fire Protection Association. This is not the complete and official position of the NFPA on the referenced subject, which is represented only by the standard in its entirety.

Most fire officers would subscribe to the risk principles in the NFPA standards. These principles can be shortened and made easy to remember with the following brief statements:

- Risk a life to save a known life.
- Perform in a predictable, practiced manner to save an unknown life or valued property.
- Take no risk to save what's lost.
- Default to a defensive strategy when risks are excessive or conditions deteriorate quickly.

These statements may seem black and white to some, but experience tells us that their interpretation is still wide open. Some incidents may present themselves with *pure* risks; that is, the conditions present make it easy to classify the NFPA risk principle that is applicable. More often, incidents present *speculative* risks that can lead to differing opinions as to which principle is appropriate. The ISO must be willing to discriminate and make judgments regarding speculative risks. For example, the first statement says, "Risk a life to save a known life." It does not say, "*Sacrifice* a life to save a known life." Throwing firefighters at a rescue may not be the answer. Likewise, the statement is not plural. At some point, the IC or ISO must determine how many firefighters should be at risk to save a life.

The phrase "perform in a practiced, predictable manner to save an unknown life or valued property" is also open to interpretation. Are there signs that a survivable space exists within a burning building? What is valued property? The ISO needs to have some sense of what constitutes survivable spaces and valued property. We will discuss survivable spaces a bit later in this chapter. For valued property, one criterion is often used: Valued property is physical property whose loss will cause harm to the community. Examples may include a hospital, a significant employer (economic base), a utility infrastructure, or a place with historical significance. In these cases, the ISO should calculate the risk (the possibility and severity of an injury), then weigh safety factors in favor of the firefighter. In other words, do what is practiced and routine within a structured plan. For example, an interior firefight at a county clerk's office may be extended because of the potential for the loss of irreplaceable records. The ISO may recommend that additional engine companies be ordered for backup or that emergency roof and ceiling bracing be installed to prolong the time the firefighters can fight or protect salvage operations.

"Take no risk to save what's lost" may seem obvious on the surface but should remind us that buildings under construction, buildings being demolished, and abandoned buildings should be approached with a defensive strategy only and should not be entered. The last risk principle, "Default to a defensive strategy when risks are excessive or conditions deteriorate quickly" is a relatively new principle developed by NFPA committee members to address the notion that some situations are "losers" from a risk-taking perspective. There are incidents that, upon arrival, are extraordinarily dangerous (risky) and warrant an obvious defensive approach (pure risk). The application for the fourth risk principle comes when well-meaning firefighters engage offensively and, usually through no fault of their own, don't appreciate rapidly deteriorating conditions (i.e., excessive risk). This is where the ISO can make a huge difference: The rate of change (rapid deterioration) is causing an imminent threat to firefighter survival and, thus, warrants an ISO intervention for emergency evacuation or an IAP adjustment suggestion for the IC.

IAFC Rules of Engagement

The Safety, Health, and Survival section of the International Association of Fire Chiefs (IAFC) developed and delivered a *Rules of Engagement* promotion as part of their mission to make sure everyone goes home following an incident. The promotion is another example of a consensus fire service risk-taking value system. The Rules of Engagement program directs firefighters and ICs to follow the NFPA risk principles and more specifically details survival edicts to manage risks. Even though the campaign is directed at firefighters and ICs, the ISO can benefit from its content TABLE 7-1 .

■ Local Risk-Taking Values

Determining the appropriateness of risk-taking (acceptable or unacceptable within defined criteria) is perhaps the most important decision that the ISO has to make at the incident scene. In most cases, the IC has already established risk boundaries for working crews using the NFPA principles. For example, the IC who has pulled interior crews and deployed exterior defensive fire streams has, in essence, determined that firefighters may no longer risk interior firefighting. In other cases, the line is not so clear. This is where the ISO has to make the *value* decision of whether the positive outcome of a specific task, strategy, or action is worth the risk of injury. Some have expressed this task as deciding *risk/benefit*. As an ISO making value decisions, you must first understand what values are in play at the local level.

Community Expectations

Most firefighters accept the notion that they may have to risk their lives to save a life. In most cases, the community served by the firefighter expects this of the firefighter and the fire department. This basic expectation is what separates the firefighting profession from many others and gives it a status of respect in the community.

Firefighters who swear (literally—when they take an oath) to uphold the values of the community (to protect life and property) must draw on courage and, in some cases, bravery to meet the community's expectations. However, the same courage and bravery must be tempered with a heavy dose of prudent judgment so that a situation does not unnecessarily harm the firefighter. The key word is "unnecessarily." When a firefighter dies protecting property (as opposed to life), the taxpayer may question whether prudent judgment prevailed. Clearly, the community does not expect firefighters to die needlessly. This is where the IC and the ISO must step in and establish a pace, an approach, or a set of guidelines to avoid an unnecessary casualty.

The proliferation of cell phone cameras and social media networks has increased the community expectations of firefighters. Anyone who captures images of firefighters taking high risks at fires, unusual rescues, and disasters quickly posts them and the news media exploits them. The stories and images make good copy and many of them go "viral." When John and Jane Doe view these events, they may mistakenly believe that their own firefighters are trained and ready to perform in the same way for their local community. This puts risk-taking pressure on a local fire service that may not be equipped or trained for such an event FIGURE 7-2 . In many ways, the fire service has become the agency of "first and last resort" for events that affect the public: It is the first agency they think to call when emergencies strike, and, when they can't think of whom to call for an unusual event, they'll call the fire department. Again, this reliance on the fire service puts risk-taking pressure on local responders.

Department Values and Skills

When determining whether a risk is acceptable or unacceptable, ISOs must consider what is commonplace and accepted by their own department in terms of risk management and compare that with a list of risk management criteria that are relative to an IAP. Influencing IAP risk factors that *should be* commonplace include:

- Routine evaluation of risk in all situations
- Well-defined strategic options
- Standard operating procedures (SOPs) or standard operating guidelines (SOGs) that are routinely followed
- Effective training and task accomplishment
- Full personal protective equipment (PPE) use
- Effective incident command structure
- The use of safe practices and safety officers
- Routine use of rapid intervention crews (RICs)
- Adequate resources
- Rest and rehabilitation
- Regular evaluation of changing conditions and IAP effectiveness
- Application of lessons learned from previous incidents, critiques, and national experiences

By comparing this list to what is practiced within the ISO's own department, a foundation for risk decision making can be created. For example, an ISO may find that risk-taking for a given incident is acceptable because his or her department routinely adjusts and communicates IAPs on a proactive basis relative to

Table 7-1	Rules of Engagement for Firefighter Survival
1.	Size up your tactical area of operation.
2.	Determine the occupant survival profile.
3.	*Do not* risk your life for lives or property that cannot be saved.
4.	Extend *limited* risk to protect savable property.
5.	Extend *vigilant* and *measured* risk to protect and rescue *savable* lives.
6.	Go in together, stay together, come out together.
7.	Maintain continuous awareness of your air supply, situation, location, and fire conditions.
8.	Constantly monitor fire ground communications for critical radio reports.
9.	You are required to report unsafe practices or conditions that can harm you. Stop, evaluate, and decide.
10.	You are required to abandon your position and retreat before deteriorating conditions can harm you.
11.	Declare a mayday as soon as you THINK you are in danger.

Reproduced from: IAFC Safety, Health, and Survival Section. (n.d.) "Rules of Engagement for Firefighter Survival." Rules you can LIVE By. http://iafcsafety.org/

FIGURE 7-2 A community may mistakenly expect their local fire department to offer a wide range of services because they saw other firefighters performing such services on the news or social media. As a result, local firefighters may feel pressured to take risks for which they are unprepared.

changing and anticipated conditions. Another ISO will find the risk-taking unacceptable because the department is typically reactive and struggles with clear IAP communication.

The ISO must evaluate an operation and decide whether the situation fits the organization's "normal" way of handling the incident. Such a decision can easily lead to conflict. The department may have embraced an inherently unsafe way of doing a task. A good example is the department that allows freelancing due to insufficient on-scene staffing or the department that routinely launches an interior fire attack before four people have assembled on scene. In these cases, the ISO must be extra alert and should try to build a bridge toward a safer operation. These cases try the ISO and present a challenge to reduce injury potential creatively. Most are solved by well-articulated observations and suggestions communicated to the IC to change the IAP.

Regarding skills, the ISO should recognize when crews are attempting to perform a skill for which they have never been trained. The rise-to-the-task attitude of most firefighters creates such situations—again, a good problem to have, but one wrought with traps. In these cases, the ISO should inform the IC that technical assistance is required and encourage a slower pace and an increase of zoning or backup for those currently on scene.

As you can see, the foundation for risk-taking is established by defined values. As professional firefighters, you have to know which values apply to you. Ask yourself this question: If an investigation is conducted, which values will I be held to: (a) consensus fire service values or (b) local risk-taking values? To apply these values, you must be keenly dialed into the conditions and activities present at a given incident. We call this situational awareness (SA).

Situational Awareness

In a simpler time, the term size-up described the continuous mental process used to evaluate incident facts and probabilities. In the 1970s, we were exposed to a simple five-step size-up. In the 1980s, it was a 13-point size-up. In the 1990s, it was

a 27-point—you get the picture. The complexities of incident handling have increased exponentially, and fire service leaders attempt to create and define size-up models to help us mentally process an incident. These efforts are to be commended.

Despite such efforts, the ISO needs a method to rapidly "get" what is happening at an incident and the threats at play. Instead of using a size-up model, consider using a SA approach that helps you "read risk." SA, in this context, is different from what firefighters are usually thinking about (task accomplishment). SA is actually a measurement of accuracy. In fact, the Naval Aviation Schools Command in Pensacola, Florida, defines SA as the degree of accuracy to which one's perception of his or her current environment mirrors reality. Similarly, ISOs' SA is their ability to accurately read potential threats and recognize factors that influence the incident outcome. There are many ways to achieve accuracy in awareness, but first you must be keenly aware of the factors that reduce your ability to stay aware and keep an open mind (see the *Awareness Tip* box).

Awareness Tip

Factors That Reduce Situational Awareness
According to the Naval Aviation Schools Command, the following factors have an adverse effect on SA:

- Insufficient communication
- Fatigue and stress
- Task overload—*too many things that need to be accomplished**
- Task underload—*overly routine tasks or boredom*
- Group mind-set and biases—*"We always do it this way."*
- "Press-on" philosophy—*"Cowboy up!"*
- Rapidly degrading operating conditions

*Italicized text has been added by the author of this text.

Keeping your mind open is a challenge. Dr. Richard Gasaway, a former fire chief, has developed numerous books, lectures, and online training programs to help firefighters understand the impediments to SA as well as solutions that can help with SA improvement (see "References and Additional Resources" at the end of this chapter). In several of his presentations, Dr. Gasaway suggests that there are three levels of SA:

1. *Capturing current cues and clues.* This level involves paying attention to what is happening now.
2. *Comprehending the situation.* At this level, you make sense of what is happening based on what you know and have experienced, or you acknowledge that you haven't experienced this situation before.
3. *Predicting the future.* This level involves projecting what is likely or possible.

The three levels of SA are used to help you make risk/benefit decisions. Each level is influenced by your mental state, training, experience, and personal values. All of these influences can be improved and, thus, so can your ability to keep an open mind (better SA) to read risks. Here are some tips for improving each influence:

- *Mental state.* One's mental state is improved through physical fitness, adequate sleep, proper nutrition and hydration, stress-reduction measures, and a "stay-calm" attitude during incidents.
- *Training.* Training is improved through a career-long commitment to seek out new information and skills and retain and practice what you've already learned.
- *Experience.* One's experience improves through actual incident handling, <u>vicarious learning</u>, and <u>recognition-primed decision-making (RPD)</u> exercises. Vicarious learning can be loosely defined as the process of observing others to develop knowledge, skill, or experience base. Reading line-of-duty death (LODD) technical investigative reports can provide a way for you to picture the actions/factors that others have experienced and to use that information to prevent a similar occurrence. RPD is a mental model that suggests that many quick decisions are made using mental templates (what works and what doesn't) from previous experiences that fit the images that you are currently witnessing.[2] Actual incident handling, full-scale response drills, and professional shadowing (ride-along with other fire departments) can help expand your RPD for quick decisions.
- *Personal values.* The values that one draws upon when assessing situations and making decisions can be improved through educational endeavors (vs. training), a commitment to reduce injuries and death, the use of consensus risk values, and a caring attitude toward others.

Clearly, the ISO needs to maintain a constant SA of the incident environment, but he or she has the additional responsibility of appraising how well firefighters and officers are gauging their own SA. All too often, the real or perceived urgency of the incident overrides prudent judgment by firefighters (and some officers), leading to a lack of SA **FIGURE 7-3**. Although the urgency of task completion is appreciated, the ISO needs to gauge whether the loss of SA by the working responders is leading to a trap or unacceptable risk-taking, or both.

FIGURE 7-3 The urgency of an incident can override prudent judgment and lead to reduced SA by firefighters.

As you can see, the ISO needs to embrace SA as an essential mental element when evaluating risk-taking. Let's now address how SA is applied in risk-reading models.

Models for Reading Risk

Over time, experienced ISOs have developed their own approach to reading risk and making judgments. For the most part, the acceptable/unacceptable decision for risk-taking will be made using a combination of established SOPs/SOGs, SA, RPD (for quick decisions), and risk modeling (when time allows or for IAP input).

As introduced above, RPD is an experience-based method that is often used to make quick decisions; that is, you see a set of conditions and you compare that to images/outcomes that you know or have experienced previously. Recognize the trap: If you are witnessing something you've never experienced, a quick, appropriate acceptable/unacceptable mental decision may not present itself. If the situation is especially complex or is rapidly deteriorating, effective decision making can collapse (as a result of stress) and leaders become more dogmatic and resort to habits or actions that may not fit the situation.[3] It is absolutely essential that the ISO self-acknowledge when an incident is unlike those experienced previously. That is the first step in preventing a collapse in decision making. Moving forward, the ISO should apply a thoughtful mental process to judge risk-taking. This is where risk-reading models can help. They are mental-processing tools that help to weigh SA inputs and get to a better-fitting decision regarding acceptable or unacceptable risk-taking.

■ The ISO Read-Risk Model

The ISO read-risk model was developed by this author, who street tested and adjusted it over a 5-year period as a duty safety officer. Since then, the model has been updated using National Institute for Occupational Safety and Health Administration (NIOSH) LODD technical reports, lessons learned from unusual and complex incidents, and input from dozens of fire officers. It can serve as a simple three-step mental process that can become "automatic" if practiced.

Step 1: Collect information—SA cues and clues.
- Read the building (or incident environment).
- Read the smoke (or the event that is causing harm).
- Read the firefighters (Do they have SA? Are they doing predictable things and getting standard outcomes?).

Step 2: Analyze.
- Define the principal hazards (the severity and likelihood of threats).
- Determine whether there are survivable spaces for potential victims.
- Compare the rate of change with the fire ground clock (Are we ahead or behind the change curve?).
- Determine what is really to be gained by fire service actions (saving people or property).

Step 3: Judge risk.
- Acceptable: If within the risk-taking values established by the department, continue to monitor, look for ways to reduce risk, and advise the IC.
- Unacceptable: If an imminent threat exists, make an immediate intervention. If not imminent, bring risk-reduction solutions to the IC.

Judgment regarding survivable spaces is perhaps the toughest (and most speculative) one to render. In many ways, this judgment is the most important input to the acceptable/unacceptable risk-taking decision. Most agree that any known victim should be given the benefit of an aggressive search and rescue operation, but there are situations in which the chance of a save is nonexistent. The ISO must be willing to use all information inputs to rate survivability. Some call this task *rescue profiling*. Rating the survivability of a space is typically derived from a combination of reading the smoke (and flames), reading the building (looking at windows, considering access/egress options, predicting collapses, etc.), and then evaluating the resources on hand and the time options for a search. From this, survivability can be rated as high (go), marginal (go with caution), and zero (no go) (see the *Getting the Job Done* box).

Getting the Job Done

Rating Survivability

The ISO can use reading smoke and reading building skills (addressed in the chapters on those topics) to help analyze the conditions present that indicate that an environment within the building is survivable, and therefore, make a judgment regarding acceptable or unacceptable risk-taking. The survivability rating can be classified using the following levels:

- *High survivability—GO.* Known victims are in danger and the conditions indicate a good possibility that a rescue effort can save the victim.
- *Marginal survivability—GO with CAUTION.* Known or unknown victims are in danger but the conditions are such that the victim may not survive; still, there is not enough information present (from the smoke or window status) to classify the situation as a zero-survivability index.
- *Zero survivability—NO GO.* The smoke, fire, building collapse, or window status is such that no victim is likely to be alive within the given space.

■ U.S. Coast Guard SPE Model

The ISO three-step read-risk model is directed primarily at fire incidents, although it can be adopted for all hazards. The U.S. Coast Guard (USCG) has developed a risk-taking model that may also be useful for determining acceptable risks at nonfire incidents. The severity, probability, and exposure (SPE) model[4] was developed as part of the USCG Operational Risk Management Program and is used to determine whether a mission should be advanced or abandoned. The model helps to assess risk potential by using numeric values. First, the severity, probability, and exposure threats are assigned a number based on the following scale:

- Severity: 1–5 (1 is minimal harm, 5 is potential death)
- Probability: 1–5 (1 is not likely, 5 is high certainty)
- Exposure: 1–4 (1 is within protective controls, 4 exceeds protection)

The ratings of each input are multiplied together, resulting in a number between 1 and 100. The USCG uses a result between 50 and 80 as a trigger to reassess risk-taking. Anything more than 80 is cause to abandon the mission. These trigger numbers may be useful for the USCG in making go/no-go decisions but perhaps needs tweaking for a fire service ISO, as an operation may already be under way. With all respect to the USCG, the model can be manipulated and reinterpreted for the fire service. Consider using this adaptation of the USCG SPE numbers:

- **Known trapped and endangered victims**
  - SPE 1–50: Acceptable risk
  - SPE 51–80: Marginal risk but monitor closely
  - SPE 81–100: Prepare for rapid withdrawal
- **Unknown or no victims (or nonsurvivable conditions)**
  - SPE 1–20: Acceptable risk
  - SPE 21–50: Marginal risk but monitor closely
  - SPE 51–100: Unacceptable risk, defensive operations only

There are other risk-reading models to choose from. The important point is that the ISO needs to use a defined mental process to bridge SA to a judgment regarding risk-taking. Once a risk-taking judgment is made—acceptable or unacceptable for various activities—some follow-through is warranted. Unacceptable risk-taking warrants an intervention based on the perceived immediacy of the threat. Imminent threats should cause the ISO to immediately stop, alter, or suspend operations and possibly to withdraw crews by direct order. Less-threatening situations, or those that are looming but not immediate, should be handled through the IC. If you are uncomfortable with the risks being taken but cannot quite justify or articulate your concern, then that in itself is reason to visit with the IC. You may find it effective to approach the IC and ask how he or she feels about the situation—sort of a "reality check." This is an example of being a consultant and a helper in the team approach to managing risks. Where the risk judgment is viewed as acceptable, that also should be communicated to the IC, and his or her thoughts and future concerns should be confirmed. As with many ISO functions, the risk-taking judgment is ongoing for the duration of the incident.

Chief Concepts

- NFPA 1561 requires that the safety officer, where appointed, be the designated risk manager for an incident and that he or she determine whether the operation falls within the criteria of acceptable risk-taking as defined in the department's risk management plan.
- *Dangerous* and *risky* have different meanings. The fire scene environment is inherently full of dangers (threats), but firefighters make decisions regarding the dangers they face; therefore, the conditions present are *dangerous*, but the act of firefighting is *risky*.
- Risk-taking values are influenced by two fire service consensus documents (NFPA standards and the IAFC Rules of Engagement), as well as two local considerations (community expectations and the local fire department's values and skills).
- NFPA standards outline four risk management principles. These principles can be paraphrased as:
  - Risk a life to save a known life.
  - Perform in a predictable, practiced manner to save valued property.
  - Take no risk to save what's lost.
  - Default to defensive tactics when conditions deteriorate quickly.
- When the threats to firefighters are deemed excessive and/or are deteriorating quickly, firefighters should view the situation as an imminent threat to survival and withdraw to defensive positions.
- There are many risk management criteria that are relative to an IAP. The ISO needs to compare his or her own department's way of handling an incident to the IAP risk factors to help determine their foundation for risk-taking. Some of the criteria include routine evaluation of risk in all situations, well-defined strategic options, full PPE use, effective ICS, adequate resources, and regular evaluation of changing conditions and IAP effectiveness.
- Situational awareness (SA) is the degree of accuracy to which one's perception of his or her current environment mirrors reality.
- Many factors can reduce one's ability to have SA. They include insufficient communication, fatigue and stress, task overload or underload, group mindset and biases, and rapidly degrading operating conditions.
- Making a judgment regarding acceptable/unacceptable risk is grounded in SA. SA is influenced by mental alertness, training, experience, and personal values—all of which can be improved.
- Many quick decisions are made using recognition-primed decision making (RPD), which is a mental model by which one matches witnessed conditions with those

that have been previously experienced. Actual incident handling, full-scale response drills, and professional shadowing (ride-along with other fire departments) can help expand RPD for quick decisions.
- Risk-reading models can help the ISO make acceptable risk judgments, especially when the ISO has not experienced a previous similar event. The ISO read-risk model and the U.S. Coast Guard severity, probability, exposure (SPE) model are notable examples, although many exist.
- The three-step ISO read-risk model is:
  1. Collect incident SA cues and clues (read the smoke, building, and firefighter actions).
  2. Analyze the SA (define the threats, judge survivable spaces, judge the rate of change, define what is being saved).
  3. Judge the risk-taking (acceptable or unacceptable based on values and principles).
- Survivable spaces should be judged and assigned a level: high = go; marginal = go with caution; or zero = no go.
- The U.S. Coast Guard's SPE model assesses risk based on a number derived from a formula. A scale value is given to severity (1–5), probability (1–5), and exposure (1–4). These values are multiplied together, creating a number between 1 and 100. A result over 80 indicates an extraordinary risk.
- Unacceptable risk-taking warrants an intervention based on the perceived immediacy of the threat. Imminent threats should cause the ISO to immediately stop, alter, or suspend operations and possibly to withdraw crews by direct order. Less-threatening situations, or those that are looming but not immediate, should be handled through the IC.

Key Terms

recognition-primed decision making (RPD) A mental model that suggests that many quick decisions are made using mental templates from previous experiences that fit the images that you are currently witnessing.

situational awareness (SA) The degree of accuracy to which one's perception of his or her current environment mirrors reality.

valued property Physical property whose loss will cause harm to the community.

vicarious learning The process of observing others to develop knowledge, skill, or experience base.

Review Questions

1. Describe the differences between *dangerous* and *risky*.

2. List the influences on risk-taking values.

3. List the four risk management principles outlined in NFPA standards.

4. What is "valued property"?

5. What is meant by situational awareness, and what factors diminish one's ability to be situationally aware?

6. What are the four influences on situational awareness that can be improved?

7. Describe the three-step ISO read-risk mental process used to judge whether a risk is acceptable or unacceptable.

8. List the three levels used to profile survivability for a given space within a building.

9. What is recognition-primed decision making, and how can it be improved?

References and Additional Resources

Centers for Disease Control and Prevention. NIOSH Alert: Preventing Deaths and Injuries of Firefighters Using Risk Management Principles at Structure Fires. July 2010. http://www.cdc.gov/niosh/docs/2010-153/. Accessed May 5, 2015.

Emery, Mark. 13 Incident Indiscretions. *Health & Safety for the Emergency Service Personnel*, 15, no. 10, FDSOA, October 2004.

Federal Emergency Management Agency, U.S. Fire Administration. Risk Management Practices in the Fire Service, FA-166. December 1996. http://www.usfa.fema.gov/downloads/pdf/publications/FA-166.pdf. Accessed May 5, 2015.

Gasaway, Richard. *Situational Awareness Matters!* Website for articles, handouts, and downloads regarding situational awareness is available at http://www.samatters.com.

International Association of Fire Chiefs (and several cosponsors). Rules of Engagement for Firefighter Survival (poster). Available at http://iafcsafety.org/rules-of-engagement-training-poster-released/.

Kipp, Jonathan D., and Murrey E. Loflin. *Emergency Risk Management*. New York: Van Nostrand Reinhold, 1996.

Klein, Gary. *Sources of Power: How People Make Decisions*. Cambridge, MA; MIT Press, 1998.

NFPA 1250, *Recommended Practice in Fire and Emergency Service Organization Risk Management*. Quincy, MA: National Fire Protection Association, 2020.

NFPA 1500, *Standard on Fire Department Occupational Safety and Health Program*. Quincy, MA: National Fire Protection Association, 2018.

NFPA 1561, *Standard on Emergency Service Incident Management System and Command Safety*. Quincy, MA: National Fire Protection Association, 2020.

Putnam, Ted. *Collapse of Decision-Making: Findings from the Wildland Firefighters Workshop*. Missoula, MT: USDA Forest Service Publication 9551-2855 MTDC, July 1996.

U.S. Department of Transportation, U.S. Coast Guard. Operation Risk Management, COMDTINST 3500.3, November 1999.

U.S. Fire Administration, National Fallen Firefighters Foundation. 16 Firefighter Life Safety Initiatives. http://www.everyonegoeshome.com/16-initiatives/. Accessed May 5, 2015.

Endnotes

1   NFPA 1561, *Standard on Emergency Service Incident Management System and Command Safety*. Quincy, MA: National Fire Protection Association, 2020.

2   Klein, Gary. *Sources of Power: How People Make Decisions*. Cambridge, MA; MIT Press, 1998.

3   Putnam, Ted. *Collapse of Decision-Making: Findings from the Wildland Firefighters Workshop*. Missoula, MT: USDA Forest Service Publication 9551-2855 MTDC, July 1996.

4   U.S. Department of Transportation, U.S. Coast Guard. Operation Risk Management, COMDTINST 3500.3, November 1999.

In 2013, a firefighter was killed and another trapped while conducting a primary search after the fourth alarm at a fire in an apartment complex. The firefighters were in an interior hallway on the first floor of the fire building, knocking on doors, when a collapse occurred. The apartment complex was built in 1980, consisted of multiple three-story buildings, and was of a lightweight, woodframe nature. There had been numerous fires at this complex in the past, so veteran fire officers were aware of the building's lightweight features. At the time of the collapse, fire had consumed the upper section of one of the apartment buildings and there were multiple master streams in operation. The incident was into the fourth alarm and second hour of operations.

The first arriving companies had previously completed a primary search of the fire building (witnessed and reported to chief officers). During the intital operations, there was some confusion regarding the command post location, who was in command, and who was the safety officer. Approximately 30 minutes into the incident, defensive operations were ordered. Within the next hour, master streams were deployed, the third and fourth alarms were ordered, and evacuation of exposed apartment buildings ensued. Due to some confusion at the command post, the initial primary search of the fire building was not logged on the incident command board and a decision was made to search the first floor again. When a company went to conduct this redundant primary search, the strategy on the fire ground was defensive. There were seven master streams in operation at the time. It was never communicated to the rest of the fire ground that a company was going into the fire building's interior and to shut down their master streams.

1. Based on the information provided here, what factors may have been present that reduced situational awareness?

2. Using the USCG severity, probability, and exposure (SPE) model, what values would you calculate for this incident?

3. Which analytical risk-taking factor from the three-step ISO read-risk model seems to have been missed at this incident?

    A. Rating survivable spaces

    B. Comparing actions with the fire ground clock

    C. Establishing a rapid intervention team

    D. Reading smoke

4. The three levels of situational awareness are influenced by all of the following, EXCEPT:

    A. personal values

    B. experience

    C. one's mental state

    D. adequate resources

Flames: © Ken LaBelle NRIFirePhotos.com

Knowledge Objectives

Upon completion of this chapter, you should be able to:

- Describe potential hazardous energy sources that could be present at an incident. (NFPA 5.2.13 , pp 105–118)
- Identify the methods used to determine the presence of potential hazardous energy sources. (NFPA 5.2.13 , pp 106–107, 112, 114, 116)
- State the methods used to communicate the presence or absence of potential hazardous energy sources. (NFPA 5.2.13 , pp 107, 118)
- Describe the methods used to establish and mark hazard zones necessary to protect members from potential hazardous energy sources. (NFPA 5.2.7 NFPA 5.2.13 , pp 107, 118–119)
- Identify the major components of an electrical grid system. (pp 107–108)
- List forms of alternative energy and their associated hazards. (pp 110–112)
- List the chemical properties of common utility gases. (p 112)
- List the hazards associated with utility water and storm sewer systems. (p 113)
- Describe examples of mechanical hazardous energy. (p 113)
- List the hazardous energy sources in vehicles. (p 114)
- Discuss weather as hazardous energy and list the warning signs that extreme weather is approaching. (pp 115–117)

Skills Objectives

Upon completion of this chapter, you should be able to:

- Demonstrate the ability to recognize indicators of potential hazardous energy sources. (NFPA 5.2.13 , pp 105–117)

A line-of-duty death report has just been issued detailing the incident report of a firefighter killed while operating at a house fire. During the incident, the firefighter came in contact with energized wires that had fallen from a service drop. You research more extensively and find that this is a much more common occurrence than you had realized. You know that you must identify hazardous energy as one of your primary responsibilities at incidents.

1. What is your department's policy for disconnection of utility service? What equipment is necessary to identify whether power is disconnected or isolated?
2. If an energy source such as a downed electrical wire or wires that can potentially fall is identified on the fire ground, how does your department secure the area and make all members operating aware of the hazard?

Introduction: Defining Hazardous Energy

Ask a firefighter to define hazardous energy and you will likely hear a list containing physically hazardous energy sources like electricity and pressurized vessels. Although these are certainly *forms* of hazardous energy, a list is not a definition. In the context of this text, <u>hazardous energy</u> is defined as the unintended, and often sudden, release of stored, residual, or potential energy that will cause harm if contacted.

The NFPA definition of hazardous energy sources is more specific:

> Electrical, mechanical, hydraulic, pneumatic, electrical, nuclear, thermal, gravitational, or any other form of energy that could cause injury due to the unintended motion energizing, start-up, or release of such stored or residual energy in machinery, equipment, piping, pipelines or process systems. (NFPA 1521, 2020 Edition; 3.3.20)

Our modern, technologically advanced society is full of hazardous energy sources—to the point that most of their associated risks are taken for granted. Although we have engineered safety into most hazardous energy sources, this sometimes only sets the stage for complacency. Emergency responders see the results of this complacency often—and unfortunately—as we rescue victims. Most firefighters appreciate and respect the dangers associated with electrical equipment, pressure vessels, and hazardous chemicals. Still, the demands of an incident may place firefighters in harm's way if one of these hazardous energy forms were to be released. This is where the incident safety officer (ISO) comes in.

To be an effective and efficient ISO, you must front-load your understanding of hazardous energy forms so that you can better identify sources, predict the energy's release potential, and intervene appropriately to protect fellow responders. If you understand the hazards associated with a hazardous energy source, you can compare what you know to the incident conditions and make a judgment about the stability of the source. Usually we do this by categorizing the energy as *stable* or *unstable*. The effective ISO further categorizes stable and unstable based on the future potential of the energy:

- Stable—not likely to change
- Stable—may change
- Unstable—may require attention
- Unstable—requires immediate attention

Once a hazardous energy source has been identified and its stability categorized, appropriate control measures or zoning communications need to take place. This chapter lists hazardous energy forms and furnishes information on them so that you can better understand their hazardous potential and take actions to prevent injuries.

Forms of Hazardous Energy

■ Electricity

The integrity of electrical systems is based on their being properly grounded, insulated, and circuit protected. A disruption of any of these components can pose a danger to firefighters. **FIGURE 8-1** outlines basic electrical terms. The ISO who is familiar with them can better communicate electrical hazards and concerns.

Fire and rescue departments are often called to incidents that involve electrical distribution equipment (cat on the pole, wires down, and the like), but rarely do firefighters solve the problem without the assistance of the local power company. Occasionally, however, firefighters must act without the power company to save lives or stop a potential loss. In these cases, a basic understanding of electrical system integrity is important. For the ISO, the understanding of these systems is *essential*. The best way to learn about them is to contact your local power company; most provide no-cost seminars and workshops for firefighters. In the scope of this text, we concentrate on a few *recognition* concepts for the ISO. We will start with general electrical hazards, then discuss distribution grid components, and finish with individual building (or user) components.

General Electrical Hazards

Electricity is always trying to seek the *path of least resistance* to ground. Incidents involving electrical equipment, such as pad transformers and downed wires, are especially dangerous. In these cases, electricity may be seeking ground through vehicles, fences, and other conductors. A victim in an energized

Electrical Terms

Ampere (Amps)
The unit of measure for VOLUME of current flow.

Continuity
The completeness of a current circuit; the ability of electricity to pass unbroken from one point to the next.

Conductivity
The measure of the ability of a material to pass electrical current.

Energized
The presence of electrical current within a material or component.

Ground
The position or portion of an electrical circuit that is at zero potential with respect to earth. The point of electrical return for an electrical circuit.

Ground Fault Interruption (GFI)
A device that will break continuity in a circuit when grounding occurs before the current returns to the distribution source.

Grounding
The act or event of creating a point for electrical current to return to zero potential.

Ohm
The unit of measure for resistance of electrical current. One ohm equals the resistance of 1 volt across a terminal at 1 ampere.

Resistance
The degree that a material holds or impedes the flow of electrical current.

Static Discharge
The release of electrical energy that has accumulated on an insulated body. Static is a stationary charge looking for a ground.

Voltage/Volt
The FORCE that causes the flow of electricity. A volt is the unit of measure for electrical force or potential.

Watt
A unit of measure for the amount of energy a specific appliance uses.

FIGURE 8-1 Common terms associated with electricity.

car is like a bird on an overhead power wire; that is, as long as the ground is not touched, no electrocution can take place. Anything or anyone who touches the vehicle and the ground at the same time can create a path for electricity to seek ground.

Firefighters attempting to approach compromised electrical equipment may feel a tingling sensation in their feet; this is a danger sign known as <u>ground gradient</u>. Ground gradient is electrical energy that has established a path to ground through the earth and is energizing it. A downed power line may be energizing the earth in a concentric ring of up to 30 feet (9 m), depending on the voltage of the source, especially if the ground is wet or snowy. The electrical field gets weaker rapidly as the distance from the point of entry increases. Therefore, there is a difference in the potential voltages at any two points within the gradient. This potential difference is known as *step potential*. The difference in potential between the points where the

firefighter's feet make contact with the ground can cause electrical current to flow through the firefighter's body. Step potential can be reduced or eliminated by wearing the appropriate nonconductive footwear, keeping the feet together, or shuffling the feet in small sliding motions. A firefighter who feels tingling in his or her boots must back away using a shuffle-foot motion to keep both feet in contact with the ground.

Firefighters working at structure fires, motor vehicle accidents, wildland fire operations, and any incident at which contact with electrical energy is a possibility need to be aware of the following electrical shock hazards:

- Electrical current can flow through the ground and extend outward for several feet, depending upon the conditions (i.e., ground gradient).
- Contact with downed power lines must be avoided, as the lines may still be energized. *Treat all downed power lines as energized.*
- Overhead power lines may fall onto and energize conductive equipment and materials located on the ground. Fences, building rain gutters, and other conductive metals are especially dangerous shock hazards.
- Smoke can become electrically charged and conduct electrical current.
- During solid-stream water applications, if the stream touches or even comes near energized power lines or equipment, the water can conduct electricity back to the nozzle.

Identifying shock hazards is not a perfect science; in most cases, the proactive recognition of shock hazards occurs through situational awareness; that is, you put "two and two together" through observations of the equipment (e.g., damaged or compromised electrical component) and the environmental conditions (e.g., wet surfaces, a metal fence). Shock hazards can also be identified through sensing equipment. Some ISOs carry specialized electrical-current–sensing equipment like a "hotstick" or "voltage alerting pen" to detect electrical current movement through objects. Thermal imaging cameras may also assist by spotting overly "hot" signatures of electrical equipment and appliances. As with any specialized tool, training and understanding of the equipment's limitations are essential. Last, there are tell-tale signs of electrical problems (shock potential):

- The odor of overheated wires (plastic melting and/or burning oil)
- Visible smoke or haze emitting from electrical equipment and wires
- A "buzzing" sound coming from wires or transformers (Note: Transformers typically "hum" in normal operation, especially older ones.)
- Tingling sensations in feet, which indicate ground gradient and step potential
- Tingling or static sensations around the face or hands, which indicate arcing potential or static discharge through air
- Metallic materials (such as electrical equipment cabinets and boxes) with discoloring or blistering paint or that are glowing

Shock Hazard Warning Procedures

Any firefighter who discovers or suspects an electrical shock hazard (e.g., wires down, damaged electrical equipment) should take immediate action to warn others. The following steps should be covered by a department standard operating procedure/guideline, and all members should be trained to perform them:

1. Immediately warn others using an urgent or priority radio message.
2. Act as guard or sentry to warn others until the incident commander (IC) can assign resources.
3. Use a box light, lantern, or other scene light to illuminate the hazard at night.
4. Flag the area as a "no-entry" zone using red-and-white diagonal-striped barrier tape (10 feet [3 m] minimum clear zone for ground hazards—greater when wet).
5. For overhead power pole fires or hanging wires, clear a zone equal to the distance between two poles.
6. Ensure that the local power company has been notified to respond.

Failure to adequately warn other firefighters of a known electrical hazard has been a contributing factor in firefighter electrocutions. Following the above steps can prevent this from happening again. The following case studies relate to this hazard:

- Volunteer Fire Fighter Electrocuted by Downed Power Line Following Severe Weather—North Carolina, available at http://www.cdc.gov/niosh/fire/reports/face201319.html
- Career Fire Captain Electrocuted After Contacting Overhead Power Line from the Platform of an Elevating Platform Fire Apparatus—Pennsylvania, available at http://www.cdc.gov/niosh/fire/reports/face200801.html
- Career Captain Electrocuted at the Scene of a Residential Structure Fire—California, available at http://www.cdc.gov/niosh/fire/reports/face200507.html
- Volunteer Fire Fighter Dies After Coming into Contact with a Downed Power Line—Arkansas, available at http://www.cdc.gov/niosh/fire/reports/face9946.html
- Downed Power Line Claims the Life of One Volunteer Fire Fighter and Critically Injures Two Fellow Fire Fighters—Missouri, available at http://www.cdc.gov/niosh/fire/reports/face9937.html
- Electrical Panel Explosion Claims the Life of a Career Assistant Fire Chief, an Electrician, and Seriously Injures an Assistant Building Engineer—Illinois, available at http://www.cdc.gov/niosh/fire/reports/face9928.html
- Volunteer Fire Fighter Electrocuted While Fighting a Grass Fire—California, available at http://www.cdc.gov/niosh/fire/reports/face9926.html

There are other steps that firefighters can take to protect themselves from electrical hazards:[1]

- Treat all downed power lines as if they are energized.
- Keep a safe distance from downed power lines—at least 10 feet (3 m).
- Wear the appropriate personal protective clothing and footwear.

- Do not attempt to move or cut downed wires.
- Avoid walking or standing in pooled water when downed power lines are present.
- When fighting fires near downed power lines, do not use solid-stream nozzles. Never apply water directly to electrical equipment that is burning or arcing. Because water is a good conductor of electricity, the current may flow back through the hose stream to the nozzle and cause a serious injury. Using a high-pressure fog nozzle to break up the water stream can help prevent electrical current from flowing back to the nozzle.

At all incidents, the ISO should evaluate the proximity and integrity of electrical systems. When a component of the total system is deemed to have lost integrity, the ISO must assume that an electrical danger exists and communicate that assumption to responders (see the *Safety Tip* box). In some cases, the electrical system can be perfectly intact and still pose a threat to firefighters, particularly when aerial apparatus and aluminum ladders are being used near power lines and electrical equipment. Although it happens rarely, power can "jump" or arc through the air if a conductor comes close to the exposed wire or power connection. **TABLE 8-1** shows the 360-degree minimum operating distances for equipment operating near electrical lines.

Distribution Grid Components

FIGURE 8-2 shows a typical electrical generation and distribution system, commonly known as the "grid." Along the chain of the grid, various equipment components handle the electricity. **FIGURE 8-3** lists some of the specific hazards associated with this equipment.

Being familiar with the arrangement of overhead equipment, cables, and power lines on power poles may help the ISO recognize potential hazards. **FIGURE 8-4** shows a typical overhead power distribution setup, although it is important to understand that there are numerous configurations.

Incident operations involving electrical grid equipment are numerous, and specialized tactics should not be attempted without consultation with the electrical power authority. A couple of specific types of grid incidents or hazards merit special attention.

Substation Fires

Electrical transmission and distribution substations use large transformers to step down voltage. Incidents involving this

TABLE 8-1	Minimum Distances for Equipment/Apparatus Operating Near Exposed Distribution Wires and Electrical Conductors	
Voltage		**Distance**
0–50,000		10 feet (3 m)
50,000–200,000		15 feet (4.5 m)
200,000–500,000		20 feet (6.1 m)
500,000–750,000		35 feet (10.7 m)
750,000–1,000,000		45 feet (13.7 m)
Source: OSHA 75 CFR 48142 (August, 2010); ASME B30.5a (1995).		

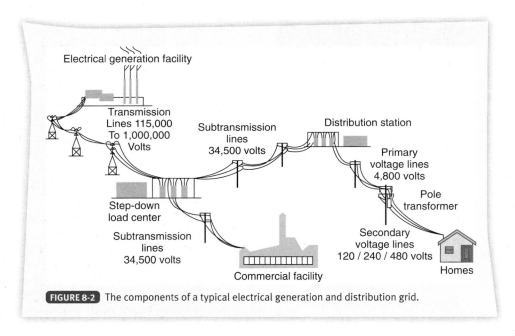

FIGURE 8-2 The components of a typical electrical generation and distribution grid.

Electrical Grid Equipment Hazards

PowerLines/Wires

Uninsulated, under tension, arc danger, difficult to know voltage/amperage, downed wires may jump/recoil, ground/fences/gutters easily energized when in contact, power feed may be from both directions.

Load Center and Distribution Stations

Usually fenced in with security measures. High voltage hazards. Can only be isolated and taken off line by the power company. Transformers and cooling fluid tanks can hold thousands of gallons of oil. Power line anchors and overhead supports can fail (steel and aluminum) leading to line recoil that can topple poles leading away from the site.

Pole-Mounted Transformers

Usually step-down type, difficult to extinguish, may drop/dangle, may cause pole damage, may cause wire failure, may drip hot oil and start ground fire, sealed transformers have been known to explode when burning or when exposed to the flames of another fire.

Pad-Mounted Transformers

Usually low-voltage/high-amperage, may energize surrounding surfaces, pooled water may conduct current, difficult to extinguish oil/pitch fire, possibility of arcing.

Ground Level Vaults

Confined space, possible O_2 deficiency, buildup of explosive gases/smoke, cable tunnels can transfer fire/heat, significant arc danger.

Subterranean Vaults

Same as ground level vaults, plus water collection hazard, difficult to ventilate and extinguish, can "launch" a manhole cover if accumulated gases ignite.

FIGURE 8-3 Hazards associated with electrical grid equipment.

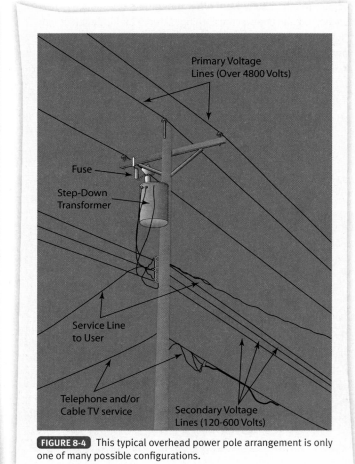

FIGURE 8-4 This typical overhead power pole arrangement is only one of many possible configurations.

equipment present several hazards to responders. Large transformers (bigger than some homes) can hold several thousand gallons of oil and may have an auxiliary cooling oil tank nearby (that holds thousands more gallons). When a transformer catches fire, it may burn for several days, and the heat generated from an oil-fed fire can cause serious damage to the high-tension wire infrastructure and other transformers. In most cases, firefighters should *let the equipment burn until the power company makes a decision* regarding what needs to be cooled or extinguished. It is likely that the power company will want to cool the high-tension wire anchors and super structure above the burning equipment with broken streams. Large transformer fires can be extinguished, but it takes copious amounts of water to bring the transformer core down below 400°F (204°C; the ignition temperature of the oil). Substations are typically well drained so water cannot pool, but make sure the drainage systems are working.

High-tension wires that separate from anchor points can "reel coil" with tremendous force; that is, when the tension is released suddenly, the wire snaps, or whips, in a direction opposite to the tension force. When that happens, two energies are at play: First is the law of motion—for every action (or force), there is an opposite and equal action (force). Second, steel retains a certain memory. Most steel cable is tightly wound onto a spool during production. When it is released from its application, it returns to its reeled state—hence the term "reel coil" or recoil. The recoil may topple one or more towers downstream before the force is dissipated. Make sure responders stay clear of "whip" pathways.

Incidents Involving Wind Turbine Power Generators

Historically, humans have harnessed wind for many purposes. These efforts continue to expand, and now, large wind farms are creating enough electrical energy to supply the needs of an entire town. Do-it-yourself residential owners are installing small wind turbines to supplement traditional energy services or to live off the utility grid entirely. Hazards associated with wind turbines are typically proportional to the size or scale of the turbine. In addition to the typical hazards associated with any electrical equipment, wind turbines add the following hazards:

- Rotating blades pose a significant hazard. Once in motion, these blades carry inertia, which makes them difficult to stop. The larger the blade, the more inertia it will have. Firefighters should give moving blades a wide berth.
- Not only is the atmospheric wind causing movement, but the rotating blades themselves create pressure gradients that include negative pressure upwind, positive pressure downwind, and pressure outflow along the circumference of the blade travel.
- The electromagnetic turbine or motor is a mechanical device that generates heat.
- The size of the wind turbine can introduce hazards associated with the height of the tower and blades. A typical wind farm tower and nacelle (the housing for the turbine) can rise hundreds of feet **FIGURE 8-5**. Individual blades can be 60 to 130 feet (18–40 m) long, creating a rotational diameter and danger zone of more than 300 feet (90 m)! These large blades may appear to be rotating rather slowly; that is an optical illusion. The tips of a 60-foot (18-m) long blade are actually traveling more than 70 miles per hour (mph; 113 kph). Technical rescue skills and safety precautions should be employed for trapped or injured personnel within or atop the towers.
- Large wind turbine nacelles contain many other moving gears and systems **FIGURE 8-6**. Many of these systems are automated in that an outside wind change will

FIGURE 8-5 Wind farm towers can rise hundreds of feet and have a rotational hazard zone of more than 300 feet (90 m).

cause changes in blade pitch, yaw direction (swivel of the nacelle), or shaft braking. While lockdown systems may also be present, firefighters should not work inside these nacelles without a technical expert present.
- Burning turbines and blades will likely produce toxic smoke from the many high-tech metals and plastics present. As a rule, a fire in a large wind turbine is best watched from the perimeter of a substantial collapse zone. Consult your local wind farm power experts for more guidance.

Building (User) Electrical Components

The electrical grid is designed to feed multiple end users, typically buildings. The electrical equipment within the building can range from simple (residential properties) to sophisticated (commercial/industrial). However, most have the following components in common:

- *Grid feed and main.* Overhead power lines or underground conduit supplies the building from the grid. Generally, an exterior power meter and main shutoff will be located where power enters the building. Multiple meters can indicate the number of separate units or tenant spaces within a building.
- *Power distribution circuit protection.* The "breaker box" may be co-located with the main shutoff and meter on residential buildings. Larger commercial buildings will typically have multiple breaker boxes on the interior.

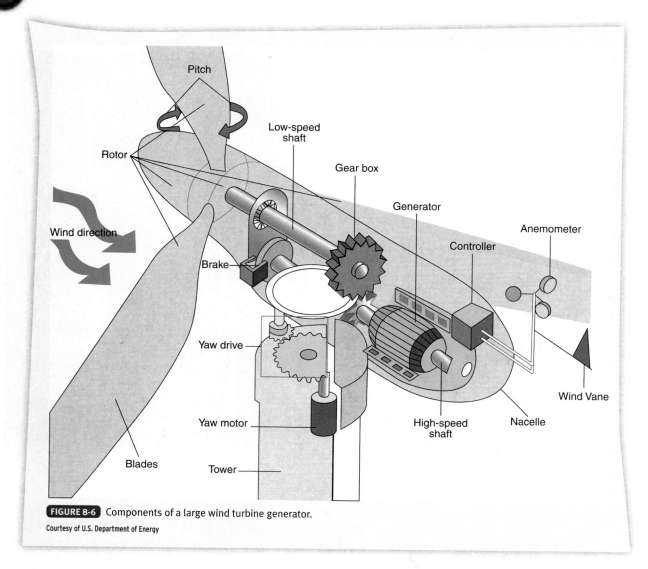

FIGURE 8-6 Components of a large wind turbine generator.

Courtesy of U.S. Department of Energy

- *Transformer(s).* Buildings that require higher voltage (220 volts, 480 volts, or higher) and/or multiphase electrical supplies will also have transformers to help regulate the voltage needs within the buildings. Transformer boxes can be located externally (pad-mounted, below ground, or rooftop) or internally (common in older buildings). *Hazard note: Older, internally located transformers may still contain poly-chlorinated biphenyls (PCBs)—a very toxic carcinogen—although most have been replaced. Regardless, interior-located transformers present a significant smoke, fire, and shock hazard.*
- *Wiring.* Electrical wires within a building are typically dual-insulated and multiwire, with a hot, neutral, and ground wire all wrapped in a single sheath. This type of wiring is often called "Romex." Romex runs are typically located behind interior wall, floor, and ceiling coverings. Permanent wire runs that are exposed are required to be within a protective conduit. Knob and tube wiring (K&T) may still be found in buildings built before 1930. K&T is easy to identify, as it includes single, minimally insulated or uninsulated parallel wire

runs supported by ceramic knobs used as spacers/supports between the wires. There is no ground wire.
- *Outlets, fixtures, and switch boxes.* Individual lights, switches, equipment, appliances, and devices access electricity through a wall- or ceiling-mounted box or by plugging into an outlet.

Most hazards associated with building electrical components can be eliminated by shutting off the power at the main—a common procedure at most building fires. This procedure can be rendered ineffective if the building has systems to achieve an uninterrupted power supply (UPS). UPS systems can include backup generators, battery banks, or alternative power supplies, such as solar systems. Shutting down commercial backup power systems is often beyond the technical abilities of firefighters. Closing individual circuit breakers can help isolate a damaged portion of the building if interior operations are under way. Otherwise, it's best to consult the building engineer or mechanical technician who represents the property.

Alternative Energy Sources

The increased reliance on computers and data transmission systems has led to an increase in the presence, and capacity of, UPS systems. Further, the "green" movement has inspired many to create cleaner ways of generating power to reduce one's "carbon footprint"; some even choose to live off the grid. These demands and desires have created a whole host of alternative energy sources that can be found in buildings of all types and sizes. Entire books have been written to cover alternative energy sources. Here, we will touch on a few and describe some notable hazards that firefighters may face when handling incidents at which alternative energy sources are present.

Battery Rooms

UPS battery rooms are commonplace in many data storage and processing centers, as well as in telecommunication switch equipment centers. In such installations, it is not uncommon to use reconditioned wet cell batteries from submarines for UPS purposes, and some of these battery banks can take up more than several thousand square feet of floor space. Emergency responders working in or around UPS rooms need to exercise extreme caution because of the presence of direct current (DC) power created by the batteries. Pooled water, battery acid, and battery rack hardware can become DC energized because there is typically no ground fault protection with DC power. A main shutoff may separate the batteries from an inverter, but the batteries will continue to be a shock hazard. Likewise, these batteries are typically on a slow charging system, creating hydrogen gas as a by-product.

Emergency Backup Generators

With the need for UPS, many businesses (and some residential properties) have chosen to install backup generators to provide electrical power when the distribution grid fails to deliver. These generators are typically fuel-powered (gasoline, diesel, natural gas, and/or propane) engines that create DC power. The DC power flows through an inverter that converts the electrical current to alternating current (AC). The inverter may be a separate device or built right into the engine. Sizes of these generators can range from a small portable engine (similar to a lawn mower engine) to a full-blown railroad locomotive. In the cases of larger, permanently installed generators, several systems are likely in place that automatically start and run the generator. Shutting down outside power feeds from the distribution grid can trigger the start-up of the generator. Likewise, the presence of a permanently installed generator may also indicate that a UPS battery room is present, along with the hazards associated with a battery room. In addition to electrical current hazards is the storage and distribution of the fuel used to power the back-up generator.

Solar Energy

Solar energy is used primarily in two ways: electrical generation and thermal heating. Solar electrical generation systems include an array of photovoltaic (PV) panels that collect light energy from the sun to stimulate electrons in a silicon crystal or film (or other evolving material). This stimulation is captured as DC current, which is fed via electrical wire to an AC inverter, optional battery storage, AC breaker panel, distribution grid meter, and eventually the electrical appliances within a building **FIGURE 8-7**. Technological advances have introduced PV modules that look like roofing tiles and are incorporated into the roof itself as a protective cover; they look like glass

tiles. Thin film PVs are flexible and look like a mat. Hazards associated with PV solar systems include:

- Additional roof dead load of 3 to 5 pounds per square foot. A standard 3- by 4-foot PV panel weighs approximately 40 pounds.
- Hot surfaces while the sun is shining on PV panels.
- Constant DC generation while daylight is striking the PV panel. This energy is flowing from the panel to the first DC disconnect, which is likely near the inverter. The use of a 100% light-blocking cover can stop the flow from the panels, but all of the panels must be covered. Note: Apparatus scene lighting and most streetlights are not likely to stimulate electrical generation in PV panels, although the evolving efficiencies in PVs may change this. Also note that PV panels that break or collapse as a result of a fire may still generate electricity if exposed to daylight.
- Potentially unstable surfaces. Firefighters should never walk on or intentionally break or move PV panels. Although glass used for solar panels is typically tempered and is difficult to break, every effort should be made to stay away from the panels. Incorporated roof-tile PVs and thin-film PV modules have a certain amount of durability. Both are quite smooth and should not be walked upon. They are, however, easy to remove or break up with prying tool.

Solar thermal systems are designed primarily to heat up a liquid that is used for heating or hot water use. The panels used to collect the sun's energy may, at first glance, look like a PV panel, but they are different. Solar thermal panels can be distinguished from PV panels by the piping that is connected to the panel. PV panels will have electrical conduit attached, whereas thermal panels will have liquid piping (copper, PVC, or other liquid tubing). Thermal solar panels may be covered by flat glass or may appear as multiple glass tubes. The flat

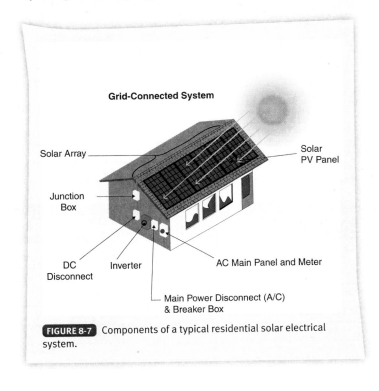

Grid-Connected System

Solar Array

Solar PV Panel

Junction Box

DC Disconnect

Inverter

AC Main Panel and Meter

Main Power Disconnect (A/C) & Breaker Box

FIGURE 8-7 Components of a typical residential solar electrical system.

glass panels look more like the PV panels but have more internal volume (depth). Within the flat glass panels or glass tube arrays are energy-absorbing tubes filled with a liquid medium (usually a food-service-grade antifreeze or gel). The liquid moves from the panels in a closed-piping system to a heat exchanger within the building. The heat exchanger is likely inside a tank or vessel that transfers heat to water (for hot water taps) and/or a radiant hot water heating system.

Hazards associated with solar thermal systems include:

- Additional roof dead load of 5 to 8 pounds per square foot
- Pressure vessel rupture due to heating of the closed-loop liquid system
- Hot fluid scalding if the piping system becomes damaged
- Development of an inhalation hazard if the antifreeze is exposed to fire

As with the PV panels, firefighters should not operate near the glass panels or tubes. Typically, the external glass used for thermal panels remains cool to the touch even under direct sunlight. However, the internal components and liquid can be extremely hot and cause contact burns.

Additional forms of alternative energy sources are being developed and include geothermal (harnessing heat energy form below the earth's surface), hydrogen fuel cell, hydromotion (tidal), and biomass sources.

Utility Gas

Firefighters who have seen a propane tank BLEVE (boiling liquid expanding vapor explosion) undoubtedly agree that the most important utility to control is gas. The most common utility gases are propane and natural gas (methane). Both are colorless and odorless in their natural form so an odorant is usually added to make leaks noticeable, although that odorant is added at supply plants. Raw propane and methane from refineries and ground wells may be transferred to pipelines that convey the gas many miles before an odorant is added. Propane and natural gas can be stored in a liquid state: liquefied petroleum gas (LPG) and liquefied natural gas (LNG).

At incidents, the ISO should evaluate the integrity of the gas fuel supply and containment vessels. The integrity of a gas system relies on a closed (air-tight) supply vessel (a tank or piping), a shutoff valve, a pressure regulation device, and a distribution system with protection at each appliance (shutoff valve and surge protection). Each component in the system is designed for a certain pressure. Pressures that exceed the maximum (such as from exposure to heat from a fire) can cause a pressure-relief device to activate, introducing expanding flammable gas into the environment. Likewise, trauma to this system can create holes, pipe separation, or container failure. Once gas escapes from the system, fires may be accelerated or toxins may be released and inhaled. Unignited gases can accumulate in confined spaces and present an explosion hazard if an ignition source and proper air mixtures are introduced. Knowing the properties of common gases can help the ISO determine the risks associated with the gas present **TABLE 8-2** .

Flame impingement on stored LPG and LNG storage vessels is extraordinarily dangerous. The tanks contain product in a liquid state. That liquid can boil and expand as the fire heats the tank, and the resulting pressure can cause tank failure and a BLEVE. Tank cooling with adequate gallons per minute (GPM) is mandatory **TABLE 8-3** . If that is not possible, evacuate all personnel according to the evacuation distances in Table 8-3.

TABLE 8-2	Properties of Common Utility Gases	
Property	**Propane (LPG)**	**Natural Gas (Methane)**
Chemical makeup	C_3H_8	CH_4
Vapor density (air = 1)	1.52 at 70°F (21°C)	.55 at 70°F (21°C)
Boiling point	−44°F (−42°C)	−259°F (−162°C)
Ignition temperature	871°F (466°C)	999°F (537°C)
Flammability Limits		
Upper	9.5	14.0
Lower	2.4	5.3

TABLE 8-3	Cooling and Evacuation Suggestions for LPG Tank Fires	
Tank Size (gallons)	**Minimum Water Cooling Rate (GPM)**	**Minimum Evacuation Distance**
0–150	50–100	1,000 feet (305 m)
150–1,000	100–150	1,500 feet (457 m)
1,000–18,000	500–1,000	4,000 feet (1,219 m)
18,000	1,000+	1 mile (1,609 m)

Source: Conservative estimates derived from the U.S. Department of Transportation's 2012 Emergency Response Guidebook.

When cooling tanks, it is important to keep from extinguishing burning product that is escaping from a pressure-relief valve.

Utility Water, Storm Sewer Systems, and Flash Floods

Some may argue that utility water and sewer systems are not hazardous energy. Anyone who has witnessed a rapidly enlarging sinkhole after a waterline break readily asserts that water is indeed hazardous energy. Water and storm sewer systems can cause safety concerns in numerous incident situations. Uncontrolled water flow can cause initial and secondary collapses in structures. A damaged sewer or storm drain system can leach water into surrounding gravel and dirt beds, undermining the earth supporting a structure, road, or other utility system.

Firefighting efforts may introduce significant quantities of water into a structure. Who evaluates where all that water will go? The ISO should investigate large quantities of water flowing in and around the incident. In one incident, the assigned ISO saw large quantities of water pooling and bubbling in the yard next to a dwelling fire, which seemed unusual. Upon investigation, the ISO determined that the piping feeding the hydrant being used for water supply had begun leaking underground. The pooling was the first visual clue that the ground underneath the firefighting operation was being undermined. The water department was notified and the water shut down, but not before the nearby street pavement began sinking next to a staged engine.

Water that collects in basements or other building areas may extinguish gas-fired equipment pilot lights, with the result that raw gas is being "bubbled" into the water. Although many gas-fired devices have fail-safe systems to stop raw gas from flowing, in a fire, after a collapse, or in waterlogged conditions these fail-safes may be damaged. Likewise, pooled water and electrical equipment do not usually play well together.

Storm sewer systems are designed to receive potentially large volumes of water, literally thousands of GPM. In flash flood situations, these systems can be overwhelmed with water and debris. The debris can get caught up at storm grates and cause localized flooding. The tremendous force of the flowing water can pull unsuspecting victims into the debris and grates, pinning them below water levels or crushing them with force. Firefighters have been killed attempting to rescue victims caught in storm sewer systems. Swift water rescue techniques and solid risk management concepts need to be employed when working around storm sewer systems during flash floods. Likewise, flash flooding of rivers, canals, and other open-conveyance water systems requires preplanning and training for potential rescue situations **FIGURE 8-8**. Personal floatation devices and rope throw bags should be mandatory safety equipment for all flash flood responders.

Mechanical Energy

Systems that include pulleys, cables, conveyers, counterweights, and springs are examples of mechanical stored energy. Lightweight high-rack storage systems can be classified as stored potential energy in the form of dead or live loads and gravity. The sudden release of mechanical systems can be caused by heat, trauma, or overloading, or all of these.

Steel cables used in pulley systems are likely to recoil violently when they fail. Simple "guide," or "guy," wires for poles, signage, and antennae are typically tensioned and can release and recoil with amazing force. Freestanding truss structures, like antennae, are nothing more than vertical cantilevered beams and are weakened quickly when exposed to the heat of a fire. These weakened towers fail quickly, and horizontal forces such as wind or blunt force accelerate the collapse.

Pressurized Systems and Vessels

Pressurized equipment comes in many forms, sizes, mediums, and arrangements. Regardless, suddenly released pressure from an enclosed container can literally slice through a person. Pressurized systems use either hydraulics (liquids) or pneumatics (air/gases) as a medium to achieve power or force. Pressurized systems typically include a reservoir for the medium, some sort of pump, tubing or conveyance system, and the tool or implement that uses the pressure to achieve a task. Connections, valves, manifolds, and pressure control devices may also be present. Heat applied to a pressurized system causes pressure to increase, which can exceed the system's design limits. Built-in pressure-relief devices may not be able to relieve pressure as fast as the pressure is being developed, leading to an explosive failure of a component.

Closed containers of various products (mainly liquids and some gases) may become "pressure vessels" when heated by fire or hot smoke. Although gaseous product containers are designed to hold certain pressures, the fire can cause the pressure to exceed design limits. Failure of the container can cause the container to become an unguided missile. Liquid containers may or may not be designed to hold pressure, yet pressure is developed within them when they are heated. Again, they may explode or become missiles or cause the liquid stored within to boil over and spill. As a rule, the stronger the container is, the more initial resistance to pressure it has and the more explosive it becomes when it fails.

FIGURE 8-8 Flash flooding of rivers and canals can cause significant damage and requires preplanning and training for technical rescue potentials.
Courtesy of Loveland Fire & Rescue Fire Authority

■ Hazardous Energy in Vehicles

In this text, all sorts of transportation equipment are classified as vehicles. Cars, trucks, boats, planes, trains, and the like are included in the classification. Vehicles contain many forms of hazardous energy. The following factors can contribute to hazardous energy when working at incidents involving vehicles:

- *Stability/position.* Rolling weight, instability, collapse (crush), failure of ground to hold the load
- *Fuel systems.* Types of fuels (including alternative), storage, pumps, fuel lines, pressurization
- *Electrical systems.* Batteries, converters, high-voltage wires, generators
- *Power generation systems.* Pulleys, belts, heat, noise, thrust, exhaust gases
- *Suspension/door systems.* Springs, shocks, gas or pneumatic struts
- *Drive/brake systems.* Pressure vessels, heat, springs, torsion, exotic metal fumes
- *Restraint safety systems (air bags).* Trigger systems, chemicals, delayed or unpredicted deployment

The list can go on, but we will describe a few specific systems and hazards in more detail as well as some hazards associated with a specific type of vehicle.

Alternative Fuel and Energy Systems

New fuel and power system technologies are being developed to address traditional fuel supply shortages and dependency on foreign oil. Alternative fuels, fuel cells, and high-voltage systems present additional hazards to responders. Many of our educational resources (books, suggested procedures, and incident experience) trail the incident potential of these hybrid systems, although progress is being made. NFPA offers an online firefighter training program for electric hybrid vehicles. The training program is free and can be accessed at the Electric Vehicle Safety Training website (http://www.evsafetytraining.org).

Identifying a vehicle with an alternative or hybrid power system can be challenging. A few "standards" are in place, but they are likely to change as manufacturers develop new and improved systems. Although the standards are subject to change, some identifiers have been adopted. Bright orange conduit or cable can indicate that high-voltage power (up to 700 volts and 125 amps) is used in the drivetrain of a vehicle **FIGURE 8-9**. These high-voltage systems typically have large battery packs to help store and feed energy. Bright blue conduits or cables can indicate medium-voltage drivetrains (less than high-voltage systems but more than typical 12- or 24-volt systems). Ethanol fuels (like E-85) are polar solvents that can render class B foam ineffective. Crews must use alcohol-resistive foams for spill and fire control. Err on the side of caution when dealing with these newer technologies.

Vehicles powered by compressed natural gas (CNG) or propane are required to have a decal or sticker on the right-rear portion of the vehicle and adjacent to refilling locations. CNG labels are standardized as a blue diamond with the words "CNG," "LNG" (a BLEVE hazard), or "un-odorized CNG" within the diamond. Propane-powered vehicles should have a black diamond with "Propane" or "LPG" within. The most common location for the stored CNG or LPG fuel tanks (compressed and liquefied gas vessels that should be treated as a BLEVE hazard at fire incidents) is within the frame rails, although they may be roof-mounted on buses (usually under a lightweight tank cover).

Drive Brake Systems: Tractors/Trailers

Semitruck tractors and cargo trailers may have two more hazardous energy sources in addition to the common automotive hazards already listed: namely, air brake chamber springs and split-rim wheels. Air brake chambers are located just behind each tire/wheel assembly attached to the axles. When exposed to fire, these chambers are likely to fail, and the heavy-duty spring within the chamber can become a projectile capable of injuring unsuspecting firefighters **FIGURE 8-10**. Split-ring wheel rims are an older method to help keep a tire secured to a wheel for heavy-duty applications. Opposing tension between the split-ring steel, tire air pressure, and the metal wheel creates a hazardous energy threat that can suddenly release from fire exposure. Although safer methods have been

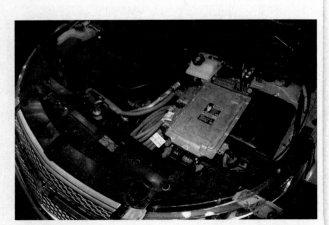

FIGURE 8-9 Bright orange conduit or cable can indicate that high-voltage power (up to 700 volts and 125 amps) is used in the drivetrain of a vehicle.

FIGURE 8-10 Notice the heavy springs that have separated from the axle-mounted air brake chambers during this tractor/trailer fire. They can release with tremendous force and easily injure firefighters during fire attack.

Courtesy of District Chief Chris E. Mickal/New Orleans Fire Department, Photo Unit

developed, many split rims are still in service. Protective cooling of air chambers and wheel rims is paramount when fighting tractor/trailer fires.

Restraint Safety Systems: Air Bags

The accidental or delayed deployment of air restraint devices can cause—and has caused—injury to rescuers. C-spine trauma, muscle sprain, contusions, and lacerations are all possible if the unsuspecting firefighter is suddenly hit by a late-deploying restraint bag while attempting to extricate victims of vehicle collisions. A victim trapped in a vehicle with an undeployed air restraint safety device is in danger; consider this an *immediate rescue environment*. In some cases, the actions of firefighters can cause the activation of an air restraint device. Although it is beyond the scope of this text to discuss all the steps to deactivate air restraint devices, responders should research and develop procedures for restraint device deactivation.

Locomotives

Modern train locomotives are actually very large electrical generators and should be treated as such. The electrical generator is typically diesel-fuel fired. The electricity developed is used to rotate the train's wheels through electric motors, power air compressors for car braking, and develop induction energy to also help in braking (magnetic). Many historical tourist districts still run steam locomotives. These are typically large-scale water boilers that are heated by coal or oil. Hazards are typical to any boiler: scalding liquid/vapors, explosive ruptures, and uncontained fuel fires. The braking systems are typically mechanical on steam locomotives.

Aircraft

Specific to aircraft are the hazards of radar, exotic metals, and the potential of munitions in military and covert-mission (i.e., CIA, mercenary) aircraft. These topics are covered in the *Technical Rescue Incidents* chapter.

■ Weather

The popular phrase "If you don't like the weather, wait 5 minutes" rings true in many regions of the United States and is often cited by locals no matter where you may travel. Weather is a dynamic, complex, and often misunderstood force that firefighters must contend with. Common firehouse talk suggests that the "big one" will hit when the weather is at its worst (perhaps another example of Murphy's law). Often, the adverse effects of weather cause the incident for which we are called. Progressive fire and rescue departments have acknowledged this possibility and have taken steps to reduce the impact of adverse weather on personnel. Once on scene, the ISO should consider weather as a form of hazardous energy and weigh the effects of weather extremes with the "behavior" of the incident and of the incident responders.

Effective ISOs study weather and understand the particulars of weather patterns found in their geographic regions. Likewise, they keep abreast of daily forecasts and weather observations as a matter of habit and readiness. From this constant attentiveness, they can make reasonable predictions on the impact of weather-related risks.

In-depth professional weather spotter and forecasting training is available from local television weather forecasters and the National Weather Service (often free of charge). The National Weather Service is part of the U.S. Commerce Department's National Oceanic and Atmospheric Administration (NOAA), which has an extensive collection of pamphlets, books, videos, and charts available for weather-related education. This section of the text highlights weather considerations essential to effective ISO awareness and should not replace more in-depth training from NOAA resources. These considerations are wind, humidity, temperature, and the potential for change.

Wind

Of all the weather considerations affecting firefighting operations, wind is by far the most important. Nothing can change a situation faster—or cause more frustration in an operation—than the effects of wind. Wind is created as air masses attempt to reach equilibrium. Warm and cold fronts cause changes in atmospheric pressure and therefore gradients in pressure. Other factors, such as the jet stream (the prevailing wind), day/night effect (the diurnal wind), upslope/downslope, and sea breeze, all influence wind. The arrival of a cold front can cause a 180-degree change in the wind direction. A falling barometer (the measure of atmospheric pressure) can indicate an approaching storm and subsequent wind. A warm, dry wind that spills down through mountain canyons or valleys can be extremely dangerous to firefighters in that it quickly changes fire behavior (whether wildland or structural). Likewise, the sudden reduction of a warm wind allows the local or prevailing wind to influence the fire. A sudden calm period may indicate an upcoming weather event, such as a thunderstorm downdraft or a wind shift.

Some indicators of wind and wind changes can be found in the patterns formed by clouds. High, fast-moving clouds may indicate a coming change, especially if the clouds are moving in a different direction than the surface wind. Lenticular clouds (very light, sail-shaped cloud formations found high aloft) indicate high winds that may produce strong downslope winds if they surface. The development and subsequent release of a thunderstorm (cumulonimbus clouds—Latin for raincloud) causes erratic winds and strong downdrafts **FIGURE 8-11**. It is not uncommon to have four or five wind direction and speed changes in the vicinity of a thunderstorm squall line.

ISOs should understand the wind patterns for their specific geographic region. Further, the ISO should evaluate existing and forecasted winds and predict their influence on fire behavior and the safety of responders. As wind velocity increases, so does the risk to firefighters (see the *Safety Tip* box). Being aware of local wind tendencies and evaluating their effects on an incident can make a difference between a high-risk and a relatively safe operation.

Humidity

The ISO working a large wildland fire has likely established communication with the fire behaviorist and weather forecaster assigned to the incident. One important piece of information to be had from this communication is the relative humidity of the air. The primary reason is obvious: Lower humidity means increased fire spread. The ISO assigned to a

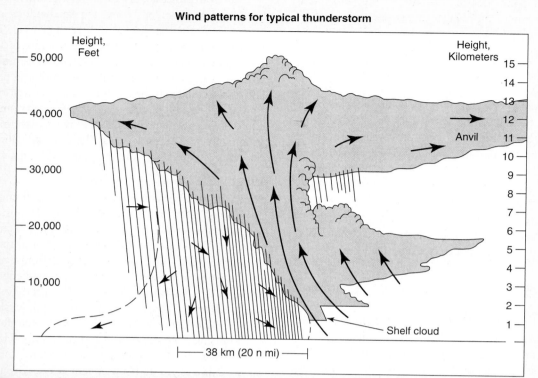

Wind patterns for typical thunderstorm

FIGURE 8-11 Erratic wind shifts are common during thunderstorms. The formation of a shelf or wall cloud at the base of a cumulonimbus cloud can indicate the potential for a tornado.

structure fire needs to look at humidity for another reason: Humidity affects firefighters in a number of ways. In especially dry environments (hot or cold), firefighters become dehydrated quickly from breathing as well as sweating. In high-temperature, high-humidity environments, the firefighter becomes dehydrated through profuse sweating in an attempt to cool the body (metabolic thermal stress). In cold-temperature, high-humidity environments, the firefighter fights penetrating cold (pain) and the effects of rapid ice buildup on surfaces and equipment. High humidity also affects structural fires by keeping smoke from dissipating in the outside air. Prolonged low humidity can cause accelerated fire spread in lumber and other wood products, such as shake shingles and plywood, and flying brands can retain their heat longer and fly farther in low humidity.

Temperature

Which is easier: launching an aggressive brushfire attack when it's 100°F (38°C) or maintaining a critical defensive fire stream between two buildings when it's 10°F (−23°C)? Answer: It depends. From the ISO's perspective, temperature needs to be evaluated relative to its effect on firefighter exposure. Acclimation is critical. Firefighters working in International Falls, Minnesota, are well acclimated to the cold. A firefight in subfreezing temperatures may even seem commonplace. Conversely, firefighters in Yuma, Arizona, are probably accustomed to aggressive operations when it's 105°F (41°C). Invert the two extremes and the stage is set for firefighter injury. The ISO needs to consider acclimation when evaluating the weather and its

effect on firefighters. Practically speaking, you need to be concerned when your firefighters are operating in an environment outside their temperature "norm" for the region. The firefighter stress effects of temperature and humidity are covered more completely in the chapter *Reading Firefighters*.

Weather Shifts

In most cases, the prevailing weather condition for a given incident is what everyone will have to endure for better or for worse. At times, however, the duration of the incident or the development of a severe storm during it causes significant danger to responders. The ISO familiar with local weather tendencies can advise the IC of weather indicators that may require a shift in the incident action plan. Simple weather spotter guidelines and instructions, like those in the following list, can help the ISO warn others.

- *Watch the sky*. This simple activity is often overlooked. If foul weather is approaching or suspected, find a close vantage point to evaluate cloud patterns and wind activity **FIGURE 8-12**. A quick phone call to mutual-aid agencies upwind may reveal useful information about an approaching storm.
- *Note 180-degree changes in wind direction in a short period of time (a few minutes)*. Be especially wary of a sudden calm. A wavering smoke column is likewise noteworthy. In most cases, each event is a sign of an unstable air mass and may point to a significant weather change.
- *Be mindful of the potential for a flash flood*. Mentally compare previous rainfall and ground saturation with current rainfall to determine the potential for a flash flood.

Safety Tip

Wind Danger at Structural Fires

Firefighters are well aware of the effect winds can have on outside or wildland fires, but their effect on structural fires is equally important. Of particular note are the choices that firefighters make for the point of fire attack and the point of ventilation. One age-old tenet of fire stream application is to attack from the unburned side. Although this principle is appropriate sometimes, we know that at other times it is not possible, practical, or desirable. Wind influences (wind speeds of 10–20 mph [16–32 kph] or greater)[2] should be considered when making attack point decisions. Open doors, sudden window failure, and other vent spaces can allow wind to penetrate the interior space of the building and influence fire behavior. All firefighters should be aware of this possibility, and the ISO should evaluate attack options to see if firefighters are in a position to be trapped, overrun, or exposed to a wind-fed acceleration of the fire inside the structure. Likewise, the influence of wind should be part of the decision-making process when locating ventilation openings and flow paths. Similarly, a strong wind can easily defeat the desired outcome of positive-pressure ventilation fan use.

The National Institute of Standards and Technology (NIST) has released an outstanding technical report on the effects of wind-driven fire behavior. *Firefighting Tactics Under Wind Driven Fire Conditions: 7-Story Building Experiments* (NIST TN 1629, April 2009) is available at the NIST website (http://fire .nist.gov/bfrlpubs/fire09/PDF/f09015.pdf). One important observation came out of almost all the experiments: Wind-driven interior fires create untenable conditions for firefighters trying to access the fire from downwind hallways and corridors: Temperatures can spike to nearly 2,000°F (1,093°C)! The following case studies highlight some outcomes of these conditions.

- Career Probationary Firefighter and Captain Die as a Result of Rapid Fire Progression in a Wind-Driven Residential Fire—Texas, available at http://www.cdc.gov/niosh/fire/reports/ face200911.html
- Career Firefighter Dies in Wind-Driven Residential Structure Fire—Virginia, available at http://www.cdc.gov/niosh/fire/ reports/face200712.html
- Volunteer Fire Fighter Dies and Three Firefighters Are Injured During Wildland Fire—Texas, available at http://www.cdc .gov/niosh/fire/reports/face201110.html

FIGURE 8-12 This cloud formation indicates an unstable air mass and developing thunderstorm. In less than 20 minutes, large hail pellets fell and a funnel cloud was formed.

Evaluate your incident location and whether you are in a low, flood-prone area or along a drainage path.

- *Developing thunderstorms can produce rapid changes.* The changes include straight-line winds (microbursts) of 100 mph (161 kph), hail, and lightning. Tornadic thunderstorms, characterized by a rotation between the rain-free base of a cloud and a forming wall (or shelf) cloud, are especially worthy of attention. A tornado forming in the vicinity of an incident or coming toward it should be immediately communicated to the IC and may be grounds for emergency abandonment and a switch to defensive operations. Protective withdrawal from the area may even be considered.

- *At night, use lightning flashes to define cloud formations that may be tornadic.* Large hail (1/4 inch or larger) can also indicate that you are near the area where a tornado is most likely to form.
- *Calculate the distance between you and the lightning.* A simple rule of thumb to calculate lightning distance is to count the seconds between the lighting flash and the thunder. Divide this by five and you have the distance in miles. Thunder that claps fewer than 5 seconds from the flash means you are in the lightning area. Although lightning can strike anywhere at any time, firefighters can protect themselves by becoming the smallest target possible. Stay away from poles, fences, and trees. Avoid standing in water. Suspend operations with aluminum ground ladders (practice solid risk management).
- *Deep snow not only makes travel difficult; it can hide hazards at the incident.* Covered swimming pools, trip hazards, and electrical equipment can be hidden under snow. The weight of snow can place stress on buildings, as well as on utility lines, poles, signs, and other structures. Ice storms may cause the collapse of buildings, power poles and lines, and signage. Hose advancement becomes difficult in deep snow, causing the rapid

fatigue of firefighters. Frequent crew rotation and rehab are of utmost importance. Driven snow can impair visibility and make smoke-reading difficult.

Miscellaneous Hazardous Energy Forms

The list of hazardous energy forms can seemingly go on indefinitely. Earthen materials (sand, gravel, stone, clay, and like materials), ice, and flowing water (streams, canals, surf, storm drainage) can present significant hazards to firefighters. Often forgotten in the list of hazardous energy forms are animals and medical facilities.

Animals

The evacuation of penned, fire-threatened animals is highly dangerous. It is essential to have a plan. Some fire/rescue departments perform animal rescues as a human lifesaving activity (a deterrent to emotional and risky Good Samaritan rescue attempts). Rescues of small animals may be within the abilities of most responders, but larger animals require more caution and, generally, more expertise. In these cases, it is best to enlist a veterinarian trained in emergency large-animal care to provide guidance and veterinary care during the rescue. Large-animal rescues can become quite technical. Where the potential exists, fire departments should preplan, train, and equip members in order to have a safe and successful rescue **FIGURE 8-13**. Do

FIGURE 8-13 Large-animal rescues can be quite technical and may require specialized training and equipment **(A)** as well as veterinarian assistance **(B)**.

Courtesy of Justin King, Loveland Fire Rescue Authority, CO

not underestimate the power (hazardous energy) of a spooked horse, cow, dog, llama, ostrich, moose, yak, wildebeest, or any other animal.

Medical Care Facilities

Medical care facilities have become increasingly high-tech, involving many devices, chemicals, and systems that can be included in the hazardous energy category. Magnetic resonance imaging (MRI) equipment uses highly energized magnets that can pose compelling hazards—and disruption of radio communications. Medical X-ray imaging, stored compressed gases, high-voltage exam, and laser surgical tools and equipment may also be found in medical facilities. As with all hazardous energy forms, be aware of their potential and expect the worst.

Preventive Actions for Hazardous Energy

This chapter has not only introduced hazardous energy sources, it has provided warnings and several conservative suggestions on how to approach, avoid, or help minimize the inherent threat they create. Compiling a list of hazardous energy abatement methods is compelling but would comprise volumes of text. Presented here are some overhead or general actions that an ISO should consider to help prevent injuries and deaths that may occur from contact with or exposure to hazardous energy sources.

Awareness—Communication

Most, if not all, hazardous energy threats should be communicated to responders. The simple act of making personnel aware of hazardous energy can help prevent unintentional contact or exposure and resulting injuries or death. Hazard awareness can be considered the *first action or priority* when dealing with hazardous energy. Imminent threats warrant immediate communication in the form of an urgent or priority radio message. That message should contain not only the type and location of the hazard, but also some direction of what personnel or activities need to be withdrawn. Urgent or priority communications should be followed by acknowledgment of the personnel in the threat area. Hazardous energy sources that don't present an immediate threat can be communicated using less urgent methods, including face-to-face interaction, general safety messages or briefings, IC information sharing, and flagging (using streamers, lights, and other signaling devices) to draw attention to the threat.

Avoidance—Zoning

The second priority preventive action is that of zoning to help achieve hazard avoidance. This chapter has provided some suggestions regarding the appropriate avoidance distances that should be considered for a given hazardous energy threat. Where no direction is given, the ISO must make some judgment regarding perceived severity of the threat and the size (distance) of the zone. The severity of the threat helps to establish the type of zone that is warranted. There are four types of control zones based on severity, required personal protective equipment (PPE), and procedures:

- No-entry zone. This is an area where no responders are allowed to enter, regardless of PPE, due to dangerous conditions. Some call no-entry zones an "exclusionary zone." A collapse zone is a specific form of a no-entry zone.
- Hot zone. This area immediately surrounds a hazardous area and can be considered an immediately dangerous to life and health (IDLH) environment requiring appropriate PPE, accountability procedures, and a standby rapid intervention crew (RIC).
- Warm zone. This area surrounds a hot zone and is where personnel, equipment, and apparatus are operating in support of an operation. Decontamination processes and RIC standby are typically found in the warm zone. Personnel working in the warm zone should have a defined level of PPE appropriate for their assignment and exposure to conditions.
- Cold zone. This area surrounding the warm zone is used for an incident command post, support-agency interfacing, and staging of personnel and equipment. The use of PPE is rarely needed in the cold zone.

Wherever possible, control zones should be identified with barrier tape, signs, cones, flashing beacons, physical barriers such as fences, or other methods to help denote their presence. The use of law enforcement, traffic control resources, or security resources may be needed to help denote the areas and discourage public entry to a defined zone. Obviously, it may not be practical to mark a control zone due to the location, distance, resources, or other considerations for a given threat.

Where barrier tape can be used to help denote a control zone, it is recommended that a color scheme be used to help differentiate the zones (see the *Getting the Job Done* box).

Defining the different control zones addresses the severity component of avoidance. Establishing the size of a given zone is also important. We've presented some suggestions throughout

Getting the Job Done

Barrier Tape Color Scheme for Zoning
The following color scheme is suggested and encouraged in NFPA 1500 for the marking of control zones at an incident **FIGURE 8-14**:
- No-entry zone: Red and white diagonal-striped or chevron
- Hot zone: Red
- Warm zone: Yellow
- Cold zone: Green

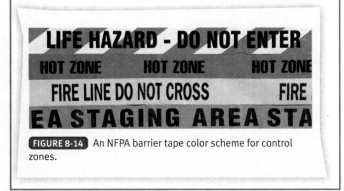

FIGURE 8-14 An NFPA barrier tape color scheme for control zones.

the chapter, but there is probably no perfect formula to calculate control zone distances. That said, the ISO is usually responsible for making a judgment regarding control zone sizes/distances. The following factors can help when making this judgment:
- The height of something that may collapse. It is best to be conservative and add at least 50% to the collapse height.
- The wind direction and speed.
- Product flammability and potential for a BLEVE or percussive unconfined vapor cloud explosion (PUVCE). A PUVCE (pronounced "poov-see") can occur from the ignition of a propane vapor cloud or other flammable gas or liquid vapor that is heavier than air.
- The length of cable or wire and the direction of recoil when it releases (add 25%).
- The potential for a domino effect.
- The direction, volume, and speed of liquid flow.
- The presence of protective barriers that can absorb the hazardous release (e.g., stout walls, fences, restraint devices).
- For electrical hazards, the pooling of firefighting water and the presence of other conductive materials.

■ Lock-Out/Tag-Out

The unintentional start-up or energizing of hazardous energy equipment can cause significant injury to those trying to rescue a worker trapped in/by the equipment. To prevent this outcome, ISOs should employ lock-out/tag-out (LOTO) safeguards. LOTO safeguards ensure that electricity remains off and valves, switches, and other initiating devices remain in an off or closed position. As the name implies, this process usually includes the application of a lock and a warning tag. Where a lock or tag is not practical, posting a sentry can help deter others from turning something on. Industry is required to use LOTO procedures, defined by the Occupational Safety and Health Administration (OSHA) under OSHA 29 CFR 1910.147, when engaged in the maintenance, servicing, and repair of hazardous energy sources and equipment. Although not specifically mentioned in the OSHA CFR, rescuers performing extrications in and around hazardous energy equipment should be familiar with the requirements and apply them. Many ISOs and heavy rescue teams carry OSHA-compliant LOTO kits for such an occurrence.

While performing a LOTO procedure is paramount, a diligent ISO will also ensure that any local isolation switches are also turned off. For example, at a machinery entrapment, the LOTO procedure may take place in the electrical room hundreds of feet away from the actual piece of machinery, which will usually have an isolation switch ("stop" button) within arm's length of the entrapment hazard.

■ Protective Cooling

Pressure vessels, conveyance systems, and other closed-loop or sealed hazardous energy systems are susceptible to rupture when exposed to external heat and flame impingement. Protective cooling can help prevent rupture in cases where water reactivity, electrical conduction, and pooling do not cause an additional hazard. Input from technical experts can help the IC/ISO make decisions regarding the appropriateness, placement, and GPM requirements of protective cooling streams.

Wrap-Up

Chief Concepts

- Hazardous energy is defined as the unintended, and often sudden, release of stored, residual, or potential energy that will cause harm if contacted. Forms of hazardous energy include electrical, mechanical, hydraulic, pneumatic, nuclear, thermal, gravitational, climatic, and natural sources like surf, streams, and large animals.

- The ISO needs to identify the presence of potential hazardous energy sources and make a judgment regarding their status: stable (not changing or likely to change) or unstable (predictable or unpredictable). Once a hazardous energy source has been identified and its stability categorized, appropriate control measures and zone isolation areas need to take place.

- Electricity is always seeking the path of least resistance to a ground, and electrical distribution systems are designed to transfer the flow from generation to the end user. The ISO needs to understand the components of a typical electrical distribution grid as well as those in the end-user building.

- When a component of the grid or end-user system is deemed to have lost integrity, the ISO must assume that an electrical danger exists and communicate that assumption to responders using an urgent or priority message. In some cases, the electrical system can be perfectly intact and still pose a threat to firefighters, particularly when aerial apparatus and aluminum ladders are being used near power lines and electrical equipment. Power can arc or jump through the air if a ladder or other piece of conductive equipment comes close to an exposed wire or power connection; a 10-foot (3-m) minimum clearance is required, and greater distances of up to 45 feet (14 m) may be required for higher voltage equipment and wires.

- Shock hazards are the greatest threat from electricity. Ground gradient is electrical energy that has established a path to ground through the earth and is energizing it. A firefighter walking on a charged surface may encounter step potential, wherein the difference in voltage between each leg causes electrical current to flow through the firefighter's body. A firefighter who feels tingling in his or her feet must back away using a shuffle-foot motion to keep both feet in contact with the ground.

- The primary method for reducing electrical shock threats is to shut off the electricity where possible. When that is not possible, shock hazards can be reduced by treating all wires as energized and keeping a 10-foot (3-m) clearance from the wires or any conductive material (including pooled water) that may be in contact with the wire or equipment.

- Substation transformer fires and incidents involving large wind farm turbines should not be engaged without technical assistance from the local electrical company or provider.

- Alternative energy systems can include uninterrupted power supply sources such as battery rooms and auxiliary generators that may provide power even though the grid power has been shut off by the fire department. Other alternative energy sources can include solar electrical and thermal panels: Photovoltaic (PV) panels provide DC electrical generation using light-collecting materials. Solar thermal panels use light to heat collection tubes that contain a fluid or gel that transfers heat to an exchanger for hot water use or radiant heat-producing tubes.

- Common utility gases include propane and natural gas. Both can be stored as a liquid—LPG and LNG. The primary hazard is found in their flammability. Storage vessels with LPG and LNG can become a BLEVE (boiling liquid expanding vapor explosion) when flame impinges on the vessel. Cooling streams of significant GPM can help prevent a BLEVE, although it is important to keep from extinguishing burning product escaping from pressure-relief valves. When high GPM flow is not available to prevent a BLEVE, immediate evacuation of the public and responders is warranted. Evacuation distances for large LPG and LNG storage tanks can exceed a mile.

- Damaged domestic water systems and overwhelmed storm sewer systems create hazardous energy in the form of powerful and uncontained flow that can cause collapses, sinkholes, and drowning of victims in its path.

- Mechanical hazardous energy includes tensioned wires, cables, springs, conveyers, and even high-rack storage (gravitational).

- Pressurized systems (hydraulic and pneumatic) typically include a reservoir for the medium, some sort of pump, tubing or a conveyance system, and the tool or implement that uses the pressure to achieve a task. Heat applied to a pressurized system can create pressures that exceed the system's design limits. Built-in pressure-relief devices may not be able to relieve pressure as fast as the pressure is being developed, leading to an explosive failure of a component.

- Vehicles contain many forms of hazardous energy, including fuel systems, heat generation, restraint systems (air bags), batteries, springs and struts, and the like. Of particular note are air bags that have failed to deploy after a crash; firefighters should treat this scenario as an immediate rescue situation for victims still in the vehicle. Other notable hazards include the evolving use of alternative fuel and energy sources (hybrid and electric vehicles). Responder safety training programs are gradually being developed to help deal with the hazards they present.

- Effective ISOs study weather and understand the particulars of weather patterns found in their geographic regions. Likewise, they keep abreast of daily forecasts and weather observations as a matter of habit and readiness. Wind is perhaps the most influential weather factor that creates hazards for firefighters, especially for structural firefighting. Winds above 10–20 mph (16–32 kph) can cause extreme fire behavior according to the latest studies. Many firefighters have been killed by wind-driven fire events in buildings.
- Other weather hazards include the potential for rapid weather change, tornadic activity, lightning, severe downpours, and hail. Free weather-spotter and severe storm recognition training is available from local news TV stations and from the National Oceanic and Atmospheric Administration.
- Miscellaneous forms of hazardous energy are numerous, but often forgotten are the hazards associated with animals and medical facilities. The evacuation of penned animals threatened by fire can be daunting. Likewise, the rescue of large animals may require technical expertise and the assistance of a large-animal veterinarian. In medical facilities you can find compressed gases, MRI equipment, and laser machines.
- Generally speaking, the ISO is responsible to develop preventive actions for all hazardous energy forms. The priority of actions includes awareness-level communications; avoidance-level zoning considerations; lock-out/tag-out; and preventive cooling.
- Control zones must be defined and, where possible, marked with color-coded barrier tape:
  - The no-entry zone is off limits to all responders due to the dangers present (red/white diagonal-striped tape).
  - The hot zone is an IDLH zone that requires full PPE, RIC, and accountability tracking (solid red tape).
  - The warm zone is a support zone for working crews, apparatus, decontamination, and RIC staging (yellow tape).
- The cold zone is where the command post is located; it also provides an area for staging unused equipment (green tape).

Key Terms

cold zone The area surrounding a warm zone that is used for an incident command post, support-agency interfacing, and staging of personnel and equipment.

ground gradient Electrical energy that has established a path to ground through the earth and that is now energizing the ground.

hazardous energy The unintended and often sudden release of stored, residual, or potential energy that will cause harm if contacted.

hot zone The area immediately surrounding a hazardous area that can be considered an immediately dangerous to life and health (IDLH) environment requiring appropriate PPE, accountability procedures, and a standby rapid intervention crew.

no-entry zone The area in which no responders are allowed to enter, regardless of PPE, due to dangerous conditions.

warm zone The area surrounding the hot zone where personnel, equipment, and apparatus are operating in support of an operation.

Review Questions

1. Define *hazardous energy*.
2. List four ways to categorize the status of hazardous energy.
3. Define *ground gradient* and *step potential*.
4. Why are UPS systems hazardous?
5. How can electrical shock hazards be detected, and what steps can be taken to avoid them?
6. List the communication steps that should be taken when anyone discovers an imminent energy threat.
7. List several forms of alternative energy and their associated hazards.
8. Name the two most common utility gases and list several hazards associated with them.
9. Describe the differences between a BLEVE and a PUVCE.
10. List the hazards associated with utility water and storm sewer systems.
11. Give examples of mechanical hazardous energy.
12. List the hazardous energy sources in vehicles.
13. What are the four considerations relating to weather as a hazardous energy that need to be evaluated?
14. List some warning signs that extreme weather is approaching.
15. List several sources of miscellaneous hazardous energy.
16. Describe the general protective actions that need to be taken when an imminent hazardous energy threat has been found.
17. Describe the methods used to establish and mark hazard zones necessary to protect members from potential hazardous energy sources.

References and Additional Resources

Backsteom, Robert, and David A. Dini. *Firefighter Safety and Photovoltaic Installations Research Project*. Northbrook, IL: Underwriters Laboratories, 2011.

Duke Energy. First Responder Beware: Electrical and Natural Gas Safety. 2012. http://www.dukesafety.com/firstresponders/. Accessed September 25, 2014.

Grant, Casey C. *Firefighter Safety and Emergency Response for Electric Drive and Hybrid Electric Vehicles*. DHS Assistance to Firefighter Grant Funded Study. Quincy, MA: Fire Protection Research Foundation, 2010.

Grant, Casey C. *Firefighter Safety and Emergency Response for Solar Power Systems*. DHS Assistance to Firefighter Grant Funded Study. Quincy, MA: Fire Protection Research Foundation, 2010.

National Institute for Occupational Safety and Health. *Electrical Safety: Safety and Health for Electrical Trades Student Manual*. NIOSH Publication No. 2009-113. Cincinnati, OH: U.S. Department of Health and Human Services, Public Health Service, Centers for Disease Control and Prevention. http://www.cdc.gov/niosh/docs/2009-113/. Accessed May 6, 2015.

National Institute of Standards and Technology. *Firefighting Tactics Under Wind Driven Fire Conditions: 7-Story Building Experiments*. NIST Technical Note 1629, Gaithersburg, MD: NIOSH, 2009.

Occupational Safety and Health Administration. *Control of Hazardous Energy (Lockout/Tagout)*. Title 29 CFR 1910.147. Washington, DC: U.S. Department of Labor. https://www.osha.gov/SLTC/controlhazardousenergy/. Accessed May 6, 2015.

The Alternative Energy website offers an introduction to alternative energy sources and new technologies and is available at http://www.altenergy.org.

The National Weather Service website contains many weather-related publications and is available at http://www.nws.noaa.gov/om/publications.shtml.

Your local power and/or gas utility company can offer electrical and gas equipment hazard information. Many offer free firefighter response training.

Endnotes

1 Duke Energy. First Responder Beware: Electrical and Natural Gas Safety webpage. 2012. http://www.dukesafety.com/first-responders/. Accessed September 25, 2014.

2 National Institute of Science and Technology, *Firefighting Tactics Under Wind Driven Fire Conditions: 7-Story Building Experiments*. NIST Technical Note 1629. Gaithersburg, MD: NIST, 2009.

The alarm for an electrical substation fire came in around 1300 hours on one of those days when you would rather stay indoors. The temperature was below zero, the wind was howling, and a stinging snow squall was just beginning. Upon arrival, crews found a large step-down transformer burning intensely. Exposures included the overhead power line grid, one other transformer, a cooling oil reservoir (approximately 500 gallons), and a small control building. The local power company field representative had just arrived, and the decision to delay the fire attack was made. It took 20 minutes to verify that the substation grid was deenergized and power rerouted. An overhead wire steel support structure was showing some signs of heat stress, and there appeared to be some heat discoloration of the oil reservoir. The power company field representative then asked the fire department to start cooling the exposures and the transformer core. The IC and the designated ISO conferred on the request with the power company representative. They discussed the following:

- The ground is frozen, and the drainage system built into the substation may not handle the cooling streams.
- The wind and temperature are likely to cause the stream spray to build up and form as ice on nearby high-tension wires, substructures, and towers. The structural integrity of these exposures was questioned.
- Failure to cool the intense fire may cause involvement of the second transformer, a boil over of oil in the reservoir, or collapse of the overhead high-tension wires and support structure.
- Power has been disrupted for several hundred users.
- The weather is expected to get worse.
- Two portable master streams flowing 500 gpm each are deemed sufficient to cool the transformer. Two large hand lines would be needed to cool the overhead wire support structure and the oil reservoir. There are adequate resources to launch the attack and an adequate water supply was nearby.

1. What are the hazardous energy forms present at this incident?

2. What firefighter safety issues should be addressed prior to fire attack?

3. How should the stability of this hazardous energy incident be categorized?
   A. Stable—Not likely to change
   B. Stable—May change
   C. Unstable—May require attention
   D. Unstable—Requires immediate attention

4. Which of the following represents the first action or priority when dealing with hazardous energy?
   A. Awareness communication
   B. Avoidance—zoning
   C. Lock-out/tag-out
   D. Protective cooling

Note: This case study is taken from an actual incident. After an additional 20 minutes of planning and a 10-minute setup, the fire was attacked. Extinguishment took 40 minutes. Ice buildup caused the overhead wires to sag, but did not collapse the steel support structure.

Flames: © Ken LaBelle NRIFirePhotos.com

Knowledge Objectives

Upon completion of this chapter, you should be able to:

- List the three factors that influence the chance of overexertion. (p 127)
- List the three ergonomic stressors that can produce injury and the three strategies to mitigate them. (pp 127–128)
- Describe the criteria used for heat and cold assessment of personnel at an incident. (NFPA 5.2.14 , pp 129–131)
- State the signs and symptoms of cardiac, heat, and cold stress. (NFPA 5.2.14 , pp 128–131)
- List the essential ingredients that help human cells perform. (p 131)
- Explain the role of hydration in preventing injuries. (pp 132–133)
- Define the conditions or activities that determine the need for rehabilitation strategies at an incident. (NFPA 5.2.14 , pp 135–138)
- Define the five elements that make up firefighter REHAB. (p 136)

Skills Objectives

Upon completion of this chapter, you should be able to:

- Demonstrate the ability to assess or develop a rehabilitation plan at an incident. (NFPA 5.2.14 , pp 135–138)

s your experience level has increased since your appointment to incident safety officer, you have gained valuable insight into your responsibilities. Even at incidents where building construction hazards are monitored, fire behavior factors are stable, and resources exceed the need, hazards still exist. Firefighters doing work can exhibit poor decision making or work beyond their physical and mental limits.

1. How many times have you worked yourself or witnessed firefighters working beyond the point of exhaustion?
2. What conditions or symptoms would you identify that would require you to stop a firefighter from operating at an incident scene?

Introduction: Is It Possible to Read Firefighters?

What is meant by *reading firefighters*? Essentially, we're talking about the ability to observe firefighters doing their thing and predict when they will make a mistake, exercise poor judgment (risk-taking), or exceed their abilities. So far in Section 2, we have focused on the need to front-load your knowledge so that you can rapidly apply knowledge at incidents through recognition and analysis (a skill). This last front-loading chapter can be viewed as the wildcard because predicting firefighter behavior is not easy given the many factors that can affect firefighter judgment while performing tasks. Let's look at this challenge more closely.

Think of an incident response in your fire service career where you made an honest human mistake that could have led (or did lead) to a serious injury. At some point after the incident, in a moment of introspection, you probably said, "What the heck was I thinking?" Believe it or not, the answer to the question is simple: You weren't. Most working firefighter judgment mistakes can be explained—even those that seem to defy explanation.

Human performance (including mental processing) is dependent on many factors. A breakdown in any of them can lead to decreased human performance—and mistakes. More often than not, overexertion causes mental (and physical) mistakes. We know that overexertion is the leading cause of injuries (and deaths) at incidents.[1] The heart of *reading firefighters* is the evaluation of factors that lead to overexertion and therefore injuries. Before we look at the prevailing overexertion factors, we have to disclose that we, the fire service, have actually set up firefighters to be injured. This may sound condemning, but I think you will accept the statement if you agree with the setup. Read on.

The Overexertion Setup

Consider the following statement: "Firefighters are the only professional athletes who need to achieve peak performance without warming up."[2] Few occupations stress the human body to the degree that firefighters experience in hostile, working-fire environments **FIGURE 9-1**. Further, firefighters are called upon

to perform such physical feats anywhere and at a moment's notice. Several celebrity athletes have even commented on the demands of the firefighter. The TV sports channel ESPN carried a special in which professional sports figures (some of whom are considered the most fit of all athletes) participated in a simulated firefighting task competition. Following the event, every interviewed athlete admitted that the physical challenges of firefighting were off the charts. They couldn't imagine doing the same thing in smoke, heat, or mentally challenging situations. We all know that the incident scene is not a competition, although the race against time occasionally presents itself as such. This is but one part of the setup.

Most firefighters accept that an incident requiring peak performance can happen at any time, although we structure our lives around the typical daily routine. Our chores, daily assignments, meals, workouts, and rest periods are mostly defined, and then we throw in an incident response when called upon. We accept this way of life, yet it is part of the setup. If you just ate, your body may not be able to operate at peak performance. If you are between meals and hungry when "the big one" hits, you will struggle to perform at peak. If you feel a cold coming on, if the alarm awakens you from a deep slumber, if you just finished a tough workout—you get

FIGURE 9-1 Few occupations stress the human body to the degree experienced by firefighters.

© Kim Fitzsimmons. Used with permission.

the point. These situations lead to rapid overexertion of the body and decreased performance of human cells—including brain cells! The conclusion to our earlier question stands: You weren't thinking, because your cell performance was stressed.

Some believe that the fire service cannot change the reality that incidents occur while we sleep, when we are hungry, or when we are least capable of performing. However, the essence of fire service professionalism is the ability to function safely at *all times*. Therefore, it is incumbent upon us to address the physical and mental demands on firefighters, regardless of *when* an incident occurs. Remember Chief Alan Brunacini's assertion: *"For 200 years we've been providing a service at the expense of those providing the service."* If we are truly dedicated to preventing injuries and deaths, we need to address the number one cause of injuries—overexertion. To prevent overexertion, and to help firefighters stay alert, strong, and clear-headed—even at those inopportune times—an understanding of the proactive and real-time overexertion influences is required.

■ Overexertion Resistance

The study of human performance and, more recently, of firefighter performance is well documented in texts, trade magazines, and medical journals. One conclusion is reached by all: Firefighting requires humans who are physically fit. Those who are not fit for duty have decreased overexertion resistance, and they present an injury risk to themselves as well as to others due to the team nature of firefighting. Conversely, the physically fit firefighter can resist overexertion longer, with minimal injury risk **FIGURE 9-2**. Nevertheless, at some point both the fit and the unfit firefighter can become fatigued, and regardless of their levels of fitness, both types of fatigued firefighters create an equal injury risk.

Enough cannot be said about the need for firefighters to proactively keep themselves strong, flexible, and aerobically fit. Efforts to improve an individual's strength, flexibility, and aerobics to help prevent overexertion at incidents can be defined as work hardening. We know that some incidents will require a high level of aerobic fitness (i.e., fitness relating to the body's efficient use of oxygen). Thus, firefighters need to include duration training in their work-hardening regime (30–40 minutes of running, cycle spinning, stair climbing, and the like). Other incidents require anaerobic fitness (i.e., fitness relating to rapid bursts of muscle and oxygen use). This type of fitness is achieved through heavy resistance training like weightlifting and core strengthening. Both forms of work hardening (aerobic and anaerobic) should include "warm-down" stretching activities to improve flexibility. A complete work-hardening regime should also include occasions in which individuals perform their "workout" in the context of incident handling; that is, they wear their full personal protective equipment (PPE) ensemble and use the tools of the trade while exercising.

Work hardening is an important part of overexertion resistance, as is efficient *fueling*. Firefighters should adopt a program of efficient fueling (feeding) that supports their metabolism. Like a professional athlete, firefighters need to dial in their individual diet needs to best support cell performance and their work-hardening efforts. Those who know how their bodies respond to stress and how to immediately address metabolic deficiencies perform better and help prevent injuries. Again, this kind of preparation is proactive prevention, and all efforts to achieve this level of preparation can help in the overall firefighting effort. Still, incidents occur in which even the most fit firefighter has not adequately warmed up or is not in a perfect cell state to perform. That is reality.

■ Overexertion Realities

The diversity in physical performance capability (and incapability) of firefighters presents difficult challenges. At any given incident, some firefighters are operating at peak physical and mental performance, while others are ready to succumb to an overexertion injury. Incident safety officers (ISOs) must deal with "here-and-now" overexertion threats to firefighters regardless of their proactive work-hardening efforts. ISOs can, however, make a difference in the way they address human overexertion factors. A solid understanding of overexertion influences can help ISOs quickly identify when they are adding to injury potential. With this foundation, they can read firefighters and help implement strategies to prevent overexertion.

FIGURE 9-2 The physically fit firefighter can resist the chance of overexertion longer but will sometimes still succumb.

© Jones & Bartlett Learning. Photographed by Glen E. Ellman.

The three stressors that influence overexertion are related to the environment, human physiology, and the quality of rehabilitation (rehab) efforts.

Environmental Stressors

The onset of overexertion is often influenced by the environment where incident activities are occurring. The environment presents ergonomic and thermal (hot and cold) challenges. Let's start with the ergonomic challenges.

Some define ergonomics as the science of adapting work or working conditions to a worker. Others define it as the study of problems associated with people adjusting to their work environment. In the fire service, both definitions are appropriate. The firefighter's workplace includes the fire station and apparatus (which can be engineered to be ergonomically friendly) and the incident scene (which creates challenges that are ergonomically problematic). The cold reality is that firefighters don't have the luxury of having an "ergonomically friendly workplace" at an emergency scene **FIGURE 9-3**. However, we can minimize injury potential by understanding ergonomic stressors and the strategies for abatement.

■ Ergonomic Stressors

Most of the activities at an incident require some sort of physical labor. Scientifically, this labor translates to human kinetics—muscular and skeletal motion—and most agree that firefighting activities require a stressful range of muscle and skeletal motions. The *aware* ISO evaluates those stresses

FIGURE 9-3 Incident scenes are rarely "ergonomically friendly."

and, when appropriate, calls a "time-out" to adjust the way firefighters have chosen to tackle the situation or to simply give them a break. At times, the awareness is simply common sense. More professionally, the awareness comes in the form of an equation:

Physical setting + Worker relationship to setting + Task requirements = *Ergonomic stress*

Each of the following ergonomic stressors should be evaluated to determine whether an injury potential exists. Often, a slight change in any of these areas can reduce injury potential.

Physical Setting

Understanding stressors related to the physical setting involves examining the surface conditions of the working area. Obviously, secure footing is essential to safe operations. Temperature variations, close proximity to equipment, and lighting integrity also need to be included here. Often, the setting stressor has to do with the space size or geometry of the work location: Anyone who has had to remove a large unconscious person from a tiny bathroom in a single-wide mobile home can attest to the ergonomic stressors involved. Distractions are also part of the work setting: noise, flashing lights, swinging mechanical equipment, and weather extremes.

Worker Relationship to the Setting

The relationship of the worker to the setting has to do with body mechanics that are being employed. Workers might be forced to tackle a task while bending down, standing, crawling, or stooping. Other times they must ascend or descend to accomplish a task. Pulling, twisting, and pushing to accomplish an objective are other relationships that may cause injury. The speed or pace at which the task is completed may also create an injury relationship. In some cases, the relationship of the worker is size dependent, which explains why responder physical diversity is viewed as a positive: Each department needs the big, robust firefighters as well as smaller and more agile ones.

Task Requirements

It is essential to consider the energy required to accomplish a task, as well as the amount of focus or attention. Sometimes, the task priority or immediacy is such that it creates additional stress. In other cases, the limited number of people (staffing) available to accomplish the task is the primary factor leading to injury. The types of tools and equipment necessary to accomplish the task also need to be considered.

■ Ergonomic Abatement Strategies

Once the physical setting, worker relationship to the setting, and task requirements have been evaluated, certain hazards become obvious. At that point, the ISO can use one or more of the following three strategies to abate or mitigate the hazard. These strategies are awareness, accommodation, and acclimation.

Awareness

The awareness strategy acknowledges that the workers are less apt to suffer an injury if they are aware of the problem, thereby heightening their cautiousness. Awareness is perhaps the most used abatement strategy and certainly the most simple. Examples include the warning of a slippery surface or a reminder to lift with a tight core and leg leverage.

Accommodation

Used here, accommodation includes efforts to alter or adjust the environment, worker relationship, or task to reduce injury potential. The use of PPE is accommodation. Placing a roof ladder on a pitched roof is a method of accommodating the worker in an environment that requires ascending and descending on a pitched surface (and the added benefit of distributing the impact load). Adding more people to help perform a task is a form of staffing accommodation. Artificial lighting is an accommodation to darkness. An obvious, but often underutilized, accommodation strategy is to reduce the pace or speed of an activity. Having a water bottle available at each apparatus riding position reminds firefighters to *prehydrate* while en route to a reported incident, another accommodation strategy.

Acclimation

This is the most difficult strategy to implement during an incident, in that most acclimation is done proactively and is typically gradual. Acclimation can be defined as an individual's gradual process of becoming accustomed to an environment. Physical fitness programs that include strength, aerobic, and flexibility training for work hardening are examples of proactive acclimation. Likewise, live-fire training and in-context drills are forms of proactive acclimation. At an incident, the incident commander (IC) or ISO may choose to have a crew practice a skill set in a safe area before members are assigned to an ergonomically challenging area. Hazardous materials technicians are known to employ this strategy when dealing with new or unusual equipment at an incident. Firefighters who have a psychological or physiological response to heights (acrophobia) can sometimes benefit from acclimation by slowly ascending and descending to various heights and spending some time getting used to the view.

Another underused acclimation strategy at an incident is to have crews perform simple stretching and warm-up exercises prior to an assignment. Although this may feel awkward, consider the statement that we are the only athletes who need to perform at peak with no warm-up. While first-due responders may have to immediately jump in and perform, there is a compelling argument that staged firefighters should do some sort of warm-up. Incidents that require crew rotation or defined work periods are perfect candidates for preassignment warm-ups (e.g., wildland incidents, hazardous materials incidents, lengthy technical rescues). The investment in an acclimation warm-up can prove very effective in preventing strains/sprains and other overexertion injuries!

■ Thermal Stressors

In the *Reading Hazardous Energy* chapter we discussed weather as a hazardous energy source, impacting fire behavior as well as firefighter safety. Here, we will more specifically detail the physiological impacts of the weather and other hot/cold influences (thermal stress) on firefighters. Thermal stress can come in the form of hot and cold (temperatures). Heat stress can be further divided into external (environmental) heat and internal (metabolic) heat. We know that the compartments of a building on fire can present significantly elevated environmental temperatures that we try to control either through ventilation or cooling. In doing so, we raise our internal metabolic temperatures through physical labor.

Exposure to hot and cold can be influenced by several environmental factors, including:

- Ambient air temperature
- Humidity
- Wind speed
- Thermal radiation (flames and reflected heat energy from surfaces)
- Direct sun exposure
- Wetness (rain, snow, fire streams)

Each of these influences can accelerate the way heat and cold stress the human body. The danger of thermal stress comes when the human body core temperature deviates from its ideal or normal range. The human body is amazing with its many systems to regulate a normal (average) core temperature of 98.6°F (37°C). For most physically fit adults, a deviation of one or two degrees can be tolerated without risk. Thermal stress and the chance of injury elevates exponentially outside of this range. Although medical literature varies, most agree that core temperatures above 101°F (38.3°C) or below 96°F (35.6°C) will start to affect physical performance—including reduced mental alertness or decision-making capability. There are some individuals who can tolerate a larger range, but most are affected outside this range.

In addition to the preceding environmental influences, the ability of an individual to maintain his or her core temperature is influenced by the following:

- Activities performed (metabolic heat)
- Duration of exposure to hot or cold
- PPE type
- PPE condition (wet, dry, wear and tear)
- Hydration level of the individual
- Physical condition of the individual (fit/unfit, body fat content)
- Acclimation to local norms
- Miscellaneous issues, such as nourishment, medications, illness, fatigue, cardiac risk factors

Heat Stress

The primary mechanism to prevent heat stress is sweating (evaporative cooling). Full structural protective clothing reduces the body's ability to evaporate heat. Likewise, high humidity reduces the benefit of sweat evaporation. The ability of the body to maintain its core temperature decreases as the worker becomes dehydrated through perspiration. Working in direct sunlight adds heat stress, especially at higher altitudes (exposure to more ultraviolet rays). Firefighters suffering from heat stress go through a series of heat-related injuries that grow progressively worse as heat builds.

Although it is important to know the different types of heat stress injuries and their warning signs/symptoms, it is more important to predict and prevent them. The ISO can use criteria from the National Oceanic and Atmospheric

Administration's (NOAA's) heat stress index and exposure charts to help assess the threat of heat stress injuries **TABLE 9-1 and 9-2**. The charts are simple to use and should be referenced by the ISO when heat stress factors are present (many ISOs keep the charts in their field kits for quick reference). If nothing else, they provide a reminder to accelerate crew rotation cycles and improve rehab efforts (employ active cooling remedies and hydration).

Predicting heat injuries using the NOAA charts is a great start to reducing the chance of injuries. However, during interior structural fire operations, the ISO should *presume* that

NOAA's National Weather Service
Heat Index
Temperature (°F)

Relative Humidity (%)	80	82	84	86	88	90	92	94	96	98	100	102	104	106	108	110
40	80	81	83	85	88	91	94	97	101	105	109	114	119	124	130	136
45	80	82	84	87	89	93	96	100	104	109	114	119	124	130	137	
50	81	83	85	88	91	95	99	103	108	113	118	124	131	137		
55	81	84	86	89	93	97	101	106	112	117	124	130	137			
60	82	84	88	91	95	100	105	110	116	123	129	137				
65	82	85	89	93	98	103	108	114	121	128	136					
70	83	86	90	95	100	105	112	119	126	134						
75	84	88	92	97	103	109	116	124	132							
80	84	89	94	100	106	113	121	129								
85	85	90	96	102	110	117	126	135								
90	86	91	98	105	113	122	131									
95	86	93	100	108	117	127										
100	87	95	103	112	121	132										

Likelihood of Heat Disorders with Prolonged Exposure or Strenuous Activity
☐ Caution ☐ Extreme Caution ■ Danger ■ Extreme Danger

TABLE 9-1 NOAA Heat Stress Index Chart
Courtesy of National Oceanic and Atmospheric Administration/Department of Commerce.

Awareness Tip

Heat Stress Injuries and Warnings

Heat stress injuries can get progressively worse as heat exposure and core temperatures rise. Here is a look at some early warning signs of heat stress and the progressive nature of injuries that can occur.

- Transient heat fatigue (THF). THF is not a true injury; rather, it is an early warning sign that the core temperature is elevated. It is characterized by the onset of physical exhaustion that is remedied by rest and hydration only to return quickly and more profoundly upon engagement with the hot environment. Thermal radiation from flames and hot surfaces is the most common cause of THF. THF is a signal that active core cooling remedies are needed.
- *Heat rash.* The development of a heat rash is another warning sign that active cooling remedies are needed. Heat rash is characterized by blotches of red skin or red bumps, especially around the neck, cheeks, underarms, and groin.
- *Heat cramps.* This is a true injury marked by painful muscle spasms. In simple terms, it is a warning that the body's electrolytes are not balanced. Dehydration may also play a role in heat cramps. Medical attention is warranted.
- *Heat exhaustion.* Progressively worse than heat cramps, heat exhaustion is actually an early form of hypovolemic shock due to the loss of water and electrolytes through sweating. It is characterized by cool and clammy skin that may be ashen or gray and may be accompanied by dizziness, nausea, and/or headaches. Medical attention is warranted.
- *Heatstroke.* Heatstroke is the most serious heat stress injury and is a result of a failure of the body's cooling mechanisms. Heatstroke is often characterized by a change in behavior (decreased mental alertness) and, in some cases, hot, dry, flushed skin (late sign). Heatstroke is considered a medical emergency, and rapid transport to a hospital is mandatory.

TABLE 9-2	Heat Exposure Injury Risks
Apparent Temperature	**Heat Stress Risk with Physical Activity and/or Prolonged Exposure**
90–105°F (32–41°C)	Heat cramps or heat exhaustion possible
105–130°F (41–54°C)	Heat cramps or heat exhaustion likely, heatstroke possible
130°F (54°C) and up	Heatstroke highly likely

Note: Add 10°F (~6°C) to the apparent temperature when working in structural PPE or when working in direct sunlight.

firefighters have elevated core temperatures, even on days when outside temperatures are comfortable. The presumption is based on the fact that the interior of the building is likely to have elevated ambient temperature, radiant energy, and humidity levels—and firefighters are working in PPE that doesn't readily dissipate metabolic heat. Studies by the International Association of Fire Fighters (IAFF), Toronto Fire Services, and Orange County (CA) Fire Authority have all shown that firefighters engaged in operations wearing full structural PPE will have elevated core temperatures, averaging over 101.5°F (39°C) after 20 minutes (one SCBA bottle) of activity.

Typically, firefighters have used passive cooling to reduce core temperatures during periods of rehab. Passive cooling includes the use of shade, air movement, and rest to bring down core temperatures. Left to their own judgment, many firefighters use perceived comfort to determine when they are ready to go back to work, leading to THF. When used, paramedics and EMTs assigned to rehab functions have typically monitored firefighters' heart rates and their perceived comfort to determine when they have recovered sufficiently enough to don PPE and resume firefighting operations. New studies have shown that this approach is *not* adequate and that heart rate and perceived comfort are not good indicators of sufficient core temperature cooling.[3]

Fire Marks

Active Cooling: The Orange County Fire Authority Study

Active cooling includes the use of external cooling methods to help lower an individual's core temperature and, thus, the chance of thermal stress injury. In 2007, the Orange County (CA) Fire Authority conducted a series of drills to better understand the performance implications of hydration, core temperature, and incident rehab practices. The study included over 100 participants, representing a wide range of ages, body types, fitness levels, and experience. Participants swallowed a capsule that allowed the researchers to document baseline, active firefighting, and recovery body core temperatures. Firefighters performed a live-fire interior attack with simulated victim rescue (15 minutes) along with another 15 minutes of fire ground support activities (e.g., stairs, overhaul tool work). This was followed by a 20-minute rest period outdoors with temperatures in the 80s and roughly 40% humidity.

The data collected from the study confirmed some long-held beliefs but also revealed the need to include active cooling as part of rehab when engaged in structural firefighting. Some highlights from the report include the following:

1. The average core temperature following the drills was 101.8°F (38.8°C); 44% of the participants had readings over 102.1°F (38.9°C), with a high of 106°F (41.1°C).
2. After five minutes of rest, the average core temperature continued to rise.
3. With no intervention, the average core temperature after 15 minutes of rest was 101.1°F (38.4°C).
4. The core-temperature capsule results were compared to a tympanic measuring device. Tympanic readings were 1.3°F lower than the capsule just after the drill but dropped to 3.8°F lower after 5 minutes of rest.
5. Cooling remedies tested included passive cooling (PPE removal), misting fan, cold/wet towels, and forearm immersion. The average drop in core temperatures after 20 minutes (as measured by the capsule) were:

- Passive = 0.75°F
- Misting fan = 0.88°F
- Cold/wet towels = 1.18°F
- Forearm immersion = 1.22°F

As you can see, active cooling (towels/forearm immersion) is 50–60% more effective than passive cooling. Further, the use of tympanic measuring devices may not be an effective tool for measuring the effectiveness of cooling methods. Use of misting fans, previously considered an active cooling technique, is marginally better than PPE removal. Using misting fans in high-humidity environments will further reduce their effectiveness. The study also measured heart rates, hydration levels, and other data to help understand the physiological impacts of structural firefighting as well as rehab strategies and should be considered a mandatory reading for the ISO.

The final report can be found at the Orange County Fire Authority website (http://www.ocfa.org) or as part of a larger study captured in a National Fire Academy Executive Fire Officer Program report titled *Firefighter Rehabilitation in the Orange County Fire Authority: Are We Meeting the Need?*

If we adopt the assumption that all firefighters working in full structural PPE have elevated core temperatures, and that any core temperature over 101°F (38°C) increases injury risks (especially mental judgments), then it makes sense to use active cooling strategies as part of rehab. Active cooling is the process of using external methods or devices (e.g., hand and forearm cold water/ice immersion, covering the head and neck with cold wet towels, and gel cooling vests) to reduce elevated body core temperatures. Several firefighter-involved studies have shown that active cooling is *best* achieved using a forearm immersion technique or the use of cold, wet towels on the head and neck (see the *Fire Marks* box).

Cold Stress

Cold stress is similar to heat stress in that a series of injuries can occur if the body's core temperature cannot be maintained (see the *Awareness Tip* box). In this case, however, the core temperature is being reduced or cooled below desirable levels. Moisture in the form of perspiration and wetness from fire streams becomes an enemy that wicks away heat for those working in frigid environments. The body reacts to cold exposure by burning more calories to stimulate metabolic heat; muscles begin to shiver. Consciously, a cold person is likely

Awareness Tip

Cold Stress Injuries and Warnings

Like heat stress, cold stress injuries can get progressively worse. Here is a look at some early warning signs of cold stress and the progressive nature of injuries that can occur.

- *Shivering.* Sudden exposure to a frigid wind will cause almost all humans to shiver. This is a natural reaction and not an injury per se. When shivering is not relieved by warmth or is uncontrollable, it could be a warning of a true cold exposure injury.
- *Frostnip.* The combination of humidity, sweating, and fire stream exposure combined with frigid wind (wind chill) can lead to frostnip, which is a local injury usually involving the fingers, nose, ears, and toes. Most people do not realize they have frostnip (no pain), which can be recognized by pale skin and loss of sensation. Rewarming is usually all that is necessary. Delayed rewarming or prolonged exposure to the cold can lead to frostbite.
- *Frostbite.* Frostbite is the actual freezing of skin tissue, and the severity is gauged by how deep the frozen tissue is. It can be characterized by cold, hardened skin and may appear as a whitish or bluish-white color. Medical attention is warranted.
- *Hypothermia.* As the name implies, hypothermia is the lowering of the body's core temperature. The onset can be rapid, especially in wet conditions. Once a core temperature drops below 95°F (35°C), symptoms and signs develop that indicate the need for medical attention. Uncontrolled shivering, actions to stimulate warmth (stomping, jumping up and down), and a loss of concentration are early warnings of mild hypothermia. As the core temperature drops further, there may be loss of coordination, confusion, lethargy, and cardiac arrhythmias. Immediate medical care and transport is required for hypothermia.

to engage in some kind of physical activity (like jumping up and down) to stimulate metabolic heat. If the cooling is not stemmed, the body will begin to vasoconstrict vessels near the skin in an effort to preserve core heat. The reduced blood flow to these peripheral tissues increases the risk of local cold stress injuries (e.g., frostnip and frostbite of ears, nose, hands, feet, exposed skin). Again, if the cooling is not reversed, the core temperature decreases to the point that vital organs start to shut down—a hypothermic medical emergency.

Firefighters are often caught off-guard when it comes to cold stress. Consider a crew that has launched an aggressive interior fire attack at a building on a very cold night. Once inside, the thermal radiation, PPE, and their metabolic heat combine to raise the core temperatures of the crew. As they rotate outside to change bottles, the frigid air may actually feel good, at least initially. Their core temperature comes back down, although the combination of the sweat evaporating and the frigid air quickly tricks their hypothalamus. The body reacts and begins to burn calories to produce heat. As they prepare to reengage, their core temperature is already rising just as they reenter the hot environment, sending another mixed signal to their heat-regulation system. They start sweating again, and as Murphy would have it, the crew is ordered to go defensive. They retreat to the cold in a dehydrated, fuel-depleted state—an insidious equation that allows cold stress to rapidly take hold.

Predicting cold stress injuries is also aided by a NOAA chart, and the ISO can use it for assessing the cold stress potential. The wind chill chart is a bit different than the heat stress index in that humidity is replaced on the matrix with wind speed **TABLE 9-3**. Exposure time to frostbite is based on uncovered,

dry skin. Wet clothing combined with the cold air and wind will accelerate frostbite and likely lead to hypothermia.

■ Fighting Thermal Stress

The prevention of thermal stress injuries can be accomplished through accommodation, rotation, and hydration.

- Accommodation starts with efforts to remove the firefighters from the environment that is creating the thermal stress. The accommodation effort continues with the use of forearm cold water submersion and/or cold towel covering for heat stress. For cold stress, body warming efforts and drying of PPE are necessary.
- Rotation is the planned action to rotate crews through rehab, heavy tasks, and light tasks to minimize the stress caused by working in extreme environments.
- Hydration cannot be overemphasized in heat stress environments, but it is also effective, and often forgotten, in cold stress environments (covered in greater detail later in this chapter).

Physiological Stressors

Physiological performance depends on the metabolic processing (cell chemistry) of the firefighter. If the cell chemistry of a firefighter is not functioning optimally, the risk of overexertion increases. Since overexertion is the leading cause of firefighter injuries and deaths, it makes sense that we do all we can to prevent it. We've already discussed the importance of being physically fit and the impacts of thermal stress on firefighters. Here, we discuss the relationship of firefighter performance (alertness, decision making, meeting physical demands) and human cell performance (performance of muscles, tendons, ligaments, brain, heart, lungs, blood). Remember that introspective question after a mistake: *What was I thinking?* And the reason you weren't: In most cases, you weren't thinking because your (brain) cell performance wasn't optimal. If we understand what contributes (and distracts) from optimal cell performance, we can help prevent those mistakes and injuries.

With all due respect to the medical community, cell performance is really as simple as providing the cell with three essential ingredients: oxygen, water, and fuel. The cell converts those ingredients into some form of energy (muscles activate, brain reasons, etc.) as well as heat and several other by-products. The human body includes a vast array of systems to help make the cell ingredient delivery and by-product

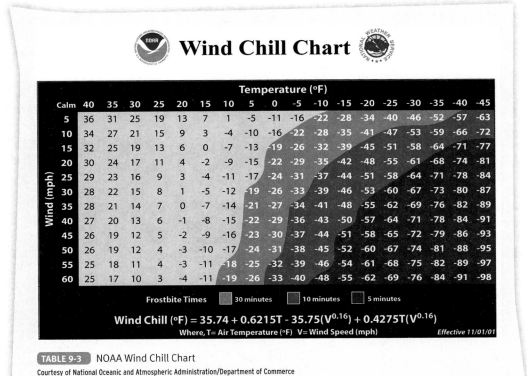

$$\text{Wind Chill (°F)} = 35.74 + 0.6215T - 35.75(V^{0.16}) + 0.4275T(V^{0.16})$$

Where, T= Air Temperature (°F) V= Wind Speed (mph)   Effective 11/01/01

TABLE 9-3 NOAA Wind Chill Chart
Courtesy of National Oceanic and Atmospheric Administration/Department of Commerce

removal process work. A quick, simple review of those systems is in order before we get to the cell ingredients:

- *Lungs.* Oxygen intake and carbon dioxide removal
- *Stomach.* Fuel/water intake and preparation
- *Digestive tract.* Transfer of prepared fuel/water to the cell supply chain, and recycling, storage, or removal of unused supplies
- *Heart.* Pump to keep the supply chain moving
- *Blood.* Cell ingredient and by-product transport medium
- *Arteries/veins.* Pathways for cell ingredient and by-product movement
- *Nervous system.* Communication system that links all components of the supply chain

This list is an oversimplification of the components involved in managing the essential cell ingredients, but a more in-depth understanding is likely unnecessary in most critical situations. The key is understanding that the nature of firefighting challenges the demand and supply of those ingredients, as well as the removal of by-products. If we dig a little deeper into each ingredient (using firefighter language as opposed to medical language), we can better prepare the ISO to read firefighters and thus prevent death, injuries, and mental lapses.

■ Oxygen

Oxygen and fuel combine in the cell to create energy. Voluntary and involuntary demands on the cell change the metabolic rate. Naturally, the higher the demand, the more oxygen/fuel is needed, causing an increase in breathing and heart rates. Cell performance is diminished if the demand can't be met. The aerobic component of work hardening improves the body's ability to meet these demands through acclimation. Those who haven't acclimatized will stress cell performance because the heart–lung relationship has not been strengthened.

The fire ground—namely the products of combustion from the fire—also stresses oxygen intake and can impact the heart. Carbon monoxide and hydrogen cyanide lead the list of hundreds (if not thousands) of contaminants that interfere with oxygen intake, although the wearing of SCBA during firefighting and overhaul makes this risk a nonissue. The reality is that carbon monoxide and hydrogen cyanide levels may still be at elevated levels outside the burning building, and the diligent ISO will ensure the correct setup of hot/warm and cold zones to alleviate this hazard. Both are clear gases (you don't know you're breathing them), and both are known to have cumulative destructive effects on the heart and lungs. Firefighters with existing or developing heart and lung ailments are especially at risk. Although studies continue, most agree that the firefighter cardiac events that occur on the fire ground are the result of a preexisting heart/lung condition *plus* an increased voluntary or involuntary demand on the heart/lungs *plus* the onset of dehydration *plus* the exposure to carbon monoxide and hydrogen cyanide. It is this final exposure to toxic gases that can be said to actually trigger the cardiac event.

■ Hydration

Water is vital to the peak operation of nearly every body system, from the transport of nutrients, to blood flow, to waste removal, to temperature regulation. Firefighters engaged in strenuous physical activity while wearing PPE ensembles can sweat out over a quart of water per hour. In addition, the body is losing water through waste removal and breathing. When the body becomes dehydrated, cells become stressed, leading to fatigue, reduced mental ability, cramps, hyper- or hypothermia, and, in extreme cases, medical emergencies such as renal (kidney) failure, shock, cardiac irritation (inadequate stroke volume), and death. The hydration of all firefighters should be paramount. In fact, firefighters need to prehydrate, hydrate, and rehydrate as a matter of routine for working incidents **FIGURE 9-4**. It is interesting to note that many firefighters (and the general population) are mildly dehydrated on a daily basis. Proactive measures to stay hydrated include the hourly consumption of water as a daily routine and the monitoring of urinary output (near-clear urine is ideal). Another proactive hydration suggestion is for firefighters to have a filled water bottle at their apparatus riding position to consume water while enroute to a potential working incident.

As a rule, firefighters should strive to drink a quart of water an hour during periods of work; this is best delivered in 4- to 8-ounce (120- to 240-mL) increments for every 15 to 20 minutes of exertion.[4] Substituting soda or other liquids for water can slow the absorption of water into the system. For this reason, water is the beverage of choice for initial incident operations. When the physical demands of an incident are prolonged, hydration with water alone is not adequate. Sweating firefighters must replace essential electrolytes (namely potassium and sodium). Failure to replace electrolytes will lead to a chemical imbalance at the cell level, which leads to overexertion and the accompanying risk of injury and mental lapses. Those electrolytes help keep the cell walls isotonic to blood and the conductivity of the nervous system sound; therefore, failure to replace electrolytes can irritate the heart.

The consumption of sports drinks (SDs), like Gatorade, is perhaps the easiest way to replace electrolytes at a working incident, although electrolytes can be found in many nutritional foods (especially green vegetables). Most of the available SDs have been scientifically developed to provide a balance of essential electrolytes, carbohydrates (less than 10%), and water (hydration) that have an osmolality that is similar to blood,

FIGURE 9-4 Working firefighters need to prehydrate, hydrate, and rehydrate early and often at working incidents.

meaning they can be more readily absorbed and are isotonic to cells. Most literature suggests that hydration strategies should start with water for the first hour of work followed by SDs after the first hour. A higher heat stress index (more sweating) suggests that SDs should be introduced sooner, but after initial water hydration. Likewise, SDs should not totally replace water but should merely be a component of the hydration plan.

It is important to note that SDs should not be confused with the increasingly popular energy drinks (EDs) like Red Bull or Rock Star. Most EDs contain a high concentration of sugar and caffeine as well as various herbs, vitamins, and extracts. This combination (and concentration) slows the absorption of the beverage into the bloodstream and can be compounded with stomach irritation for a dehydrated person. Additionally, EDs can have the detrimental effect of reducing intracellular hydration as well as increasing urinary output, blood pressure, and heart rate, according to many medical and fire service studies. As a result, some fire departments, and incident management teams, have banned ED consumption from the workplace.[5]

Coffee is a staple of the firehouse and many peoples' lives. Most know that coffee is a diuretic (increases urinary output and dehydration), although many argue there is compelling evidence that a single cup of coffee can actually improve mental alertness and physical performance. Both arguments are supported by science. The issue here is in acclimation. Those who routinely drink one or two cups of black coffee in the morning do benefit (psychologically and physiologically) from the beverage and there will be negligible dehydrating effects as long as water is also consumed. The addition of sugar and cream to coffee starts to eliminate the caffeine benefit and slows absorption. The point here is not to support coffee consumption for working firefighters but merely to acknowledge that a single, daily cup of black coffee isn't going to throw off one's body chemistry. Plus, there is the enjoyment of a cup of coffee during that predawn extended incident! Remember to keep the volume down and continue to drink water and electrolytes if you've been sweating.

■ Fuel Replacement

While dehydration and thermal stress can lead to energy depletion, most firefighters associate energy depletion with the need for food. The nutrition, or "fueling," of firefighters can either rejuvenate them or put them to sleep. Too often, rehab feeding efforts accomplish the latter effect. The ISO should view firefighter fueling from the perspective that a properly nourished firefighter works smarter and safer. The *improperly* fed firefighter not only experiences a "crash" but likely makes sluggish mental calculations, leading to injury. (Remember that question, *What was I thinking?*)

So what is the proper way for firefighters to nourish themselves? A study of essential nourishment theory can answer this question. Remember, however, that individual metabolisms are diverse. Metabolic rates are influenced by lifestyles, fitness, illnesses, over-the-counter and prescription drugs, and circadian rhythm. Circadian rhythm refers to a person's physiological response to the 24-hour clock, which includes sleep, energy peaks, and necessary body functions. The following explanation of cell fueling is meant to maximize cell performance for hardworking responders at an incident.

Cell-Fueling Theory

For a cell to work efficiently, it must use a balance of the three essential ingredients: oxygen, water, and fuel. While the meaning of *oxygen* and *water* is self-explanatory, *fuel* is a more complex concept to understand. The fuel component is glucose (from food), balanced and made usable by insulin. When this balance is present, other essential hormones and enzymes are created that make for a well-running human machine. True, *all* food is fuel for the firefighter. That is, all carbohydrates from food are converted to glucose (a sugar) when the muscles and brain are being worked hard. Insulin is released into the bloodstream to help regulate the glucose use. The entry rate of carbohydrates into the bloodstream is known as the glycemic index. A high-glycemic-index food source quickly spikes blood sugar (and the resulting insulin surge). The sugar is rapidly burned at the cell level, leaving too much insulin in the system, which is telling the body to store carbohydrates, not use them. This is dangerous! It makes sense, then, to fuel firefighters with a low-glycemic-index food so that the blood sugar levels and insulin are stable, gradual, and consistent TABLE 9-4 . The glycemic rate is but one part of the equation. The cell-fueling strategy must also include other essential elements to keep energy levels high.

Cell-Fueling Strategy

The key to providing quick energy that optimizes cell performance is to consume a balance of foods from the three food

TABLE 9-4	Glycemic Index Comparisons for Common Carbohydrates
Carbohydrates with a High Glycemic Index (Avoid)	
White sandwich bread	Pizza crust
Hamburger/hot dog buns	Durum spaghetti
Rice cakes	French fries
White flour tortillas	Chips/pretzels
Orange juice	Most cereals
Watermelon	Cookies
Grapes	Cakes
Raisins	Donuts
Bananas	Bagels
Potatoes	Pastries
Carrots	Macaroni
Pumpkin	Most candy
Sweet Corn	
Carbohydrates with a Low to Moderate Glycemic Index (Better Choices)	
Most vegetables	Non-instant oatmeal
Apple juice	Oat bran breads
Apples (whole)	Mixed whole-grain breads
Oranges (whole)	Whole-grain tortillas
Pears	Low-fat yogurt
Peaches	Soups without pasta
Most beans	Most nuts
Milk	

groups of low-glycemic carbohydrates, protein, and fat. Ideally, this balance should be a 40/30/30 mix—that is, 40% carbohydrate, 30% protein, and 30% fat.[6] Some experts argue that the 40/30/30 balance is a bit weak on carbohydrates. The argument is fair, especially for athletic firefighters who are accustomed to demands of their workouts and physical activities. Simply stated, their metabolic rate is high and very efficient at utilizing carbohydrates. Whether it is 40% carbohydrates or 50% carbohydrates, a balanced approach achieves the following benefits:

- The low-glycemic carbohydrates stabilize insulin release into the bloodstream, helping to reduce the roller coaster of blood sugar levels that often leads to sporadic activity, chemical imbalance, and fatigue.
- The protein helps cells rejuvenate and facilitates the building of new cells (amino acids are the building blocks of cells).
- Dietary fat helps essential hormones form and stabilizes the carbohydrate entry rate. Additionally, dietary fat can signal the body that "enough" food has been eaten and helps limit overeating by working crews. Overeating can tax the digestive system, causing blood to shift from the muscles and the brain; this is one of the reasons why we sometimes feel like sleeping after a meal.

Choosing the best carbohydrate, protein, and fat also promotes steady, sustained performance. Carbohydrates are the toughest to balance correctly because so many of the foods we typically find on the fire scene are rich with unfavorable carbohydrates. Donuts, fries, candy, bread, potatoes, bananas, citrus fruit juice, and chips/crackers all have moderate to high glycemic indexes. Good carbohydrates include green vegetables, apples, whole-grain breads, tomatoes, and oatmeal. Protein can be derived from low-fat meats such as turkey, chicken, and fish. Eggs, low-fat cheese, and beans also offer protein. Fats should be of the monounsaturated type like olive oil, nuts, and low-sugar nut butters.

The department that preplans nourishment for rehab efforts can achieve the desired balance. Many take the easy path and use commercially available sport/energy bars—a decent choice assuming that the bars are low-glycemic index and contain good protein and fat (preplan your choices!). Menu planning with a nutrition or fitness expert is another way to preplan rehab. When a department has not preplanned rehab nourishment, the ISO can make recommendations based on the 40/30/30 balanced approach. A sampling of nourishment for rehab is offered in the *Awareness Tip* box.

Deciding *when* firefighters should eat is subject to debate. One solution is simply to fuel the firefighters when they are hungry. This aim sounds simple, although its application may be subjective and reliant on perception. A further examination of some fueling theory (minus the medical speak) provides good planning insight on when to feed firefighters.

When a firefighter is working hard, cells pull glucose from that already present in the muscles and bloodstream. Once that glucose is used (it goes quickly), the body searches the digestive tract for readily available glucose. If it has been more than an hour or two since food has been eaten, there will be no readily available glucose. The body then takes stored glycogen from the liver—a very limited supply. When this supply is used up, the

---

Awareness Tip

Sample of Balanced Nutrition for Firefighters Working at an Incident

Morning: Breakfast burritos, apples, and water. Breakfast burritos can be made with scrambled eggs (egg white only is best), chopped Canadian bacon, chopped green peppers, and low-fat cheese, all wrapped up in a whole-grain flour tortilla. Fresh salsa can also be added.

Afternoon/evening: Turkey and cheese sandwich or wraps (thinly sliced whole-grain bread or tortillas), water, apples, and small portions of almonds or cashews. It is better to have high-piled turkey than to eat two or three sandwiches; the increase in bread intake eventually throws off the balance. Mustard, hot sauce, pickles, lettuce, and tomatoes will not hurt the balance, whereas catsup, sandwich spreads, and mayonnaise will.

Any time: Commercially available energy bars and water. It is important to make sure the energy bars have the 40/30/30 balance or state they are "low glycemic index." At extended incidents, water can be supplemented by electrolyte replacement sports drinks. Almonds and whole apples provide a great snack anytime. Your body uses them both very efficiently.

---

body attempts to gain glucose by breaking down muscle, but those muscles are being used. So, the body resorts to the long, difficult process of breaking down body fat stores. Although that sounds like a great way to lose a few pounds of unwanted fat, we must understand that the incident scene may be the *worst* place to fat-burn. First, fat-burning takes time (longer than 30 minutes to start) and it is accompanied by the need for more water and oxygen. As the fat is transformed to usable blood sugar, unwanted by-products are created in the form of added metabolic heat and the creation of lactic acid. These by-products interfere with optimal cell performance. From this explanation, we can derive a good rule of thumb for feeding firefighters:

- Feed *now* if it's been more than 2 hours since the last food intake.
- Feed *every 2 to 3 hours* when physical and mental demands remain.
- Feed more often when firefighters are working in cold environments (they are burning more carbohydrates).

Granted, the more physical the task is, the more important feeding becomes. Often, fire department members performing less physical tasks (ICs, ISOs, staging managers, and others with such assignments) get left out of the fueling cycle. Remember, optimal thinking requires optimal cell performance also. Those assigned to less physical tasks require balanced fueling—just at a lower *volume* than those physically working hard.

One last comment on firefighter fueling: Many fire departments resort to convenient food choices (fast foods, coffee/donuts) or the well-meaning offerings of citizens or volunteer civic groups (cold-cut sandwiches, soda pop, chips, hot dogs) at incidents. Although all food is fuel, these choices can reduce sustained energy levels and optimal mental performance. If the incident is almost over, or the need for sustained energy and mental acuity has been lowered, the acceptance of these choices may be a minor issue. If the goal of rehab is to sustain energy and mental levels at a given incident, these choices are detrimental. Again, preplanning efficient rehab food choices—and educating those who generously assist firefighters with rehab—is time well spent.

Rehabilitation Efforts

The rehab effort that includes balanced nutrition, substantive hydration, and relief from thermal stress keeps firefighters performing well, both mentally and physically. This effort in itself reduces mental lapses and injury potential caused by overexertion. The NFPA has acknowledged this in several documents and, frankly, has tasked the SOFR with ensuring that rehab efforts are efficient. Starting with the NFPA requirements, we will address rehab efforts and give you some more suggestions on how to evaluate and improve them.

■ NFPA Rehabilitation Requirements

The NFPA defines emergency incident scene rehab as

> An intervention designed to mitigate against the physical, physiological, and emotional stress of firefighting in order to sustain a member's energy, improve performance, and decrease the likelihood of on-scene injury or death.

Several NFPA documents address the various roles, responsibilities, and requirements for incident scene rehab, and the ISO should have an awareness of those:

NFPA 1584. *Standard on the Rehabilitation Process for Members During Emergency Operations and Training Exercises* (2015). This document addresses the responsibilities of a fire department to develop an effective rehab program. Other highlights include:
- Specific rehab standard operating procedures/guidelines (SOPs/SOGs) that should be developed
- Responsibilities of the IC, company officer, rehab manager, and individual members regarding rehab (the ISO is not specifically mentioned)
- Rehab site characteristics
- Specific procedures and guidelines for rest, hydration, nourishment, EMS, cooling/warming, and so on
- Annex material that provides useful details that can be used as an educational tool

NFPA 1561. *Standard on Emergency Service Incident Management System and Command Safety* (2020). Of note in this standard is the requirement that the SO "shall ensure that the incident scene rehabilitation area has been established." An annex note that supports that requirement states that on-scene rehab should address:
- Rest
- Hydration
- Basic life support observation and care
- Energy nutrition (food and electrolyte replacement)
- Accommodations for weather conditions

The ISO requirement and areas to address are like marching orders: They not only assign responsibility but they outline the subject objectives that the ISO needs to address.

NFPA 1521. *Standard on Fire Department Safety Officer Professional Qualifications* (2020). As you've come to know, the 1521 standard goes on to further outline the SOFR's job performance requirements in carrying out the marching orders (see the *NFPA 1521 Objectives* box).

NFPA 1521 Objectives: Rehabilitation

5.2.14 Monitor conditions, including weather, fire fighter activities, and work cycle durations, given an incident or planned event, so that the need for rehabilitation can be determined, communicated to the IC, and implemented to ensure fire fighter health and safety.

(A) Requisite Knowledge. Comprehensive knowledge of heat and cold assessment criteria, rehabilitation strategies, including NFPA 1584, *Standard on the Rehabilitation Process for Members During Emergency Operations and Training Exercises*, SOP/Gs and training materials; available resources that can be used for rehabilitation, signs and symptoms of cardiac stress, and heat and cold stress.

(B) Requisite Skills. Ability to recognize signs of cardiac, heat, and cold stress; set up a rehab area and ensure that members use it as designed.

Reproduced with permission from NFPA 1521, *Standard for Fire Department Safety Officer Professional Qualifications*; © 2020, National Fire Protection Association. This is not the complete and official position of the NFPA on the referenced subject, which is represented only by the standard in its entirety.

A casual observation of these NFPA citations might lead one to believe that the ISO must set up and manage the rehab effort. That literal view was *never* the intent of those of us who helped to write and pass the standards. NFPA 1584 supports this view, as the ISO is not even mentioned. In fact, 1584 clearly states that the department is responsible for preplanning a rehab program and training its members on how to best achieve rehab. Further, it states that the IC is responsible for making sure the important function of rehab is established. Therefore, the ISO's role is to make sure the IC has addressed the function and then evaluate its effectiveness.

It is interesting to note that NFPA 1584 is fairly vague on when to establish a rehab unit. It merely says that "Rehab shall commence whenever emergency operations or training exercises pose a potential safety or health risk to members." From the ISO's perspective, that could be interpreted as every incident and exercise! In fairness, though, the supporting annex material provides a list of factors that may cause a fire department to initiate rehab, including incident time, complexity, intensity, and weather issues (NOAA heat/cold reference charts).

■ Evaluating Rehabilitation

We've established that the ISO is responsible for evaluating the rehab effort. To do so, the ISO needs to know when and what needs to be done to prevent overexertion and sustain alertness and physical energy. We've addressed some of those issues throughout this chapter. Formally, however, the ISO needs to front-load some logistical considerations that help achieve the desired outcome. Several valuable resources exist that can help the ISO better understand the logistical components of an effective rehab system. These resources continue to be updated as the fire service attempts to address the number one cause of on-scene injuries: overexertion. The ISO should stay up to date with best rehab practices and make the effort to incorporate those when evaluating rehab. As a starting point, the information in NFPA 1584 provides reasonable criteria to help the ISO do an evaluation. Further education and evaluation criteria regarding rehab logistics can be found in the following documents:

- *Rehabilitation and Medical Monitoring: A Guide for Best Practices* by the International Association of Fire Chiefs (IAFC). This 50-page guide includes sample SOPs, lots of best-practice suggestions, and useful background information.
- *Emergency Incident Rehabilitation* by the U.S. Fire Administration (USFA). This revised publication was developed in cooperation with the IAFF and includes over 170 pages of in-depth science, suggestions, and procedures to help maximize firefighter rehab efforts.

Using information gleaned from these resources, several important points can be made that will help the ISO evaluate rehab efforts. First, rehab, at minimum, should include five components (listed in priority order): rest, hydration, basic life support (BLS) monitoring and care, energy nutrition, and accommodation for weather conditions. Second, the rehab effort should be scalable based on the incident activities, duration, complexity, weather conditions, and number of responders that are involved.

Five Components of Rehabilitation

If we take the priority order of the rehab components and shuffle them, the word REHAB becomes an acronym that helps the ISO remember the components that should be included and evaluated:

- **R = Rest.** This is a time-out to help firefighters stabilize their vital signs. Pulse rate, blood pressure, and breathing rate should return to normal. The most important vital sign to stabilize is core temperature. It needs to be reduced through active cooling after exertion in PPE and/or heat stress environments. Part of the rest component is to provide a break from the mental/psychological issues that may be associated with the incident.
- **E = Energy nutrition.** Make sure provisions are made for balanced food nourishment to improve sustainable energy and mental acuity.
- **H = Hydration.** The drinking of water cannot be overemphasized as part of rehab. Ensure that sanitary drinking water is available and remind firefighters of the need to drink 4 to 8 ounces (120 to 240 mL) for every 15 to 20 minutes of work. Electrolytes need to be replaced after an hour of sweating (or sooner). Other beverages such as energy drinks, coffee, soda, and fruit juices should not replace water and sports drinks for hydration and electrolyte replacement.

- **A = Accommodation for weather.** Removing or shielding firefighters from weather conditions is especially important for heat and cold stress environments. Precipitation (snow/rain) and the effects of wind should be avoided, especially in the cold.
- **B = BLS observation and care.** At minimum, BLS care and transport resources should be assigned at firefighter working incidents. Given the high risk (and history) of cardiac-related injuries in the fire service, ALS resources are more desirable. Paramedics (preferably) or EMT-basics should make a judgment on whether a firefighter can return to incident duties based on their best judgment and the vital signs acquired from the individual firefighter.

Scaling the Rehabilitation Effort

For initial operations and small-scale or short-duration incidents, rehab can take place at a SCBA bottle changing station and may take on an informal atmosphere. Some call this approach *self-rehab* **FIGURE 9-5**. As an incident evolves, becomes more complex, is extended, or requires significant resources, the rehab effort should become formal. Formal rehab requires the IC to form a rehab unit or group and assign personnel and logistical support to carry out the five REHAB components in a preplanned and trained manner. Let's look at both self-rehab and more formal rehab approaches a bit more critically.

Self-Rehabilitation

Self-rehab relies on preestablished SOPs/SOGs, training, and accountability of members to meet the five REHAB components. As a basic rule (from NFPA 1584), self-rehab should take place after each bottle of SCBA used (30- to 45-minute bottles or equivalent workload). The first break should be at least 10 minutes long. Subsequent bottle breaks should be 20 minutes (1-hour bottles require 20 minutes of rest starting with the first). Let's call this the 10/20 rule.

Most firefighters understand the need to rest and hydrate during periods of self-rehab and will likely self-initiate.

FIGURE 9-5 Small-scale or short-duration incidents typically rely on *self-rehab* to address energy revitalization.

Courtesy of Frisco (TX) Fire Department

Logistical support is usually readily available and includes fresh drinking water and a place to doff PPE and sit and relax out of harm's way. Most firefighters will cut short their rest time if left to their own discretion, especially if "fun" tasks still need to be accomplished. Although the 10/20 rule is reasonable for most firefighters, the ISO should know it takes at least 20 minutes for vital signs to resume normal levels in an average firefighter (fitness wise). It will take more time in heat stress environments or for those who are not work hardened. If staffing levels are such that a 20-minute break is an unobtainable luxury, then the ISO needs to advise the IC and help revise the incident action plan to match the resources—and rehab time—available.

When self-rehab is being practiced, some firefighters may not fully appreciate or initiate the other three components of REHAB: energy nourishment, BLS monitoring, and weather accommodation. This potential oversight is a drawback to self-rehab, and the ISO should be mindful of it and add reminders. Regarding BLS medical monitoring and care, many fire departments that offer that level of service may discount the need to have a separate BLS resource assigned when self-rehab is being used—figuring that the BLS- or ALS-trained firefighters will keep an eye on each other. We have to be honest that firefighters working at the so-called "bread and butter" incident may have a mind-set that the incident is no big deal, which becomes a setup for masking medical needs. Couple that with the "can-do" attitude that is desired and expected in the service and you have a formula for missing the need for extended rehab and/or medical care for a given individual. The bottom line is that we shouldn't rely on working firefighters to carry out the medical monitoring and care portion of self-rehab.

A reoccurring recommendation in many firefighter line-of-duty death reports is the need to assign, at minimum, a BLS resource for firefighters. The rate of cardiac events experienced by firefighters is unusually high compared to the general public. As such, ALS resources are preferred. The ISO needs to remind the IC of this need even for small-scale or short-duration incidents where self-rehab is used. When assigned, paramedics/EMTs should not sit idly by; they should get involved at the self-rehab station and monitor the firefighters as their primary task.

Formal Rehabilitation

Incidents that are large, of long duration, resource intensive, and/or complex require a formal rehab effort. Fire departments should have SOPs/SOGs and member training that outlines the requirements for the establishment and operation of a formal rehab unit: *when* and *how*. History (and federal law) suggests that formal rehab should be established *when*:

- Incidents require physical labor over 2 hours
- High-rise fires are not controlled by a single company
- Wildland fires require more than an initial attack company to control
- Hazardous materials incidents require technician-level response (OSHA CFR requirement)
- Confined spaces are involved (OSHA CFR requirement)
- Incidents expose responders to weather extremes for longer than an hour (subjective)

- Incidents require excessive physical labor and the rotation of crews
- Incidents involve defined hourly work periods (e.g., urban search and rescue, disaster responses, searches)

These are realistic guidelines for *when* formal rehab should be established, but the *how* part of the equation should also be addressed. At minimum, the formal rehab effort should clearly address the five REHAB components and include assigned resources and personnel to make it function in an effective manner. When it comes to evaluating the formal rehab effort, the ISO should ask the IC to assign an assistant safety officer (ASO). The mere fact that a formal level of rehab has been established also indicates that the size, duration, or complexity of the incident will require the ISO to focus on the hazard environment, risk-taking, planning, and action plan elements that are ongoing.

Some fundamental formal rehab issues need to be addressed, whether rehab efforts are being evaluated by the ISO or the ASO. While not all-inclusive, these issues are directly related to rehab effectiveness or have been historically problematic.

- Formal rehab should take place in an area that is removed from the action area of the incident. An area should be chosen that allows firefighters to remove their PPE and still be safe/protected from environmental hazards.
- Likewise, the rehab area should be divided into segments: a gross decontamination area (wash off smoke contaminates); an area to doff PPE and exchange bottles; a face/hands wash area; medical care area; and an area to relax that provides for hydration, nutrition, and active cooling/warming as needed. **FIGURE 9-6**.
- The relaxation area can achieve better results if the gathered firefighters are out of the sight line of the evolving incident and working firefighters (mental break).
- Adequate staffing for the rehab function is paramount to effective operations. Starting with medical observation and care, it is best to have two individuals who are paramedic trained and have the authority to give a go/no-go verdict on whether a firefighter should return for an assignment (see the *Safety Tip* box). If an individual requires medical intervention and transport, the medical care function will cease for other firefighters. Have a plan and be proactive.
- Regarding PPE/SCBA and tools, having assigned staff to help firefighters doff their gear and prepare it for reassignment is valuable. When PPE is doffed at fires, it is best to have a manner to help air the equipment out and assist in drying. Persons handling this gear should use nitrile gloves and particulate masks. Smoke contaminants in today's fires are toxic and cancer-causing.
- Consider taking carbon monoxide and oxygen level readings in and around the rehab area. Vehicle exhaust fumes may be presenting a silent threat that is defeating rehab benefits (especially prevalent in cold environments).
- Listen to firefighters in the relaxing area. Try to sense their general mood and attitude.

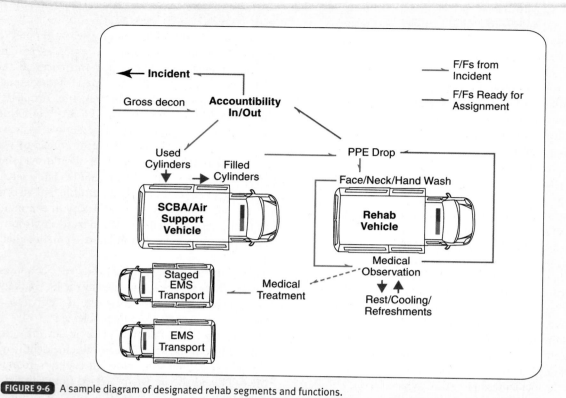

FIGURE 9-6 A sample diagram of designated rehab segments and functions.

Medical Go/No-Go Criteria for Firefighters

All responders should be aware of the medical criteria that can assist in deciding whether a firefighter in rehab should return to staging for an assignment. Additionally, paramedics and EMTs assigned to the medical monitoring and care rehab task should have the authority to keep firefighters in rehab or initiate care and transport as warranted. The USFA *Emergency Incident Rehabilitation* document uses the following as criteria to determine the need for further rest or care *following* 20 minutes of rest, cooling, and hydration:

- Pulse rate in excess of 120 beats per minute
- Body temperature above 100.5°F (38°C) (Tympanic measurements might be 2°F lower than actual core temperature.)
- Diastolic blood pressure above 90 mmHg
- Systolic blood pressure over 130 mmHg

Signs and Symptoms of Cardiac Distress or Injury Potential

Additionally, firefighters, rehab ALS/BLS attendants, and the ISO should be aware of the other signs and symptoms that could indicate injury potential or cardiac distress:

- Difficulty breathing/shortness of breath
- Chest pains
- Extreme fatigue/sluggishness
- Altered level of consciousness
- Dizziness/loss of balance
- Poor skin color
- Slurred speech

It is not the duty of the ISO (or ASO) to help rehab attendants accomplish their task. The ISO should offer suggestions when an essential part of rehab is not being addressed or offer suggestions to improve effectiveness.

An incident that involves unusually gruesome situations, serious firefighter injury, firefighter death, or other potential psychological stresses (now called an <u>atypically stressful incident</u>) is cause for the ISO to initiate, through the IC, a process of rotation and/or crew removal. NFPA 1584 requires that a team or crew that has lost one of its members to injury or death be released from the incident and returned to quarters for follow-up actions. We will discuss this further in the chapter *Postincident Responsibilities and Mishap Investigations*, although for now, it is important that the ISO be especially alert for firefighter behavioral issues that can lead to an injury during or following an atypically stressful incident. Likewise, any firefighter who is removed from an incident for medical reasons (but not transported to a care facility) should *not* be allowed to go home or be alone. This is especially important for the firefighter who says, "I'm just not feeling good." He or she should return to a fire station and be monitored for a reasonable time frame as determined by a medically trained person and the medical condition that led to incident removal.

Finally, rehab efforts should continue after crews demobilize and leave the scene. Although we are all different, true recovery from an overly strenuous firefight or other incident can take 24 hours. Postincident rehab should include PPE cleaning/exchange, personal showering, continued hydration/nourishment, and a nap!

Chief Concepts

- Reading firefighters refers to the observation of firefighters and predicting when they will make a mistake, exercise poor judgment (risk-taking), and/or overextend their abilities.
- Overexertion is the number one cause of injuries and deaths in the fire service, often leading to mistakes in judgment or decreased mental awareness.
- Avoiding overexertion requires firefighters to proactively keep themselves strong, flexible, and aerobically fit. Efforts to improve an individual's strength, flexibility, and aerobic fitness to help prevent overexertion can be defined as *work hardening*.
- The stressors that influence the chance of overexertion include the environment, human physiology, and the quality of rehab efforts.
- Environmental stressors include ergonomic issues and thermal (hot and cold) challenges.
- Ergonomic stressors are related to the physical setting, the relationship of the worker to the setting, and the task requirements. An improvement to any of these dynamics can reduce the chance of overexertion or other injury and is accomplished using the mitigation strategies of awareness, accommodation, and/or acclimation.
- Thermal stress can come in the form of heat and cold (temperatures). Heat stress can be further divided into internal (metabolic) heat and external (environmental) heat.
- The ISO can use charts (heat index and wind chill) from the National Oceanic and Atmospheric Administration (NOAA) as criteria to proactively assess heat and cold thermal stressors for incident personnel.
- Heat stress injuries get progressively worse as body core temperature rises. Signs and symptoms of heat injury potential include transient heat fatigue, heat rash, cramps, cold and clammy skin, dizziness, nausea, and, in severe cases, altered state of consciousness and loss of perspiration.
- ISOs should presume that firefighters working in full PPE and SCBA have elevated core temperatures. The remedy is active cooling through cold/wet towels on the head and neck, or forearm immersion in cold water. Passive cooling includes shade, air movement, and the removal of PPE. In most cases, passive cooling is relatively ineffective for core temperature reduction.
- Cold stress injuries can be local (frostnip or frostbite) or systemic (hypothermia). Signs and symptoms include uncontrolled shivering, loss of sensation in peripheral tissues, pale/ashen/hardened skin, and, in severe cases, lethargy and altered state of consciousness.
- Optimal cell performance is required for arduous firefighting tasks. Cells require the ingredients of oxygen, water, and fuel (blood glucose), delivered in a balanced way, to work optimally. The fire ground challenges the uptake and delivery of these ingredients.
- Firefighters can lose a quart of water an hour through sweating, breathing, and cell waste removal. Hydration of 4 to 8 ounces (120 to 240 mL) for every 20 minutes of physical labor is paramount to prevent dehydration. Failure to hydrate will cause rapid overexertion and the chance of injury or death. After initial hydration, the replacement of electrolytes lost through sweating is essential. Sports drinks, as opposed to energy drinks, can help replace electrolytes.
- Fueling (feeding) of firefighters should begin immediately if it has been more than 2 hours since they have eaten. Otherwise, fueling should take place every 2 to 3 hours for working firefighters.
- Food choices for working firefighters should include a balance of 40/30/30—that is 40% low-glycemic carbohydrates, 30% protein, and 30% dietary fat. Several nutritious sports/energy bar brands can offer this balanced fueling, but they should be researched. Convenient fast foods and other high-fat/high-carbohydrate prepared foods should be avoided.
- NFPA requires individual departments to develop SOPs/SOGs that define when and how rehab efforts are to commence. As a rule, the incident duration, complexity, intensity, and weather issues should be considered when developing the need to activate rehab efforts.
- The acronym REHAB helps the ISO remember the five essential elements of the incident rehab effort:
  R = Rest (remove and relax)
  E = Energy nutrition
  H = Hydration
  A = Accommodation for weather conditions
  B = BLS medical observation and care (ALS is recommended)
- Rehab can be scaled as *self-rehab* for short-duration/small-scale incidents or as a *formal rehab unit or group* when the size, duration, complexity, or strenuous nature of the incident demands.
- When a formal rehab unit or group has been established, the ISO should request the help of an assistant safety officer (ASO) to help evaluate rehab efforts. The active part of incident handling will likely demand ISO attention.
- All firefighters, rehab managers/attendants, and the ISO should know the signs of cardiac stress or injury potential: difficulty breathing/shortness of breath, chest pains, extreme fatigue/sluggishness, altered level of consciousness, dizziness/loss of balance, poor skin color, and slurred speech.

Wrap-Up, continued

Key Terms

<u>acclimation</u> An individual's gradual process of becoming accustomed to an environment.

<u>accommodation</u> The efforts to alter or adjust the environment, worker relationship, or task to reduce injury potential.

<u>active cooling</u> The process of using external methods or devices (e.g., hand and forearm cold water/ice immersion, covering the head and neck with cold wet towels, and gel cooling vests) to reduce elevated body core temperatures.

<u>atypically stressful incident</u> An incident that involves an unusually gruesome situation, serious firefighter injury, firefighter death, or other potential psychological stress.

<u>circadian rhythm</u> A person's physiological response to the 24-hour clock, which includes sleep, energy peaks, and necessary body functions.

<u>ergonomics</u> The science of adapting work or working conditions to a worker and the study of problems associated with people adjusting to their work environment.

<u>passive cooling</u> The use of shade, air movement, and rest to bring down core temperatures.

<u>transient heat fatigue (THF)</u> An early warning sign that the core temperature is elevated, as characterized by the onset of physical exhaustion that is remedied by rest and hydration only to return quickly and more profoundly upon engagement with the hot environment.

<u>work hardening</u> Efforts to improve an individual's strength, flexibility, and aerobics to help prevent overexertion at incidents.

Review Questions

1. What is *work hardening* and how can it be accomplished?

2. What are the three ergonomic stressors that can lead to overexertion?

3. What are the three "A's" to help mitigate ergonomic hazards?

4. List the types of heat stress and cold stress injuries, and give some signs and symptoms of each.

5. What criteria should an ISO use to proactively prevent heat and cold injuries?

6. List two examples of passive cooling and two methods of active cooling.

7. At minimum, how much water should working firefighters drink at an incident?

8. What three ingredients need to be balanced to help human cell performance?

9. Commercially available energy bars should ideally contain what percentage of each of the three food groups?

10. What is the difference between self-rehab and formal rehab?

11. List the types of incidents that should require formal rehab.

12. List the signs and symptoms of cardiac distress or injury potential.

References and Additional Resources

Boyd, Michael E. *Firefighter Rehabilitation in the Orange County Fire Authority: Are We Meeting the Need?* NFA EFO Paper. http://www.usfa.fema.gov/pdf/efop/efo42096.pdf. Accessed May 7, 2015.

International Association of Fire Chiefs. *Rehabilitation and Medical Monitoring: A Guide for Best Practices.* Midlothian, TX: Cielo Azul Publications, 2009.

McLellan, Thomas, and Glen Selkirk. *The Management of Heat Stress for the Firefighter.* Toronto, ON: Defense Research and Development Council, 2005.

NFPA 1584, *Standard on the Rehabilitation Process for Members During Emergency Operations and Training Exercises.* 2015 ed. Quincy, MA: National Fire Protection Association, 2015.

Ross, David, Peter McBride, and Gerald Tracy. Rehabilitation: Standards, Tools, and Traps. *Fire Engineering*, May 2004.

Sears, Barry. *Enter the Zone.* New York: HarperCollins, 1995.

The FireRehab.com website features many articles, resources, and products addressing firefighter incident rehab. It is available at http://www.firerehab.com.

The National Weather Service website offers access to NOAA weather heat index and wind chill charts. It is available at http://www.nws.noaa.gov.

U.S. Fire Administration. *Emergency Incident Rehabilitation.* 2008. Available as a download at https://www.usfa.dhs.gov/applications/publications.

Endnotes

1. Firefighter injury and death reports from the NFPA, USFA, and IAFF unanimously conclude that this statement is factual.

2. Variations of this statement have been presented by numerous fire service speakers over the past several years. The original source is not known.

3. McLellan, Thomas. *Safe Work Limits While Wearing Firefighting Protective Clothing, Toronto Fire Department Grant Study Report.* Toronto, ON: Defense Research and Development Council, 2002.

4. U.S. Fire Administration. *Emergency Incident Rehabilitation.* FA-114. Washington, DC: USFA Publications, February 2008.

5. The author has found numerous examples of EDs being banned from the workplace, including several TYPE 1 Incident Management Teams, Yates Construction, and Phelps Construction. CALFIRE and the military are considering a similar ban of EDs for crews working the field (wildfires/combat). EDs have been outlawed in several countries due to their cardiac irritation potential.

6. Sears, Barry. *Enter the Zone.* New York: HarperCollins, 1995.

INCIDENT SAFETY OFFICER
in action

Lieutenant James and Driver/Firefighter Chan made up a two-person engine crew assigned to the outlying station of a combination fire department. They had a productive morning that included their vehicle readiness checks, usual morning workout, a minor vehicle extrication incident, and the weekly task of mowing the grass in front of the station. The summer day was beautiful: sunny, a deep blue sky, 85°F, and humidity around 60% with no wind. After mowing and trimming the lawn, James and Chan decided to take a break and address their lunchtime hunger pangs. Just as they headed for the kitchen, the station alarm alerted them to a single-engine response to a rubbish fire. Upon arrival at the scene, they found a small deep-seated fire in an illegal dump down in a farmer's ravine. They reported that they could handle the incident.

Training/Safety Officer O'Connor was busy in her office, casually monitoring the radio while James and Chan worked the dump fire. A series of radio transmissions over the next 20 minutes caught her attention. The first asked for a volunteer manpower response. The next asked for a county backhoe to assist with digging out the trash. Finally, O'Connor heard a request for an EMS response for an injured firefighter—which came from Chan—not James. O'Connor grabbed her radio and responded to the incident in accordance with her responsibilities as the duty safety officer.

Upon arrival, O'Connor was informed that Lt. James had experienced a blackout and was being treated with IV therapy in the back of the ambulance and would likely be transported to the hospital for heat exhaustion.

1. What ergonomic factors contributed to Lieutenant James's injury?

2. What thermal stress factors (and specific heat index) contributed to this incident?

3. A proactive strategy for hydration should include which of the following?

   A. Pre-incident consumption of sports drinks

   B. Hourly consumption of water as a daily routine

   C. Drinking 4 to 8 ounces of water for every hour of hard work

   D. Reducing the quantity of sugar and cream added to coffee

4. When should firefighter fueling (feeding) take place as part of a working incident rehab plan?

   A. After incident stabilization efforts are completed

   B. Following periods in which firefighters have transitioned to fat-burning

   C. Any time that insulin levels are suspected to be elevated

   D. Right away if it has been more than two hours since last food intake

Note: This case study is taken from an actual injury investigation report. Names have been changed to protect the guilty.

SECTION 3

The Incident Safety Officer On-Scene: A Reference Guide

In Section 2 you were encouraged to pursue incident safety officer (ISO) mastery: the unconscious ability to be efficient and effective. The path to mastery includes learning (front-loading) and performance. While Section 2 focused on front-loading, Section 3 provides suggestions to help you perform the ISO role for various types of incidents and events. This is not to say that Section 3 is without front-loading information; it addresses many knowledge objectives that satisfy the NFPA job performance requirements. More importantly, this section addresses skill (or performance) objectives.

Section 3 has eight chapters. The chapters *Triggers, Traps, and Working with the Incident Commander*; *A Systematic Approach to the ISO Role*; and *Postincident Responsibilities and Mishap Investigations* are applicable to all incidents; the other chapters address specific incident types and events. You may be tempted to skip the first two chapters in this section (thinking, for example, "I'm familiar with ICS and Chapter 11 sounds pretty generic"), but please don't. The information contained in *Triggers, Traps, and Working with the Incident Commander* is all about *your* behaviors as the ISO and the way you interact with others at an incident. You can undo your very goal of making a difference by the *way* you perform your duties. The author of this text has made those mistakes and hopes to share with you ways to prevent doing the same.

A Systematic Approach to the ISO Role creates a "template" for the chapters addressing specific incident types. It is suggested that you read that chapter prior to the ones following it simply because it introduces a mental-processing format that is repeated for each type of incident.

As with most fire service knowledge and skill sets, personnel must practice their application on a recurring basis. So it is with the ISO skill sets. Therefore, Section 3 becomes a recurring proficiency training *reference guide*. That is, ISOs can refer to an individual chapter as part of a scheduled training regimen that helps maintain skills.

The chapter *The ISO at Training Drills and Events* is new to this text and can also be used as a planning reference.

Voices of Experience

The National Institute for Occupational Safety and Health (NIOSH) has been tracking line-of-duty death (LODD) statistics for years. Despite our training and technological advances, we continue to lose approximately 100 fire-fighters a year. While the NIOSH LODD pie charts include the usual suspects (heart attack, motor vehicle accidents, trauma, asphyxiation), they show only the physiological causes of death to firefighters. The more compelling

statistic is the *operational* causes of death, or *causal* factors. In other words, I don't care as much about the fact that firefighters died of trauma as I do about the operational issues that led to the trauma. What are the operational trends, decisions, issues, and mistakes that cause firefighters to die?

A different look at the NIOSH LODD reports over the last 20 years reveals that five factors are repeatedly to blame for LODDs, civilian losses, near misses, and significant injuries. These factors have been dubbed by fire service leaders as the "NIOSH 5." They include inadequate or improper:

1. Risk assessment
2. Incident command
3. Accountability
4. Communications
5. Standard operating guidelines (SOGs), or failure to follow SOGs

Think back to your last dysfunctional fire ground operation. Chances are that some or all of these factors were present. To ensure the most safe, effective, and efficient fire ground operation, we must set up the operation to prevent the NIOSH 5 from coming into alignment. The more factors align, the more likely that bad things will happen.

We must look at the fire ground in a different way in order to manage it and, hopefully, save citizens and ourselves from it. The fire ground has three levels: strategic, tactical, and task. The strategic level is typically greater than 50 feet (15 m) from the building. This distance renders the incident commander (IC) too far away; moreover, due to the IC's stationary position, he or she is likely to see only one side or corner from the outside of the structure.

In contrast to the strategic level, the task level is where companies get the work done. It is characterized by smoke, heat, noise, tunnel vision, and auditory exclusion. No matter what the individual's experience level, it's simply harder to see, hear, speak, and think in the task arena. In addition to these stressors, the physical and emotional demands of the fire ground create additional psychological and physiological hindrances to decision making. The radius of situational awareness is usually only 0 to 6 feet (0–2 m) in the task arena—zero feet for a smoke-filled environment and 6 feet for typical tasks such as donning self-contained breathing apparatus (SCBA), throwing a ladder, searching a room, stretching a line, cutting a hole, etc.

In between the task (company/individual) level and strategy (command post) level is the tactical level. This is the 6- to 50-foot (2- to 15-m) distance between the companies working in the hazard zone and the IC at the command post. The officer in the tactical space is far away enough to see the bigger picture but close enough to see/anticipate changing fire and building conditions. It is in this arena that the division/group/sector officer lives. This officer is charged with coordinating multiple companies working together to accomplish one or two tactical objectives (fire attack, search, ventilation, exposure control, etc.). The tactical officer must be highly mobile and highly agile in a highly hostile environment. In other words, this individual is a field commander.

The tactical level is the space that is often overlooked on the structural fire ground. We all know that the task level is where the companies get the work done. We also know that the strategic level is where the IC lives. In between, however, is the tactical space. This sweet spot is where true tactical decision making, accountability, communications, and safety should come together for success.

When companies and command are talking directly to each other and both are operating within the "fog of war," meaning that neither party *really* knows what's going on, a *tactical gap* is created. The IC is basing all of his or her intelligence on radio traffic and one or (at best) two sides of a building.

The crews, conversely, do not see deteriorating building conditions, collapse potential, fire above/beneath them, accountability issues, or the risk/gain paradox. More NIOSH 5 factors come into alignment and the tactical gap widens. The more the tactical gap widens, the more likely that firefighters and civilians will fall into it.

Here is an example from a recent fire I responded to. A house fire was dispatched midmorning on a weekday. The house was approximately 2200 square feet (200 m²), two stories, modern, and vacant. First-arriving companies reported heavy fire. In fact, fire gutted most of the house, and collapse began in the garage. Within minutes, four engines, two truck companies, a fire medic unit, and two battalion chiefs arrived. The first-arriving battalion chief initially had a 13:1 span of

control: four engines, two trucks, one fire medic, one additional chief, two radio channels, mobile data terminal updates, a GPS map, and code 3 driving while responding. Needless to say, accountability was an issue from the start.

The crews performed interior attack, search, and vertical ventilation. In addition, utilities were secured and exposures protected. As the second chief, I arrived approximately 10 minutes after the first chief. While the main fire was knocked down, a secondary search was not completed, and shortly after my arrival, the first-floor attic fire flared up significantly, with fire blowing out the roof vent and throughout the attic. Radio traffic began to escalate, and command was having increasing difficulty with feedback of multiple crews talking and muffled transmissions through SCBA. The NIOSH 5 were coming into alignment.

I entered the house as Division A, in full personal protective equipment (PPE), to fill the tactical gap between the companies working and command. My objective was to coordinate the fire attack in the attic while ensuring crew accountability and safety. I immediately told all captains, face to face, that I was their boss as Division A and that they now reported to me. The radio traffic immediately fell silent as all communications were face to face between me (the tactical-level field commander) and the companies at the task level.

The disturbing (and all-too-common) reality was that few companies were intact. Some captains were outside while their crews were inside. Some captains were inside while their crews were outside. Some didn't know where their captains or crews were. Many had inaccurate or no company designators on their PPE. In addition, an interior stairwell was weakened by the fire, and multiple crews were moving up and down it simultaneously. All the while, crews were having a hard time coordinating the fire attack and extension of a line to knock down the attic. Also, individuals were in the aforementioned garage, under a severely sagging roof *after* command stated that no one was to enter the space. Now, the NIOSH 5 factors of risk (garage), incident command (crew members not following orders), accountability (crew continuity), communications (feedback and SCBA), and SOGs (improper company identifiers) were all aligned.

These issues were all cleaned up with aggressive, face-to-face tactical field-command presence. I was simply the sergeant at arms for the IC, ensuring that his "leader's intent" was being fulfilled safely, effectively, and efficiently.

The division/group/sector supervisor's main mission is to prevent the NIOSH 5 from aligning. Doing so requires him or her to fulfill the following standing objectives simultaneously (a job description, if you will):

- Risk/gain assessor
- Safety officer
- Accountability officer
- Communication tsar
- SOG enforcer
- Tactical commander

In many ways, the division/group supervisor is an "embedded" safety officer that bridges the tactical gap. When assigned, a separate and independent ISO working for the IC must also be aware of the relationship of the NIOSH 5 and the tactical gap. Although the ISO may not be physically present in the tactical gap arena, he or she must evaluate the environment and actions and determine whether the tactical gap is incubating the NIOSH 5. Proactive use of the incident command system (including the use of an ISO and an embedded safety officer as the division/group supervisor) to prevent the NIOSH 5 from aligning will create much safer, more effective, and efficient fire ground operations.

Anthony Kastros
Battalion Chief (ret.), Sacramento Metro Fire District
Sacramento California

Triggers, Traps, and Working with the Incident Commander

Flames: © Ken LaBelle NRIFirePhotos.com

Knowledge Objectives

Upon completion of this chapter, you should be able to:

- List four options that can help the incident safety officer (ISO) trigger safe behaviors. (p 146)
- Identify imminent threats to firefighter safety. (NFPA 5.2.4 , p 148)
- List the three ISO "traps" and discuss how each can render the ISO ineffective. (pp 149–151)
- Describe elements of an incident command structure and the position of the ISO. (NFPA 5.2.1 , p 151)
- Identify the components of an incident action plan (IAP). (NFPA 5.2.1 , pp 154–155)
- Describe the components of an IAP that the ISO should monitor. (NFPA 5.2.2 , pp 154–155)
- Describe methods used by the ISO to alter, terminate, or suspend activities or operations identified as hazardous. (NFPA 5.2.2 , pp 147–148)
- Describe how an incident commander would be notified by the ISO of any actions taken to alter the IAP or activities during an operation or planned event. (NFPA 5.2.2 , NFPA 5.2.4 , pp 151–152)
- Describe the radio transmissions that are monitored to indicate communication barriers or incident hazards. (NFPA 5.2.9 , pp 152–153)
- Describe the types of incidents at which additional ISOs or technical specialists are required due to corresponding hazards of the incident type. (NFPA 5.2.10 , pp 156–158)
- Define the process for making recommendations to an incident commander to request additional ISOs or a technical specialist. (NFPA 5.2.10 , p 155)
- Describe the reasons that the duties of an ISO would be transferred from one individual to another. (NFPA 5.2.3 , p 156)
- Define the information required during a transfer of ISO duties. (NFPA 5.2.3 , p 156)

Skills Objectives

Upon completion of this chapter, you should be able to:

- Review an IAP and available information to determine hazardous conditions. (NFPA 5.2.1 , p 153–155)
- Demonstrate the ability to monitor an IAP. (NFPA 5.2.2 , pp 154–155)
- Demonstrate the ability to communicate within an incident command structure any actions taken that result in modifying an IAP. (NFPA 5.2.2 , NFPA 5.2.4 , pp 152–155)
- Demonstrate the ability to transfer IAP information and other relative incident factors to another individual during an incident. (NFPA 5.2.3 , p 156)
- Demonstrate the ability to identify radio transmissions that could indicate communication barriers. (NFPA 5.2.9 , pp 152–153)

After a recent multifamily dwelling fire where you were assigned the ISO role, many members have been heard complaining about some of your actions and functions at the scene. Some have mentioned that they thought you were functioning more as a sector/division/group officer as you worked in the incident command structure and ordered lines and ground ladders to be repositioned. Others have said you did a good job in letting them know of potential hazards and were clear in communicating hazard control zones and actions. You are preparing to write up your after-action report for submission to the operations chief.

1. What examples of ISO interventions have you seen at incident scenes in which you were involved?
2. If you were a company officer or firefighter operating at an incident scene, how would you react to an ISO intervention involving you or your crew? What if you did not agree or see the same conditions being identified by the ISO?

Introduction: It's All in the Way You Deliver the Message

Front-loading knowledge and skills (described in Section 2) can help the incident safety officer (ISO) become technically competent but does not by itself create an ISO who will make a difference. The element that tips the scale toward making a difference is the ability to communicate clearly and appeal to the safety sense that can be innocently sidestepped during working incidents.

Firefighters by and large are a proud, strong, and reasonable group who thrive on competition, adrenaline, and a "can-do" attitude toward challenges. The ISO can trigger favorable or unfavorable responses when confronted by a group of firefighters driven by challenge. Face it, there are times when the ISO must deliver a message that is unpopular or counters the desires of driven firefighters. Often, the new or inexperienced ISO falls into some common traps as crews try to discredit or argue the ISO's message. Although these traps may seem trivial, they thwart the ultimate goal of making the incident safer. Given a well-meaning ISO and a group of respectful, can-do firefighters, it might be hard to imagine anyone wishing to subvert the ISO's efforts, but incident-handling experiences are full of examples of the ISO's message falling on deaf ears. In these cases, often it's not the message that was off but the way the message was delivered, usually because of an ISO trap. The ISO who does not address the traps becomes only minimally effective in the long term.

The successful ISO uses a palette of options to deliver the safety message to working crews. There are passive and active options to trigger higher levels of safety awareness and behaviors. Knowing these triggers and applying them in an appropriate manner are the keys to making a difference.

Likewise, the ISO must present concerns and solutions in a way that is appealing to the incident commander (IC). Failure to work in harmony with the IC is also a failure in firefighter safety. Harmony is achieved when the ISO understands and works within a defined incident command structure (ICS), is knowledgeable when evaluating the incident action plan (IAP), and addresses issues respectfully with the IC.

This chapter looks at ways to "get along" with everyone through favorable triggers and to avoid harmful traps. Additionally, it examines ways to work within a scaling ICS and to evaluate IAPs.

Safety Triggers

Training and safety officers love the adage "Train like we play and play like we train" and use that ideology to reinforce safe behaviors. The training officer facilitates the "train like we play" portion while the safety officer addresses "play like we train" at incidents. The difference is that the training officer can easily stop or interrupt the training activity to remind the players of the safe behaviors associated with a given task or evolution (perform in a safe and predictable manner). The safety officer working an incident doesn't always have the option to stop the players when they are not playing like we train. Granted, overtly dangerous/immediate hazards should trigger the safety officer to stop unsafe activities or withdraw the players. (For that matter, any fire officer/firefighter should stop unsafe activities because we are *all* responsible for safety.) However, most of the time, the players are playing like they train (in a safe and predictable way). Occasionally, the urgency, duration, complexity, routineness, newness, evolution, or stressfulness of an incident will cause a player (or players) to operate in an unsafe way or unknowingly engage a threat. Further, unsafe behaviors and conditional hazards are not all created equal and come on a sliding scale (i.e., risk a lot to save a lot). Given all of these dynamics, the ISO must have a range of options to help "trigger" situational awareness and safe behaviors in the players. A safety trigger can be defined as the option or approach that an ISO uses to help remind firefighters to be situationally aware and to perform in a safe and predictable manner. Some triggers are subtle (passive), whereas others are direct (active). Safety triggers include:

- Being visible (passive)
- Setting an example (passive)
- Soft intervention (active)
- Firm intervention (active)

Let's start with the subtle and work toward the more direct.

■ Being Visible

The ISO should wear a high-visibility vest that clearly states "SAFETY" **FIGURE 10-1**. In many ways, use of this vest is like the "power of suggestion" employed by many advertisers. Upon seeing the safety vest, firefighting crews often stop an unsafe action, withdraw from an unsafe position, or quickly put their gloves back on. This self-correction is desirable on scene, as it allows the ISO to concentrate on other items of concern. Using

FIGURE 10-1 Be visible! The use of a safety vest can help trigger safe behavior through the power of suggestion.

this trigger may seem like an elementary or even childish suggestion; however, experience has proven this trigger effective and, therefore, useful for the ISO. NFPA 1561 (2020 edition) requires that the SO and any assistant safety officers (ASOs) be readily identifiable at the incident scene (5.9.6.13).

To maximize this trigger, the safety vest should be instantly recognizable from a distance. Many vests look alike from a distance, and it can be difficult to differentiate the ISO vest from the staging manager or even IC. As stated in the chapter *Designing an Incident Safety Officer System*, using a different color with highly reflective trim seems to be the best solution for safety officer vests. While no standard exists, most people associate the color *green* with safety. The National Safety Council uses a green cross as their logo and the Fire Department Safety Officer's Association uses a green Maltese cross as its symbol. ISOs wearing a self-contained breathing apparatus (SCBA) may be unable to display a unique vest; in these cases, a different-colored helmet with a "safety" emblem or title and/or the use of commercially available SCBA bottle identification sleeves can help maintain the ISO visibility trigger. A low-cost option that serves a similar purpose includes the use of bright green arm bands worn on each sleeve around the biceps area; they are easy to make.

Another advantage of having a unique visibility identifier is the ease and inclination of working crews to share their hazard observations with the ISO. For example, a firefighter team was applying defensive water to an exposure and upon seeing the ISO in the green vest turn the corner, waved him over and shared an observation that night was approaching and there was an unlighted, empty swimming pool nearby. The ISO took steps to get the pool taped off and illuminated.

■ Setting an Example

Philosopher, theologian, and Noble Peace Prize winner Albert Schweitzer once said, "Example is not the main thing in influencing others. It is the only thing."[1] Effective ISOs always try to set a good example while performing their duties. Often, it is the habits and self-discipline that an ISO displays that influence others—a passive trigger to safe behavior. To illustrate,

ISOs should always, without fail, participate in the crew accountability system. Likewise, they should use appropriate personal protective equipment (PPE), follow department policies, and obey zone markers.

Because ISOs often work alone—contrary to most fire service tenets—they are usually outside the hot zone or immediately dangerous to life and health (IDLH) hazard area and looking in. This practice is acceptable. However, to set a good example, ISOs working alone should follow some basic guidelines:

- Always be in sight of another responder.
- Always be within shouting distance of another responder.
- Always have a personal alert safety system (PASS) device turned on.
- Always participate in established accountability systems and personnel accountability reports (PARs).
- Let someone know where you are going if you are taking a tour of the incident scene.
- On larger incidents, the practice of communicating by radio to the division/group supervisor that you (as the ISO) are entering his or her area is a customary and recommended practice. Likewise, let the supervisor know when you depart; doing so assists with accountability should a PAR be necessary.
- Don't walk into, or breathe, smoke.

These guidelines may seem a bit absolute and impractical, but the intent is to set the example and act safely. Although it happens rarely, an ISO may be injured or lost and nobody knows it happened. At a recent event in Baltimore, Maryland, an ISO was performing a recon function in an uninvolved exposure and suffered an injury. The fire department cleared the scene only to be called back when a citizen reported an abandoned fire department vehicle—the safety officer's. They found him unresponsive in the exposure building.[2] The ISO died.

It is counterproductive for the ISO to be viewed as working unsafely. Think about the irony: It is rich fodder for teasing to catch the safety officer doing something unsafe! Do not give them the chance. ISOs should evaluate their own environment and exposure, making the adjustments necessary to act safely. Here are some suggestions to help become the safe example:

- In cases where the ISO needs to go into an IDLH environment or into a hot zone, a partner should be requested and the ISO "team" should use appropriate PPE and SCBA and be tracked just like any other assigned crew.
- When performing reconnaissance around a building, ISOs are likely to walk into an area where no responder is visible (if you cannot see anyone, then nobody can see you). In these cases, ask for a partner to go with you. If no one is available, walk to the corner of sides B and C, then walk back around the A side to go look at sides C and D, all the while having someone on side A keep you in view.
- ISOs need to self-monitor their rehab needs: Stay hydrated and eat something if you have been on scene longer than 2 hours. Take steps to minimize the effects of thermal stress on your thinking ability (see the *Reading Firefighters* chapter).

It is commonplace in some departments for the ISO to simply don his or her safety vest over the work uniform (no PPE). This may be acceptable for an ISO who is stationary and working at the command post (cold zone), using ASOs (in PPE) for field recon. This author strongly suggests that any ISO/ASO doing recon in and around the incident wear PPE appropriate for the incident type. Doing so sets a good example (passive trigger) and places the ISO in a stronger readiness posture should a hands-on, injury-prevention intervention become necessary.

■ Soft Intervention

Being readily identifiable and setting a good example are subtle triggers that can be classified as passive. To *actively* address safety issues, the ISO may need to intervene in a direct way. Safety issues requiring intervention can be classified as an *imminent threat* or a *potential threat*. An <u>imminent threat</u> is best defined as an act or condition that is judged to present a danger to persons or property that is so urgent and severe that it requires immediate corrective or preventive action[3] (a firm intervention, covered later). A <u>potential threat</u> is an activity, condition, or inaction that is judged to have the capability, but not immediacy, to cause harm to persons or property and thus warrants monitoring and/or operational modification. Most potential threats can be addressed by soft interventions. A <u>soft intervention</u> can be defined as awareness or suggestive communications made to crews or command staff that cause them to modify their observations and activities to prevent injury from a potential threat **FIGURE 10-2**. The intent of this intervention is to achieve positive safety changes in behaviors, operations, and actions within the framework of the incident management system and the IAP. The use of humor, subtle reminders, information sharing, and "peer-talk" are examples of soft interventions. These tactics are employed as a way to trigger safe behavior when a witnessed action or environment is concerning but not necessarily life threatening.

Soft interventions don't necessarily direct or order changes; they merely share observations or offer suggestions to modify activities. In most cases, the person receiving the message makes adjustments to the hazard within the framework of an ICS. This

is especially true if the ISO acknowledges the wisdom or choice made previously and then introduces a third interpretation or additional information that underscores the safety concern.

Injecting or introducing humor can be effective when making soft interventions—but be careful with it. The humor must not trivialize the safety concern. To make a safety point, humor is best used when the environment allows face-to-face communication and centers on the circumstance, not on the actions of firefighters. Self-deprecating humor is rarely offensive to others and can be effective.

The use of information-sharing or discovery questions can also prove to be an effective soft intervention technique. For example:

- Did any of you notice . . . ?
- What do you think about . . . ?
- How can we avoid . . . ?
- Is it me, or does that . . . ?

These suggestive questions should be delivered in a sincere, curious manner to help keep the conversation nonconfrontational. When possible, keep the questions open ended to encourage critical thinking.

Soft intervention should not be used when an imminent threat exists. The reasons are obvious: First, most recognize that the ISO has the authority to stop, alter, or suspend activities and operations *only* if an imminent threat to firefighters is present. Second, you would not want to *suggest* or downplay a directive to stop, alter, or suspend activities if the threat presents an immediacy or severity concern.

Realistically, hazards and corrective needs at an incident may fall between the need for a soft and a firm intervention. In these cases, the ISO should try a soft intervention first. If the soft intervention is ineffective, the ISO may choose to use a stern advisory. In essence, the ISO is stating in clear, direct language that the hazard or behavior is alarming. The use of a stern advisory should be reserved for the few unfortunate occasions when an individual or crew is acting without discipline or with a perceived disregard for the safety of themselves or others. Matching the intervention to the degree of concern is essential to achieving buy-in with the person or crew in question. If the intervention is viewed as irrelevant or demeaning to the crew, change may not occur. When the stern advisory does not achieve change, consider taking the issue (and solution) to an operations section chief (when assigned) or to the IC.

■ Firm Intervention

A <u>firm intervention</u> can be defined as a direct order to immediately stop, alter, suspend, or withdraw personnel, activities, and operations due to an imminent threat. The firm intervention by an ISO is an official order, with all the authority of the IC to back it up **FIGURE 10-3**. For example, an ISO who witnesses a crew operating in an imminent collapse zone could relay via radio:

> *Urgent Message.* Attack Team 3, from Safety: Evacuate that position immediately. The Charlie side is an imminent collapse zone. Move to the Bravo side. Acknowledge.

Any time a firm intervention is used, the ISO should relay the order to the IC (and operations section chief, if

FIGURE 10-2 Often, a simple reminder (soft intervention) is all that is needed to prevent an injury.

FIGURE 10-3 A firm intervention is a direct order to immediately stop, alter, suspend, or withdraw personnel, activities, and operations due to an imminent threat.

assigned) as soon as the situation allows (i.e., personnel are clear of the threat). The authority to use and report firm interventions is reflected in the NFPA 1561 standard (2020 edition):

> **5.9.6.11** *At an emergency incident where activities are judged by the safety officer as posing an imminent threat to responder safety, the safety officer shall have the authority to stop, alter, or suspend those activities.*

> **5.9.6.11.1** *The safety officer shall immediately inform the incident commander of any actions taken to correct imminent hazards at the emergency scene.*

ISO Traps

ISOs can find themselves getting trapped into operational modes and activities that render them (and the ISO program) ineffective. Specifically, the ISO's general approach may cause responders to dismiss the ISO's efforts. Often, it is the well-intentioned, inexperienced ISO who wanders into the following types of traps.

■ The Bunker Cop

The *bunker cop* syndrome is evident in the ISO who spends too much time looking for missing, damaged, or inappropriate use of PPE. For example, the ISO may be always asking working firefighters, "Where are your gloves?" or "Why isn't your helmet on?" Like a radar traffic cop (focused only on speeding), the bunker cop is focused on one specific component of safety: PPE.

Unfortunately, this focus causes the ISO to miss the big picture of incident safety **FIGURE 10-4**. Further, it eventually alienates the ISO and undermines an ISO program in the eyes of firefighters; most firefighters know what the PPE expectations are and they resent the "babysitting." Occasionally, firefighters need to be reminded of PPE safety expectations, but the reminder is best delivered by the company officer, team leader, or group supervisor. If these leaders take care of PPE issues, the ISO does not have to be on patrol for them. As mentioned previously, leading by example is an excellent way of encouraging proper on-scene behavior. Company officers, division/group supervisors, and ASOs must all exemplify leadership traits focused on safety.

On the other hand, ISOs could be considered negligent if they fail to recognize situations in which firefighters are not using appropriate PPE. In these cases, they should report the infraction to the person's team supervisor as a soft intervention. If PPE issues keep resurfacing at incidents, the department values and standard operating procedure (SOP) enforcement are delinquent. If so, the ISO needs to work within the framework of departmental change and, unfortunately, is forced to be a bunker cop in the short term.

A similar, yet subtle deviation of the bunker cop syndrome is the ISO who focuses on firefighter skill proficiency at incidents. It is easy to watch a crew throwing ground ladders and determine whether they are doing it "by the book," but the ISO who brings up skill deficiencies during an incident may

FIGURE 10-4 The ISO who takes the *bunker cop* approach misses the big picture.

not be received well by working crews. An ISO once made the following mistake: A veteran fire officer who was operating as a ventilation group supervisor found the ISO skill-performance intervention very annoying. He promptly put the ladder down and said, "Here, if you're so smart, you do it!" As you can see, the ISO didn't make a difference. In some cases, perhaps the tone of the ISO's intervention is too dictatorial, thus explaining an emotional reaction. If, however, the tone of the ISO intervention is friendly, the ISO must look at the emotional reaction as an indicator of something deeper, such as a sign of stress, a rehab need, or a personality conflict.

New or inexperienced ISOs often fall into this trap because of their familiarity and comfort with basic PPE use and skill evolutions. When the new ISO is (or is acting as) a company officer, PPE and crew safety are comfort zones, and nobody on the crew questions the company officer's authority to address the issues.

Safety Tip

See the Big Picture

While PPE use and skill proficiency are critical to safe, efficient incident response, the ISO needs to always have an eye on the bigger picture, including factors such as fire behavior, building construction, and risk/benefit evaluations.

■ The CYA Mode

The well-known abbreviation *CYA* best describes ISOs who spend an inordinate amount of time ensuring that they are not held personally accountable for incident scene actions. Although the label is earthy, the behavior must be acknowledged. The proliferation of NFPA standards, the Occupational Safety and Health Administration's (OSHA) Code of Federal Regulations (CFR), and local requirements (SOPs) have placed ISOs in a position of liability by mandating their compliance with the due diligence requirements outlined in those codes and standards. It is easy to imagine an ISO being charged for a significant firefighter injury or duty death if he or she did not enforce those written requirements. In fact, a case can be made that no death should occur when an effective IC and an ISO are on scene. This "legal liability" environment can lead to a CYA approach by the ISO.

An ISO who is caught in the CYA trap is constantly citing CFRs, standards, SOPs/SOGs, and other numbered requirements as the reasons for bringing up safety concerns: "You can't do it that way because OSHA 1910.134 says so." A worst case is one in which the ISO tries to "wash his or her hands" of a safety infraction. Usually, this is a result of an ISO's inability to get an IC or other officer to change an operation or task: "If someone gets hurt, it's not my fault." Perhaps less severe but equally guilty of the CYA syndrome is the ISO who says, "Do you realize how much paperwork I have to do if someone gets hurt?" That phrase is only funny once.

In all cases, ISOs who use CYA tactics are destined to fail. Firefighting crews see right through their interventions and simply dismiss their recommendations or orders as pure

self-preservation. To avoid this outcome, ISOs must display a genuine concern for everyone's safety. They must take personal responsibility for each and every firefighter's safety. Although codes, regulations, and SOPs/SOGs may tell them safe ways of doing things, successful ISOs meet requirements not because of the requirements, but rather because they believe in doing things the safe way—to reduce the threat of injury. To avoid the CYA label, practice "good intent" and "personal concern." In other words, DTRT—*do the right thing*—because it is the right thing to do for the circumstances.

■ The Worker

The ISO who pitches in and helps crews with their tasks falls into the *worker* trap **FIGURE 10-5**. To be effective, ISOs must stay mobile and observant; they must not allow themselves to get involved assisting with a tactical assignment. Too often, this is easier said than done. In many cases, ISOs may feel it is necessary to help move hose lines or throw ladders or make tactical (command) decisions because there simply are not enough people to do all the assigned tasks. If this is the case, the IC should consider retaining the safety officer responsibility and allow the former ISO to lead and work as part of the assigned

FIGURE 10-5 The ISO must avoid getting drawn into performing tasks.

crew. Perhaps more effectively, the tasks and risks should be reevaluated and prioritized to match the number of personnel available (change the IAP!). If there are too few people to do assigned tasks, then the ISO role becomes even more critical, and the need to see the big picture and not get caught up in activities is all the more important.

Before his retirement, Division Chief Gene Chantler of the Poudre (Colorado) Fire Authority said it best:

> If an officer assigned as SO [safety officer] becomes directly involved with suppression, . . . there are two people who made a mistake. The first is the individual assigned to the position; it is clear they are unsuited to fill that role. The second person to make a serious mistake is the IC for appointing someone to the position who doesn't clearly understand what the job entails.[4]

These three traps (bunker cop, CYA, and worker) can, and have, wrecked many ISO programs. The ISO must look past them and be constantly vigilant to avoid them. The bottom line is that an ISO needs to be viewed as a valuable consultant who works with a genuine concern for safety.

Working Within Command Systems

The valued ISO must work within the framework of an established ICS. Organizationally, the ISO typically reports directly to the IC, even though there are still variations of command systems within the fire service. NFPA 1561 and the National Incident Management System (NIMS) are clear that the safety officer, when appointed, shall be integrated within the ICS as a command staff member who reports directly to the IC. Nationally, efforts are being made to unify the various ICS systems (NIMS compliance) so that all jurisdictions will use a common language and expansion processes. With this in mind, we need to address several issues related to the ISO and the commonality or compliance with NIMS. These issues include:

1. IC relations and communications
2. The ISO's role regarding IAPs
3. Local expansion of the ISO function in incidents where a single ISO cannot perform all the required duties
4. The role of the ISO within the NIMS typing scheme

■ IC Relations and Communications

Nearly all incident management systems hold the IC responsible for the safety of responders. This responsibility may not be "dumped off" on an ISO; the IC is ultimately responsible. This responsibility is reinforced in NFPA 1561 (2020 edition):

> **5.9.6.10** *At an emergency incident, the incident commander shall be responsible for the overall management of the incident and the safety of all members involved at the scene.*

When using an ISO, the IC is putting trust in the ISO to handle the safety function as an extension of the IC's responsibility **FIGURE 10-6**.

Making a difference as an ISO is codependent on the ISO's support of the IC and an IC's faith in the ISO. When the two

FIGURE 10-6 The IC is ultimately responsible for the safety of all responders; the ISO is trusted to help meet that responsibility.

elements come together positively, the injury and death potential is reduced. To achieve this harmony, the ISO must strive to present tangible, well-articulated hazard observations and steer clear of any statement, feedback, or action that subverts the IC. The relationship with the IC is strengthened when the ISO embraces and supports several key components of an ICS—namely authorities, communications, and a solution orientation.

Authorities

The assigned ISO must yield to the IC's authority and not pursue any argumentative approach to correcting tactics or strategies that the IC has implemented. The ISO who feels strongly about a request to change a situation has the responsibility to speak clearly, professionally, and rationally to the IC. Once the IC has shared his or her decision, the ISO must accept it and move forward. To belabor a disagreement can be damaging to relationships and does not assist in resolving the concern. Where conflict exists, the ISO may consider a return to information gathering. Incidents are dynamic, and perhaps the new information gleaned can help the IC buy into the ISO's suggestions.

The safety responsibility and authorities that the IC and ISO work to achieve are similar but differ slightly. The best expression of this concept is in a quotation from the U.S. Fire Administration's (USFA) *Risk Management Practices in the Fire Service:*[5]

> Incident Commander's View: *"Get the job done and operate safely"*

> Incident Safety Officer's View: *"Operate safely and still get the job done"*

As already stated, the ISO has the authority of an IC only when an *imminent threat* is present. In all other cases, the ISO needs to work out issues in the framework of the ICS. That means all proposed changes and concerns go through the IC. When the IC delegates the operations function to an operations section chief, the ISO may establish a link with the operations chief so that changes can be made at that level, and the IC should approve the communication link.

Communications

A majority of the National Institute for Occupational Safety and Health Administration (NIOSH) investigations into multi-firefighter fatality incidents fault communication failures, most of which are caused by three barriers: equipment (radio) issues, human factors, and administrative controls (SOPs/training).

The ISO's role in supporting ICS communications is two-fold: (1) Monitor overall radio communications, and (2) use various communications methods to maintain contact with the IC and working crews. For the latter, the two primary means of communication typically used by the ISO are radio and face-to-face.

When using the radio to communicate, be mindful of the limitations and congestion that are often present. Although local procedures and policies are likely to guide your radio use, some suggestions for the ISO are warranted (see the *Getting the Job Done* box).

While radio communication is essential, face-to-face is the preferred (and most effective) method for communications with the IC and outside crews. This method allows for dialogue and feedback, and it eliminates most of the barriers imposed by radio transmissions. Further, face-to-face communications allows both parties the opportunity to visually confirm that the message is understood and encourages spontaneous dialogue. As a rule, the ISO should have face-to-face communication with the IC every 15 minutes at routine incidents and more frequently if conditions or factors change **FIGURE 10-7**. At prolonged incidents, the 15-minute face-to-face rule may not be practical or warranted. In these cases, the ISO and IC should agree on a face-to-face communication schedule.

As mentioned earlier, the ISO must also support the IC through the monitoring of communications. Knowing the common communication barriers—radio issues, human factors, and administrative controls—is a starting place. The accompanying *Awareness Tip* outlines several indicators that a communication barrier or breakdown exists.

Actively listening to radio communications for barrier indicators is one part of monitoring communications. The second—and perhaps more important—portion of communications monitoring has to do with evaluating *what* is said, *how*

FIGURE 10-7 For routine incidents, the ISO should have face-to-face communication with the IC every 15 minutes and more frequently for rapidly changing events.

Getting the Job Done

Radio Use Suggestions for the ISO

- *Use the identifier "Safety" when speaking on the radio.* This is congruent with NIMS and has the potential impact of gaining attention. The use of day-to-day readiness identifiers for the ISO is not consistent with NIMS practice. For example, by using a day-to-day readiness radio designator of, say, "Battalion 101" when filling the ISO role at an incident, the battalion chief can cause confusion as to his or her role. At an incident, the best practice is to use *functional assignment* radio labels. For example:

  Battalion 101 as the ISO: **Ventilation Group from Safety.**

  Truck 3 Captain as Ventilation Group Supervisor: **Safety from Vent Group, go ahead.**

  ISO: **Vent Group from Safety, your secondary escape ladder is being moved to the Bravo–Charlie corner because of heat exposure. Acknowledge?**

  Ventilation Group: **Safety from Vent Group, we copy that the secondary escape ladder is now at Bravo–Charlie.**

- *Limit radio use to the communication of significant safety messages, hazards, and firm interventions.* Limiting your radio use accomplishes two things: First, it allows you to hear more of what everyone else is saying. Second, others who rarely hear from "Safety" are more apt to listen if and when you do break in. If you find that you've had to deliver more than a handful of radio firm interventions, start thinking that the IAP doesn't match conditions and go suggest a tactical time-out through the IC.
- *An order from the ISO to immediately evacuate a building or area should be prefaced by "emergency traffic"; all other firm interventions should be prefaced by "urgent" or "priority message."* Local SOPs have to be considered here. For departments that haven't addressed this issue, the language suggested here should yield the desired outcome.
- *Follow any firm interventions with a radio communication to the IC.* The use of a firm intervention will likely impact the IAP, and the IC needs to acknowledge the change. Waiting for a face-to-face update may not be timely.

it is said, and the *acknowledgment*. Further, the ISO is listening for what is *not* being said. These aren't just radio-listening evaluations; the ISO keeps an ear toward the face-to-face conversations heard among crews. By doing so, the ISO judges not just the effectiveness of communications but also the potential for an injury or the need for an intervention. Let's consider these dynamics more closely:

- **What** is said. At minimum, the ISO should expect to hear the following: the IC communicating a brief IAP, crews giving their CAN report (conditions, actions, needs), the achievement of benchmarks (primary search all clear, water on fire, etc.), and PARs. Failure to hear these should cause the ISO to visit the IC. Conversations between working crews should be job-focused, relative to the task at hand, and supportive of the IAP. ISOs observing face-to-face conversations have uncovered numerous cases in which the crews were not on the same page as the IC—an issue worthy of exploration.

- *How it is said.* In an ideal world, there would be no fire-fighters yelling at an incident scene. It's not our emergency. Reality happens, though. There are many reasons that firefighters might be yelling; perhaps they just need to be heard over loud ambient noise. The key for ISOs is to listen to *how* working crews deliver messages on the radio and face-to-face and compare it to the environment and tasks that are under way. You would hope to hear calm, clear, and concise communications in most cases. Loud, emotionally charged messages can be symptomatic of inexperience or an adrenaline rush but can also indicate an urgent need or firefighter emergency. Message senders who sound out of breath, weak, or pained

should also raise concern. Any emotional or pained voice inflection that is out of place or unusual should warrant immediate ISO attention to investigate further.

- *Acknowledgment.* Many communication issues can be resolved through the use of standard message formats and a communications model that includes a message-repeating acknowledgment:

Command: Engine 2 from Command.

Engine 2: Engine 2, go ahead Command.

Command: Engine 2 from Command, stretch a wye-pack line to side Charlie to prevent exterior extension. You will be Division Charlie.

Engine 2: Engine 2 copies Command, stretching a wye-pack to side Charlie to prevent exterior extension and assuming Division Charlie with a crew of three. Engine 2 D/O will remain Engine 2 Operator.

Command: Command copies Division Charlie is a crew of three. Engine 2 D/O will remain Engine 2 Operator.

Many fire departments use this type of command–order model, which makes it easier for the ISO to pick up missed acknowledgments. Models that don't repeat messages, or those urgent situations in which someone short-cuts the model, make it more difficult for the ISO to catch acknowledgments. An unanswered message can be a flag for an intervention. Consider the risk, crew, location, and conditions present. In some cases, the activation of a rapid intervention crew (RIC) may be necessary based on unacknowledged messages alone.

- *What is not being said.* We've already discussed what should be communicated or what to do if there seems to be insufficient communication. Here we are addressing the interaction between individual crews and command. Face-to-face communication should be taking place regularly. The communication might even be spirited; working incidents are what we train for and the can-do attitude should be felt. Granted, people get tired as an incident drags on, but even fatigue should be felt in the pulse of communications. When the ISO senses that face-to-face communications seem off or nonexistent, some attention to the rehab effort or other stress-related issues may be required.

Solution Orientation

For an IC, problem solving is not only expected, it is required. Too often, however, subordinate officers bring problems to the IC to be solved. While the authority and responsibility to solve these problems rest with the IC, it seems reasonable to bring the problem *and* a solution to the IC for consideration. This *solution-orientation approach* accomplishes two things: First, the IC gets a head start on problem solving—one solution is already drafted. Second, the IC can troubleshoot the solution and offer guidance to others or simply adjust the overall IAP accordingly.

ISOs should embrace the solution-orientation approach, not only to present themselves as partners to the IC, but also to enhance their credibility. In many ways, ISOs are consultants to the IC in that they ask questions, define problems and strengths, draft solutions, and offer recommendations for them to make a decision **FIGURE 10-8**.

Awareness Tip

Communication Barrier Indicators

Communication failures are often cited in firefighter line-of-duty deaths (LODDs), injuries, and near misses. The ISO needs to actively listen to communications as part of their incident-monitoring function. The following list of indicators can help the ISO judge whether a communication failure potential requires an intervention.

- *Incomplete or fractured radio messages.* This issue can relate to equipment problems or human factors. Certain buildings can contribute to the problem (high-rises, malls, large public assembly venues, etc.). Runners, talk-around radio options, and tactical time-outs for face-to-face updates can help.
- *Too much communication.* This issue most likely relates to human factors or administrative control. The use of a tiered message priority system and training can help. On scene, the ISO may have to suggest to the IC to gate-keep by telling all units to standby and to then direct the radio use one by one. The declaration of "priority messages only" or "essential traffic only" can also help.
- *Not enough communication.* In most cases this issue relates to human factors and can be corrected through the IC by asking for status reports/PARs. If that doesn't help, there may be an equipment issue; consider using runners to make sure crews are on the correct tactical channel. Talk-around options may need to be explored.
- *Excessive feedback or interference.* This issue most likely relates to a human factor. Remote microphones are quite sensitive, and voice amplification systems (for SCBA) and multiple radios on high volume near the communicator can increase the potential for feedback. Loud equipment (apparatus, PPV fans, etc.) may be part of the problem. Self-correction, headset use, and step-away reminders can help.

FIGURE 10-8 The ISO should be a consultant who is part of the solution path, as opposed to simply bringing problems to the table.

When a concern arises that requires attention (*not* an imminent threat), the ISO can use a standard solution-orientation approach to communicate it. The approach consists of three shared statements and a question:

1. Here's what I see (a factual observation).
2. Here's what I think it means (your judgment about the hazard).
3. This is what I would do (your solution).
4. What do you think?

By ending with a question, you are acknowledging the IC's authority. This approach is very effective and can work in the most difficult of situations.

The ISO's Role Regarding IAPs

ICs are responsible for developing and communicating an IAP that reflects the overall strategy, tactics, risk management, and responder safety objectives (NFPA 1561, 5.3.16). For initial operations, the IAP is rarely captured in written form, although good commanders use a memory-jogging template to help them better communicate a brief radio IAP in a standard format. An example of this template is found in **FIGURE 10-9** and shows the components of a good IAP for day-to-day incident handling.

It cannot be overemphasized that the ISO must be dialed into the IC's IAP! In fact, once assigned the ISO function at an incident, the first order of business is to confirm the IAP with the IC. The ISO uses the IAP as a foundation for the evaluation of risk-taking as he or she surveys and monitors incident conditions and activities (recon). Of particular interest to the ISO is the risk management level established by the IAP and the responder safety objectives. The ISO needs to monitor these IAP components and provide input when conditions or activities warrant an update or change.

In cases where the first-arriving companies and/or the IC have not communicated an IAP, the ISO should initiate the conversation to do so with the IC. Failure to develop and communicate an IAP has been cited in dozens of NIOSH LODD investigative reports, most coming from single-family residential fires.

A formal written IAP is beneficial for certain incidents and may, in fact, be required by OSHA CFR or other mandatory directives. These incidents include:

- Hazardous materials (hazmat) technician–level incidents (required)
- Technician-level technical rescues (required)
- Complex incidents
- Incidents that span beyond one operational period
- Incidents involving multiple agencies or jurisdictions
- Incidents at which the plans and/or logistics sections have been formalized

A written IAP can range from a simple incident briefing form (ICS 201) to a multipage packet containing dozens of forms prepared by various personnel who submit them for IC approval. The ISO's role for a simple IAP is to review it, make suggestions, and sign off on the IAP. In more complex situations, the ISO will be required to prepare specific forms that are included in the IAP packet. Namely, the ISO may have to prepare the following:

- ICS 208: Safety message/plan
- ICS 215A: IAP safety (risk) analysis
- ICS 214: Activity log
- Site safety plan (hazmat incidents and technical rescues)
- ICS 206: Medical plan (reviewed by the ISO)

Examples of these forms can be found in the appendix of this text. Incidents that require the ISO to prepare and review various forms signal the need to expand the ISO role to include ASOs to maintain a field monitoring presence. The request for ASOs should be made through the IC.

Local-Level ISO—Function Expansion

When a single ISO cannot perform all the safety functions at an incident, the ISO should request, from the IC, an ASO. NFPA 1561 addresses this issue:

> **5.9.6.4** *Assistant safety officers shall be assigned when activities, incident size, incident complexity, or other needs warrant extra personnel to ensure the achievement of safety functions.*

This citation gives some clues regarding the circumstances that suggest the need for an ASO or ASOs (e.g., activities, size, complexity). To be more specific, the ISO should request an ASO from command when one or more of the following incident circumstances are present:

- An incident that covers a large geographic area and includes numerous branches, groups, or divisions. The challenge of monitoring multiple crews over a widespread area is more than a single person can reasonably achieve.
- Firefighter emergencies where the RIC has been activated. This is an incident within an incident, and it is beneficial for the ISO to support command while an ASO monitors the RIC operation and rescue.
- An incident at which crew risk-taking is considered extreme and the need exists for focused monitoring. This allows the ISO to continue safety functions for the whole incident while the ASO focuses on the committed crew(s).
- Incidents that require the input of a safety officer for an established plans section.

Incident Action Plan (Fire) - Briefing Sheet

IAP Briefing Sheet (circle):   **Initial**   **Second**   **Third**

Incident Name:	Location:

Overall Strategy:		Benchmarks
☐ Transitional (default)	☐ Offensive	☐ 360 Complete
☐ Defensive	☐ Marginal	☐ All-Clear, Primary Complete
Risk Profile		☐ Under Control
☐ **HIGH** – Life ☐ **MEDIUM** – Control ☐ **LOW** – Property		☐ Loss Stopped

Tactical Priorities/Objectives:	Companies Assigned: (circle Group/Div Sup)	CAN Report	PARs	
1st:				
2nd:				
3rd:				
4th:				
5th:				

Radio Channel Assigned: _____ Secondary Ops Channel: _____

Hazard Issues/Status	Safety Systems	
		Reminders
Utilities _____	☐ ISO _____	☐ Communicate IAP
Access/Egress _____	☐ RIC _____	☐ Update IAP
Bldg. Integrity: _____	☐ IC Aide_____	☐ Time Prompts
Heat/Smoke Flow: _____	☐ REHAB _____	☐ CAN/PARS
Other: _____	☐ EMS Standby_____	☐ Control Zones
Other: _____	☐	☐

IC: _____   Date: _____   Time: _____

FIGURE 10-9 A sample incident action plan.

■ Incidents that require a certain technical expertise to assist the ISO with safety functions (e.g., county health official, building engineer, specialty team member).

■ Incidents that involve the interface with local, state, or federal health or safety representatives.

■ Multiagency incidents using a unified command structure or incidents at which area command is established.

Certain types of incidents, specifically hazmat incidents (technician level), confined space rescues, and trench rescues, mandate the appointment of an ISO with the appropriate training/certification. When the designated ISO does not possess technician-level training or certification for the type of incident in question, appointing a technician-level–trained ASO as part of the ISO staff helps satisfy the safety needs of the technician-level teams. The technician-trained ASO can be titled assistant safety officer–hazmat (ASO-HM) or simply hazmat safety officer in hazmat incidents, and assistant safety officer–rescue tech (ASO-RT) for a confined space incident. When ASOs are utilized, the ISO should be located in the same area as the IC, becoming a central contact point for the

ASOs. This structure allows the ISO to manage overall safety functions for all responders (including non-technician-level responders) and all the concerns that shape the IAP.

When ASOs are used, the ISO must take on some supervisory responsibilities. Namely, the ASO or ASOs need to be clear on their functions and communication channels. At a minimum, the ISO who is granted additional ASO resources needs to brief them prior to deployment **FIGURE 10-10**. That briefing should include discussion of the following details:

- The IAP
- Assigned communication channels, such as tactical and command
- Geographic area or area of focus for each ASO (Example: "Go to the West Division and monitor and intervene as necessary.")
- Dot-line reporting (Example: "West Division is the Engine 10 Captain and they have three companies assigned. Check in with E-10 Captain.")
- Desired reporting schedule, such as every 20 minutes
- Documentation requirements (Example: "Maintain a unit log and report any firm interventions to me via radio. I will forward them to the IC.")

Transferring the ISO Function

As an incident escalates or is prolonged, it is sometimes necessary for the originally assigned ISO to transfer the safety function to another person. The reasons that the duties of an ISO would be transferred from one individual to another are myriad:

- The incident has become complex and requires a more experienced/trained ISO.
- The IC wishes the original ISO to stay in the field as an ASO (already dialed into the activities taking place) and wants a formal ISO to stay at the command post.
- The duration of the incident requires the ISO to take a break or requires multiple operational periods.

FIGURE 10-10 ASOs should be requested when the incident warrants extra personnel to ensure the achievement of safety functions. Those ASOs need to be briefed by the ISO.

- The incident is being turned over to another agency or to a preformed incident management team (discussed later).

Transferring the safety function is much like transferring command—a formal briefing needs to take place. Using the MEDIC acronym can help the ISO remember what information needs to be transferred to the incoming ISO during the formal briefing:

- **M = Monitoring issues:** Give the new ISO an overview of the incident scene and let him or her know when the last recon trip was made. Describe the established risk-taking level so that the incoming ISO can get a handle on the appropriate monitoring approach (*risk a life to save a life* requires more attention to monitoring).
- **E = Evaluations:** Note any hazards that have been evaluated. Also describe your impression of the rehab effort and any concerns regarding crew fatigue.
- **D = Developed preventive measures:** Share details regarding any safety zones, especially "no-entry" zones. Likewise, note any mitigation countermeasures (scene lighting, protection from thermal stress, crew rotation schedule, etc.).
- **I = Interventions:** Note any firm interventions that took place. History may repeat itself!
- **C = Communications:** Obviously, the new ISO needs to know the radio channels that have been established. Likewise, any assigned ASOs need to be informed that the transfer is being made. Perhaps less obvious is the need to share any safety messages, site-safety plans, or threat analysis forms that have been developed.

You may have noticed that no mention was made to share the IAP with the incoming ISO. Each incident will have its own circumstances, but generally, the incoming ISO needs to close the IAP loop through the IC.

The Role of the ISO for Regional, State, and National Incident Types

Some incidents will present complexities that surpass the resource and management capabilities of the local response. As an incident response effort expands, the basic duties and responsibilities of the ISO also become more complex. A system of expansion is in order, and in fact, a template has been created to help guide that expansion. NIMS is part of the National Response Framework and is designed to help transition the management of local incidents into one that uses regional, state, and national resources. Although the scope of this text involves the ISO at the local level, some discussion is in order to help the ISO understand how the position expands and transitions into a system that helps achieve safe operations for the largest of incidents. We can begin with a look at a system of classifying incident magnitude: the national incident typing scheme.

The National Incident Typing Scheme

The lessons of recent disasters have motivated emergency response agencies to adopt "typing systems" for state and interstate mutual aid resources. The National Framework NIMS document has given responsibility to the NIMS Integration Center (NIC) to develop a resource typing protocol. NIC is responsible for the ongoing development and refinement of various NIMS activities and programs. To meet their charge, NIC has created a system of incident "typing" for NIMS all-hazards incident response. The USFA has created some measurements or

Fire Marks

The National Incident Typing Scheme
The following represents an overview of the national incident typing scheme and includes some of the measurements or triggers that the USFA has addressed to help clarify the typing intent.

- **Type 5 (local event):** The incident can be handled with local resources and without the activation of general staff positions of the ICS (e.g., planning, operations). Incident duration is one operational period that rarely exceeds a few hours. No written IAP is required.
- **Type 4 (local event):** The incident requires more than several resources (original jurisdiction plus mutual aid), and the IC activates command staff and general staff functions as needed. The incident is limited to one operational period. No written IAP is required, but documentation should begin for operational briefings and priorities.
- **Type 3 (regional event):** Capability needs have exceeded the abilities of initial resources, and IC positions are being added to match the complexity of the incident. Command staff, general staff, division/group supervisors, and unit leaders are being used. A preorganized and trained Type 3 incident management team (IMT) assumes the management of initial actions where they exist. Multiple operational periods are being considered or developed. A written IAP is required.
- **Type 2 (multiregion, state, or initial federal event):** The incident requires the response of resources from out of the area, multiple operational periods are involved, and most or all of the ICS positions are being filled. A preorganized and trained Type 2 IMT is assigned. Guidelines suggest an incident involving less than 200 personnel per operational period or 500 total. A written IAP is required for each operational period.
- **Type 1 (federal event):** A Type 1 incident is the most complex. All the ICS positions are activated and branches are established. A preorganized and trained Type 1 IMT manages the incident.

Sourced and paraphrased from the U.S. Fire Administration.

triggers based on ICS systems that can help clarify the NIMS typing scheme (see the *Fire Marks* box).

The typing system is supported by pre-established incident management teams (IMTs) for Type 1 and 2 incidents. There are roughly 16 Type 1 teams assigned to 11 geographic areas in the United States.[6] These teams comprise 30 to 50 personnel who respond to incidents on a rotational and/or geographic basis. The majority of these team members are employees of other agencies. For example, the Type 1 safety officer for a given team serves as the career fire chief of a local fire department.

In addition to the Type 1 and 2 IMTs, there are four National Incident Management Organization (NIMO) teams that consist of 5 to 8 full-time persons (federal government employees) that assist the larger teams. The United States and Alaska are divided into 11 geographic areas for the purpose of incident management and mobilization of resources (people, aircraft, and ground equipment). Within each area, an interagency Geographic Area Coordinating Group (GACG), made up of fire directors from each of the federal and state land management agencies from within the area, is established. Working collaboratively, the GACG's mission

is to provide leadership and support not only for wildland fire emergencies but also for other emergency incidents (e.g., earthquakes, floods, hurricanes, tornadoes), as necessary. Authority for establishment of the GACG is through departmental policy and interagency agreements. Cost-effective sharing of resources among public agencies is a key component of the GACG mission and is expected by the public, Congress, and states.

All agencies and geographic areas work together under the auspices and direction of the National Interagency Fire Center (NIFC). The National Interagency Coordination Center (NICC) is the focal point for coordinating the mobilization of resources for wildland fire and other incidents throughout the United States using a Geographic Area Coordination Center (GACC), which serves as an overhead "dispatching" center for the teams.

Some regions (state or multiple regions within a state) have seen the benefits of IMTs and have created and trained Type 3 IMTs to fill the void between local events and state- or national-level events. Many of these teams have been "certified" and are recognized through the Federal Emergency Management Agency. Taken further, training programs are available through the USFA to help local entities develop Type 4 and 5 IMTs, usually consisting of a pool of trained personnel who can serve in various ICS positions.

All the Type 1, 2, and 3 IMTs have designated personnel who are trained and on call to fill the safety officer position. The IMTs use the acronym *SOF* for the position we've been labeling ISO. The SOF designator is followed by the type level; for example, a Type 1 SOF is labeled SOF1. Before we look at the role of the SOF, let's cover the time line that can be expected for the expansion and transition of incident types.

IMT Time line for Response

As mentioned earlier, Type 4 and 5 incidents are those that only last one operational period for control actions. An incident that lasts longer than one operational period is a candidate for the response of a Type 3 IMT if such teams exist for the area. On-call members of the IMT need time to deploy and set up. Such is the case for Type 1 and 2 IMTs. **FIGURE 10-11** shows a typical time line for the IMT response and operations.

Transitioning to Type 1 or 2 Incidents[7]

As an incident expands, the previous role of the ISO transitions from one with lots of time spent surveying incident conditions/operations ("boots on the ground" time) to one that is quite strategic with lots of time spent meeting, planning, and communicating at the command post. (The field assessment component of the safety function is accomplished mainly by ASOs once the incident gets to the Type 1 or 2 level.) Part of this transition will include a face-to-face session where the IC (and likely the initial ISO) will meet with the incoming IMT to discuss expectations, authorities, and objectives. If and when the IMT assumes management of the incident, the former ISO is likely to receive another assignment or may actually need a break!

Moving forward, the IMT SOF will attend a separate command staff meeting and is then given time to conduct an initial assessment of hazards and risks in order to prepare an incident risk assessment worksheet **FIGURE 10-12**. The worksheet is developed using information from the previous ISO, field observations, meetings with operations/planning sections (current and upcoming missions or objectives), and historic

IMT Timeline for Response and Operations

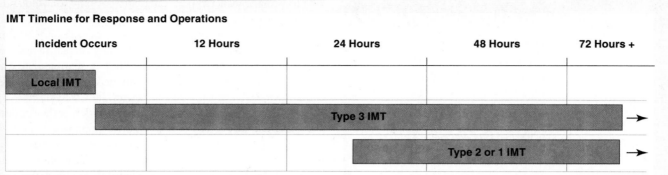

Incident Occurs	12 Hours	24 Hours	48 Hours	72 Hours +
Local IMT				
	Type 3 IMT			→
			Type 2 or 1 IMT	→

Upon arrival, the IMT's incident commander (IC) will meet with the local IC and the agency administrator (county executive, city manager or mayor, etc.) to determine what they desire/expect from the IMT and to obtain any necessary delegation of authority. The incoming IC will then brief the rest of the incoming IMT about its role/expectations. The IMT will then integrate as requested into the current ICS structure.

FIGURE 10-11 A typical time line for the request and response of a predesignated IMT.

FIGURE 10-12 The SOF1 tracks hazards and control measures on an incident risk assessment worksheet.

or previous experience data that are relevant to the current incident. The SOF then begins a cyclic process of meetings, briefing preparations, further risk/hazard analysis, posted safety message development, IAP and medical plan review, and preparation for the next operational period **FIGURE 10-13**.

The SOF1 or SOF2 relies on line safety officers (titled *SOFRs*), who serve as ASOs. The SOFRs monitor working teams, deploy safety measures, and make interventions. This is not to say that the SOF1 or 2 is spending all his or her time doing paperwork; he or she is constantly on the move, seeking input, making site (facility) inspections, communicating with/directing SOFRs, maintaining relationships with other IMT members, attending meetings **FIGURE 10-14**, and correcting safety hazards as they arise. If an emergency or unforeseen situation arises, the SOF will treat it as "an incident within an incident" and likely dispatch an SOFR to the occurrence location—the point being that the SOF1 tries to remain strategic and available to the IC.

The SOF1 uses many more communications tools than an ISO uses for a typical local event. Obviously, face-to-face and radio methods are used, but the SOF1 expands those options and uses cell phones (voice and text), emails, and developed briefings and posted messages to address and share hazard information.

Pacific Northwest Incident Management Team 2
Safety Officer Responsibilities

SOF 1 – Command Staff	SOF 1 – Strategic/Field
0530 OPS/Safety Briefing0530	OPS/Safety Briefing or
0600/1800 Briefing at ICP	Brief at Spike Camp
0830 Risk Analysis 24/48/72	Supervision of Line SOFs
1200 C & G Meeting	DIV/GRP assignments for SOFs
Pre-Strat (Day-Night)	Order/Release Line SOFs
Planning Meeting (Day-Night)	Performance Evaluations-SOFs
Conf. Call w Safety & Health	Work w Ops on Strategy & Missions
Safety Message & 9 Line	Camp Inspections including Spikes
IAP Inputs due	Medical Response Map Verified
Chronology & TOC	Incident within an Incident
Night on call (alternate)	Night on call (alternate)
Safety Summary & Narrative	Accident Investigations and Reports
Input to Safety Narrative	Post Safety Signs at ICP Base Camp
Lessons Learned	
DMOB Plan	

FIGURE 10-13 The responsibilities of a Type 1 SOF include both command staff and field commitments.

FIGURE 10-14 A Type 1 IMT gathers information for an IC and general staff briefing at the Deception Complex fire in Oregon (2014).

Chief Concepts

- The ISO can trigger safe behaviors using a palette of options. Being visible by wearing a unique vest or other identifier and practicing safety behaviors are examples of passive triggers. Using soft and firm interventions are examples of active or direct triggers.

- Firm interventions should be saved for those occurrences at which an imminent threat exists. An imminent threat is one that is judged to present a danger to persons or property that is so urgent and severe that it requires immediate corrective or preventive action. A firm intervention should be considered a direct order with all the authority of the IC.

- ISOs can fall into traps that may render them ineffective. These traps are (1) acting as a "bunker cop," (2) adopting a "CYA" approach, and (3) performing tactical activities (becoming a "worker").

- NFPA standards and NIMS require the ISO to be integrated into the incident management system as part of the command staff, reporting directly to the IC.

- Elements that must be addressed in the IAP include the overall strategic goal, risk-taking level, tactical assignments, incident priorities, communications channels, and identified safety concerns.

- The ISO focuses on risk management levels and the safety hazard components of an IAP when monitoring an incident.

- A firm intervention by an ISO (stop/alter/suspend/withdraw) requires the ISO to immediately notify the IC of the actions taken so that the IAP can be adjusted. The use of a radio to report the intervention is appropriate to avoid any delay in adjusting the IAP.

- The ISO uses face-to-face as the primary communications tool and should meet with the IC every 15 minutes for typical incidents. The ISO should reserve the radio for communicating firm interventions. When using the radio at incidents, the designated ISO should use the functional identifier of "Safety" versus his or her day-to-day readiness identifier (e.g., Battalion 101).

- The ISO should monitor radio transmissions and actively listen for signs of communication barriers such as incomplete/fractured messages, excessive communication (overload/lack of tiered or communication priority), insufficient communication, and excessive feedback interference. Barriers can be caused by equipment issues, human factors, or administrative controls (SOPs/training).

- The ISO should use a *solution-oriented* approach when bringing issues to the IC. The approach includes three statements and a question:
  1. This is what I see.
  2. This is what it means to me.
  3. Here's my solution.
  4. What do you think?

- The ISO can expand the safety function needs by requesting ASOs through the IC. Circumstances that would require ASOs can include geographic size, complexity, RIC activation, or risk level of the incident. Likewise, an ASO would be useful when the incident requires the ISO to participate in the plans section or when multiple agencies or area command is involved.

- ISO duties could be transferred from one individual to another for myriad reasons:
  - The safety function requires a higher level of training/experience.
  - The ISO needs a break or there are multiple operational periods.
  - The ISO needs to become an ASO to retain focus on a given activity.
  - The incident is being transferred to another agency or IMT.

- When transferring ISO duties, the ISO can use the MEDIC acronym to remember the information required to be shared: **m**onitoring issues, **e**valuations, **d**eveloped preventive measures (zones, etc.), **i**nterventions, and **c**ommunications.

- The national "typing scheme" is a way to categorize incidents based on their resource needs:
  - Type 5 = Local incident that is easily handled
  - Type 4 = Local incident requiring additional resources
  - Type 3 = Regional incident
  - Type 2 = Multiregional, state, or initial federal incident needing less than 200 personnel per operational period
  - Type 1 = Most complex type of incident

- Type 1 and 2 incidents will likely include the response of preorganized, trained IMTs with predesignated individuals to fill ICS functions. The ISO function is filled by an SOF1 or SOF2, who will likely assign line safety officers (SOFRs) to fill the ASO or field safety function.

Key Terms

firm intervention A direct order to immediately stop, alter, suspend, or withdraw personnel, activities, and operations due to an imminent threat.

imminent threat An act or condition that is judged to present a danger to persons or property that is so urgent and severe that it requires immediate corrective or preventive action.

potential threat An activity, condition, or inaction that is judged to have the capability, but not immediacy, to cause harm to persons or property and thus warrants monitoring and/or operational modification.

safety trigger An approach that an ISO uses to help remind firefighters to be situationally aware and to perform in a safe and predictable manner.

soft intervention Awareness or suggestive communication made to crews or command staff that causes them to modify their observations and activities to prevent injury from a potential threat.

Review Questions

1. List two *passive* triggers that an ISO can employ to achieve safe behaviors.

2. List two *active* triggers that an ISO can employ to increase safety.

3. List the three ISO traps and discuss how each can render the ISO ineffective.

4. Who is ultimately responsible for incident safety?

5. Describe the organizational relationship of the ISO within the ICS.

6. What are the two primary communication tools the ISO uses?

7. As a rule, how often should the ISO communicate with the IC at routine incidents?

8. List the three barriers that cause communications to break down and the four indicators that a barrier is present.

9. What is meant by a *solution-orientation* approach to problem solving? How is it accomplished?

10. List the components of an incident action plan. Which components should the ISO focus on?

11. What type of ISO actions can impact the incident action plan? How should the ISO process those actions?

12. List the circumstances that may cause the ISO role to be transferred to another individual. What information should be transferred to the incoming ISO?

13. When should an ISO expand the safety function to include ASOs?

References and Additional Resources

Brunacini, Alan V. *Fire Command*. 2nd ed. Quincy, MA: National Fire Protection Association, 2002.

Emery, Mark. The 4-C Communication Model. *Health & Safety for the Emergency Service Personnel*, 16, no. 9, September 2005.

The Federal Emergency Management Agency website offers resources for the National Incident Management System, including online training, ICS forms, graphics, and sample procedures. It is available at https://www.fema.gov/national-incident-management-system.

NFPA 1561, *Standard on Emergency Service Incident Management System and Command Safety*. Quincy, MA: National Fire Protection Association, 2020.

U.S. Fire Administration. *Voice Radio Communications Guide for the Fire Service*. Washington, DC: Department of Homeland Security, 2008.

The U.S. Fire Administration website describes training programs for local IMTs. It is available at http://www.usfa.fema.gov/training/imt/.

Endnotes

1 Anderson, Peggy. *Great Quotes from Great Leaders*. Albert Schweitzer. Lombard, IL: Great Quotations Publishing Company, 1990.

2 Lt. James Bethea of the Baltimore City Fire Department, assigned as a duty safety officer, fell through a floor opening and died of smoke inhalation in the basement of a row-home while performing recon in an exposure. The full NIOSH investigative report and details of the ISO LODD were not available at the time of this printing.

3 Definition taken from NFPA 1561, *Standard on Emergency Service Incident Management System and Command Safety*. Quincy, MA: National Fire Protection Association, 2020.

4 Chantler, Gene. The Safety Officer: A Roundtable. *Fire Chief Magazine*, February 1993.

5 *Risk Management Practices for the Fire Service, FA-166*. Washington, DC: United States Fire Administration, 1996.

6 The number of teams is dynamic and flexes based on availability, capability, and training needs. Further information can be found at www.nifc.gov/nicc/.

7 A special thanks to Type 1 Safety Officer Sam Phillips, Pacific Northwest National Incident Management Team 2 and Fire Chief for the Clallam County Fire District No. 2 (WA), for his assistance in developing this content.

Florida has seen its share of tourists, snowbirds, hurricanes, and wildfires, and Florida fire departments have had to adapt to the challenges presented by each of those. Over time, firefighters and fire officers have learned to work especially well with each other across city, district, and county lines. The statewide response system in Florida has evolved into what is now considered one of the nation's best in responding to disaster. It was not always that way. When Florida started experiencing significant interface wildfires (1990s), area fire officers were burdened by the lack of training, equipment, and organizational procedure needed to deal with such large fires. Several chief officers reported that it was an amazingly difficult experience. Federal IMTs and wildfire-trained crews responded to Florida in large numbers, bringing their significant experience gained from fighting wildfires in the dry, western United States. Even with their experience, Florida wildfires presented interesting challenges that the outside teams never had to deal with: swamps, high ambient temperatures, and *alligators*.

During one wildfire, an Oregon fire crew experienced a close call with an alligator. Appropriately, the safety officer assigned to the fire decided to include alligator awareness training as part of the safety briefing for outside crews. However, that led to another problem. Armed with their new alligator information, outside crews became more comfortable in the alligator habitat. Word then got out that teams from other states were actually trying to catch an alligator and have their picture taken with it as a keepsake of their Florida experience. Reportedly, bets were placed on which team would get the first picture. Can you imagine the headache that presented for a safety officer? The IMT SOF was now facing serious safety *and* legal issues!

1. Whom would you consult to develop a strategy that addresses the alligator-related issues?

2. What forms of interventions should be initiated by an ISO to address alligator (or other wild animal) encounters?

3. Under the national incident typing scheme, an incident that escalates past the local level becomes what type of incident?
   - **A.** A mutual aid incident
   - **B.** Type 5
   - **C.** Type 4
   - **D.** Type 3

4. The letter "D" in the ISO's hazard MEDIC acronym stands for what?
   - **A.** Develop preventive measures
   - **B.** Direct tactical activities
   - **C.** Determine risk acceptability
   - **D.** Don PPE

Note: Thanks are due to the numerous Florida fire officers who shared their anecdotal experiences for this case study. Special thanks go to Assistant Chief Jeff Money, Brevard County Fire/Rescue, Florida. Any misrepresentations are the fault of the author.

Smoke: © Greg Henry/ShutterStock, Inc.

A Systematic Approach to the ISO Role

Flames: © Ken LaBelle NRIFirePhotos.com

Knowledge Objectives

Upon completion of this chapter, you should be able to:

- Describe the four steps that help an incident safety officer (ISO) become integrated into an incident. (pp 167–168)
- Define the roles of an ISO at planned and unplanned events. (NFPA 5.2.1 , pp 165–167)
- List the five ISO general duties applicable to all incident types. (p 165)
- Describe the incident scene conditions that are monitored as part of an ongoing incident. (NFPA 5.2.5 , p 166)
- List the components that make up a firefighter safety system. (pp 169–171)
- Describe the risk management criteria that would cause a threat to be imminent to firefighter survival. (NFPA 5.2.4 , p 168)
- Describe methods used to ensure member accountability at an incident scene. (NFPA 5.2.6 , pp 169–170)
- Identify the consequences and hazards that the lack of accountability at an incident scene could create. (NFPA 5.2.6 , pp 169–170)
- Identify hazardous incident conditions that would require the establishment of a hazard control zone. (NFPA 5.2.7 , p 171)
- Describe how hazard control zones can be identified and communicated at an incident scene. (NFPA 5.2.7 , p 171)
- Describe how entry into the hazard zone area can be controlled. (NFPA 5.2.7 , p 171)

Skills Objectives

Upon completion of this chapter, you should be able to:

- Demonstrate the ability to evaluate hazards and assign the relative degree of risk to the hazard. (NFPA 5.2.4 , p 166)
- Demonstrate communications necessary to alert members of imminent threats. (NFPA 5.2.4 , p 167)
- Demonstrate the ability to monitor an accountability system. (NFPA 5.2.6 , pp 169–170)
- Establish hazard control zones at an incident. (NFPA 5.2.7 , p 171)

A multitude of tasks that need to be accomplished, rechecked, and monitored are common among many fire service operations. Incident checklists, computer software, and even dispatch-driven time voice prompts assist officers and members in making sure things do not get overlooked. The ISO role is no exception, and many incident scene and operational factors need to be assessed and continually monitored as the incident progresses.

1. Does your department utilize any form(s) of checklists or other prompting tools to help make sure tasks are completed and assessed?
2. If you were given the responsibility of creating an ISO incident scene checklist, what would it include? How many different types of incident checklists would you develop?

Introduction: Getting Started

The list of duties and responsibilities that the incident safety officer (ISO) is tasked with is quite lengthy. Without a systematic approach to accomplish all the ISO-related mental and physical tasks, the safety officer is destined for mediocrity. Experienced ISOs typically develop a uniform approach that is applicable to many types of incidents. Yet certain issues and frustrations challenge the ISO who is trying to develop the best approach to incident duties.

Perhaps the biggest issue (and frustration) facing the ISO is the prioritization of necessary physical and mental functions. Often, an ISO arrives at the incident scene after the initial fire attack or setup. After checking in with the incident commander (IC), the ISO hopes to get a quick briefing on what has happened, what is planned, and what the IC needs. What's next? Often the IC requests a 360-degree scene survey; sometimes, the IC may have a specific question or concern (such as, "When is this building going to collapse?"). Other times, the IC leaves the options up to the ISO by saying, "I don't have the full picture here. Find out what's going on and report back." Worse yet, the IC may just assume that the ISO knows what needs to be done and may give no indication of priorities. It is not at all uncommon for an IC to have the ISO draft a quick action plan based on the current and predicted situation status (sitstat) and resources (restat). Given all these variables, the ISO may find it hard to develop a starting place for the many items that need to be addressed—which may lead to confusion, stress, or frustration.

This chapter looks at several methods or approaches that help the ISO address all the required functions. In the early 1990s, approximately 500 feedback/review sheets from the Fire Department Safety Officers Association course "Preparing the Fire Ground Safety Officer" were reviewed. In response to the question "What can we add to the class to better address your needs?" respondents listed numerous frustrations and asked for tools to address them. Sample criticisms and frustrations were as follows:

- There are no clear starting places for ISO duties.
- Existing ISO checklists are too short and/or too general to provide what is needed to be effective.
- Typical checklists imply that once an item is checked off, it no longer needs attention—a dangerous assumption for an ISO. A checklist that includes everything that an ISO should address would be very long and unrealistic to apply to every incident.
- The ISO must stay flexible and not be sidetracked with details that can obscure the "big picture." How do we do this?
- The expectation is that the ISO needs to "see all" and "know all" to be effective. We need a tool to help us accomplish this.
- The ISO must be *reactive* to the needs of the IC as well as *proactive* in the prevention of injuries to firefighters. These opposing expectations create a priority clash at times.

The last point is compelling: The ISO role can be viewed as both reactionary and proactive. On one hand, the ISO must look at what has already happened and offer solutions to correct unacceptable hazards or risk situations. On the other hand, the ISO has the opportunity to predict future events and make suggestions to minimize the effect of the events on the firefighter. Most ICs and working crews are receptive to the reactionary component; the situation is likely to be visual and explainable. Being proactive, however, is more subjective and places the ISO in a position to "sell" something that is not so obvious. In these cases, the ISO has to deal with opinions regarding the "likelihood" that the hazard in question will affect firefighters.

Addressing all the ISO functions and balancing reactive and proactive needs can be achieved by using a systematic ISO approach to incidents. The two most common approaches to ISO duties include the use of checklists and the use of action models. To be effective, checklists and action models must have certain qualities that make them usable: flexibility, adaptability, cyclicity (the revisiting of items), proactivity, and reactiveness.

The remainder of this chapter looks at an arrival process to integrate the ISO into an incident. We finish the chapter with a list of some specific ISO focus areas that are applicable at all incidents, regardless of type.

A Simple and Effective Tool: Checklists

One thing is certain in the fire service: *We love our checklists!* Nearly every area of the fire service has adopted them because of all the benefits they provide. Some benefits are readily apparent:

- They provide a quick reminder of things that need to be done.
- When you are distracted, they help you get back on track.
- They lend themselves to uniformity (from person to person doing the same task).
- Archiving is relatively simple.
- Changing the checklist is relatively simple within the framework of a fire department.
- Most formats are easy to understand.

It is easy to see why the fire service has embraced the checklist. The flip side of the coin is that checklists can hamper the ISO's effectiveness. The following are some disadvantages:

- Checklists might imply that there is only one right way to perform the functions of the ISO.
- Checklists have a tendency to be either overly simple or overly lengthy.
- Once an item is checked off a list, the ISO may forget to revisit it.
- To cover a multitude of incident types, the ISO would have to carry a filing cabinet stocked with every conceivable checklist.
- Checklists often imply an order for task completion, especially for the new or inexperienced ISO. It is rare that ISO tasks are done in order—or even fully completed. Most ISO tasks are cyclic in nature.
- Checklists may be subject to subpoena in legal matters.

Many templates and sample forms are available to help create a checklist designed for use in a systematic approach to ISO functions. Even when a jurisdiction (especially at large, multiagency incidents) requires predesigned checklists and forms, the ISO can easily develop a helper checklist or notepad that makes up for the inadequacies of defined forms. When *required* checklists and forms are not an issue, the ISO is encouraged to create a checklist that takes into account local variables. Appendix B of this text includes checklists that have been developed by working ISOs and may be helpful when designing your own.

Action Models

Basically, an <u>action model</u> is a template that outlines a mental or physical process that considers inputs that lead to an output or outcomes. Some ISOs prefer the flexibility and adaptability of action models as opposed to checklists. The biggest advantage of action models is that they furnish a template in which to process multiple inputs in a logical, cyclic, and meaningful way.

The first and second editions of this text presented an action model developed by this author and then Training/Safety Officer Terry Vavra of the Lisle-Woodridge (IL) Fire District (most recently Fire Chief of the Buffalo Grove [IL] Fire Department). The model arose from the frustrations experienced with designing checklists for the ISO. We believed that a simple, easy-to-apply model could be created that was adaptable to most incidents, overcoming those frustrations. One of the main ingredients in designing the action model was the need to remind ISOs to be "cyclic" in their thinking and to stay open to changing inputs.

■ Cyclic Thinking

All of us use a linear thinking process to handle incidents—that is, a process having a defined starting point and a desired ending point (goal accomplishment). During the linear process, inputs are considered that may or may not change the path to accomplishing a goal. At some point the person making decisions may reach his or her maximum capacity for input, causing stress. Combine this point of maximum input with the focus to reach an ending point, and it is easy to see that some hazards may be overlooked. The situation awareness gurus call this "input saturation." This point was highlighted following the tragic Storm King Mountain fatalities during the South Canyon fire in Colorado (see the *Fire Marks* box).

Fire Marks

Overload at the South Canyon, Colorado, Fire

Fourteen firefighters died in a sudden blowup of the South Canyon fire in 1994. Following the incident and subsequent investigations, human error, decision-making overload, and communication breakdown were cited as contributors to the tragedy. Ted Putnam, PhD, specialist for the Forest Service's Missoula Technology and Development Center, wrote an article titled "Collapse of Decision Making"[1] that discusses the psychological elements that lead to failed leadership during wildland incidents. In the article, Dr. Putnam makes a case for the traps of linear thinking by discussing decision-making models and conclusions based on numerous studies. Dr. Putnam concludes that "our linear thinking tends to underestimate hazards, particularly if the hazard is increasing at an exponential rate. Under stress, leadership becomes more dogmatic and self-centered. It regresses to more habituated behavior." In other words, increasing incident stress actually leads the decision maker to minimize the number of inputs being considered, and the person resorts to what may have worked in the past—not what is necessary for the conditions currently faced. The decision makers on Storm King Mountain were simply overloaded. Dr. Putnam concludes that we need to pay more attention to the psychological and societal processes of our people and the way inputs are managed.

The IC needs to think in a linear fashion, establishing a path and working toward a positive conclusion. Still, there need to be checks and balances to that thinking; that is, the IC needs to be curious, question overconfidence, and be observant. These are not linear qualities; rather, they are cyclic ones requiring openness. Unfortunately, the IC needs to keep things moving forward. An ISO can help. It is imperative that ISOs create an environment in which they can stay open to multiple inputs and maintain a high degree of situation scanning and awareness—cyclic thinking. In many ways, the ISO is the "what-if" thinker who tends to overestimate hazards. With this in mind, the ISO Action Model developed by Dodson and Vavra used a circular image—a reminder to stay open to the hazards that may eventually cause injury to firefighters. A system of cyclic, or recurring, evaluation by the ISO can help eliminate the trap of linear thinking and underestimating hazards.

■ The Dodson/Vavra ISO Action Model

The Dodson/Vavra ISO (DVISO) Action Model is depicted as a cyclic, four-arena model that allows the ISO to mentally process the surveying and analysis of typical incident activities and concerns. As shown in **FIGURE 11-1**, the model calls for the ISO to analyze four general arenas. Neither a starting place for the model nor a direction of flow should be inferred. Upon the ISO's check-in with the IC, a starting place may be assigned. If this is the case, the ISO simply jumps into the cycle as directed. If the IC does not select a starting place, the ISO can start where he or she feels attention is warranted. Once in the cycle, the ISO should conduct an initial survey of each arena and then monitor the applicable concerns in each. In essence, the ISO performs a mental evaluation of the conditions, activities, operations, or probabilities in each arena. The DVISO action model components requiring evaluation and attention can be thought of as the *Four Rs.*

The DVISO action model has served the fire service well, mostly as a means to help local fire departments develop their own ISO checklists/action models and to assist ISO training and certification programs. While the DVISO action model provides an easy-to-remember and simple approach to ISO duties, it is woefully short and perhaps too simplistic for today's rapidly changing and complex incidents. Further, the DVISO model does not meet the specific ISO functions/duties that are outlined in various standards and laws—namely, NFPA standards and the Occupational Safety and Health Administration's (OSHA) Code of Federal Regulations (CFR). Using constructive criticism from many practicing ISOs and the duties outlined by NFPA/OSHA, a more appropriate action model can be derived: the hazard "MEDIC" approach introduced in the chapter *The Safety Officer Role.*

■ The Hazard "MEDIC" Action Model

In several chapters we have used the mnemonic MEDIC as a memory aid for ISO responsibilities and the information that needs to be passed when transferring the ISO function to another person. It can also serve as an action model, especially when expanded. Likewise, it serves as a reminder of the five

ISO general duties that are expected of ISOs at all incidents. As a starting place, recall what each letter represents:

M = Monitor
E = Evaluate
D = Develop
I = Intervene
C = Communicate

As an action model, the *monitoring* portion of MEDIC is an "input." The *evaluation* becomes an "analysis," and *developing, intervening,* and *communicating* become actions, or "outputs." Let's now expand these elements to show more specifically what each represents in helping an ISO fulfill his or her duties.

Monitor

Most ICs agree that having an ISO available to help with ongoing 360-degree scene surveys is essential to improving scene safety and staying abreast of changes that can influence the incident action plan (IAP). The ongoing surveys, called <u>reconnaissance</u>, or "recon" for short, are an exploratory examination of the incident scene conditions and activities **FIGURE 11-2**. Recon becomes a crucial part in identifying conditions and

FIGURE 11-2 ISOs perform *recon* as a monitoring function to identify incident conditions and activities that need to be evaluated.

Resources

Report ◁ ACT ◁ Recon

Risk

Action leads to "Report"

FIGURE 11-1 The Dodson/Vavra ISO Action Model.

activities that need to be evaluated (see the *Getting the Job Done* box). The recognition skill sets we covered in Section 2 (reading smoke, buildings, and hazardous energy) come into play here. As the ISO performs recon, he or she must also actively listen to communications as presented in the chapter *Triggers, Traps, and Working with the Incident Commander*.

FIGURE 11-3 The ISO uses the front-loaded skills of *reading risk* and *reading firefighters* to help evaluate incident activities and conditions.
Courtesy of Forest Reeder

Getting the Job Done

ISO Monitoring Issues at Incidents

Incident conditions that need to be monitored include, but are not limited to:

- Weather
- Hazardous energy (utilities, etc.)
- Building construction and collapse potential
- Fire and smoke conditions
- Access and egress options

Activities that need to be monitored include, but are not limited to:

- Firefighter actions
- The actions of victims and bystanders
- Apparatus placement and operations
- Other vehicle movement (traffic)
- Firefighter support functions (rehab, rapid intervention crews, staging, etc.)

Communications that need to be monitored include, but are not limited to:

- Dispatch to command messages
- Command to crew messages
- Crew to crew messages
- Radio barriers
- Face-to-face interaction

Evaluate

The evaluation component is perhaps the most important, and difficult, one to execute. Face it, your evaluations—judgments—lay the foundation for interventions and the design of measures to prevent injuries or death. Judgments are often *perceptive*, and others may not perceive things the same way you do. In some cases, the judgments you make are spontaneous and based on incomplete information gathering. Regardless, the ISO *must* make judgments. The ISO can form better judgments when evaluating activities and conditions when the front-loaded knowledge and skill sets of *reading risk* and *reading firefighters* are used **FIGURE 11-3**.

Just as incidents are dynamic, so must be the evaluation of conditions and activities. For example, the exposure of a working crew to smoke and fire conditions may be judged as acceptable as rescues are being performed, but the same exposure might be deemed unacceptable when there is nothing to be gained (the building will be torn down after the incident). The *Getting the Job Done* box outlines some specific items that the ISO must evaluate (analyze).

As you can see, the evaluation portion of the MEDIC action model is crucial and will become the springboard for developing preventive measures (proactive needs) and interventions (reactive needs).

Getting the Job Done

ISO Evaluation Issues at Incidents

- *Risk-taking level of the firefighters*. Does it match the IAP and is that appropriate for the conditions?
- *Safety systems for firefighters*. Considerations include level and use of personal protective equipment (PPE), accountability systems, rapid-intervention readiness, rehab efforts, and communications.
- *Incident hazard status and rate of change*. Are hazards imminent or potentially threatening, predictable or unpredictable, and changing quickly or slowly?
- *IAP accomplishment*. Are there enough resources (people) to meet the IAP? Are crews performing in a safe and standard manner, and are there standard and predictable outcomes for their actions?
- *IAP components*. The initial IAP established by first-arriving crews or the one communicated by the IC may not fit the incident conditions discovered during recon. The ISO needs to evaluate that fit. Likewise, certain incidents require an ISO to contribute to, or review, an IAP through the *planning* process.

Develop Preventive Measures

Within the context of the MEDIC action model, developing preventive measures is different than intervening. Think of interventions as those spontaneous and *reactive* triggers you use to directly address evaluated hazards (recall soft and firm interventions from the *Triggers, Traps, and Working with the Incident Commander* chapter). Preventive measures are *proactive* actions you implement for forecasted hazards. To better show the difference, picture a situation in which overhead power lines to a building are not exposed to fire but may become that way. The ISO would be wise to offer a soft intervention to crews nearby (reactive): "Did you see the power line feed? We should probably keep an eye on it." The ISO then develops a preventive measure (proactive) by zoning the drop zone with no-entry tape and follows up with the IC to share the observation and inquire about power company notification. If the power lines do become exposed, the ISO might make a firm intervention (reactive) to withdraw crews from the area and then post a sentry (proactive) to warn others.

Preventive measures can take on many forms. Recall some of the mitigation strategies you learned in the chapter on reading firefighters: awareness, accommodation, and acclimation. Most preventive measures will come in the form of adjusting or improving the standard way things are being achieved by crews doing what they always do. Adding light, zoning the incident, increasing rehab capabilities, and incorporating more safety briefings or messages are specific countermeasures to achieve the preventive mitigation strategies of awareness, accommodation, and acclimation.

One last point regarding the development of preventive measures: It never ends! The ISO is *always* considering and devising ways to make things safer—from the moment of incident assignment to its conclusion or ISO release. Granted, there needs to be a sense of priority and reasonableness here, but the point is that the ISO is always mentally engaged in the concept of injury prevention.

Intervene

As just mentioned, interventions are those deliberate, spontaneous, and reactive triggers you use to directly address a hazard that you've evaluated. *Imminent* threats require a firm intervention—a direct order—to stop, alter, suspend, or withdraw responders to prevent almost certain harm. The IC must be notified (via radio) of any firm intervention, as an adjustment to the IAP may be necessary. Soft interventions are used for hazards presenting *potential* harm or damage and are typically communicated through face-to-face methods during the course of the incident.

Communicate

Obviously, interventions require communication to deliver urgent and advisory messages. At times, the communication of hazards can be accomplished using warning signs, a whistle (yes, some ISOs use a whistle as an attention-getter), or safety briefings and written safety messages. Zoning tape is a protective measure but also serves as a hazard communication tool.

The act of communicating includes the ability to listen; thus the ISO should actively listen to what is being said by other responders. The "pulse" and status of working crews can be realized through simple listening and may indicate the need to improve rehab needs or visit the IC to discuss issues and solutions **FIGURE 11-4**. Likewise, the ISO needs to contribute to the conversations in a friendly, caring way. Overly directive, trivial, or negative contributions serve no purpose (even if you're just trying to be funny). As always, the ISO needs to maintain cyclic communication with the IC, including the communicated request for ASOs to help the ISO perform all the safety functions.

The communicate portion of the MEDIC action model is also a reminder that certain forms or documentation may need to be completed: activity log, site safety plans, safety messages, and the like.

FIGURE 11-4 The "pulse" of an incident can be realized through the act of simply listening to the conversations of crews.
Courtesy of Superstition Fire and Medical District

The ISO Arrival Process

Whether you choose a checklist or an action model, you still need a process to become integrated into an ongoing incident. Rarely does a predesignated duty ISO arrive on scene first. If the duty ISO does arrive first, it makes sense for the ISO to take on the role of initial IC if the situation warrants. More often than not, a first-due company arrives, gives a status report, establishes command, and declares an initial action plan or mode of operation. If the duty ISO arrives after the first-due company but before a designated or incoming command officer, the ISO should probably assume command after connecting with the first-due (working) company officer. Granted, this action is subject to local policy but is consistent with model IC systems. In most cases, the ISO arrives (or is appointed) after the establishment of a command post and stationary IC. When the duty ISO does arrive (or the ISO assignment is given), a process of integration into the incident should be mandated to ensure that the ISO is dialed into the situation. The following steps can help the ISO get dialed in:

1. *Confirm the ISO assignment.* Upon arrival at an incident, a predesignated duty ISO should meet with the IC to confirm that the IC wants and/or needs the ISO position filled. Some fire departments have a policy that the IC does not have a choice—assigning an ISO is a must. Occasionally, the IC may need the prospective ISO to take another command staff assignment or act as a section chief (operations, planning) or a group/division supervisor. In these cases, the IC retains safety functions.

2. *Collect information.* Upon confirmation, the ISO needs to gather information. It is hoped that an IC would brief the ISO on the overall IAP. In addition to the IAP, the ISO should inquire about the status of the situation and resources. This is also a good opportunity to ask about known hazards, any established control zones, and the status of rapid intervention crews (RICs) and a rehab system.

3. *Confirm communication links.* Take the time to confirm assigned tactical radio channels and, if policy does not address the issue, ask the IC which radio frequency he or she prefers to be used. Additionally, offer a face-to-face communication schedule so that the IC knows what to expect from you. It is desirable for the IC to announce over the tactical radio channel that "Safety" has been assigned (e.g., "All fire ground personnel from Fourth Street Command, Battalion 101 is now "Safety.""). If the IC doesn't make the announcement, the incoming ISO should.

4. *Don appropriate PPE and position identifier.* The ISO should don PPE that is appropriate for the likely potential hazards, as well as an ISO-identifying vest or other identifier, and check into the personnel accountability system. From here, the ISO begins his or her duties using a systematic approach (MEDIC action model).

Getting the Job Done

The ISO Arrival Sequence

The following sequence should commence when a predesignated ISO arrives on scene or when an individual is delegated the ISO position by the IC:

1. Confirm the ISO assignment (for responding duty ISOs).
2. Collect information: IAP, known hazards, sitstat/restat, zones, RIC.
3. Confirm communications: radio channels, face-to-face schedule, establishment of "Safety" on the radio.
4. Don appropriate PPE and position identifiers (vest, helmet, SCBA identification sleeve), and check into the accountability system.

Once the arrival sequence is completed, the IC is likely to ask for a more complete recon (360) of the incident. At other times, the IC may request the ISO to address a specific need, such as developing a quick IAP, checking the accountability system, etc. Where no IC direction is given, the ISO should start the MEDIC action model (monitor and recon) if no imminent threat needs an intervention (one may have been discovered during the arrival sequence).

ISO Focus Areas for All Incidents

The remaining chapters of this text will address the ISO at various types of incidents (structure fires, wildfires, hazardous materials incidents, and technical rescue), the ISO's role at training drills and special events, and the ISO's postincident duties. Each of those chapters will list focus areas that are applicable to the chapter topic. However, there are some focus areas that are common to all incidents: risk-taking level, personnel safety systems, and control zones.

■ Risk-Taking Level

It cannot be overemphasized: The ISO must confirm with the IC the risk-taking level (some call it the risk profile) that is at play for a given incident. Most use simple language (low,

medium, high) to quickly communicate a corresponding risk management principle from NFPA 1500:

- *Low.* No risk to the safety of members shall be acceptable when there is no possibility to save lives or property (take no risk for that lost).
- *Medium.* Activities that are routinely employed to protect property shall be recognized as inherent risks to the safety of members, and actions shall be taken to reduce or avoid those risks (minimize risk for savable property).
- *High.* Activities that present a significant risk to the safety of members shall be limited to situations in which there is a potential to save endangered lives (risk a lot to save people).

The declared risk profile gives the ISO a foundation from which to monitor and evaluate the conditions and activities taking place. During initial incident activities, the information necessary to make a risk profile decision may not be fully revealed or understood, creating a dilemma for the IC and ISO. This is why so many fire departments have included "360 complete" as an incident benchmark. Incident benchmarks can be defined as key phrases that signal the completion of significant tactical objectives. Common incident benchmarks include the following:

- *360 complete.* All four sides of the incident environment have been observed and notable factors have been communicated to the IC.
- *All clear.* A primary search of the incident environment has been completed.
- *Under control.* The destructive incident force (usually the fire) has been controlled but not eliminated.
- *Loss stopped.* Fire has been extinguished and overhaul and ventilation have been completed. No further loss is expected.

Risk profiles are often tied to the incident benchmarks and help responders understand what risk levels are acceptable for a given phase of the incident, especially when communicated by the IC:

- *360 complete.* IC establishes the risk profile as low, medium, or high.
- *All clear.* High risk-taking moves to medium.
- *Under control.* Medium risk-taking moves to low.

Going back to the initial stage of the incident, many crews start off operating in a high risk-taking mode (default) even though a 360 hasn't been completed. This default may have been acceptable in another time (predictable fires, stronger buildings, simple smoke). In today's rapidly changing incident environment (complex societal influences), using a high risk-taking default may not be the best practice. In fact, many National Institute for Occupational Safety and Health Administration (NIOSH) line-of-duty death (LODD) investigative reports have cited the high risk-taking default as a contributing factor and recommend that a 360 be completed and high risk-taking commence only when incident conditions indicate a high potential that savable victims need to be rescued (default to low risk-taking until a 360 is complete).

While the incident benchmarks provide a guide, the IC drives the operational risk-taking decision and must communicate it. Taking that decision, the ISO evaluates activities and conditions and may conclude that the risk profile established may no longer be appropriate. When this occurs, the ISO can either make a firm intervention (if an imminent threat is present) or visit the IC and share the concern.

Some may read the preceding description of risk profiling and say that their department does the same thing by just declaring an offensive or defensive mode. That is not the same! Offensive or defensive modes are *strategic* approaches to an incident. Some defensive strategies include high risk-taking. For example, placing a defensive fire stream to protect exposures prior to understanding the evacuation needs of the exposure (as well as collapse and utility threats) can be very risky. Some offensive strategies are done at low risk. The effective IC declares the overall strategy (offensive, defensive, transitional, or marginal) as well as the accepted risk profile for operations (high, medium, low).

Personnel Safety Systems

Many safety systems have been developed to help fire departments reduce responder injuries and deaths at incidents. The ISO's role is to monitor and evaluate the systems and take steps to make sure they are working as intended. The overall safety system for individual firefighters includes PPE, accountability systems, RICs, and responder rehab. We've addressed several of these systems throughout the text. In the context of this chapter, the individual safety systems that require ISO focus at *all* incidents are accountability systems and rehab. We will look at the RIC and PPE safety systems as focus elements in the next chapters.

Personnel Accountability Systems

The ISO needs to check to see that the fire department's personnel accountability system is being used effectively. The measurement of effectiveness is highly dependent on the type of system a local fire department uses. Obviously, the ISO needs to be intimately familiar with the accountability system (SOPs, training, strengths and weaknesses) that is in place locally. It may be surprising to some readers, but there are fire departments that still don't have a formalized accountability system, as evidenced by the repeated recommendation by NIOSH in recent LODD reports: "Fire departments should ensure that a personnel accountability system is established early and utilized at all incidents."[2]

Some fire departments rely merely on the discipline of company officers and crews as their accountability system (no formal charting or tracking). These departments are using an *informal* accountability system. History has shown that the reliance on an informal system is dangerous at best and fatal in too many cases. In cases where no formal accountability system exists, the ISO should consider assisting the IC with the declaration of a tactical time-out and a one-by-one charting of each crew, its assignment, and its location. Lives may depend on it.

There are many formal types of accountability systems with various strengths and weaknesses. Some examples include:

- Tactical worksheets and command boards
- Apparatus riding lists/cards
- Electronic bar-code systems
- Accountability tags, keys, or passports
- Radio-based electronic tracking receivers and software

Regardless of the system used, the intent of each method is to track responders by number, function, and location and provide a rapid method to do a personnel accountability report (PAR). Likewise, the system should be expandable as an incident grows in size or complexity **FIGURE 11-5**. Given these needs, the ISO should inquire (evaluate) whether the accountability system is working as intended. It is not suggested that the ISO manage the accountability system; the system should be initiated and managed by others as routine on every scene. If the ISO is tied down to initiating and updating the accountability system, he or she is an *accountability manager*, not an ISO.

The effective use of an accountability system does not end with a check on the passport status board (crews appear to be tracked by number, assignment, and location). You must also evaluate crews to make sure they are working within the framework of the action plan. Failure to work within the framework of the action plan is a form of <u>freelancing</u>. In the fire service, freelancing has typically been defined as a firefighter working alone and is viewed as a dangerous and deadly enterprise. Both infractions should be included in the definition of freelancing: (1) the act of working outside the parameters of an IAP and (2) an individual performing incident functions out of the sight or voice range of others.

The solo firefighter is probably the deadlier of the two scenarios because of the *what-if* potential **FIGURE 11-6**. What if the firefighter falls through the floor? What if the firefighter suffers a heart attack? What if the firefighter gets lost? What if the firefighter disrupts ventilation flow paths in an undesirable way? The what-ifs are endless and lead to one conclusion: Nobody will know that the firefighter experienced an emergency or injury. This is all in addition to the possibility that lone firefighters may get sucked into a situation that requires skills they may not possess or that may require more than one person to accomplish the task.

ISOs should keep a close eye on working crews and develop an eye for catching lone workers. They can also apply

FIGURE 11-5 A formal personnel accountability system should include a method to track the number, function, and location of responders and is expandable based on incident size or complexity.
Courtesy of Mike Legeros

FIGURE 11-6 Firefighters working out of sight or voice range of others is a form of freelancing—a dangerous situation.

their basic knowledge of fire ground operations and predict situations that lead to freelancing. For example, a crew is performing an assigned task and it needs additional equipment. Often, the crew breaks up and one member (a gofer) is sent for the missing tool. The incident turns tragic as the gofer becomes lost, is trapped, or is distracted by a more demanding need. Likewise, the person waiting for the gofer to return may look to find ways to be productive, leading to more what-if hazards. As another example, a firefighter, especially an inexperienced one, who is assigned to a seemingly "trivial" task, such as monitoring an exposure line or a positive-pressure fan, may search for any excuse to get where the action is. Similarly, members of a crew or team may have completed a task but do not want to rotate to staging; they want to stay involved so they look for other tasks that need to be addressed in the action area. This is well intentioned but outside the action plan. Unfortunately, the IC may or may not be aware of the need for the crew's self-assigned task and may have higher priorities in mind for the crew. In these cases, you can catch the freelancing only if you have an intimate knowledge of the IAP.

More than once, an ISO has been criticized for not having a partner or not being tracked by the accountability system. As with the IC, the ISO should be tracked by the accountability system. When performing a roving assignment or entering a collapse or hot zone, the ISO should wear appropriate PPE and have an assigned partner, just as we expect firefighters to work in teams or pairs. This is especially important when the IC wishes the ISO to get a close-up look at a fire floor, potential collapse zone, or other assignment in which a certain risk is taken to collect information. Failure to track the ISO or, worse yet, failure of the ISO to work as a team member at an emergency incident will do grave damage to the safety efforts of the department, in addition to increasing the chance of injury or death to the ISO.

Rehabilitation

NFPA 1500 requires that personnel undergo rehab in accordance with NFPA 1584 (*Standard on the Rehabilitation Process for Members During Emergency Operations and Training Exercises*). At an incident, the IC is responsible for considering the circumstances of the incident and initiating rehab as needed. Individual responders are responsible for communicating their rehab needs to their supervisor. Taken further, the ISO (where assigned) must ensure that a rehab area has been established (NFPA 1561, 5.9.7.5) and that the area addresses rest, hydration, active cooling, basic life support (BLS) observation and care, energy replacement (food and electrolytes), and accommodations for the weather. Recall from the chapter *Reading Firefighters* that it is not the task of the ISO to set up the rehab area (never intended by the NFPA chain of responsibilities); the ISO does, however, need to monitor and evaluate the effectiveness of the rehab effort.

Practically speaking, the can-do nature of firefighters and the linear strategic focus of the IC can set up a situation in which rehab is not being addressed as it should. Similarly, the routine "self-rehab" that might be automated in firefighter behavior (hydrate during every SCBA cylinder change) can set the stage for incomplete or ineffective rehab. In either case, the judgment of rehab inadequacy should cause the ISO to make recommendations to the IC **FIGURE 11-7**. The mnemonic REHAB reminds the ISO of the components of an effective rehab effort:

R = Rest (recover stable vital signs)
E = Energy nutrition and electrolytes
H = Hydration
A = Accommodation from thermal stress/weather
B = BLS medical observation and care (advanced life support [ALS] preferred)

FIGURE 11-7 The ISO should judge rehab efforts and request the IC to bump up rehab functions where it is deemed necessary.

■ Control Zones

Can you list the special emergency items that a fire engine and a police cruiser carry that are nearly identical? Your list is sure to include radios, warning lights, siren, high-visibility traffic vest, maybe a first-aid kit, and—wait for it—*yellow barricade tape!*

Next question: Have you ever seen a police officer or fire-fighter yield to yellow barrier tape at an incident? Probably not. The barrier tape is used primarily to discourage *public* entry to an established zone—not for our responders. Therein lays the issue. Incident scenes, especially those that firefighters deal with, require a hierarchy of zones that should be recognizable and understood by responders (and law enforcement officers).

The ISO is responsible for establishing control zones (or adjusting established zones) at incidents and identifying them with colored hazard tape, signage, cones, flashing beacons, fences, sentries, law enforcement or security personnel, or other measures. Further, the measures used to help identify the control zones should have some sense of priority or audience applicability to be effective. For this reason, fire departments should adopt a control zone scheme that helps to identify the hazardous conditions that are found at incidents. A common, and increasingly popular, means to categorize hazard zones includes a four-level hierarchy based on severity and required PPE and procedures.

No-Entry Zone

Nobody is allowed to enter this area, regardless of PPE, due to extremely dangerous conditions. Nobody means nobody: no responders, no police officers, no ISOs, no civilians, and no media personnel. Some call no-entry zones an "exclusionary zone." A collapse zone is a specific form of a no-entry zone. Granted, there may be a need for someone to access the no-entry area, such as a power company technician who will secure or eliminate the hazard. When warranted, the fire department is wise to make sure a rescue crew is on standby for the person or persons it has *allowed* to be in the no-entry zone. When establishing a no-entry zone, the ISO should:

1. Announce the presence of the zone and the nature of the hazard using a "priority" or "urgent" radio message.
2. Mark the area with a unique barrier tape. Red and white chevron or red and white diagonally striped tape with the words *DO NOT ENTER* or *DANGER* is the national standard (NFPA 1561) for marking no-entry zones.
3. Consider posting a person to monitor the zone and to warn others who approach it (i.e., post a sentry).

Hot Zone

Hot zones are those operational areas that are judged to be IDLH (immediately dangerous to life and health) yet are within parameters of acceptable fire department risk-taking and/or established safety systems (PPE, SCBA, etc.). When establishing a hot zone at a structure fire, the ISO must consider that carbon monoxide and hydrogen carbon gases are likely present (invisible) outside the structure. Likewise, radiant heat from flames, dirty smoke, window breakage, dropping debris, and ground ladder/fire stream dangers may require a hot zone that extends well past the walls of the involved building. Established hot zones should include:

1. A communicated level of required PPE.
2. Two-in/two-out for initial entry.
3. The use of a formal accountability system for personnel tracking.
4. Assignment and readiness of a RIC.
5. A cyclic or planned process for PARs.
6. Marking or flagging with solid red barrier tape with the words "Hot Zone" printed repeatedly on it. At times, it is not reasonable to tape the entire hot zone. In these cases, the use of cones and solid red tape at declared entry/exit points is reasonable. Also consider using an access/egress monitor to help track crews.

Warm Zone

The warm zone includes the area surrounding a hot zone or action area where personnel, equipment, hose lines, and apparatus are operating in support of an incident. Decontamination processes and RIC standby are typically found in the warm zone. Personnel working in the warm zone should have a defined level of PPE appropriate for their assignment and exposure to conditions (the ISO is usually tasked with establishing the level of PPE required for warm zones). The classic solid yellow barrier tape with the words *Caution* or *Fire Line* is standard for marking a warm zone.

Cold Zone

The area surrounding a warm zone that is used for an incident command post, support-agency interfacing, rehab, and staging of personnel and equipment should be categorized as a cold zone. Cold zones should be designated by the use of solid green barrier tape with the words *Support Zone* or *Staging Area*. The use of PPE is rarely needed in the cold zone.

Various control zones should be identified with barrier tape, signs, cones, flashing beacons, physical barriers such as fences, or other methods to help denote their presence. Obviously, it may not be practical to mark a control zone due to the location, distance, resources, or other considerations for a given threat. In these cases, the ISO is encouraged to be inventive or resourceful: The use of law enforcement, traffic control resources, or security resources may be needed to help denote and discourage public entry to a defined zone.

Wrap-Up

Chief Concepts

- The ISO assignment can be overwhelming due to the multitude of issues that an ISO must address. The expansive and dynamic nature of the job can create frustrations without a systematic approach to the assignment. The two most common systematic approaches to the ISO function make use of checklists and action models.

- Checklists are probably used more often because of their numerous advantages. The ISO, however, understands the limitations of checklists and may prefer an action model.

- The Dodson/Vavra ISO Action Model presented in previous editions of this text has been helpful but does not fully conform to established standards and laws for ISO duties. The MEDIC mnemonic is a better ISO action model.

- When expanded, the MEDIC ISO action model defines the roles of an ISO at planned and unplanned events and can serve as a reminder of the five ISO general duties applicable to all incident types.

- The incident scene conditions that are monitored as part of an ongoing incident include, but are not limited to, weather, hazardous energy (utilities, etc.), building construction and collapse potential, fire and smoke conditions, and access and egress options.

- Incident activities that warrant monitoring at all incidents include, but are not limited to, firefighter actions, the actions of victims and bystanders, apparatus placement and operations, vehicle movement (traffic), firefighter support functions (e.g., rehab, RIC, staging), and communications.

- The following sequence should commence when a pre-designated ISO arrives on scene or when an individual is delegated the ISO position by the IC.
  - Confirm the ISO assignment (for responding duty ISOs).
  - Collect information: IAP, known hazards, sitstat/restat, zones, and RIC.
  - Confirm communications: radio channels, face-to-face schedule, establishment of "Safety" on the radio.
  - Don appropriate PPE and position identifiers (vest, helmet, SCBA identification sleeve), and check into the accountability system.

- The five components that make up a firefighter safety system include level and use of PPE, accountability systems, rapid-intervention readiness, rehab efforts, and communications.

- The ISO must weigh risk management criteria and established risk profiling from the IC and compare that with incident conditions and activities. If the two don't match, the ISO may have to make a firm intervention if there exists an imminent threat to firefighter survival.

- Risk profiles are often tied to the incident benchmarks and help responders understand what risk levels are acceptable for a given phase of the incident, especially when communicated by the IC:
  - 360 complete: IC establishes the risk profile as low, medium, or high.
  - All clear: High risk-taking moves to medium.
  - Under control: medium risk-taking moves to low.

- Methods used to ensure member accountability at an incident scene can include tactical worksheets and command boards; apparatus riding lists/cards; electronic bar-code systems; accountability tags, keys, or passports; and radio-based electronic tracking receivers and software.

- Some fire departments rely merely on the discipline of company officers and crews as their accountability system (no formal charting or tracking). These departments are using an informal accountability system. History has shown that the reliance on an informal system is dangerous at best and fatal in too many cases.

- Hazardous incident conditions require a hierarchy of control zones based on their potential to do harm and the PPE and procedures that need to be in place while working in a given zone. That hierarchy should include no-entry, hot (IDLH), warm, and cold zones.

- Zones should be marked and communicated in a standard way: No-entry zones should use red and white diagonally striped barrier tape and be announced on the radio as a priority message. Hot zones (IDLH) require solid red barrier tape and the application of PPE, RIC, and formal personnel accountability measures. Warm zones can be marked with solid yellow tape, while cold zones can be marked with solid green tape.

- No-entry zones should further be controlled by signage, cones, flashing beacons, fences, sentries, law enforcement or security personnel, or other measures. Hot zones can be controlled by defined entry/egress points using tape and cones and perhaps an entry/egress monitor.

Key Terms

action model A template that outlines a mental or physical process that considers inputs that lead to an output or outcomes.

freelancing (1) The act of working outside the parameters of an IAP, or (2) an individual performing incident functions out of the sight or voice range of others.

incident benchmark A key phrase that signals the completion of significant tactical objectives. Common examples include *360 complete*, *all clear*, *under control*, and *loss stopped*.

reconnaissance An exploratory examination of the incident scene conditions and activities.

Review Questions

1. What two methods help the ISO achieve a systematic approach in addressing ISO duties?
2. List several advantages and disadvantages of using checklists.
3. What is one of the biggest traps of linear thinking?
4. List the five general duties of the ISO at all incidents (MEDIC).
5. Describe the four steps that help an ISO become integrated at an incident.
6. What are the five components of the firefighter safety system used for most incidents?
7. Identify the consequences and hazards that the lack of accountability at an incident scene could create.
8. What are the two forms of freelancing?
9. Describe the relationship of risk management criteria and common incident benchmarks.
10. An effective rehab component should include what elements?
11. List the four-tier system of establishing and marking control zones.

References and Additional Resources

NFPA 1500, *Standard on Fire Department Occupational Safety and Health Program*. Quincy, MA: National Fire Protection Association, 2018.

NFPA 1561, *Standard on Emergency Service Incident Management System and Command Safety*. Quincy, MA: National Fire Protection Association, 2020.

NFPA 1584, *Standard on the Rehabilitation Process for Members During Emergency Operations and Training Exercises*. 2015 ed. Quincy, MA: National Fire Protection Association, 2015.

NIOSH Alert, Preventing Deaths and Injuries of Firefighters Using Risk Management Principles at Structure Fires, July 2010. Available at http://www.cdc.gov/niosh/docs/2010-153/.

NIOSH firefighter fatality reports are available at http://www.cdc.gov/niosh/fire.

Putnam, Ted. The Collapse of Decisionmaking and Organizational Structure on Storm King Mountain. Findings from the Wildland Firefighters Workshop. Missoula, MT: USDA Forest Service Publication 9551-2855 MTDC, July 1996.

The Rules of Engagement for Firefighter Survival poster, by the International Association of Fire Chiefs (and several cosponsors), is available at http://www.iafc.org/files/rulesofengagementroe_poster.pdf.

Endnotes

1   Putnam, Ted. The Collapse of Decisionmaking and Organizational Structure on Storm King Mountain. Findings from the Wildland Firefighters Workshop. Missoula, MT: USDA Forest Service Publication 9551-2855 MTDC, July 1996.

2   The author found this recommendation repeated in dozens of NIOSH LODD investigation reports between 2011 and 2018. They can be found at http://www.cdc.gov/niosh/fire.

INCIDENT SAFETY OFFICER
in action

Captain Franks was recently reassigned from his truck company officer position to that of a dedicated shift safety officer. His department has only recently adopted a duty shift safety officer program, and the department chose Franks and two other captains to attend an ISO training academy at another department to learn their new assignment. Captain Franks enjoyed the academy and felt like a dedicated ISO position could make a difference. After a few weeks of learning his non-incident (HSO) responsibilities, Franks began longing for an opportunity to practice his newly acquired ISO skills. His wishful thinking was soon realized when his shift was dispatched to a working fire in a commercial building. Upon arrival at the incident, Franks was overcome with a feeling that the incident was not going well: lots of radio chatter, firefighters rushing around, thick black smoke pushing from the second floor of a large storage building, and vehicle traffic congestion. Franks dressed in his PPE, grabbed his radio, clipboard, and safety officer vest, and proceeded to review the ISO checklist, which reminded him to check in with the IC and ascertain the incident action plan. The IC briefly acknowledged Captain Franks' arrival and told him, "We have a good fire ripping on the second floor and all hell is breaking loose. Engine 2 is doing an interior attack and Truck 1 is going to the roof for vertical ventilation." The IC then rapidly listed a multitude of items that he needed Franks to accomplish: Do a 360, check the status of the attack crew who haven't been heard from, do something about the traffic congestion, find out why the truck company hasn't gotten the roof opened, and give a read on collapse potential. Captain Franks felt completely overwhelmed, but took a deep breath and began to process his responsibilities.

1. What information should Captain Franks obtain from the IC before addressing his requests?

2. What can Captain Franks do to better manage his overwhelmed feeling?

3. Based on the information presented in this scenario, which personnel safety system seems deficient?
   A. PPE
   B. Accountability
   C. Rehabilitation
   D. Zoning

4. Which incident benchmark (key phrase) should cause incident risk-taking to switch from high to medium?
   A. Under control
   B. Loss stopped
   C. All clear
   D. 360 complete

Smoke: © Greg Henry/ShutterStock, Inc.

The ISO at Structure Fires

Flames: © Ken LaBelle NRIFirePhotos.com

Knowledge Objectives

Upon completion of this chapter, you should be able to:

- Discuss the relationship of risk-taking to incident benchmarks. (pp 181–182)
- Name the three communication ingredients to an effective personnel accountability report (PAR). (pp 182–183)
- Describe the actions to be taken if a member is not accounted for at an incident scene. (**NFPA 5.2.6** , p 183)
- Identify the consequences and hazards that the lack of accountability at an incident scene could create. (**NFPA 5.2.6** , p 183)
- Identify imminent threats to firefighter safety. (**NFPA 5.2.4** , pp 177–178, 181–182)
- Classify types of imminent hazards into major categories. (**NFPA 5.2.4** , p 177)
- Describe the incident scene conditions that are monitored as part of an ongoing incident. (**NFPA 5.2.5** , pp 181–182)
- Describe the types of incidents at which ASOs or technical specialists are required due to corresponding hazards of the incident type. (**NFPA 5.2.10** , p 185)
- Describe how fire behavior affects firefighting efforts. (**NFPA 5.3.4** , pp 180–181)
- Describe the characteristics of suitable entry and egress options at building fires. (**NFPA 5.3.6** , pp 178–180)
- Describe what is meant by "rescue profile." (p 181)
- Identify the environmental and operational conditions present at an incident that assist in evaluating the capability of a rapid intervention crew (RIC). (**NFPA 5.3.1** , pp 184–185)
- Describe the standards or organizational standard operating procedures/guidelines (SOPs/SOGs) that would require the use of a RIC at an incident or planned event. (**NFPA 5.3.1** , pp 184–185)
- Identify factors that would require the capabilities of a RIC to be increased. (**NFPA 5.3.1** , p 185)
- Describe how the need for additional RIC capability is communicated to the IC. (**NFPA 5.3.1** , p 185)
- Identify fire behavior issues that should be communicated to the RIC by the ISO. (**NFPA 5.3.2** , p 185)
- Identify building access and egress issues that should be communicated to the RIC by the ISO. (**NFPA 5.3.2** , p 185)
- Identify the structural collapse issues that should be communicated to the RIC by the ISO. (**NFPA 5.3.2** , p 185
- Identify the hazardous energy issues that should be communicated to the RIC by the ISO. (**NFPA 5.3.2** , p 185)
- List several unique hazards at strip mall structure fires. (p 186)
- List four ISO functions and six assistant safety officer (ASO) functions at high-rise fires. (pp 186–187)
- Identify fire environment conditions and contaminates that necessitate the need for contamination reduction processes that help with cancer prevention. (**NFPA 5.2.15** , (pp. 187)
- Describe the process used for smoke residue gross contamination reduction at the fire scene. (**NFPA 5.2.15** , (pp.188)

Skills Objectives

Upon completion of this chapter, you should be able to:

- Demonstrate the ability to prioritize tasks at emergency scenes. (**NFPA 5.2.1** , pp 176–177, 181–182)
- Demonstrate the ability to make decisions in an environment of large unknowns. (**NFPA 5.2.1** , p 181)
- Evaluate resource needs that, if deficient, may affect emergency personnel safety and survival. (**NFPA 5.2.1** , pp 178–180)
- Demonstrate the ability to take proactive measures to ensure emergency responder safety and health. (**NFPA 5.2.1** , p 183 187)
- Demonstrate the ability to alter, terminate, or suspend operations based on anticipated fire behavior, fire dynamic, or building construction hazards. (**NFPA 5.2.2** , pp 178–181
- Demonstrate the ability to monitor an accountability system (**NFPA 5.2.6** , pp 182–183)
- Prioritize risks at incidents. (**NFPA 5.2.5** , pp 181–182)
- Demonstrate the ability to communicate issues identified at an incident scene to the RIC. (**NFPA 5.3.2** , p 185)
- Communicate access and egress hazards that are identified at a building fire. (**NFPA 5.3.6** , p 185)
- Evaluate and communicate contamination judgements to the IC and tactical work members. (**NFPA 5.2.15** , pp 187–188)
- Judge contamination reduction efforts and develop further exposure-prevention measures where necessary. (**NFPA 5.2.15** , pp 187–188)

You Are the Incident Safety Officer

At the scene of a structure fire, the incident commander (IC) has briefed you on the operational plan and locations of crews working the incident. The IC's first order for you is to establish control and access zones at the incident. You scan the fire ground and observe a defensive fire with many personnel and apparatus already assigned to hose and master stream operations along with companies who are working on protecting exposures both externally with lines and internally by checking for extension.

1. How does your department restrict access to fire areas? Are there different levels of access granted to different crews or officers?

2. How are zones marked or communicated to the crews operating at the scene? If a collapse zone is established early in the incident, how are later arriving crews informed of the zone?

Introduction: Specific Incident Types

Starting with this chapter, we look at the role of the incident safety officer (ISO) at specific types of incidents. We will devote a chapter to structure fires, hazardous materials (hazmat) incidents, technical rescues, and wildland–urban interface fires. In each, we will address several general issues associated with that type of incident and then look at unique considerations for each. You will notice that there are many NFPA-referenced job performance requirements (JPRs) (knowledge and skill) for this and the remaining chapters. Many of these JPRs are repeated because the information necessary to satisfy a generally worded JPR requires elements from each of those referenced chapters.

If you are using this chapter as a refresher or proficiency training reference, recall that the previous chapter covered umbrella issues associated with *all* types of incidents—including structure fires.

General ISO Functions at Structure Fires

Of all the incident types to which fire and rescue departments respond, structure fires can be considered the most risky for numerous reasons, the greatest of which is the compressed time window that a fire department has to make a difference. In the chapters *Reading Buildings* and *Reading Smoke*, we discuss how societal influences have changed the fire and the building. Growth-stage fires in a building can change in minutes, if not seconds, due to newer building materials and contents (e.g., plastics) and lightweight construction methods. The combination is deadly. The well-prepared ISO uses a front-loaded skill set of Read 3 to help evaluate the most prevalent risks at a structure fire: *read* smoke, *read* the building, *read* hazardous energy. Combine this approach with the ability to judge firefighter actions and effectiveness, and one has the formula to help discover potential threats and, thus, prevent firefighter injury or death.

Before we dig in, two important points need to be made regarding structure fires. First, many firefighters have sworn (taken an oath) to risk their life to save a life (a known potential that comes with the badge). Your author did the same. Second, the ISO's job is to protect the firefighters and, along with the IC, continually make judgments as to when it is no longer acceptable to take risks that can lead to firefighter injury or death. As you read this chapter, you might get the sense that the content has forgotten the first point—nope, we get it. It's the primary role of the ISO to *always* be monitoring/evaluating and devising ways to make response efforts safer. The general ISO functions at structure fires involve issues associated with reconnaissance (or simply "recon") efforts, risk-taking evaluation, and safety system effectiveness.

■ Reconnaissance Efforts at Structure Fires

We learned in the chapter *A Systematic Approach to the ISO Role* that the *monitoring* and *evaluating* portions of the MEDIC action model occur during recon. The recon effort at a structure fire should occur early and be repeated 5, 10, or even 20 times as the incident progresses (or regresses). Experience, knowledge, theory, study, and wisdom allow the ISO to use the visual image collected during the recon effort and make a determination, judgment, or advisement on the potential for firefighter injury. As a starting place, the ISO uses reading smoke, building, and hazardous energy skill sets to define the incident environment in three dimensions: the principal hazard, environmental integrity, and the effects of the surrounding elements **FIGURE 12-1**.

In addition to evaluating the environment, the ISO should watch crew activities to evaluate effectiveness. The ISO is challenged not only with evaluating current conditions but also with anticipating changes that will occur and the impact of these changes on the incident. This whole monitoring–evaluation relationship helps the ISO assess risk-taking and can be expressed in the following formula:

Principal hazards ± Integrity + Other hazards ± Resource effectiveness = Risk-taking

Let's break down the equation and show how it can work at a structure fire. Each of the formula components can add to risk-taking. Two components, environmental integrity and resource effectiveness, can help minimize (subtract)

FIGURE 12-1 The ISO uses the *Read 3* skill set (read smoke, read building, and read hazardous energy) to define the principal hazard, scene integrity, and surrounding elements.
Courtesy of Frisco (TX) Fire Department

FIGURE 12-2 Building geometry includes issues associated with a building's layout, size, stories, and other features.

risk-taking. The goal is to continually subtract. When the incident is deemed stable (no more change) and the resources have effectively controlled hazards, risk-taking should no longer be part of the equation.

An individual ISO could assign values to each equation component and work the equation like a scientist. Practically speaking, most just understand the equation principle and work it based on intellect and wisdom. We'll take the practical approach as we dive into each component.

Defining the Principal Hazards

When you ask firefighters what the principal hazard of a structure fire is, they might look at you and say (in a sarcastic tone), "The building is on fire!" This reply is 100% correct. From the ISO's perspective, though, the principal hazards need to be defined in a much more specific way (so that they can be communicated). For example, a building may be on fire, but, more specifically, the ISO sees that thick, black, turbulent smoke is traveling down the center hallway of an apartment building and a well-defined air track is being pulled back into the hallway exterior door. The ISO knows that these signs indicate a ventilation-limited fire and the potential for explosive growth. No firefighter should be in that smoke for any reason.

Principal hazards at structure fires typically fall into general categories, adding significant risk-taking (imminent) threats:

- *Hostile fire events.* Included here are flashover (turbulent smoke fills a box), backdraft (closed box with signs of extreme heat), and explosive growth (ventilation-limited fire with a developing air track).
- *Building geometry.* We're not necessarily talking about math here. In context, building geometry includes issues associated with the building's layout, size, number of floors, access options, and other features **FIGURE 12-2**. Center hall construction, confusing floor plans, large vaulted ceiling spaces and great rooms, minimal access/egress options, large void spaces, non-permitted additions/alterations, and the like are notorious threats that have contributed to firefighters' deaths.

- *Collapse potential.* Fires in void spaces, in basements, and above drop ceilings pose rapid collapse threats. Any pressurized, tan, brown, or darkening smoke from structural areas of modern, lightweight wood buildings is a collapse warning. Certain roof styles (tied-arch, bowstring truss, and exposed lightweight steel/combination trusses) collapse quickly at building fires. Unreinforced masonry load-bearing walls, parapets, cornices, and facades are all big-ticket collapse threats.
- *Hazardous energy.* Although all forms of hazardous energy can become an imminent threat, electrocutions and explosions seem to lead the list at structure fires. Power line feeds to buildings that are exposed to fire should warrant control zoning and priority communication. Likewise, the presence of pressure vessels, grain storage, and hazardous materials requires similar actions.

Environmental Integrity

Closely related to the principal hazard is the integrity of the environment. Environmental integrity can be defined as the status of the building, conditions, and hazards in terms of stability (change potential) and time (rate of change). The ISO is responsible for determining environmental integrity, considering factors such as weather, smoke, flame spread, and hazardous energy, and assesses how the building (host) is reacting to those conditions as time passes. At the structure fire, the ISO can define environmental integrity using the following terminology:

- Stable and not likely to change (subtracts from risk-taking)
- Stable but changing slowly (may or may not add to risk-taking)
- Unstable and changing slowly (adds to risk-taking)
- Unstable and changing quickly (exponentially adds to risk-taking)

Classifying environmental integrity as stable or unstable is useful, but knowing the potential *rate of change* is critical. Judging the rate of change (is it getting better or getting worse—and how fast?) is the ISO's foundation for making effective judgments on environmental integrity. For initial operations, the fire department is trying to make an incident stable—that is understood. However, the passage of time can be lost on crews (and the IC) who are trying to achieve that stability. Time is often a forgotten player. From the ISO's perspective, time is an important consideration that dictates hazard priorities, permits recon efforts, or makes the difference between an imminent threat to life and a minor operational concern. ISOs must have a keen sense of time as a resource: Time cannot speed up and it cannot slow down. You cannot trade it or buy it. You cannot keep it or give it away; yet, time does tend to "slip away."

Unfortunately, our perception of time changes at working incidents. Most fire officers can recall an event at which they would swear they were working for only 10 minutes when in fact it was 30. The inverse is equally true: You swear it was 30 minutes but it was only 10 (such as when waiting for relief to show up). How can the ISO evaluate and use time effectively? One suggestion is to preplan the use of dispatch or the communication center to broadcast elapsed incident time ("Lincoln Street Command from Dispatch, you are 20 minutes into your incident."). If such a procedure does not exist, the ISO should use a stopwatch, smartphone, or other device to monitor the passage of time.

The ISO should also project the likely on-scene time. Clearly, a 2-hour firefighting effort has different needs than an 18-hour conflagration. Short-duration incidents (less than 2 hours) may require just rest, active cooling, health monitoring, and hydration. Longer incidents may indicate the need for nourishment, extended rest, fresh crews, and sanitation logistics. Projecting on-scene time can also remind the ISO that darkness, rush-hour traffic, or the usual afternoon thunderstorm is coming.

Additional resource response time is also a consideration. One truism seems to fit the fire service:

If you need it, and it is not there, it is too late.

The ISO should look to see whether staged crews are available and ready for an assignment and project their reflex time if an unexpected event were to upset the incident action plan (IAP). Even though the ordering and assignment of resources is an IC function, the ISO can initiate a subtle reminder if necessary.

Defining Other Hazards

During recon, the ISO must assess the possible impact of any physical or atmospheric hazards on firefighters. The effective ISO evaluates any physical item, including terrain, foliage, curbs, posts, fences, drainage, signs, antennae, hazardous energy, barriers, and so on and decides whether the item could affect the operation. Likewise, the weather or other atmospheric conditions need to be weighed (day/night, incoming storm front, the flaring of a cosmic sun spot causing radio interference, Jupiter aligning with Mars, etc.).

FIGURE 12-3 Sloping grades may cause dangerous miscommunications. In this figure, both teams may believe they are on the first floor.
Courtesy of Lt. Rob Gandee

One physical feature—a building built on a sloping grade—has been cited as a significant factor in multiple firefighter deaths. In sloping-grade incidents, a crew that thinks they are working on the first floor may in fact be working on the second **FIGURE 12-3**. Usually, the presence of a fence or other barrier masks the situation from the IC or advancing crews. When observing sloping-grade situations, the ISO should relay the information to the IC and suggest that it be relayed to all responders. The IC should relay how the varying levels are to be labeled and communicated to improve clarity for those who need to report their location.

Fences, shrubbery, parked vehicles, security hardware, and other barriers may impede incident operations by restricting access or impairing an adequate size-up. The ISO should determine the relevance of the barriers to the incident as a whole. As a rule, the ISO should survey the entire area of incident impact and ensure that each crew has at least two escape areas (safe havens), as well as a clear corridor for egress.

Resource Effectiveness

The principal hazard, integrity, and other hazards are all environmental "conditions" that are discovered and evaluated during the recon effort. To complete the risk-taking equation, the recon effort must evaluate resource effectiveness. Resource effectiveness is tied to resource allocations, task applications, and the operational effectiveness of crews. Many times, the crews themselves don't see the big picture of the equation (principal hazard ± integrity + other hazards ± resource effectiveness = risk-taking) because they are in a linear-thinking mode. The ISO realizes and appreciates the linear-thinking mode of crews and serves as the wingman to help keep risk-taking in an acceptable range.

Resource Allocation

The ability to evaluate, manage, and assign resources is often used to measure the effectiveness of the IC. Budgetary concerns, response times, politics, and compatibility issues combine to limit the availability of resources to overcome an incident. While having unlimited resources to handle a serious incident is ideal, it is unrealistic to expect such a scenario in most communities. The ISO must look at what *is* available and determine whether it is adequate. Obviously, the ISO must know the IC's intended IAP as a starting place to evaluate resource effectiveness. Generally, personnel and equipment are the key resource considerations. With each of them, the ISO's basic question is, "Do we have enough to do what we're trying to do?" Answering this question acts as a check and balance that can help the IC. Any evaluation of "enough personnel" usually leads to the political, emotional, and practical discussion of staffing. From the ISO's perspective, the staffing present at an incident *is what it is*. Judging adequacy comes from a pre-established understanding of what a given staffing scheme can reasonably accomplish (see the *Awareness Tip* box).

There are several established reference guides that can help the ISO judge staffing adequacy for structure fires:

- NFPA 1710, *Standard for the Organization and Deployment of Fire Suppression Operations, Emergency Medical Operations, and Special Operations to the Public by Career Fire Departments* (2020)
- NFPA 1720, *Standard for the Organization and Deployment of Fire Suppression Operations, Emergency Medical Operations, and Special Operations to the Public by Volunteer Fire Departments* (2020)
- The NFPA *Fire Protection Handbook*, Volume II, Section 13 (2008)

The two NFPA standards (1710 and 1720) outline deployment standards for career and volunteer departments, respectively, and address items such as initial arrival staffing (four persons assembled) prior to initial attack as well as other tactical needs and support functions. NFPA 1710 summarizes that 14 persons are needed for a simple low-hazard single-family-dwelling fire. The NFPA *Fire Protection Handbook* is a bit more detailed and gives some recommended staffing numbers (initial response) for various occupancies based on their hazard potential. While far from absolute, the ISO can use the handbook numbers as a good starting place to judge staffing adequacy but needs to address some staffing issues that the book does not specifically quantify (e.g., rehab, staging, EMS, rapid intervention crew [RIC]).

Given the various reference materials, and the support needs required for a structure fire, a staffing number can be generated to help the ISO gauge needs. **TABLE 12-1** uses the *Fire Protection Handbook* numbers but extends them to include support needs as well as three-person staged companies (no driver/operator). The staffing number assumes a working fire requiring the tactical needs of fire attack, backup line, search, and flow path/access management (ventilation/forcing openings). Remember, this table is only a guide to help judge initial adequacy.

In addition to staffing, equipment allocation and adequacy need to be evaluated. In most cases, the equipment issue is

Awareness Tip

Staffing Trends and Issues

With increasing demands on fire departments (e.g., record call volume, EMS, technical rescue services, hazmat incidents, disaster response, fire behavior changes, customer-service focus) come many challenges for adequate incident staffing. Local politics, emotions, economic factors, and national standards/regulations further the challenge to adequately staff for incident operations. As it relates to structure fire attack staffing, most fire departments use the concept that firefighters form into companies, under the direction of an officer, to accomplish tasks.

The effectiveness and safety of the company is directly proportional to the staffing level (number of members) and capabilities of the company. The capabilities are determined by preincident training. Granted, effective training and proficiency are important, but with all capabilities being equal, the *number of responders* assigned to any given crew has the *greatest impact* on effectiveness. To prove this, the National Institute of Standards and Technology (NIST) released Technical Note 1661: *Report on Residential Fireground Field Experiments* (April, 2010). The report captures the results of a multiphase study of resource deployment for residential fire ground operations (basic interior fire attack with search, ventilation, water supply, etc.). More than 60 full-scale fire experiments were conducted (by well-trained, proficient firefighters) to determine the impact of crew size, first-due engine arrival time, and subsequent apparatus arrival times on firefighter safety and effectiveness. Highlights of the study concluded the following:

- Four-person crews completed essential fire ground tasks (attack, search, ventilation, etc.) 30% faster than two-person crews and 25% faster than three-person crews.
- A four-person crew can put "water on the fire" 6% faster than a three-person crew and 16% faster than a two-person crew.
- The NFPA 1710 standard of 14 assembled firefighters for basic residential fire attack cannot be met with two- and three-person crews using a typical fire apparatus deployment strategy.
- In today's structure fire environment, conditions change much faster than in yesteryear. Adequate staffing of companies is a safety issue as it relates to time.

Staffing below recommended levels can lead to higher physical stress for firefighters as they attempt to complete essential tasks for fire attack. Numerous studies (Johns Hopkins University, Dallas Fire Department, Seattle Fire Department, National Fire Academy, etc.) bear this truth out. In the Johns Hopkins study, it was found that jurisdictions operating with less than four-person companies had injury rates nearly twice the percentage of those with four.[1]

For the ISO, adequate staffing for company formation to accomplish various fire ground tasks is an issue that should be addressed. When too few resources exist for safe company staffing, perhaps the IAP needs to change.

Source: Jahnke, Sara A., Walker S. C. Poston, Nattinee Jitnarin, and C. Keith Haddock. Health Concerns of the U.S. Fire Service: Perspectives from the Firehouse. *American Journal of Health Promotion*, November/December 2012, vol. 27, no. 2, pp. 111–118.

TABLE 12-1	Staffing Guidelines for Initial Fire Ground Operations			
Occupancy	Examples	Resources	Staffing	Total
High hazard	High-rises Hospitals Public assemblies Schools Nursing homes Heavy industrial	4 Engines 2 Trucks (or similar) 2 Chiefs Accountability ISO RIC Rehab EMS Staged	16 8 2 1 1 4 2 2 6	42
Medium hazard	Apartments Offices Mercantile Light industrial	3 Engines 1 Truck 1 Chief Accountability ISO RIC Rehab EMS Staged	12 4 1 1 1 4 2 2 3	30
Low hazard	1- to 3-family dwellings Small businesses	2 Engines 1 Truck 1 Chief Accountability ISO RIC Rehab EMS Staged	8 4 1 0 (by Chief) 1 4 0 (self-rehab) 2 3	23

one of "fit." Does the equipment being used fit the action plan? Does it all reach? Reach can be applied not only to ladders and hose lines, but also to radio communications, fire streams, and building geometry. We typically look at reach as a height issue, but it also applies to the distance firefighters must travel into a building. Operations taking place well within a building (over 300 feet [91 m]) present unique concerns regarding SCBA air management, hose line deployment, equipment shuttling, lag time, and rapid egress.

Task Application

Often, *what* the crews are doing, *how* they are doing it, and *where* they are doing it make the risk equation lean toward unacceptable. Here are some considerations:

1. *Tool versus task.* Are tools being used correctly? Are they the right ones for the task? Are they being stressed past their intended design limits? Firefighters are creative and results-oriented. These attributes are admirable but can lead to injury.
2. *Team versus task.* Are there too few firefighters for the task(s) being attempted? If so, the IAP doesn't match the resources; an intervention or visit to the IC is warranted. Are there too many firefighters involved? If too many people are involved in a task (a "magnet" task—one that everyone wants to be involved in), injury may result because of congestion and team immobility, especially if power tools are involved (see the *Getting the Job Done* box).
3. *Rapid withdrawal options.* Does the crew have an escape path if things go wrong? Are there multiple escape

options? How far does the team have to travel to escape? Using a RIC to help increase rapid-escape options is a preventive safety measure.

4. *Rapid intervention options.* Is a RIC assigned/ready? Does the RIC have a solid understanding of scene hazards, ingress options, and hazards? A RIC is required for all immediately dangerous to life and health (IDLH) structure fires. The ISO should continually relay access, egress, and hazard information to the RIC officer as a matter of routine. Most of us realize that rapid intervention is *rarely rapid*. Account for this reality. We'll discuss RICs a bit later in this chapter.
5. *Trip/fall/struck-by hazards.* Evaluate the trip and fall hazards you find. Strains and sprains are the number one fire ground injury and are a result of slips, trips, and falls. Being struck is the number two cause. Can something suddenly let go and strike a crew? Awareness reminders can help.

Getting the Job Done

The Tennis Ball

Here is an example of the creativity of firefighters. A veteran safety officer of 10 years working in a busy suburban fire department shared this creative tip: He carries a fluorescent-colored tennis ball in his PPE pants pocket and uses it to get firefighters' attention. Typically, the ISO is performing recon just outside the IDLH or activity zone. When crews are focused on a task, they have a tendency to crowd each other (especially when performing forcible entry or ventilation tasks with power tools). Rather than adding to the congestion to make an awareness reminder, he simply throws the tennis ball through the site-line of the firefighters to get their attention. It works.

The veteran ISO mentioned that he has to have a ready supply of tennis balls stored in his locker because firefighters take thrown balls and chop them up, throw them into the fire, or hide them. A few of the balls make it back to the ISO at the annual awards banquet, adorned in all manner of creative graffiti.

Taken further, the ISO can use some sort of signaling device to help keep the ISO out of the hazard zone and still get a message to the firefighters, whether it's a reminder to look up, separate, or simply take a time-out, step back, and come visit the ISO for a moment. Some signaling device options include:

- Whistle or other noise-maker
- Weighted penalty flag (like a sports referee uses)
- Fist-sized bean bag
- Laser pointer (be careful with this one)
- Rapid on/off flashing of a pocket flashlight

You are sure to get some teasing regardless of the method you use—and you may just get your item thrown back at you with the velocity of a fastball. Smile and give the thumbs up.

Operational Effectiveness

The ISO needs to make a judgment that the action plan is making progress toward achieving the objectives of the IAP and leading to a stable, no-change environment. If the desired results are not being achieved, or if incident conditions are

FIGURE 12-4 Smoke that is becoming thicker, darker, or more turbulent during interior attack is cause for the ISO to make a firm intervention.
Courtesy of Frisco (TX) Fire Department

deteriorating faster than positive results can be achieved, the ISO should inform the IC. At most fires, however, ISOs are outside the building, performing recon. How can they tell whether crew efforts are succeeding? To answer, they must use their developed skills for reading smoke and reading buildings to make judgments about operational effectiveness. For most fires, a positive outcome is likely when the fire is adequately ventilated (smoke density is being reduced, built-up pressure is reduced) and fire flow is being met (applied water is quenching the red stuff and smoke is changing to clean white). The ISO should observe signs that a fire attack is succeeding. When no progress is evident, the ISO should determine whether the attack is failing due to inappropriate fire flow to the fire seat or due to the lack of adequate heat flow paths. Failure to make this evaluation may lead to a situation in which firefighters are overrun by the fire. Pockets of steam that pale in comparison to the velocity, density, and color of smoke generated may mean that the fire is releasing more heat than the stream can match. Smoke conditions that change rapidly for the worse (thicker, darker, more turbulent) during interior attack are cause for the ISO to make a firm intervention **FIGURE 12-4**. Remember, the introduction of air to a ventilation-controlled fire will cause explosive growth. In these cases, the ISO should be quick to communicate to the IC that an explosive growth is about to occur.

This brings us back to our equation: principal hazard ± integrity + other hazards ± resource effectiveness = risk-taking.

■ Risk-Taking Evaluation at Structure Fires

You may look at the preceding pages of this chapter and say, "Wow! That is way too much to remember, let alone evaluate on the fire ground." Your point is well taken, and Deputy Chief John Sullivan of the Worcester (MA) Fire Department agrees

when he says, "We often task our ISO with more responsibilities than any other human could possibly accomplish on the fire ground."[2]

The key is to continually front-load knowledge (prior to the incident), then apply a sense of priority thinking (experience and wisdom) while on scene. In actuality, the equation process can be quite fast—mostly spontaneous judgments you make when you combine observations and perceptions with your sense of priority. The bottom line: The ISO needs to make a risk-taking judgment. Even when the risk-taking equation adds up to an extraordinarily high number, you may still let the risk slide! Why? Because imperiled victims are being rescued or still need to be saved. Making the rescue determination may be obvious from an outside recon view—people on balconies, victims streaming out of a smoke-filled building, and so on. However, most times, the rescue need is a judgment based on incident conditions. Some call this rescue profiling, survivability profiling, or survivable indexing (see the *Getting the Job Done* box).

Getting the Job Done

Rescue Profiling
The ISO/IC needs to be willing to make a judgment regarding the likelihood that a known but unlocated victim can survive within a building on fire. To make the judgment, the ISO can evaluate fire, smoke, and building conditions to form an opinion regarding rescue profile.

The rescue profile is a classification given to the probability that a victim will survive a given environment within a given building space or compartment. Typical classifications are high, marginal, or zero for any given compartment within a building. While several definitions exist, the following indicators can help the ISO make rescue profile judgments:

High rescue profile. Spaces within a building where a victim would be in danger of harm from the incident and the likelihood of survival is good with a rescue effort. Some visual indicators of a high rescue profile include:

- Clear windows above or near the active fire
- Minimally smoke-stained windows and visible interior blinds
- Windows that show rivulets of water slowly moving down the inside of the glass (indicates condensation; interior temperature below 160°F [71°C], which can be survivable)
- Spaces that have obvious smoke thermal balance (smoke above, clear visibility below)

Marginal rescue profile. Spaces within the building that present a danger to occupants and rescuers but there is no compelling evidence to indicate a high or zero rescue profile. Simply put, the areas look really bad—but not bad enough to say it is a zero rescue profile. Indicators include:

- Fast, thick, dark smoke leaving a space (heat flow path) that has not gone turbulent and has not totally filled the space

- Intact, dark-stained windows that have *no* heat stress cracks
- Spaces that have already experienced a collapse but did not become fire-involved

Search and rescue efforts in marginal areas should be made with high situational awareness and a close eye toward the rate of change (fire growth, smoke worsening, and warnings of collapse/further collapse). Likewise, the ability to do rapid cooling (fire streams) and to have rapid escape options is mandatory for searches in marginal rescue profile spaces.

<u>Zero rescue profile:</u> Spaces or areas where conditions are simply not survivable. The most obvious indicator of unsurvivable space is a fully developed fire within the space (post-flashover). You can also read smoke to determine a zero rescue profile. While not absolute, the following can indicate a zero rescue profile:

- Turbulent, black, and super-dense smoke that has filled a given space within a building (black fire)
- Deeply stained windows with long, smooth heat stress cracks
- Areas that have floor-level temperatures over 300°F (149°C; beyond human life threshold)
- Areas that have experienced a collapse and rapid fire involvement

On a lighter note, spaces within a building that pose no threat to occupants (no potential for smoke, fire, collapse, or injury from firefighting operations) can also indicate a zero rescue profile; the occupants are simply not in danger.

The risk-taking associated with structure fires should be tied to tactical priorities and incident benchmarks (see the chapter titled *A Systematic Approach to the ISO Role*). The ISO looks at the tactical priorities and determines whether the risks being taken match the department's preestablished risk-taking criteria. Obviously, firefighters engaged in active victim rescue operations may have to face significant risk to their lives. Conversely, risk-taking should be reduced once an all clear (primary search done) is declared. This people-versus-property risk-taking issue may seem obvious here in print. Although such differences may seem obvious, injury and death statistics suggest that firefighters do not always practice the obvious. Unfortunately, the fire service has a poor history of making people-versus-property risk-taking decisions. Too many firefighters have died at structure fires where searches were complete, victims were all accounted for, or spaces were simply not survivable. The ISO needs to serve as the "risk cop" to make sure responders are following established risk guidelines.

Another risk evaluation can be made regarding the "pace" of the incident. Because most of what firefighters do at a structure fire takes place in a short time window, they tend to "up the pace" for task accomplishment. A quickened pace minimizes the time to think and increases pulse rates, blood pressure, and other bodily responses. A quick pace can actually minimize

the reaction time to a surprise hazard. As Chief Vincent Dunn (Deputy Chief, FDNY, retired) has said at numerous conferences: "Speed kills. Do not get caught up in the excitement of the fire ground."

Following an all clear, firefighters should reduce their pace. The ISO should watch for crews or individual firefighters who are out of synch with the incident pace; they are taking an inappropriate risk. At times, the ISO needs simply to state, "We are not going sixty-five miles per hour here—back it off to forty."

■ Safety System Effectiveness at Structure Fires

Accountability Systems

When checking into the personnel accountability system, the ISO should ask the accountability manager if things are going well. The reply can help the ISO evaluate whether the system is tracking firefighters effectively and the system is in harmony with the IAP. If not, the ISO may consider requesting that the IC call for a <u>personnel accountability report (PAR)</u>. The PAR is an organized reporting activity (usually via radio communication) designed to account for all personnel working at an incident. To be truly effective, a PAR should include

---

Safety Tip

Effective PAR Communications
Too often, a PAR takes place with incomplete communications, mostly because working crews are trying to get a task completed and they short-cut quality reporting. The ISO should be mindful of this when he or she hears PARs being reported on the radio.

Examples of *Ineffective* PAR Communication:

"Fire Attack Team 1 from Accountability: Call for PARs."

"Accountability from Fire Attack 1: We have PARs."

Examples of *Effective* PAR Communication:

"Fire Attack Team 1 from Accountability: Call for PARs."

"Accountability from Fire Attack 1: We have PARs with three, performing fire control on Division 2, Bravo side, no issues."

"Fire Attack 1 from Accountability: Copy PARs of three, performing fire control on the second floor, Bravo side."

In cases where the ISO hears a completed PAR, but the communication was ineffective, the ISO should consider visiting the IC and/or accountability manager and doing a check-up (compare the PARs, formal record, and personal recon observations) to form an adequacy opinion. In cases where the PAR is incomplete, a RIC should be deployed to find unaccounted-for responders. The ISO can best support a RIC engagement by assisting the IC and assigning an assistant safety officer for recon.

radio transmissions that include confirmation of the assignment, location, and number of people in each assignment (see the *Safety Tip* box).

Failure to account for all personnel operating at an incident scene has been cited in many line-of-duty death (LODD) investigations. In too many cases, missing firefighters have gone undetected until it was too late to launch a rescue. For that reason, PARs should be accomplished periodically and routinely at structure fires. Some fire departments mandate PARs every 15 minutes while firefighters are working in an IDLH environment or other high-risk environments. Additionally, some fire departments outline certain situations or changes that should trigger a PAR. Examples include:

- Anytime the operational mode has changed (i.e., switch from offensive to defensive)
- Anytime an incident benchmark has been achieved (e.g., all clear, fire under control)
- Following the report or witnessing of a flashover or collapse
- After the report of missing or trapped firefighters (Mayday)

When it becomes apparent by PARs or other communication that a member or crew is not accounted for, the ISO should go to the command post and assist the IC with immediate actions:

1. Verify that a person or crew is unaccounted for and its last known position/assignment.
2. Declare "emergency traffic" over command and tactical channels and communicate the designation or name of the missing.
3. Activate a RIC and direct it to the last known incident area of the missing.
4. Assign an assistant safety officer (ASO) to assist in RIC monitoring.
5. Use runners to search the non-IDLH incident operational areas (rehab, apparatus, staging, and incident perimeter).
6. Depending on immediate feedback, the ISO may suggest to the IC that a backup RIC be assigned and may request further resource response (second alarm, mutual aid).

Rehabilitation

To evaluate rehabilitation effectiveness at structure fires, the ISO needs to focus primarily on the effects of heat, physical exertion, and weather exposure. Firefighters have been known to disregard these effects if there are still "fun" things left to accomplish. Look for signs that the rehab effort is not being effective **FIGURE 12-5**. Firefighters leaving the building for a bottle change likely need active core temperature cooling, hydration, and food (if they have not eaten in the previous 2 hours). Department policies should dictate a mandatory rehab cycle for firefighters working at structure fires. When a cycle is not directed by policy, the ISO should encourage mandatory rehab after every air cylinder use or equivalent work period.

While interior crews typically receive the most rehab effort, the ISO cannot forget to check on the outside responders. We rarely consider the rehab needs of apparatus operators, RICs,

FIGURE 12-5 The ISO should watch crews for signs that the rehab effort may not be effective.

command staff, and (ironically) rehab attendants when fighting structure fires. If the outside weather is extreme, responders exposed to the elements also need relief from heat or cold and dehydration.

Radio Transmissions

The fire ground is filled with significant noise, an obvious barrier to effective radio communications that leads to missed, incomplete, or confusing communications. Additionally, crews are performing laborious tasks that require dexterity and concentration, another barrier to radio transmissions. Even so, the ISO needs to listen for unanswered radio calls and make a judgment whether the unanswered call indicates communication barriers or the need for rapid intervention to locate the crew not responding. Radio message priority is also important at the structure fire. Crews engaged in IDLH environments should have priority to report conditions, needs, and progress. The ISO needs to listen to radio traffic and, when the sense of priority is not congruent with the risks being taken, intervene if necessary. When listening, pay attention for trigger words or phrases such as the following that indicate a developing problem:

- "Mayday"
- "Urgent"
- "Emergency traffic"

- Unintelligible yelling or pained messages
- "I'm lost"
- "Look out!"

Unique Considerations for Structure Fires

■ Traffic

For the most part, the traffic around structure fires comes under control quickly, as law enforcement officers block streets where hose lines are laid. The "invasion" of large apparatus at the structure fires also helps control traffic issues. The greatest traffic risk to firefighters at structure fires is when they are arriving or moving apparatus, especially when water-tender shuttle operations are under way. For shuttle operations, the ISO should see that a traffic flow plan is communicated to all responders. The ISO should also evaluate apparatus placement, ongoing apparatus movement, and traffic lanes to determine whether hazards exist and to develop preventive measures. Using the soft intervention of "awareness" is likely the best method to address such situations. Remind apparatus operators to slow down, use spotters (in congested areas or when backing up), and expect the worst. Remind firefighters to "look both ways" and never to run or jog around apparatus movement areas.

■ The Need for an Assistant Safety Officer

In certain situations, the ISO should request an ASO (or multiple ASOs) at structure fires. Situations that warrant ASO consideration include, but are not limited to:

- *Large buildings with significant fire involvement.* The ISO is likely to need help monitoring the building, fire conditions, crew effectiveness, and/or apparatus issues. The declaration of collapse or no-entry zones may require constant monitoring by an ASO.
- *Situations in which a "plans section" is established at the fire.* The ISO needs to provide input on action plan changes and strategic-level issues. ASOs are needed to accomplish the ISO "field component" (recon, rehab evaluations, and similar functions).
- *Fires in buildings with unusual or unique hazards.* The ISO may want an ASO to handle the typical ISO functions while the ISO meets with building representatives or technical specialists.
- *Situations in which the ISO is requested to go into an IDLH environment.* In these cases, to retain an outside viewpoint, the ISO should request an ASO (and additional partner) to make the entry. If this is not acceptable to the IC, the ISO should request the IC to take back overall safety functions while the ISO enters the IDLH environment with a partner.
- *Situations in which a Mayday has been declared or the RIC is pressed into action.* A firefighter emergency can add tremendous stress on all responders. Think of it as an incident within an incident. Assigning an ASO for the RIC operation is beneficial, and it is usually best for the ISO to assist the IC at the command post.

■ Interface with Rapid Intervention Crews

By and large, fire departments have embraced the concept and use of RICs for structure fires as well as hazmat and technical rescue events. Various acronyms have been developed to identify RICs, but whatever the crews are called, they serve the same purpose. Variations include:

- RIT = rapid intervention team
- RIG = rapid intervention group
- RAT = rapid action team
- FAST = firefighter assistance and safety team

OSHA requires a minimum of two trained and equipped personnel to serve as a RIC (29 CFR 1910.134, *Respiratory Protection*) whenever SCBA is used for an IDLH environment (interior firefighting operations)—also known as the "two-in, two-out rule." NFPA 1561 states that the IC is responsible for designating and assigning an RIC to initiate the immediate rescue of injured, lost, or trapped responders (8.8) and goes on to define that an RIC must include a minimum of two trained and equipped responders to serve that purpose.

History has shown that two responders serving as the RIC is woefully inadequate to rescue a lost/trapped/injured firefighter, leading to the understanding that the two-person RIC is really intended as a requirement for *initiating* an interior attack. Most agree that a minimum of four responders are necessary for an RIC, with those four members dedicated to the function (as opposed to performing other assignments). The ISO's role with regard to a RIC is threefold:

1. Evaluate incident environmental conditions and operations to determine the need for RIC standby, and communicate the need to the IC.
2. Ensure that adequate resources are dedicated to the RIC.
3. Communicate ongoing information to the RIC leader/officer.

The Need for RIC Standby

The first item in the preceding list should be easy, assuming the department has preestablished standard operating procedures (SOPs) for RIC deployment and it is accomplished as part of the structure fire routine. Likewise, the IC is responsible for making sure the need is addressed, and the ISO should confirm with the IC that it has, in fact, been accomplished. In those instances where a RIC has not been established, the ISO can make a quick assessment of the environment and operations to see if a RIC should be established. Simply speaking, any environment that presents IDLH threats (e.g., smoke, flame, collapse potential, hazardous energy sources exposed) is a candidate for RIC allocation. The ISO then weighs that IDLH environment against the operational potential for responders to be lost, trapped, or injured in such a way as to require a risky rescue. If the risk is deemed sufficient, a RIC should be established. The key phrase here is "risky rescue." For example, the response to a fully involved car fire might satisfy the environmental need for RIC standby (smoke, fire, hazardous energy), but the location of the car fire could be such that any firefighter injured by the event can be easily retrieved and cared for (no RIC required). Take the same car fire and place it in a garage (collapse, flashover), in a tunnel (lost, limited air), or off an embankment (tricky rescue), and placing a RIC on standby becomes prudent.

Adequate RIC Resources and Capabilities

There are many training books and programs available to help prepare RIC members for their function and to improve their capabilities. This type of in-depth review is not the intent here. The objective in this discussion is to prepare the ISO to judge whether adequate capabilities have been established for the conditions at hand. We've already made the point that the two-in, two-out rule is for initial entry into a burning building and that a dedicated crew of four persons should be established for structure fires. The crew of four should assemble appropriate tools and other equipment that might be required for a firefighter rescue and take up a standby position for rapid engagement. The ISO should check with the RIC officer to see that this has been accomplished as a minimum and whether the minimum is adequate. Certain incident conditions may require the RIC capabilities to be increased (see the *Getting the Job Done* box). In these cases, the ISO should bring the issues and a solution to the IC for consideration.

Getting the Job Done

Evaluating RIC Capabilities Versus Incident Conditions

An ISO needs to offer judgment as to the adequacy of RIC capabilities with regard to incident conditions. Some fire departments have made investments to create well-trained and equipped RICs; others have not. Regardless, the starting place for judging adequacy starts with an understanding of what an established RIC is routinely capable of accomplishing. Often, RIC inadequacy exists because of environmental and operational conditions that are present at a given incident. The following are examples of conditions that may require the ISO to request an increase in RIC capabilities:

- Environmental
  - Geographical distance
  - Building geometry (number of floors, layout)
  - Access/egress options
  - Type of construction encountered
- Operational
  - Multiple/separated hot zones with many engaged firefighters
  - RIC readiness tools that won't reach (ladder/ropes, SCBA air management)
  - Inferior RIC readiness tools (hand vs. power tools, need for heavy lifting)
  - Existing RIC that is activated for firefighter assistance
  - Need for RIC rotation/rehab (climate, duration)

Where present, these conditions should be communicated to the IC to confirm the need for an increased RIC capability. This communication is typically done through face-to-face discussion (ISO to IC). The use of radio communication to request an increase in RIC capability is not advisable, as it may be construed as a request for RIC activation to an emergency. If the need to increase RIC capability is one of immediacy, then perhaps an imminent threat is the real issue and the ISO should make a firm intervention to working crews until the RIC capability issue is addressed.

ISO–RIC Communications

The ISO and RIC leader/officer share some similar responsibilities; that is, they both should perform incident recon and monitor/evaluate incident activities and conditions. Both are making a judgment regarding hazard potentials. Granted, the ISO's goal is to *prevent* a firefighter emergency, whereas the RIC leader/officer is focused on *preparing* for the emergency. Regardless, it makes sense that the two communicate on a regular basis and share perspectives **FIGURE 12-6** .

Some departments have seen the value of the ISO–RIC shared responsibilities and have organized their system and SOPs so that the RIC officer actually reports to the ISO when in standby mode. Upon RIC engagement for a firefighter emergency, the RIC reports to an operational function (IC, operations, or group/division). This reporting structure is not necessarily compliant with the National Incident Management System (NIMS), but the benefits are compelling. Local SOPs aside, this text recommends that the ISO routinely share observations with the RIC leader. Doing so is beneficial to the situational awareness of both the ISO and the RIC. At a minimum, the ISO–RIC communication partnership should share observations regarding the following:

- *Location of interior crews.* Communicating crew location ensures proper tracking and matching of tools and resources to that location.
- *Fire behavior issues.* These details include smoke and fire conditions, heat and airflow paths, potential for hostile fire events, and fire stream effectiveness observations.
- *Access/egress options.* The ISO and RIC must apprise one another of the number and locations of current access/egress points, barriers to rapid egress, forcible entry challenges, and the desire for more options.
- *Building construction collapse threats.* Collapse threats relate to building type/era/use/size considerations, established collapse zones, weak links, and likelihood and severity issues.
- *Hazardous energy issues.* The ISO and RIC can help one another monitor the status of building utilities and can alert one another to shut-off options and the presence of hazmat/pressure vessels.

■ Residential Versus Commercial Fires

Unfortunately, classifying building fires as residential or commercial can set up firefighters with mistaken expectations. In reality, some residential properties can exceed the size and fire load of a commercial building, and some commercial buildings present minimal hazards to firefighters. A new "starter mansion" or "McMansion" can exceed 6,000 square feet (560 m²) of lightweight construction and contain a fire load that exceeds the quenching ability of all the preconnected attack lines on the engine!

Size is not the only issue. Building use is equally important. Many buildings were built for residential use but are now used as commercial structures—and the converse is also true. Recall from the chapter *Reading Buildings* that the effective ISO uses the "type/era/use/size" method to classify buildings when making a "read." Further, most causes of LODDs (flashover, rapid growth, collapse, lost/trapped) occur in residential

FIGURE 12-6 The ISO and RIC officer can work as partners to share perspectives that increase situational awareness for each.

FIGURE 12-7 Strip malls often have facades that pose a rapid collapse potential. Note the fire spread path and collapse of this facade.

properties. Don't fall into the trap of classifying a building as residential or commercial.

Buildings with Central Hallways and Stairwells

A center-hall-configured building or one with central shared stairwells creates unique problems for firefighters. From the ISO perspective, these buildings present a tremendous risk for the advancing firefighter: The flow paths for ventilation and fire spread are the very same paths that firefighters use to perform search and fire attack. Dense smoke in hallways and stairwells become another risk: Smoke is fuel that ignites when the right mixture and temperature combine, causing a rapid fire spread that is faster than firefighters can react. The key is smoke cooling and flow path management: The *number one tactical priority* must be controlling smoke and heat (for search, fire attack, and firefighter safety) in buildings with central halls or stairways. These are IC tactical issues, but if they are not adequate (read the smoke and read the building), the ISO needs to intervene.

Strip Malls/Fast-Food Buildings

The neighborhood strip mall can be a firefighter killer. Specifically, high fire loads, common ceiling spaces, long

open-span trusses, and decorative facades can turn the simple neighborhood beauty salon into a surprisingly nasty firefight. Watching smoke can pay huge dividends for the ISO; be the eyes for the IC on strip mall fires. As firefighters attempt to confine the fire, watch for brown smoke, under pressure, coming from the facade distal to the fire attack. Fire can spread rapidly in facades, accompanied by rapid collapse. Often, crews use the space below a shared facade for attack access; pay particular attention to these **FIGURE 12-7**. Make sure the occupancies on either side of the involved unit have their concealed spaces exposed and preventilated. If these tactical priorities are not in place, share a solution with the IC. Strip malls typically have heavy roof loads, like HVAC systems, grease hoods, signs, satellite dishes, and other equipment. Stand-alone fast-food restaurants can share the same hazards as a strip mall.

Remember also that strip malls and stand-alone fast-food restaurants are relatively inexpensive to build. A fire that warps the bar trusses in a single unit may lead to a decision to completely demolish the building. The developer can "scrap" the entire strip mall for less than it costs to fix a few units damaged by the fire/smoke. In their preparation for fire loss, many fast-food chains have entire replacement structures ready to be rebuilt on the existing pad. Practice sound risk management at strip mall fires and fast-food restaurants.

High-Rise Building Incidents

Fighting high-rise fires is a science unto itself, and the ISO-related concerns could take up another whole book. The intent here is to give some general direction for the ISO at these complex incidents.[3] First, and perhaps foremost, the ISO should request one or more ASOs for growth-stage fires in high-rise buildings. Let's examine some other considerations.

ISO Functions at High-Rise Incidents

The ISO should take up a position in or at the command post. If the IC is in the lobby, be there. If command is two floors below the fire, be there. If command is on the street, be there. The reason is simple: ISOs need to be in on action planning and able to hear what building representatives and technical specialists have to say. They should be ready, however, to bounce between command, liaison, and plans. In all cases, they need to remain "strategic" in their safety thinking. Of particular interest to ISOs should be:

- The overall action plan
- Control of building systems, such as elevators, HVAC, and stairways
- Shelter-in-place or occupant evacuation issues (a safety concern with firefighters going one direction and occupants the other)
- Risk versus gain
- Communication and feedback from ASOs
- Outside hazards (falling glass/victims, apparatus issues, congestion)

ASO Functions at High-Rise Incidents

When command takes a position in the lobby or street, an operations section is typically set up on the upper floors below the fire. Assign an ASO to go work with the operations section, possibly bouncing between the operations section chief location and the floors below, where rehab and forward staging are located. You may also need an ASO to perform recon outside the building, monitoring access, traffic, hose, and glass/debris issues. The ASOs need to deal with typical safety functions associated with recon, tactical-level risk, and rehab. Of particular interest to the ASO should be:

- *The physical demands on firefighters.* Shuttling equipment in itself taxes the firefighters. Firefighters on the fire floor experience more rapid signs of heat stress. Rehab is hugely important: Active cooling, hydration, and electrolyte and nourishment replacement are vital right away!
- *Internal traffic control.* A plan should be in place and communicated for firefighter travel routes and accountability check-in.
- *Compartment integrity.* The ASO should encourage firefighters to be the eyes and ears for compartmental collapse possibilities. Dropped ceiling grids collapse early. More importantly, crews need to report the condition of stairwell cracks, stuck or bound exit corridor doorways, and window glass.
- *Establishment of no-entry zones around lost windows.* Air (and smoke) pressure variances between the building and the outside atmosphere may create a jet vortex near lost windows that can pick up and move an unsuspecting firefighter.
- *Development and delivery of safety briefings.* This may be an inside and/or outside ASO task.
- *Outside issues.* The ASO outside the building needs to be particularly observant of traffic issues; exposure to dropping glass, debris, or humans; and maintenance of established zones. Likewise, the use of helicopters for roof-top evacuation or firefighter/equipment shuttling may be requested or under way. An ASO can help monitor/evaluate the helicopter interface.

Contamination Reduction and Cancer Prevention

The cancer rate of firefighters is twice that of the general population[4], and we know the causes: firefighters crawling through zero-visibility smoke, prematurely doffing SCBA during overhaul, and continuing to wear contaminated gear after a fire (not to mention taking it back to their stations, cars, homes, and families). What we are talking about here the numerous carcinogens, irritants, polynuclear aromatic hydrocarbons, and particulates found in smoke (and "clear" air) during and after fires. These contaminates bind with soot and water vapor that permeates firefighter protective equipment (PPE) ensembles. Of particular concern are the particulate and aromatic contaminants in PPE clothing that expose firefighters to cancerous threats through inhalation and skin absorption for hours, and days, following a structure fire[5]. Even in "clear" air, the gases, minute particulates, and aerosols may be present-though invisible-and firefighters are exposed.

A sad irony exists in that firefighters will readily deploy a full-blown hazmat technician-level decon evolution for a suspected illicit kitchen drug lab, yet they are unlikely to provide any decontamination measures for smoke residues at a structure, car, or dumpster fire! This needs to change.

Leading fire officers and the NFPA have recognized the cancer threats and are aggressively developing SOPs and standards to help reduce cancers threats through contamination-reduction processes whenever firefighters are exposed to smoke environments. Likewise, our fire service leaders are insisting that SCBAs and cover-all PPE are utilized throughout overhaul and during the initial steps of fire cause determination and investigations. Fire Departments that have not developed smoke contamination-reduction measures are behind the cancer-prevention curve and need to so immediately. Where no contamination-reduction measures are taking place, the SOFR should step-up and champion the cause by suggesting some simple solutions (see *Getting the Job Done box*).

Strategy and Tactics

Firefighters and fire officers enjoy talking (and arguing) tactics and strategy—especially for structure fires. Although we have emphasized judging risk-taking as it relates to the IAP, we haven't instructed the ISO to judge the IC's chosen tactics and overall strategy. The reason is simple: That is the IC's responsibility. When the ISO gets involved in tactics and strategy, he or she is circumventing the IC, which can lead to command dysfunction and/or broken IC relationships. The ISO acts as the eyes and ears of the IC by gaining a view of the safety details of conditions and activities that the IC can't see. This safety-support role may sound ideal, but it is essential and easier said than done. The hallmark of good fire officers is their ability to choose effective and efficient tactics to support an overall strategy for a given incident. You could say that it is in our DNA to always develop, analyze, and alter tactics and strategies. Chief of Safety Stephen Raynis of FDNY notes, "If a fire officer is not usually assigned as an ISO, it is very difficult to remove oneself

Getting the Job Done

Contamination Reduction at Structure Fires

Studies show that the smoke from a typical structure fire (as well as car and dumpster fires) contains chemicals that are carcinogenic, and the smoke can form an oily residue and film on firefighter PPE and equipment. These chemicals can pose a cancer threat to firefighters directly through inhalation and skin absorption as they handle their gear after smoke exposure.

While on scene, firefighters need to take steps to avoid smoke residue exposure. The following techniques can help with contamination reduction at fires:

1. Prior to disconnecting the SCBA mask air supply, perform gross head-to-toe contamination reduction using a garden hose or booster line, bristle brush, and spray bottle with a mild soap solution. Those assisting this contamination reduction should be wearing particulate masks and disposable EMS gloves and some sort of cover-all clothing.

2. Remove structural gloves and don EMS gloves to disconnect air mask supply, remove helmet, and doff the remainder of the SCBA. Remove PPE and place them in a "soiled" drop zone.

3. Personnel who are refilling SCBA bottles should use disposable EMS gloves and mask to help minimize exposure to any remaining contaminates.

4. Prior to entering the rehab or rest area, use sanitary wipes to help cleanse sweat and soot from skin surfaces of the hands, neck, and face (at minimum).

5. Crews rotating back into the hazard environment need to don clean EMS gloves to re-don their PPE and SCBA. Best-practice is to also use a clean head/neck hood for return trips to the hazard zone.

6. Before leaving the scene, place contaminated PPE and SCBA in a disposable trash bag (clear is best so others know what is in the bag) for transport pack to the station for further cleaning.

7. At the station, clean smoke residue from all equipment and SCBA (using disposable EMS gloves). Process soiled PPE and undergarments per department policy (machine washing or packaging for a contract cleaner). Prepare a second set of PPE for incident readiness (many departments now issue two sets of PPE for this reason).

8. As soon as possible, take a shower (shower within the hour), before eating, relaxing, or sleeping.

Some "old-schoolers" may find this process impractical or unrealistic, and to those we must ask: "How committed to cancer prevention are you?"

---

from the thought process of being a tactical officer and concentrate on safety concerns only" (see endnote 2). The key for the ISO is to have a supportive tactics and strategy attitude and focus on safety issues (risk-taking).

A conundrum is created when the ISO feels deep down that the basic safety issue is that the IC's action plan (tactics and strategy) doesn't fit the incident. It could well be that the IC is the dysfunctional component leading to inappropriate risk-taking. Commander dysfunction is usually the result of one or more of the following:

- Incomplete or inaccurate incident intelligence
- Failure to update the IAP to address changing conditions
- Inexperience of the IC

When any of these dysfunctional triggers occur, it is best for the ISO to avoid attacking the IC's tactics and strategy. Instead, gather some factual observations and develop a solution that focuses on risk reduction—not the IC's action plan. Recall the solution-driven approach from the chapter titled *A Systematic Approach to the ISO Role*: "Here's what I see. This is what it means to me. . . . We could do this. . . . What do you think?"

Wrap-Up

Chief Concepts

- The well-prepared ISO uses a front-loaded skill set of "Read 3" to help evaluate the most prevalent risks at a structure fire: *read* smoke, *read* the building, *read* hazardous energy. Combine this approach with the ability to judge firefighter actions and effectiveness, and one has the formula to help discover potential threats and, thus, prevent firefighter injury or death.

- Risk-taking at a structure fire can be expressed using the following formula:

  Principal hazards ± Integrity + Other hazards ± Resource effectiveness = Risk-taking

- The inputs for the risk-taking formula are derived from the observations and judgments made during the ISO's recon efforts.

- Principal hazards at a structure fire that can present imminent threats include hostile fire events, building geometry, collapse potential, and hazardous energy.

- Environmental integrity can be defined as the status of the building, conditions, and hazards in terms of stability (change potential) and time (rate of change).

- Time considerations at an incident include rate of change, projected on-scene time, and reflex time.

- Other hazards can include issues related to the weather, sloping-grade buildings (which is the first floor?), vehicles, or other barriers.
- Resource effectiveness includes issues associated with staffing, equipment reach, task applications, and operational accomplishment. The ISO needs to judge the adequacy of resources and see if it matches the IC's IAP. As a guideline, adequate staffing for initial fire ground operations for various occupancy hazards is 42 for high-hazard buildings, 30 for medium, and 23 for low.
- The ISO should watch the smoke to help gauge operational accomplishment. Smoke conditions that change rapidly for the worse (thicker, darker, more turbulent) during interior attack are cause for the ISO to make a firm intervention.
- Rescue profiling can help the ISO compare IAP activities with the potential to save victims. Using visual clues, the ISO can profile the likelihood of victim survivability as high, marginal, or zero. Too often, firefighters have died at structure fires where searches were complete, victims were all accounted for, or spaces were simply not survivable.
- Effective accountability systems have a method for reporting (personnel accountability report [PAR]) that includes the verification of personnel assignments, location, and number of people in the assignment. Incomplete PARs may indicate the need for RIC engagement to assist lost, injured, or trapped firefighters. A PAR should take place on a regular basis (defined locally), anytime an operational mode has changed (offensive/defensive), following a collapse or flashover, or following a Mayday.
- The greatest traffic risk to firefighters at structure fires is when they are arriving and moving apparatus, especially when water-tender shuttle operations are under way. For shuttle operations, the ISO should see that a traffic flow plan is communicated to all responders.
- The ISO should consider the request for ASOs when responding to fires in large, unusual, or hazardous buildings; when a "plans section" is established; when an ISO is needed within an IDLH environment; or when a Mayday is declared.
- While the IC is responsible for assigning resources for a RIC, the ISO needs to confirm the RIC assignment, ensure that the RIC is adequate for the incident, and provide ongoing communication of incident observations to the RIC leader/officer. Issues that should be communicated between the ISO and RIC officer include those concerning fire behavior, access/egress options, building collapse threats, and hazardous energy potential.
- Other unique ISO considerations for structure fires include hazards associated with residential versus commercial fires, buildings with central hallways and stairwells, strip malls and fast-food restaurants, and high-rise buildings.
- The cancer rate of firefighters is twice that of the general population, and we know the causes: firefighters crawling through zero-visibility smoke, prematurely doffing SCBA during overhaul, and continuing to wear contaminated gear after a fire. Fire departments need to develop procedures and decontamination measures that are routinely practiced at all structure fires. The ISO may have to create ad hoc recommendations where decontamination is not practiced at structure fires. Essential parts of structure fire decontamination should include:
  1. Gross decontamination of gear prior to mask removal, using a brush and small hose water stream
  2. Use of disposable EMS gloves for doffing and reapplying contaminated gear
  3. Use of sanitary wipes to clean hands, neck, and face prior to rehab
  4. Bagging and transport of contaminated gear back to the station for machine cleaning
  5. Showering after equipment cleaning
- The ISO is the eyes and ears of the IC at structure fires and should not focus on tactics and strategy. Rather, the ISO should focus on safety details and the reduction of risk-taking. If the basic incident issue is inappropriate tactics and strategy, it could be that the IC has incomplete/inaccurate incident intelligence, hasn't adjusted the IAP to changing conditions, or lacks experience. In all cases, the ISO should bring recon observations and a solution to the IC and should not attack the IC's tactics and strategy.

Key Terms

building geometry The constellation of elements that must be taken into consideration at a structure fire, including issues associated with the building's layout, size, number of floors, access options, and other features.

high rescue profile A classification indicating a high likelihood that a person located within a fire-involved structure can be successfully rescued.

marginal rescue profile A classification indicating a moderate likelihood that a person located within a fire-involved structure can be successfully rescued; in this scenario, there is clear danger to occupants and rescuers but no compelling evidence to indicate a high or zero rescue profile.

personnel accountability report (PAR) An organized reporting activity (usually via radio communication) designed to account for all personnel working an incident. To be truly effective, it should include radio transmissions that include confirmation of the assignment, location, and number of people in each assignment.

rescue profile A classification of the probability that a victim will survive a given environment within a given building space or compartment. Classifications include *high rescue profile*, *marginal rescue profile*, and *zero rescue profile*.

Wrap-Up, continued

zero rescue profile A classification indicating no likelihood that a person located within a fire-involved structure can be successfully rescued; in this scenario, conditions are simply not survivable.

Review Questions

1. An effective PAR should include the communication of what three elements?
2. What are the top three actions that an ISO should take if a member is not accounted for at an incident?
3. What are four principal hazards (imminent threats) that are found at most structure fires?
4. What is meant by incident integrity?
5. List four instances in which an ISO should request ASO assistance at structure fires.
6. What is meant by a zero rescue profile?
7. List the environmental and operational factors that can signal the need for increased RIC capabilities.
8. What are the critical observations that the ISO and RIC leader/officer should share with each other?
9. What is a magnet task? What can the ISO do to help address magnet tasks?
10. List the three resource considerations at structure fires.
11. List several unique hazards at strip mall structure fires.
12. List four ISO functions unique to high-rise fires.
13. List six ASO functions unique to high-rise fires.
14. List the steps that should be taken for contamination reduction at structure fires.
15. What are the typical causes of IC dysfunction (inappropriate tactics and strategy) at incidents?

References and Additional Resources

Clark, William E. *Firefighting Principles and Practices.* 2nd ed. Tulsa, OK: Fire Engineering Books and Videos, PennWell, 1991.

Emery, Mark. 13 Incident Indiscretions. *Health & Safety for the Emergency Service Personnel,* 15, no. 10, FDSOA, October 2004.

Emery, Mark. The Ten Commandments of Intelligent and Safe Fireground Operations. *Health & Safety for the Emergency Service Personnel,* 15, no. 11, FDSOA, November 2004.

The Firefighter Close Calls website contains fire ground stories, near-miss reports, and lessons learned at structure fires. It is available at http://www.firefighterclosecalls.com.

Mittendorf, John W. *Truck Company Operations.* 2nd ed. Tulsa, OK: Fire Engineering Books and Videos, PennWell, 2010.

National Fire Protection Association. *Fire Protection Handbook.* 20th ed. Vol. 2, Section 13. Quincy, MA: NFPA, 2008.

National Institute for Occupational Safety and Health. *NIOSH Alert: Preventing Deaths and Injuries of Firefighters Using Risk Management Principles at Structure Fires.* July 2010. http://www.cdc.gov/niosh/docs/2010-153/pdfs/2010-153.pdf. Accessed May 22, 2015.

National Institute of Standards and Technology. *Technical Note 1661: Report on Residential Fireground Field Experiments.* April 2010. http://www.nist.gov/el/fire_research/upload/Report-on-Residential-Fireground-Field-Experiments.pdf. Accessed May 22, 2015.

Nedder, Joe. *Fire Service Rapid Intervention Crews: Principles and Practice.* Burlington, MA: Jones & Bartlett Learning, 2015.

NFPA 1710, *Standard for the Organization and Deployment of Fire Suppression Operations, Emergency Medical Operations, and Special Operations to the Public by Career Fire Departments.* Quincy, MA: National Fire Protection Association, 2020.

NFPA 1720, *Standard for the Organization and Deployment of Fire Suppression Operations, Emergency Medical Operations, and Special Operations to the Public by Volunteer Fire Departments.* Quincy, MA: National Fire Protection Association, 2020.

Norman, John. *Fire Officer's Handbook of Tactics.* 3rd ed. Tulsa, OK: Fire Engineering Books and Videos, PennWell, 2005.

The Rules of Engagement for Firefighter Survival poster, by the International Association of Fire Chiefs (and several cosponsors), is available at http://www.iafc.org/files/rulesofengagement-roe_poster.pdf.

U.S. Fire Administration. *Study of Cancer Among Firefighters.* Firefighter Safety and Health. Washington, DC: USFA, November 2013. Links to the report can be found at http://www.usfa.ffema.gov/operations/ops_wellness_fitness.html. Accessed June 2015.

Endnotes

1 Jahnke, Sara A., Walker S. C. Poston, Nattinee Jitnarin, and C. Keith Haddock. Health Concerns of the U.S. Fire Service: Perspectives from the Firehouse. *American Journal of Health Promotion,* November/December 2012, Vol. 27, No. 2, pp. 111–118.

2 Author's note: Quotations are taken from NIOSH LODD Report F2012-13, which is one of the most complete, well-written LODD reports and should be mandatory reading for all suppression firefighters. It is available at http://www.cdc.gov/niosh/fire/reports/face201213.html.

3 Author's note: A special thanks to the late Safety Chief David Ross (Toronto Fire Services) and to Battalion Chief Gerald Tracy (FDNY, retired) for sharing their wisdom on high-rise fires. The author takes all responsibility for interpreting their input.

4 U.S. Fire Administration. *Study of Cancer Among Firefighters.* Firefighter Safety and Health. Washington, DC: USFA, November 2013. Links to the report can be found at http://www.usfa.fema.gov/operations/ops_wellness_fitness.html. Accessed June 2015.

5 Underwriters Laboratories. *UL Study on Firefighter Exposure to Smoke Particles.* Northbrook, IL: UL, PennWell, 2010.

On September 14, 2002, a 53-year-old male career firefighter died after falling through a roof following roof ventilation operations at a house fire.

The structure was a 96-year-old, 2½-story wood frame dwelling with balloon-frame construction. The roof was steeply pitched and had intersecting gables consisting of 2 × 4-inch timbers covered with 1-inch wood planks and four or five layers of asphalt shingles. The building was located on a corner lot on a grade. The second floor was accessed at street level from the rear. The first-arriving engine reported smoke showing under the rafters and established command per department policy. The IC and engine crew conducted an interior size-up to determine the location of the fire and whether there were any occupants (there were none). Because of the heat and smoke encountered as they tried to climb to the top floor, the crew had to back down the stairs. The IC then exited the structure and called for roof ventilation. A firefighter from a rescue squad replaced the IC on the second interior attack attempt. A second interior fire attack was begun through the rear (second floor) of the building, which initially was relatively free of smoke. As this attack crew tried to climb the stairs to the top floor, they too encountered heavy smoke and extreme heat, causing them to back down the stairs to wait for ventilation.

After positioning the aerial platform over the roof reportedly "as far as it would go," the victim and his partner, carrying only a chain saw, exited the aerial platform and walked approximately 15 feet to the area to be ventilated. The victim's partner had donned his self-contained breathing apparatus (SCBA) before leaving the ground and, upon exiting the platform, went on air due to heavy smoke. The victim, who was not wearing SCBA, was observing his partner making the ventilation cuts. After the last cut was made, the victim, who had been covering his face with his hands (presumably because of the thick smoke), told his partner that they had to leave immediately. The firefighters retreated toward the aerial platform, but the victim stopped a few feet from the platform, saying he could not continue. Seconds later, the area of the roof under the victim failed, and he fell through the roof into the structure and the fire. Within minutes the interior attack crew found the victim and, with the help of the rapid intervention crew (RIC), removed him. He was transported to a local hospital where he was pronounced dead.

The victim was on the roof for less than 7 minutes. He did not fall through the hole cut by the saw; rather, he fell through an area of the roof between the cut hole and the aerial platform. According to the fire department report, this area presumably failed due to direct exposure to the fire below and the weight of the victim above. The victim fell approximately 10 feet.

1. From the information given, what are some of the structural (building) factors that contributed to the collapse?

2. What are the key information elements presented in this case that could signal an ISO that the risk-taking was out of balance?

3. The environmental integrity for this case study could be best described as which of the following?

   A. Stable but changing slowly
   B. Unstable and changing slowly
   C. Unstable and changing quickly
   D. Unknown

4. When a Mayday has been declared and the RIC is pressed into action, it is best for the ISO to:

   A. communicate a risk vs. benefit judgment to the IC.
   B. perform a rapid recon of the incident and judge the rescue profile.
   C. deploy with the RIC to help monitor hazards associated with the incident within an incident.
   D. report to the command post to assist the IC and request an ASO for RIC monitoring.

Note: This case study was taken from NIOSH Firefighter Fatality report #2002-40, available at www.cdc.gov/niosh/fire.

The ISO at Wildland and I-Zone Fires

Flames: © Ken LaBelle NRIFirePhotos.com

Knowledge Objectives

Upon completion of this chapter, you should be able to:

- Identify actions or operations that will be altered, terminated, or suspended to protect members' heath safety if identified by the ISO. (NFPA 5.2.2 , pp 197–198)
- Identify imminent threats to firefighter safety. (NFPA 5.2.4 , pp 195–196)
- Classify types of imminent hazards into major categories. (NFPA 5.2.4 , pp 194–195)
- Describe the incident scene conditions that are monitored as part of an ongoing incident. (NFPA 5.2.5 , pp 195–198)
- Define the wildland fire terminology relevant to fire growth and behavior. (NFPA 5.3.5 , pp 194–195)
- Describe fire growth and blowup factors in wildland and cultivated vegetation fires. (NFPA 5.3.5 , pp 194–195)
- List six situations that may require the appointment of an assistant safety officer at the wildland fire. (p 199)
- Define *LCES*. (p 197)

Skills Objectives

Upon completion of this chapter, you should be able to:

- Analyze the fuel, weather, and environmental effects on vegetation fires that lead to hostile fire conditions. (NFPA 5.3.5 , p 194)

An unusually dry spring has prompted a series of small brush fires in an outlying area in your community. You are concerned about the potential for these fires to become a major incident should they spread toward some newly developed areas as well as some well-established subdivisions that have not maintained adequate clearance between homes and vegetation. You voice your concerns at the monthly staff meeting, and a working group is established to tackle this challenge.

1. What capabilities does your department currently have in place to respond to a fire where vegetation and multiple buildings are involved?
2. As the ISO, what additional training or information beyond structural operations should you begin to learn and be ready to apply?

Introduction: Into the Wild

The phrase "wildland and I-zone fires" in the chapter title is not meant to be restrictive. A farmer's corn stubble field fire is not necessarily a wildland fire by definition; it is a cultivated vegetation fire. Likewise, not all wildland fires are I-zone fires. The term "I-zone" is another term for wildland–urban interface (WUI) and applies to areas where homes and businesses have minimal separation from, or are interspersed with, natural-growing wildland areas. The use of the word *urban* in WUI shouldn't be interpreted literally; it can refer to suburban or rural buildings also. For brevity, in this chapter we will use the generic terms "wildland fire" or "wildfire" for all types of vegetation or I-zone fires unless there is an issue specific to the I-zone.

At times, structural firefighters treat small brush fires as a sort of "nuisance" event that lacks the challenge and excitement of structure fires. Too often, the attitude that accompanies the nuisance incident can result in reduced attention to personal protective equipment (PPE), adequate fire flow, and sound attack strategies. Conversely, trained wildland firefighters believe that wildfires have more challenges and excitement than structure fires and, thus, put more emphasis on PPE and sound attack strategies, even for small brush fires. When the two groups come together for a common fire, the differences present interesting conflicts. Thankfully, most of us learned the lessons of the California firestorms; the Long Island (New York) wildfires; the vast wildfires in Florida, Oklahoma, and Texas; and most recently, the largest wildland fires in the histories of Colorado and Washington. More than ever, wildland and structural firefighters are sharing important perspectives and finding common ground in operations.

The scope of this chapter is to provide the structurally oriented fire officer with incident safety officer (ISO) insight as part of the initial response to a wildland fire. As a wildland fire grows, so does its resource demand. At some stage, the fire may grow beyond the resources of the local fire department and its mutual aid assistance. Once this occurs, the ISO functions are likely to be transferred to a trained safety officer who is part of an incident management team (IMT). In this chapter, therefore, we address the *initial* ISO duties and focus areas for Type 5 and Type 4 wildland fire incidents (see the chapter titled *Triggers, Traps, and Working with the Incident Commander* for definitions of the fire types).

Before we get into the ISO issues at wildland fires, we need to front-load some language and concepts that are associated with these events.

Wildland Fire Language

The language used to communicate specific details of wildland fires may not be commonplace for the typical municipal firefighter. In some cases, the wildfire community (in particular, the National Wildfire Coordinating Group [NWCG]) uses language and phrases that might conflict with municipal firefighters. For example, if you tell a structural firefighter that you "need a line over there," he or she would go grab a preconnected hose line and deploy it. Tell the same thing to a wildland-trained firefighter and the response would be, "Scratch, fire, or wet?" In the wildland fire community, a *line* is a barrier to fire spread. More specifically, a control line is the all-inclusive term for all constructed or natural barriers and treated fire edges used to control a wildland fire. A fire line is the part of a control line that is scraped or dug to mineral soil. A scratch line is a preliminary control line hastily constructed as an emergency measure to check fire spread. A wet line is water or a water agent sprayed on the ground as a temporary control line from which to ignite or stop a low-intensity fire.

The wildland fire suppression community rarely uses typical control zone language (hot, warm, and so on) for the wildland fire. Many of the "zones" used for wildfires are based on the descriptive parts of the fire or features in the fire area **FIGURE 13-1**. Words such as "head," "flanks," "origin," and "spots" are used to describe parts of the wildfire. Occasionally, geographical feature names, like "saddle," "chimney," and "WUI," are used to describe wildfire zones.

Wildland Fire Concepts

The basic approach to structure fire control is to manage heat flow paths and "put wet stuff on the red stuff." Not so for the wildland fire suppression professionals. Their approach is to create a 360-degree control line around the fire—the reason they use phrases such as, "The fire is 18% contained." There

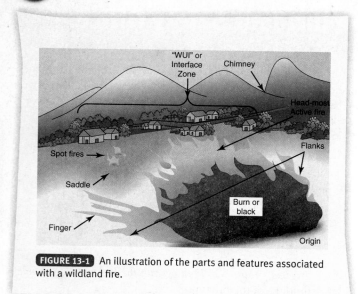

FIGURE 13-1 An illustration of the parts and features associated with a wildland fire.

are times when water is used to knock down or extinguish wildfires, but that doesn't count; the wildland fire gurus view that as a temporary measure to check or stall growth until a full control line is complete.

Threatened buildings in the I-zone are typically classified as defensible or indefensible. The difference in the two typically deals with the distance between the building and the flame lengths of burning vegetation. Other factors such as combustible siding/roofing material (shingles) are also considered. Those that are deemed defensible are often "treated" by a structural protection crew that attempts to quickly remove combustibles (vegetation and man-made stuff) around the structure and then apply water or water-based agents on or around the structure. Agents include:

- A solution of Class A foam and water
- Compressed air foam with water
- Fire retardant (like the pink slurry dropped from aircraft)

The methods of attack employed for wildland fires are numerous but are typically classified as *direct* or *indirect*. As a simple example, the local community fire department responds to a wildland fire and uses a direct attack to knock down the "flanks" and "head" of the fire (wet stuff on the red stuff). Dedicated wildfire crews will typically choose an anchor point (a nonthreatened natural feature like a river, lake, or rock outcropping or a man-made element like a wide roadway) and start to create a fire line (indirect attack).

The language and concepts presented here might be a bit oversimplified; regardless, they are important to start with as they set the stage for the ISO's ability to judge risk-taking and understand how a simple wildfire can turn into a firefighter line-of-duty death (LODD) potential.

General ISO Functions at Wildland Fires

Upon arrival and assignment at a wildland fire, the local fire department ISO should quickly grasp the potential for firefighters being overrun by the fire. The well-known factors

that affect wildland fire spread are weather, topography, and fuels **FIGURE 13-2**. Some general considerations for each are as follows:

- *Weather.* Temperature, relative humidity, barometric pressure, winds, and weather "events" such as microbursts and tornadic activity need to be considered. Also, the arrival or departure of high- or low-pressure fronts can cause a complete change in fire behavior.
- *Topography.* Factors such as slope (degree), aspect (relationship to the sun), and physical features (chimneys, saddles, barriers, etc.) all influence wildland fire behavior.
- *Fuels.* Wildland fuels are affected by moisture content, fuel type (ground, aerial, etc.), and the continuity of fuel (sparse, fractured, dense).

These factors may combine in such a way that firefighters could be overrun, just like rapid fire spread in a structure fire. Also, just like a flashover in a structure, wildland fires can experience hostile events:

- Blowup. A wildland fire term used to describe the sudden advancement and increase in fire intensity due to wind, prewarmed fuels, or a topographical feature such as a narrow canyon or "chimney." Sometimes the term "blowup" is used when a ground or surface fire becomes an aerial or crown fire. Blowup is so named because it often disrupts or changes control efforts. Because of the sudden increase in fire intensity or rate of spread, a powerful convection column can form, which can create its own wind, causing additional concerns.
- Fire storm. A violent convection column caused by a large continuous area of intense fire. Fire storms are characterized by violent surface in-drafts near and beyond the perimeter and occasional tornado-like whirls. A fire storm is considered *severe* fire behavior and direct attacks are to be avoided.
- Flare-up. A sudden, but short-lived rise in rate of spread or fire intensity that is usually attributed to wind, fuel, or topographical changes. Flaring is a warning sign of an upcoming blowup.

FIGURE 13-2 Wildland fire spread is affected by weather, topography, and fuels.

- Torching. The burning of the foliage of a single tree, or small bunch of trees, from the bottom up (sometimes called "candling"). Torching isn't necessarily a hostile event but can lead to larger flare-ups and blowups.

The ISO needs to evaluate the many factors related to fire behavior and then attempt to classify the fire in terms of firefighter safety. Evaluating flame length can also help the ISO determine if initial attack crews are at risk (see the *Awareness Tip* box).

Awareness Tip

Wildfire Flame Length Interpretations
- Less than 4 feet (1.2 m): The fire can generally be attacked directly using hand lines and tools.
- 4 to 8 feet (1.2 to 2.4 m): The fire is too intense for a direct attack on the head. A flanking attack with increased gallons per minute may be effective. Indirect firebreaks and wet lines are advisable.
- 8 to 11 feet (2.4 to 3.4 m): The fire presents serious control problems. Direct fire attacks are dangerous.
- More than 11 feet (3.4 m): Major fire runs are likely. Defensive measures are required.

By evaluating wildland fire behavior (potential for blowup, signs of flaring, and flame length) the ISO can form a judgment about incident potential and risks. The observation of flaring with flame lengths over 8 feet (2.4 m) can be considered an imminent threat if initial attack crews are attempting a head-on fire attack. A firm intervention may be warranted. In all cases, initial fire behavior observations and concerns should be shared with the incident commander (IC). If it has not been done, the IC absolutely needs to get a weather forecast and judge fire growth potential based on fuels, the topography, and the current and forecasted weather. Although the ISO needs to focus on crew risk-taking, it may be a good first step to help the IC gather this information and agree on an incident action plan (IAP) that fits predicted behavior. For initial attack efforts, this fuel–topography–weather relationship to an IAP can be accomplished face-to-face in just a few minutes.

As with structure fires, the general functions of the ISO involve issues associated with reconnaissance efforts, risk-taking evaluation, and safety system effectiveness.

■ Reconnaissance Efforts at Wildland Fires

The reconnaissance (recon) effort at a wildland fire is not as simple as walking around the building. Recon vehicles, helicopters, and assistant safety officer (ASO) field reports all help. ASOs or lookouts (trained in ISO functions) can help with recon; their reports are important not only to the ISO but also to the IC (more on ASOs later). Coordination is critical. *Recon caution*: Climbing to high ground to get a good look at the fire can prove fatal. If an elevated feature near the heel or flanks or a position upwind of the fire affords a good sight line, take advantage of it as long as there is no remote possibility that the

fire will go there. Using a vehicle to drive around the fire perimeter for recon could be dangerous around the head or advancing fingers. The use of a topographical map can be useful in understanding the geographical features of the fire area, and most have roadways included. The map can help the ISO (and IC) understand the fire area land and access issues, making it a recon tool. Reading a topographical map is a developed skill—one that ISOs should pursue if they haven't already done so.[1]

Keep in mind that the purposes of recon are to define the principal hazard, judge the potential for environmental change (integrity), define the impact of the physical surrounding, and equate the exposure of crews to help define risk-taking. The formula covered in the chapter *The ISO at Structure Fires* is also applicable at wildland fires:

Principal hazards ± Integrity + Other hazards ± Resource effectiveness = Risk-taking

Defining the Principal Hazards

The principal hazard (imminent threat) in most wildland environments is usually associated with one or more of four factors:
- Rapid fire spread (flare-ups, blowups, and fire storms)
- Physical exertion (prolonged aerobic exertion)
- Traffic issues (smoke obscuration, off-roading, congestion on narrow unpaved roads)
- Hazardous energy (power lines, transformers, fences, rivers, falling trees, wild animals)

Intervention is usually required in each area to prevent injuries, especially for structural firefighters who are not trained for, or routinely involved at, wildland fires.

Environmental Integrity

Defining environmental integrity involves the consideration of factors such as weather, smoke, flame spread, and hazardous energy. Just like the structure fire, the ISO can define environmental integrity using the following system:
- Stable and not likely to change (subtracts from risk-taking)
- Stable but changing slowly (may or may not add to risk-taking)
- Unstable and changing slowly (adds to risk-taking)
- Unstable and changing quickly (exponentially adds to risk-taking)

Recall that the passage of time is a factor when judging integrity. The presence of high winds and dry fuels can create a very unstable environment with a change rate measured in seconds. Add in a sun-side aspect and an upward slope and you have a formula for disaster. Unfortunately, the time window for initial effectiveness at a wildland fire is usually measured in minutes and hours, as opposed to seconds and minutes, as at the structure fire.

Projecting on-scene time can remind the ISO of many considerations, especially the advent of nightfall and predicted weather changes. Make sure you have a good weather forecast and use your weather-watching skills. Travel distances can affect reflex time for immediate resource needs; having multiple staging points for equipment at a geographically widespread incident is wise, and the ISO may have to bring this suggestion to the IC.

Defining Other Hazards

The wildland fire environment is as diverse as nature itself. Numerous trip-and-fall hazards top the list, especially where there are no roads, trails, or other established pathways for firefighters to work from. Animals, vermin, and noxious vegetation may present hazards not routinely experienced by structural firefighters. Burned-out trees and dead limbs, known as snags, may have been weakened and are a real hazard to firefighters; many have been killed by falling snags. "Look up and live" applies to snags and power lines. Do not downplay the potential of hazardous energy: Downed wires and ground transformers can create a ground gradient or direct electrocution hazard that can zap unsuspecting firefighters. Common fencing in the wildland environment includes the use of metal wire (such as barbed-wire) and can extend an electrocution potential when downed wires are present (the downed wire may have caused the fire!).

Active firefighting operations are typically suspended or become conservatively defensive at dark because there are too many unseen hazards at night; in fact, many fire departments have standing orders that an initial wildfire attack at night can be attempted only from established roads or trails. The ISO should have that discussion with the IC and perhaps broadcast a priority safety message to all responders for night operations.

Resource Effectiveness

At most wildland fires, the ISO is not in a good position to evaluate operational effectiveness because of distance, terrain, and smoke. The use of roving ASOs may be helpful where the geographical layout of the incident minimizes ISO effectiveness. Still, the ISO needs to make a judgment that the actions under way are making progress toward achieving the objectives of the IAP. Taking a tour in a vehicle (keep away from the fire head) or helicopter can help. The tour should include an evaluation of resource allocation, task application, and operational effectiveness.

Resource Allocation

The ISO should evaluate whether enough personnel are available for the IAP and whether enough responders are in staging should an immediate need evolve. If not, the ISO should bring the observation and a suggested solution to the IC. Making a judgment regarding sufficient staffing for a wildfire is difficult to quantify because of the many factors that have to be considered. Those factors include geographical dispersal, method(s) of attack chosen, the need to protect structures, and the physically taxing challenges that wildfires present for responders. Unfortunately, the ISO is in the position of using gut judgment to determine whether enough people are present to achieve the IAP. Some assistance for making the judgment can be found in some staffing awareness tips (see the *Awareness Tip* box).

Some may argue the point, but most firefighters *live* for the chance to fight fires, structural or otherwise. They know that threatening wildfires are physically demanding and require concentration and awareness. Once the initial "rush" of adrenaline has worn off, firefighters should slide into a routine. If rehab is effective, the routine *should still be energetic* (with all the positive indicators). If not, rehab (and staffing) is likely failing and needs adjustments. Making this observation is a critical duty of the ISO when evaluating staffing adequacy at the wildland fire.

Awareness Tip

Staffing Awareness Tips for Wildland Fires

Adequate staffing for wildland fires is hard to quantify; however, there are some general performance guidelines that can help an ISO judge staffing adequacy. These guidelines are not absolute and serve to help the ISO uncover potential staffing issues that can lead to injury.

- Vehicle-based direct attacks (brush trucks/patrols using pump-and-roll spray nozzles) should have a minimum of two personnel in the vehicle.
- "Boots on the ground" direct attack with small hose lines and hand tools should be done with no less than a three-person crew, with one being an officer/leader/supervisor (four is better). This total assumes they are near a vehicle/roadway. When two or more crews are working an area, add another two persons to serve as lookouts/spotters.
- "Hike-in" hand crews should have a minimum of five personnel, with one being an officer.
- A structural protection crew should include four persons: a driver/operator and firefighter for hose work and an officer and firefighter for direction/lookout/debris tasks. A trained, four-person crew can reasonably treat a defensible structure in 20 minutes.
- A water shuttle operation (ground-based tenders/tankers) should include a support team of five plus the driver/operators. Two members are at the fill station and two are at the drop point. The last person is the water supply officer, who should develop, communicate, and monitor a traffic plan. If multiple fill stations are being used, consider having two persons on each tender in lieu of the fill station crew.
- Staffing for support functions such as accountability, staging, and rehab must be increased at wildland incidents.

Task Application

Often, considering all relevant factors—*what* the crews are doing, *how* they are doing it, and *where* they are doing it—makes the risk equation lean toward unacceptable. Here are some considerations:

- *Tools.* Chain saws are invaluable, but they create dangers to those not accustomed to their operation for clearing trees, brush, and other obstacles. A structural firefighter may be quite accomplished at using a chain saw for cutting roofs, but a 3-foot-diameter tree or snag requires different skills. Structural firefighters may not be accustomed to lengthy hose lays, water supply issues, and the use of Class A foam in wildland environments. Pump calculations may be off, putting firefighters at risk. Some apparatus may not have hose bed protection, allowing flying embers to ignite stored hose. Again, include reminders in safety briefings and interface with crews during recon to create awareness of these safety issues.

■ *Team versus task.* Crews protecting structures may get sucked into structural firefighting if the wildland fire penetrates a building. The ignition of drapes or light combustibles inside a building may occur and is easy to control. Ignition and quick fire spread of soffit spaces, decking, and exterior siding/roofing is a different story. This is usually a dangerous proposal and is considered high risk. Often it is best for the crew to grab and go; that is, grab any civilians (homeowner), throw hoses up on the apparatus (for their own protection, if needed), and get out of the area.

Digging a fire line with hand tools requires not only physical effort but *space.* Hand line crews need to spread out but stay in voice distance of each other. The use of a compressed air foam system (CAFS) for fire control and structural treatment requires different hose and nozzle techniques that could catch the unprepared firefighter off-guard. Built-up pressure at a closed CAFS nozzle can literally snap a firefighter's wrist if he or she isn't ready for it when opening the discharge. CAFS hose lines are notorious for kinking. Reminders can help.

■ *Trip/fall/struck-by hazards.* The wildland fire environment is basically one big, spread-out trip-and-fall hazard. It bears repeating that strains and sprains are the number one fire ground injury and are a result of slips, trips, and falls. Being struck is the number two cause. The classic rubber firefighter boot is poorly suited for wildland operations, as are lightweight trail-hiking boots. Firmly laced, leather, over-the-ankle boots with a rigid sole are best for footing in uneven and rough terrain. We know the ISO can't make people change their footwear on the spot, but awareness reminders—and perhaps a suggestion to stay on roads and trails—can help. Likewise, a reminder to look up, look down, and slow down the pace can help prevent a strain, sprain, or struck-by injury.

■ *Rapid withdrawal options.* The acronym LCES is used to address rapid withdrawal options. LCES stands for lookouts, communication methods, escape routes, and safety zones and is pronounced "laces," as in *boot laces.* Individual crews should be constantly pointing out LCES options. Some LCES advocates have altered the acronym by adding *awareness*—making the acronym LACES. If you do not hear LACES chat on a regular basis, the crew leaders are not doing their job, and a friendly (and private) reminder is warranted. Once the friendly reminder is delivered, gauge the ongoing actions of the leader. If there is no change in behavior, this may indicate that the leader needs to be rotated to the rehab sector and that rehab efforts are behind the curve.

Backing apparatus into dead-ends, driveways, and structural protection assignments is an important rapid withdrawal option. This is not common practice for municipal-oriented firefighters, but it is essential for rapid retreat in the wildfire environment.

Rapid withdrawal for those performing a direct attack on a ground-fuel fire is best achieved through communicated escape routes and safety zones away from the fire. For initial operations and unexpected flare-ups or erratic wind shifts, the crews can use the "black" or burned area as an emergency withdrawal area. While not ideal (heat, carbon monoxide, embers), it may be safer than trying to escape through burnable fuels.

Operational Effectiveness

The ISO needs to make a judgment that the action plan is making progress toward achieving the objectives of the IAP and leading to a stable, no-change environment. If the desired results are not being achieved, or if incident conditions are deteriorating faster than positive results can be achieved, the ISO should inform the IC. At wildland fires, however, the ISO will rarely have all the inputs needed to judge operational effectiveness like one would at a structure fire. The reading smoke skills used at a building fire are minimally helpful, mostly because the wildland fire produces so much whitish smoke (light fuel pyrolysis, moisture, dust, and ash). So how can they tell whether the crews are being effective? The answer comes in the form of crew-report monitoring, flame length observations (they have reduced or subsided), and the attainment of a stable, slow-change (but predictable), or no-change environment.

Most local department fire attack control efforts at wildfires are based on direct flame quenching, wet lines, and structural protection. While this operational accomplishment is considered positive, it is also temporary. Recall from the preceding wildland concepts that an incident is considered contained only when a control line surrounds the perimeter of what was burning. Overhaul within the control line could take hours, even weeks, depending on the size of the control zone perimeter and the fuel status remaining within.

■ Risk-Taking Evaluation at Wildland Fires

The often unpredictable nature of wildland fires tends to cause victims to self-rescue/evacuate. That is not to say there will not be victims, and this possibility needs to be evaluated. There are some people who demand the right to stay and protect their property and/or livestock. Once the victim evacuation question is resolved, the prevailing life risk is to firefighters. The IAP should reflect this change in a reduced risk profile. It is easy to say that no risk should be taken if evacuation and search efforts have eliminated rescue needs, but do you write off a farmer's field or let a watershed area burn? Value must be considered. Good fire control efforts can actually prevent millions of dollars in mud-slide damage when the rains come after a wildfire or save hundreds of thousands of dollars in agricultural revenue. The effective ISO understands these concerns and works with the IC to achieve a "calculated" risk-taking environment that favors firefighter safety and is still aggressive in fire control efforts. In other words: intellectual aggressiveness.

Judging the pace of the incident is part of risk monitoring. Structural firefighters engaged in wildland fires should slow their typical pace. This statement is certainly arguable if structural protection, with its typical short time window, is necessary in I-zone fires. Another viewpoint offers a variant: Making a structure defensible against an advancing fire could take 20 to 30 minutes for an experienced four-person WUI-trained crew. If that time is not available, given flame lengths and speed of fire spread, then the structure is *not* defensible.

If the fire is progressing faster than crews are effective, the IC may become trapped in linear thinking and regress to habitual behaviors in crisis mode. The ISO who senses this development must shore up the IC function with cyclical thinking, suggestions (based on observations and evidence), and support.

■ Safety System Effectiveness at Wildland Fires

Accountability Systems

The collection of accountability passports may prove to be troublesome in the initial stages of the wildfire. During wildfire attack operations, firefighters may spread out and lose the typical crew integrity found at structure fires. The initial response to a wildfire in a forested I-zone is not conducive to typical accountability system use. Consider the following: The apparatus assigned to a wildfire incident are usually dispatched to an "area" versus a given address. As the first-due engine approaches the area, it may find a threatened structure and start an attack while another apparatus finds a flank or head and starts a different attack—and the two are a mile apart by road. When this occurs, the responding IC (with help from the duty ISO) needs to establish a fixed command post, orchestrate a radio roll call, and plot initial accountability on a map of the area (they may never collect accountability passport or tags, or not for an hour or more into the incident). The ISO may have to help the IC accomplish the initial plotting as they gather not only locations but CAN reports (conditions, actions, needs) from the crews that engaged prior to fixed IC establishment. From that point forward, a quick IAP can be formed after the IC and ISO try to gauge the fire spread potential (weather, fuels, topography). Additional resources can then be tracked as they arrive and are assigned.

Rehabilitation

Physical exertion and exposure to heat and smoke lead the list of rehabilitation factors that require significant attention at wildfires. Firefighters using structural PPE for wildland incidents are at extreme risk for heat stress. Many structure fire departments now supply wildland-appropriate gear for their firefighters. If not, the department is willingly setting firefighters up for a heat stress injury when structural PPE is used (more on this later). In all cases, rapid hydration, electrolyte replacement, and food are essential; the firefighters are more than likely using aerobic output (requires more hydration and carbohydrates for a longer duration) versus the anaerobic effort commonly required for a structure fire.

Medical monitoring should also be prioritized for rehab; heart attacks remain the number one killer of firefighters. Firefighters should not be allowed to go back to work based on their perceived comfort; a qualified medical attendant should make the determination based on core temperature and other vital signs. Physically stressed firefighters *should*

not be allowed to leave the incident; they should remain in rehab until they have been medically cleared or are transported to a definitive care facility. Rehab attendants should consider cardiac monitoring for any firefighter who exhibits even remote signs of heat stress. If advanced life support (ALS) medical observation is not available at rehab, consider asking the IC to order ALS services to monitor those with signs of heat stress.

When the wildfire incident is spread over a large geographical area, a central rehab station may not be effective or practical. Rehab resources (and EMS standby) may have to be shuttled and staged to safe zones around the fire's perimeter. The rehab personnel are entrusted by the IC to make these determinations, and it is the ISO's job to check that they are effective—give it the effort.

Personal Protection Systems

The classic structure fire personal protection system consists of a structural PPE ensemble, self-contained breathing apparatus (SCBA) with personal alert safety system (PASS), accountability system, and rapid intervention crew (RIC). The personal protection system at a wildfire is quite different **FIGURE 13-3** and includes:

- Lightweight brush coat and pants (Nomex or similar fire-resistive material) with cotton undergarments
- Over-the-ankle, lace-up leather boots with traction sole
- Lightweight hard hat with ear/neck fire-brand protection
- Goggles
- Single-layer leather gloves
- Mask or scarf-like nose/mouth smoke filter
- Web gear with fire shelter and water bottle
- LCES + 10 standing orders + 18 watch-outs (mental safety system)
- Crew integrity (accountability)

Local "town" fire departments that respond to fires in the wildland may not have all of the items from this list; they use their structural PPE and systems. This can set the stage for rapid physical exertion and injury potential. Too many structural firefighters shortcut their PPE use when fighting the local brush fire, and history is full of instances in which the local brush fire took the lives of firefighters simply because they didn't have appropriate PPE.

Fire departments should strive to have wildland gear for their members (grants are available for this purpose). The ISO at a working wildfire where responders are in structural gear (or street clothes) must address the issue. Solutions include ensuring rapid rotation and rehab, restricting direct attack locations, or simply getting the IC to switch the IAP to an indirect attack only. It may go without saying, but structural gear is better than street clothes for anyone doing structural protection or working the fire perimeter.

Unique Considerations for Wildland Fires

■ Traffic

Smoke obscuration is the leading traffic concern at wildland incidents. Efforts to divert traffic away from smoky areas are warranted. Small fire apparatus dispersed over a wide

FIGURE 13-3 The wildfire PPE ensemble is quite different from a structural PPE ensemble.

geographical area can create daunting concerns. Drivers of small, mobile brush patrol vehicles may not be able to see or hear firefighters working on foot in the area; they should keep headlights and flashing lights on at all times in smoky conditions. A general safety message broadcast over the radio frequency can remind drivers to use spotters (especially when backing) and to reduce speed. As with structure fires, remind firefighters to "look both ways" and never to run or jog in apparatus movement areas.

Traffic from evacuees can jam up fire apparatus, especially where narrow, rural roads are encountered. The use of law enforcement may help, and a traffic plan may have to be developed. The IC could ask the ISO to develop a master safety plan that includes escape routes (for vehicles) and safety zones. In such cases, the ISO should ask the IC for an ASO.

■ The Need for an Assistant Safety Officer

The ISO should request an ASO at wildland fires in certain circumstances:

- The fire affects a widespread geographical area.
- A plans section has been established. The ISO needs to provide input on IAP changes and strategic-level

issues, and ASOs are needed to accomplish the ISO field component (e.g., recon, rehab evaluations).

- The fire has been (or is expected to be) active for more than 4 hours.
- A base camp has been established. The ISO needs someone to communicate the IAP and safety briefings to incoming crews. The base camp may also be the rehab area, where food service and sanitation needs are established.
- The IC has asked the ISO to develop safety plans (briefings, escape routes, safety zones, or traffic management maps).
- The fire response involves air resources (helicopter attack, fixed-wing tankers).

The use of ASOs signals the ISO to become more stationary so that he or she can manage incoming information from the ASO(s) and relay/consult with the IC. That is not to say the ISO can't leave the command post; he or she will likely need to meet with other support personnel, do some overhead recon (e.g., observation points, air-recon), and visit evolving base or staging areas. At a minimum, the ISO should develop a safety briefing sheet that can be passed to the staging manager for multiagency responses, longer-duration incidents, and unusual fires (see the *Getting the Job Done* box). Safety briefings are routine for wildland-trained firefighters. A good crew boss repeats safety messages throughout the incident. This behavior should be encouraged at every chance.

Taking the time to diagram and document hazard and safety issues is well received if (or when) the incident is passed to the IMT's SOF (incident management team safety officer). Using standard ICS forms will be appreciated by the incoming IMT (included in this text's appendix).

Getting the Job Done

Preparing a Safety Briefing for Wildland Incidents
Wildland fires that require the use of resources from multiple jurisdictions or those that last 4 hours (or more) signal the ISO to use ASOs and to develop safety briefings for incoming and rotated crews. The standard National Incident Management System (NIMS) form 208 (Safety Message/Plan) provides a starting place for developing briefings. Unfortunately, the form gives only a rudimentary prompt—not specifics—for development. For the wildland fire safety briefing, the ISO should address, at a minimum, the following:

- Brief overview of the IAP (if not communicated by others)
- Active fire areas and anticipated weather/fire spread issues
- Known safety hazards (hazardous energy, animals, noxious plants, aircraft use)
- Specific precautions to be observed and practiced
- Established safe zones and escape routes (traffic plan)
- Established rehab/EMS standby locations
- LCES reminders

Early in the incident, the ISO may choose to deliver quick oral safety briefings (face-to-face or via radio) as crews arrive. As the incident evolves, the briefings should be in writing and may be delivered by the ISO, ASOs, or staging managers.

■ Interface with Aircraft

At a typical Type 5 or Type 4 fire, you rarely see helicopters and fixed-winged aircraft conducting water and retardant drops, although more fire departments are using this valuable resource on a quick-call basis during fire season. Being familiar with the hazards associated with these resources is essential to the ISO **FIGURE 13-4**. If you have never had training on interfacing with contracted wildland aircraft, seek it out (contractors are usually willing during the off-season). You can also access aircraft interface training online at the Interagency Aviation Training website (http://iat.nifc.gov).

Inexperienced firefighters or crews working a wildfire may seek out an opportunity to get "slimed" by a fire-retardant drop from aircraft; this warped mentality is like that of trying to melt a helmet at a structure fire. The accuracy of these drops is measured in terms within hundreds of feet of the target area. During safety briefings you should discuss the inherent dangers of being slimed, as injuries are likely. However, the message should also cover situations where firefighters are in danger of being overrun by the fire and being slimed is the safer alternative (of course, the preference is to stay out of such situations proactively). The trapped firefighters will welcome the drop and accept the risk of a slime injury versus being overrun by fire.

If an aircraft refueling or resupply location is established near the incident (within the fire department's jurisdiction), the ISO or an ASO should make a site safety visit, especially if local fire crews are assisting with the air contractor (like providing an engine and hoses to help with aircraft water resupply).

■ Incident Escalation

Once a fire becomes a Type 3, the ISO function may be transferred to a regional or state SOF who is trained for and experienced in the role. Having good notes that include the chronological order of events helps the transition. During the transition, a series of briefings and communication is likely; be present and offer your observations to bring the IMT up to

FIGURE 13-4 The use of aircraft at wildland fires introduces unique hazards. If you do not have training in aircraft interface, seek it out!

speed. The incoming SOF3 may choose to use you in an ASO role (SOFR—line safety officer), or you may be assigned to the resource pool. The IMT is likely to bring a focus to the incident that may be different from yours or the IC's, but remember that the goal is the same.

Wrap-Up

Chief Concepts

- Wildland fires are those with natural or cultivated vegetation and may include the exposure of homes or businesses. The wildland–urban interface (WUI) areas are those in which homes, isolated structures, and businesses have minimal separation from, or are interspersed with, natural-growing wildland areas. WUI areas are often called the I-zone.

- The language used by wildland-oriented firefighters is slightly different than that used by structure-oriented firefighters. A *control line* is not a hose line; rather, it is an all-inclusive term for all constructed or natural barriers and treated fire edges used to control a wildland. A *fire line* is the part of a control line that is scraped or dug to mineral soil. A *wet line* is water or a water agent sprayed on the ground as a temporary control line for a low-intensity fire.

- The "zones" for a wildfire are usually labeled as descriptive parts of a fire (e.g., head, flanks, fingers) or by geographical features (e.g., saddles, WUI).

- Fire control measures for wildland fires can be direct (attacking flames) or indirect (control line creation away from the fire). Threatened structures are classified as defensible or indefensible, depending on distance and exposure combustibility.

- The ISO at Type 5 and Type 4 wildfire incidents needs to evaluate, more than anything, the potential of responders being overrun by fire. The ISO uses factors of weather, topography, and fuels as a starting place to evaluate overrun threats. He or she must then look at flame lengths and hostile fire event potential to make the overall judgment.

- Hostile wildfire warnings include flare-ups and torching, which may lead to an event such as a blowup (event) and fire storm (most severe event). The observation of flare-ups with flame lengths over 8 feet (2.4 m) can be considered an imminent threat if initial attack crews are attempting a head-on fire attack.

- Recon at wildfires can be difficult and dangerous due to geographical issues and exposure to advancing fires. Finding an unthreatened vantage point and using aircraft, ASOs, or crew reports can help.

- The principal threats at a wildfire include rapid fire spread (overrun), physical exertion, traffic issues, and hazardous energy. Standing dead or fire-damaged trees are called *snags*. Snags and downed or damaged power lines are significant hazardous energy threats. Physical exertion threats can be reduced through frequent and effective rehab and reminders for responders to slow their pace. Initial wildfire attack at night should be attempted only from established roads and trails due to the potential for many unseen hazards.

- At wildland fires, the ISO is not in a good position to evaluate operational effectiveness because of distance, terrain, and smoke. The use of roving ASOs may be helpful where the geographical layout of the incident minimizes ISO effectiveness. Staffing for a wildfire is difficult to quantify because of the many factors that have to be considered. Those factors include geographical dispersal, method(s) of attack chosen, the need to protect structures, and the physically taxing challenges that wildfires present for responders. Unfortunately, the ISO is in the position of using gut judgment to determine whether enough people are present to achieve the IAP. Some guidelines can help: two-person crews for each brush truck, three-person minimum for hand crews near roads, five-person crews if hiking in, four-person crews for structural protection, five-person water shuttle support crew plus a driver for each tender, increased staffing for rehab/staging/accountability, and so on.

- All responders at a wildfire should use LCES to address rapid-withdrawal options: lookouts, communication methods, escape routes, and safety zones. The ISO needs to reinforce and remind leaders/officers/supervisors that they are responsible for constantly reinforcing LCES. Rapid withdrawal includes the practice of backing apparatus into dead-ends, narrow roads, and driveways so rapid retreat can be achieved. The "black" or burned area of a fire can be used as an emergency escape area, although it is not ideal.

- Operational accomplishment can be judged by crew-report monitoring, flame observations (they have laid down), and attainment of a stable, slow-change (but predictable), or no-change environment. Flame extinguishment is a temporary accomplishment; wildfire control is deemed complete only when a control line has been established around the fire perimeter.

- Civilians are likely to self-evacuate or do so when advised by law enforcement. Some civilians may choose to stay and protect their property. Once civilian evacuation has been addressed, the only human life threat at a wildfire is to the firefighters. Risk-taking should be reduced, although it cannot be totally eliminated. Protecting livestock and structures, reducing damage to cultivated vegetation, and controlling watershed areas are still important to prevent large financial losses. Making a structure defensible against an advancing fire could take 20 to 30 minutes for an experienced four-person crew. If that time is not available, given flame lengths and speed of fire spread, then the structure is not defensible and risks should be reduced.

- Initial accountability at a wildfire may be difficult because crews are dispatched to an *area* as opposed to an address. Responders may attack fires at various locations that are separated, requiring the IC and ISO to quickly plot crew locations on a map to begin accountability. When plotting, the IC and ISO can use CAN reports (crew conditions, actions, and needs) to help develop the IAP and predict overrun potential.

- The ISO is responsible for evaluating and addressing rehab needs at all incidents. The wildfire incident may require multiple rehab/EMS standby locations and resources to address the need. The physical exertion at a wildfire is usually aerobic, requiring more attention to hydration and nourishment as well as medical monitoring (for heat stress and heart attacks).

- Firefighters without PPE specific to wildland suppression might use structural PPE, incomplete PPE, or street clothes. All of these cases are hazardous to firefighters. Solutions include ensuring rapid rotation and rehab, restricting direct attack locations, or simply getting the IC to switch the IAP to an indirect attack only.

- Traffic issues at wildfires can be challenging. The ISO may have to assist the IC in developing a safety plan that includes traffic issues (escape routes, safety zones, evacuee issues).

- The ISO should request ASO assistance for wildfires when:
  1. The fire impacts a large geographical area.
  2. A plans section has been established.
  3. The fire is active for more than 4 hours.
  4. A base camp (staging area) has been established that includes rehab, food services, and sanitation needs.
  5. Situations require the development of safety plans and briefings.
  6. Air-attack resources are being used.
- The use of aircraft for Type 5 and Type 4 incidents is becoming more common. Where used, the ISO needs to address issues of inexperienced firefighters wanting to get "slimed." Also, the ISO (or ASO) needs to do a site safety visit for aircraft resupply areas if they are within the jurisdiction of the fire.

Key Terms

blowup A wildland fire term used to describe the sudden advancement and increase in fire intensity due to wind, prewarmed fuels, or a topographical feature such as a narrow canyon or "chimney."

control line All constructed or natural barriers and treated fire edges used to control a wildland fire.

fire line The part of a control line that is scraped or dug to mineral soil.

fire storm A violent convection caused by a large continuous area of intense fire. They are characterized by violent surface in-drafts near the fire perimeter and occasional tornado-like whirls.

flare-up A sudden, but short-lived, rise in wildland fire intensity that is usually attributed to wind, fuel, or topographical changes.

scratch line A preliminary control line hastily constructed as an emergency measure to check fire spread.

torching The burning of the foliage of a single tree, or small bunch of trees, from the bottom up (sometimes called "candling").

wet line Water or a water agent sprayed on the ground as a temporary control line for a low-intensity fire or to ignite a burn-out.

wildland–urban interface (WUI) Also referred to as the I-zone; areas where homes and businesses have minimal separation from, or are interspersed with, natural-growing wildland areas.

Review Questions

1. What is an I-zone?
2. What is the difference between a control line, fire line, and wet line?
3. Define the following terms: flare-up, torching, blowup, and fire storm.
4. List the factors that influence fire spread in the wildland environment.
5. What are the four principal hazards (imminent threats) that are present at most wildfires?
6. The ISO should remind responders to look up and look down for specific hazards. What are they looking for?
7. Describe the minimum staffing guidelines for the following assignments:
   - Vehicle-based brush patrol
   - Hand crews near roads
   - Hike-in hand crews
   - Structural protection crews
8. List six situations that may require the appointment of an ASO at wildland fires.
9. List at least five issues that should be addressed when the ISO develops a wildland fire safety briefing.
10. Discuss a troubling issue that may arise when ground firefighters interface with aircraft.

References and Additional Resources

Fire Operations in the Urban Interface. S-205 course, NFES 2171. Boise, ID: National Fire Interagency Fire Center, updated periodically.

Intermediate Wildland Fire Behavior. S-290 course, NFES 2387. Boise, ID: National Fire Interagency Fire Center, updated periodically.

National Wildfire Coordinating Group. *Glossary of Wildland Fire Terminology.* PMS 205. Boise, ID: NWCG, 2014. http://www.nwcg.gov/pms/pubs/glossary/index.htm. Accessed May 23, 2015.

The National Wildland Fire Training website offers wildfire training resources. It is available at http://www.nationalfiretraining.net.

Endnotes

1. Topographical map reading skills can be learned from many sources. There are several online sources, many of which are free. Likewise, training classes offered by the U.S. Forest Service and wilderness organizations can be found for little or no cost.

Electrical hazards are among the various hazards firefighters face during wildland fire suppression activities. Firefighters performing fire ground operations near downed power lines may be exposed to electric shock hazards through the following means:

- Electrical currents that flow through the ground and extend several feet (ground gradient)
- Contact with downed power lines that are still energized
- Overhead power lines that fall onto and energize conductive equipment and materials located on the fire ground (fences)
- Solid-stream water applications on or around energized, downed power lines or equipment

Let's look at two cases that illustrate these hazards.

Case 1

On June 23, 1999, a 20-year-old male volunteer firefighter was electrocuted while fighting a grass fire. The volunteer firefighter was one of a crew dispatched to a grass fire where a power line was reported to be down. The volunteer firefighter arrived and immediately helped the deputy chief and a firefighter/paramedic extinguish the fire on the east flank. The volunteer firefighter then walked toward a smoldering pile of brush near the downed power line. As he pulled a charged, 1-inch line over the uneven terrain, he apparently tripped and fell onto the 6,700-volt, downed power line. Other firefighters on the fire ground used a nonconductive tool to pull the line from under the victim. He was moved to the street, received cardiopulmonary resuscitation (CPR), and was then taken to a local hospital, where he was pronounced dead.

Case 2

On October 4, 1999, a 20-year-old male volunteer firefighter was electrocuted and two other firefighters were injured when they contacted an energized electric fence while fighting a grass fire. Central dispatch notified the fire department of a fire that was started when a downed power line ignited the surrounding grass. The chief arrived first, followed by Engine 1 and two firefighters. The chief indicated to central dispatch

and to the responding firefighters that the electric fence bordering the area was energized by the downed power line. The driver of Engine 1 and the three firefighters crawled underneath the bottom wire of the electric fence. They positioned themselves approximately 50 feet from the downed power line and attacked the primary fire. After the fire was extinguished, the three firefighters crawled under the fence a second time. It is believed that, when one of the firefighters was crawling on his or her back under the electric fence, a hook from the firefighter's bunker coat might have contacted the bottom wire of the fence. It is believed that the other two firefighters were shocked while trying to help the firefighter who was still energized. All three were removed from the energized area, and basic first aid procedures were administered until the ambulance arrived. One of the injured was transported by helicopter to an area hospital, and another was transported by ambulance to the local hospital and later to the burn unit of an area hospital. The third firefighter was pronounced dead on arrival at a local hospital.

NIOSH recommends that fire departments take the following precautions with regard to electrical hazards at wildland fires:

- Keep firefighters a minimum distance away from downed power lines until the line is de-energized. The minimum distance is equal to the span between two poles.
- Ensure that the incident commander conveys strategic decisions related to power line location to all suppression crews on the fire ground and continually reevaluates fire conditions.
- Establish, implement, and enforce standard operating procedures (SOPs) that address the safety of firefighters when they work near downed power lines or energized electrical equipment. For example, assign one of the fire ground personnel to serve as a sentry to ensure that the location of the downed line is communicated to all fire ground personnel.
- Do not apply solid-stream water applications on or around energized, downed power lines or equipment.
- Train firefighters in safety-related work practices when working around electrical energy. For example, treat all downed power lines as energized and make firefighters aware of hazards related to ground gradients.

1. What are some of the specific similarities in the two cases that led to the fatalities?
2. What are some of the obstacles that may interfere with implementing the NIOSH precautions? (Be specific.)
3. All firefighters should use the mental safety system of "LACES" at a wildland fire. "LACES" is an acronym for:
   A. Lookouts, awareness, communications, escape routes, and safety zones.
   B. Lookouts, awareness, control lines, energized equipment, and safety zones.
   C. Lookouts, anchor points, communications, energized equipment, and snags.
   D. Lookouts, anchor points, control lines, escape routes, and snags.

4. Initial wildland fire attack operations at night are extremely dangerous because of the potential for many unseen hazards (like downed power lines); therefore, firefighters should:
   A. don reflective traffic vests over wildland PPE.
   B. limit initial attack options to positions from established roads and trails.
   C. increase staffing to include more lookouts.
   D. attack only from the "black" or burned area.

CHAPTER 14

The ISO at Hazardous Materials Incidents

Flames: © Ken LaBelle NRIFirePhotos.com

Knowledge Objectives

Upon completion of this chapter, you should be able to:

- Define the reporting structure for an assistant safety officer–hazmat at a hazardous materials (hazmat) technician-level incident. (p 205)
- Identify imminent threats to firefighter safety. (NFPA 5.2.4 , pp 206, 208, 212)
- Describe the incident scene conditions that are monitored as part of an ongoing incident. (NFPA 5.2.5 , pp 206–208)
- Describe the types of incidents at which additional ISOs or technical specialists are required due to corresponding hazards of the incident type. (NFPA 5.2.10 , pp 209–211)
- Define the process for making recommendations to an incident commander to request additional ASOs or a technical specialist. (NFPA 5.2.10 , pp 210–211)
- Identify the types of hazmat incidents that would require the assignment of an ISO or ASO trained as a hazmat technician. (NFPA 5.5.1 , pp 205–206)
- Identify the standards and regulations that would require the assignment of an ISO or ASO trained for a hazmat incident. (NFPA 5.5.1 , p 205)
- Describe the elements of an incident action plan (IAP) related to safety plan development at a hazmat incident. (NFPA 5.5.2 , pp 208, 211)
- Describe the elements of a hazmat safety briefing that are based on an IAP. (NFPA 5.5.3 , p 211)
- Describe risk management principles used in the safety component of an IAP to identify corrective or preventive actions at a hazmat incident. (NFPA 5.5.2 , pp 208–209)
- Describe the strategies used to establish the necessary hazmat control zones. (NFPA 5.5.4 , pp 209–210)
- Describe the methods used to identify and communicate the various types of hazmat control zones to personnel operating at the incident. (NFPA 5.5.4 , pp 209–210)
- Describe the factors that would cause an ISO to adjust established control zones. (NFPA 5.5.4 , p 209)
- List five or more hazards that may be encountered at a clandestine drug lab incident. (pp 211–212)
- List and describe the three strategic goals for the safety section at a weapons-of-mass-destruction or terrorist incident. (p 212)

Skills Objectives

Upon completion of this chapter, you should be able to:

- Perform a review of a hazardous materials incident action plan. (NFPA 5.5.2 , p 211)
- Conduct a hazardous materials safety briefing. (NFPA 5.5.3 , p 211)
- Establish control zones at a hazardous materials incident. (NFPA 5.5.4 , pp 209–210)

A chemical spill from a punctured 55-gallon barrel has occurred at a loading dock of a chemical manufacturing facility. A technician-level response is on the scene and planning to mitigate the incident after recon and research have identified the product and potential hazards. An officer from the hazmat team has met you at the incident and informed you that he will be taking over the ISO role at this incident, as you are trained only to the operations level.

1. How does your department determine the credentials or certification levels necessary to function as an ISO at a special response incident such as a hazmat or technical rescue incident?

2. What functions can you perform at this incident if the hazmat team ISO takes responsibility for the technician-level duties?

Introduction: Hazmat Homework

Hazardous materials (hazmat) incidents create an enormous potential for short- and long-term disruption of our public areas, infrastructure, and environment (groundwater, habitats, farms, and other resources). Environmental consciousness has driven society to develop specific goals, procedures, and laws to govern the production, distribution, and use of hazardous materials. The realization of terrorist threats and the use of weapons of mass destruction (WMD) have further heightened societal awareness and expectations for proactive threat response and mitigation. The production and distribution of illegal drugs and substances have spawned an epidemic of dangerous drug labs and indiscriminate toxic waste disposal. The legalization of marijuana and hemp-based products in some states is challenging fire departments as grow-houses and oil-extraction processes create unique hazards.

As public awareness of hazmat issues increases, so does the expectation for fire departments to properly handle hazmat incidents. As a result, hazmat incidents have become the most regulated of all the incidents to which fire departments might respond. Regulations, the threat of acute and chronic health issues, and the potential damage to the environment have led fire departments to develop a hazmat response system that includes procedures, equipment, and training to help protect firefighters and the community they serve. The appointment of an incident safety officer (ISO) at a hazmat incident is an integral part of this response system. Never has it been more important for a duty ISO or potential ASOs to do their homework. Even if an ISO is not trained to the hazmat technician level, there are many facets of the hazmat incident that he or she must understand when performing safety functions.

The assignment of an ISO at a hazmat technician-level incident is not discretionary; it is mandated by law.[1] As a starting place, the ISO needs to be aware of the many federal regulations—defined in the Occupational Safety and Health Administration's (OSHA) Code of Federal Regulations (CFR)—that affect the role of the ISO at hazmat incidents:

- 29 CFR 1910.95, *Occupational Noise Exposure Limits*
- 29 CFR 1910.120, *Hazardous Waste Operations and Emergency Response Solutions*
- 29 CFR 1910.134, *Respiratory Protection*
- 29 CFR 1910.146, *Permit-Required Confined Spaces*
- 29 CFR 1910.1030, *Blood-Borne Pathogens*
- 29 CFR 1910.1200, *Hazard Communication*

Where there is regulation, there is liability, thus putting tremendous pressure on the fire department—and on the ISO. At the hazmat incident, the ISO should have the professional competencies for the level of incident involved. These competencies are defined in NFPA 472 *Standard for Competence of Responders to Hazardous Materials/Weapons of Mass Destruction Incidents* (2018). The standard outlines awareness-, operations-, technician-, and command-level training requirements. For example, if a fire department offers technician-level response, the ISO (by law *and* to be effective) needs to have that level of competency as well as some specific requirements for the ISO function. Where the ISO does not have a technician-level certification, an <u>assistant safety officer–hazmat (ASO-HM)</u> should be appointed to help with technician safety functions. An ASO-HM is someone who meets or exceeds the NFPA 472 requirements for a hazardous materials technician and is trained in the responsibilities of the ISO position as it relates to hazmat response. When an ASO-HM is appointed, the ISO retains an overall safety function responsibility while the ASO-HM works with the technician-level group or branch. Still, the overhead ISO needs to have the competencies outlined for hazmat command level.

Efforts are under way to standardize language and titles. Still, the ASO-HM may be titled differently based on how the incident management system has been scaled or by local standard operating procedures (SOPs). Some of these titles are:

- Hazmat safety officer
- Hazmat group safety officer
- Hazmat branch safety officer

In this text, the designation ASO-HM is used to indicate the person fulfilling safety functions for the technician-level components at an incident. Organizationally, the ASO-HM should report to and work with the ISO. In reality, the ASO-HM works with three or more people: the ISO, the hazmat branch director (or hazmat group supervisor), and any technical specialists or industry representatives that the plans section chief (or incident commander [IC]) has assigned to assist **FIGURE 14-1**.

The purpose of this chapter is to address specific ISO/ASO-HM challenges that need to be considered in performing the assigned safety functions at the hazmat incident. This chapter is *not* designed to replace hazmat-technician professional

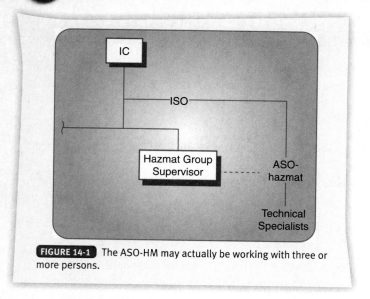

FIGURE 14-1 The ASO-HM may actually be working with three or more persons.

competencies. The intent is to describe the general duties of the ISO, followed by unique considerations for hazmat incidents. In this discussion, it is assumed that an ISO is in an overhead role and an ASO-HM is assigned for technician-level components.

General ISO Functions at Hazmat Incidents

Upon arrival and assignment at a hazmat incident, the ISO should ascertain from the IC the level of control actions that the fire department is attempting (e.g., mass decontamination and victim assistance, zone and isolate, or initial control measures).

When the fire department is engaging in immediate mass decontamination and/or victim assistance, the ISO can help assist with the establishment and marking of corridors and zones for the victims if this hasn't been addressed. Next, the ISO needs to evaluate the initial control zone and isolation efforts as well as the exposure of crews (distance, personal protective equipment [PPE]).

From there, the ISO should meet with the IC and discuss the tactics and overall strategy that are being attempted. The ISO may have to respectfully challenge the tactics and strategy if it is out of synch with the established level of control actions. For example, the IC has declared an awareness-level approach (zone and isolate only), but tactics are assigned that are technician-level in nature. Recognizing this problem and challenging the IC's judgment are founded in observations and hazmat understanding (again, the ISO must do some hazmat homework—the liability is too great). Some factors that might indicate that the tactics and strategy are out of sync include:

- Type of material involved (solids, liquids, gases) and the known or unknown chemical properties
- PPE inadequacy
- Size of the problem (puddle vs. pool, area covered or exposed, containment, potential for escalation)
- Guidelines from the Department of Transportation's *Emergency Response Guidebook*[2]
- Immediacy of civilian life threat and evacuation needs

In cases where the fire department has decided to "sit" on the situation and wait for other resources, the ISO should brief standby crews on zones and safety issues and then perform recon to ensure that all the incident hazard potentials have been considered. Follow-up can include product research and inquiries with the responsible party or technical experts (government officials, industry representatives, or hazmat technicians).

The balance of this chapter addresses ISO/ASO-HM issues for incidents where the fire department is engaged in operations- or technician-level functions. First, though, it is important to repeat that the ISO must be trained and qualified as a hazmat technician and hazmat safety officer to perform ISO functions on those occasions. If the assigned ISO does not have that qualification, an ASO with those qualifications *must* be assigned.[3] Even if the primary ISO has the required training, the safety function focus for technicians performing entry/control/abatement can be demanding. It is often best to have an ISO who is strategic and assists with overhead and support personnel (non-technician) safety functions and an ISO dedicated to the hazmat team (ASO-HM). The wise ISO will also request a third-party technical expert to assist with the safety function. Many fire departments maintain a resource call list that includes technical experts that have agreed to assist in case a technical need arises (i.e., structural engineer, chemist, and information technology [IT] guru). Whether preplanned or spontaneous, all requests for additional ISO assistance should go through the IC.

■ Reconnaissance Efforts at Hazmat Incidents

Upon arrival and assignment at a hazmat incident, the ISO should confirm that initial responders have appropriately zoned and isolated the hazard. The ISO and ASO-HM must verify that the defined zones and gateways are appropriate and communicated. In doing so, the ASO-HM should seek input from the technical reference specialist (TECHREF) assigned to the hazmat team or from the technical reference library, and not rely on previous experiences (more on zones later). The TECHREF can also help the ISO/ASO-HM define the principal hazard, environmental integrity (and threat), the effects of the surrounding elements, and the exposure threats to responders **FIGURE 14-2**. The majority of recon issues deal directly with the information garnered through the TECHREF. The ISO can use that information and apply his or her own observations gathered from a perimeter walk to determine the level of risk-taking. The formula for risk-taking changes slightly for the hazmat incident:

Chemical properties ± Integrity + Other hazards ± Resource effectiveness = Risk-taking

Chemical Properties

Typically, the chemical properties of the commodity involved dictate the principal hazard. When defining the principal hazard, the ISO/ASO-HM must understand the potential for detonation, ground leaching, gas plume spread, and liquid flow and the acute and chronic health hazards associated with exposure to the chemical. The hazmat team TECHREF is usually responsible for finding this information and sharing it with the team and the IC/ISO/ASO-HM for review.

Resource Effectiveness

The ASO-HM is in the best position to evaluate issues regarding operational effectiveness. The ISO uses the CAN report (conditions, actions, needs) from the ASO-HM and combines it with recon observations from the surrounding environment and support responders. As ever, the ISO needs to get to risk-taking. The resource effectiveness part of the equation includes resource allocation, task application, and operational effectiveness.

Resource Allocation

Many fire departments use interagency agreements to assemble enough hazmat technicians to pull off a mitigation action plan. Small fire departments may rely on a county or regional hazmat team. The assembly of interagency/regional teams takes time. From the ISO/ASO-HM perspective, the mitigation effort should not commence until adequate teams are assembled, equipped, and briefed. The history of working together that these interagency teams have is important: Do they train together often? Are there equipment compatibility issues? Do all teams use the same incident command structure (ICS) and language? Answers to these questions can help the ISO make judgments about the potential for positive or negative risk-taking. Of particular concern is the training of personnel to perform their assigned task. Responders who are assigned tasks that exceed their training can create an injury potential, along with liability should an injury occur. When a person's level of training is not known or confirmed, the ISO/ASO-HM should relay the concern to the hazmat group supervisor and/or IC. Non-technician-level responders assigned to support functions may not understand the operational and legal liability ramifications of taking shortcuts. The ISO and ASOs must reinforce the incident action plan (IAP) and remind those less-trained that shortcuts are not permissible.

A mass decon effort for contaminated victims can require significant staffing to provide assistance, direction, and care. Resource considerations will almost always include the need for provisions like disposable towels, gowns, blankets, and shelter. Modesty issues may also require additional staffing and separation barriers. The focus of the IC and engaged workers might miss these resource needs; the ISO or ASO is likely to discover the need during recon.

Understanding the types of equipment present, what is needed, and what is still coming can help the ISO/ASO-HM evaluate adequacy. At times, specialized equipment may be required to stabilize the hazard. If the hazmat team has never worked with the specialized equipment, the ASO-HM may need to encourage additional on-the-spot training time to ensure that the technicians can operate the equipment in fully encapsulating ensembles. In one example, a hazmat team and industrial technical consultant agreed that a large sewer vacuum truck would be the best tool to collect a spill and help prevent its spread into a groundwater source. The truck arrived, and the hazmat entry team spent an hour learning the operation of the truck and practicing the use of collection hoses while in their Level A suits—all before they actually started the cleanup.

Operations taking place well within a building (over 300 feet [91 m]) present unique concerns regarding SCBA air

FIGURE 14-2 The hazmat team's technical reference specialist (TECHREF) is a key player for supplying the ISO and ASO-HM with critical incident threat information.

Environmental Integrity

Environmental integrity considerations at the hazmat incident typically include the weather, infrastructure stability, container condition, and hazardous energy. Assigning a value to the integrity of each of these factors is essential for the ISO/ASO-HM team to classify the event:

- Stable and not likely to change (subtracts from risk-taking)
- Stable but changing slowly (may or may not add to risk-taking)
- Unstable and changing slowly (adds to risk-taking)
- Unstable and changing quickly (exponentially adds to risk-taking)

Other Hazards

The physical location of the hazmat incident helps the ISO define the effect of the surroundings. For example, if the incident is taking place on a public roadway, features such as terrain, foliage, curbs, posts, fences, drainages, barriers, and population density need to be evaluated. Conversely, an incident at a fixed processing or distribution facility presents physical hazards such as the presence of complex equipment, access restrictions, security features, and hazardous energy.

management, equipment shuttling, lag time, and rapid egress. The same can be said for outdoor hazmat incidents with a large immediately dangerous to life and health (IDLH) area or numerous no-entry zones. The use of golf carts or other shuttle vehicles can maximize air usage and minimize escape times. Using wagons, sleds, and hand carts can ease the work load of technicians carrying equipment. Obviously, equipment decon time increases with the creative use of these implements.

Task Application

Often, *what* the crews are doing, *how* they are doing it, and *where* they are doing it make the risk equation lean toward excessive. Here are some considerations:

- *Tool versus task.* The tools being used by hazmat technicians range from overly simple (rubber plug) to amazingly complex (chlorine "C" kit or a mass-gas chromatograph). Reliance on initial and recurrent training is essential to their effective use. At times, the technicians may request practice time if a tool is unfamiliar to them or as a quick refresher before going into the hot zone.
- *Team versus task.* The primary concern here is the level of PPE required for a given team to perform their task. The ASO-HM should dictate these requirements based on TECHREF input and standard practices.
- *Rapid withdrawal options.* Hazmat technician crews are required by law (the OSHA CFR) to preplan rapid withdrawal paths and escape areas that are specific to a given incident. Likewise, a standby hazmat rapid intervention team (HAZRIT), dressed to the same PPE level as the IDLH entry team, must be available for rapid engagement. Support personnel who are assisting with decon should have a <u>contaminated safe refuge area</u> for their withdrawal if warranted.

Operational Effectiveness

Judging operational effectiveness may be difficult if the ISO is maintaining a strategic profile at the command post. The ISO must therefore rely on the ASO-HM to judge the effectiveness of the technicians' operation. Other assigned ASOs can help with effectiveness judgments of support activities for those not directly involved with the technicians. Most hazmat operations are centered on an action plan that is created, discussed, and communicated prior to actual operations. This helps preplan operational effectiveness. The ASO-HM can troubleshoot the action plan to make sure the operation will be effective and to address contingencies should an unplanned event occur.

■ Risk-Taking Evaluation at Hazmat Incidents

The principle of "risk a life to save a life" may not be appropriate for the hazmat incident. Other than initial responder actions and civilian evacuation efforts, the control efforts of the hazmat responders should be highly calculated and heavily weighted in favor of safety. The ISO and ASO-HM must strive to agree on the overall risk profile of the incident, and the IC must include that risk-management profile in the IAP.

Two additional risk issues are at play for hazmat: liability and risk communication. Liability (the legal responsibility) is tied directly to the training levels of responders and further

stabilization (entry) objectives. The basic risk liability question is whether a separate, contracted environmental cleanup team can achieve the same entry objectives. Sometimes a fire department hazmat team has to make an entry to verify that the incident is stabilized and savable victims are removed and decontaminated. Once it is decided that a fire department hazmat team entry is warranted, the risk issues associated with hazmat incidents are usually tied to risk communication.

Most firefighters understand that the hazmat incident requires a slow, calculated approach and are likely to back off on risk-taking, but others may not understand this need or practice it. The ISO may have to communicate established risk guidelines to other emergency response affiliates or industry representatives at the incident. Some may not understand the magnitude of the event or the potential of the chemicals involved. Industry representatives who have cost responsibility for the incident may front a "no big deal" approach to minimize media coverage or implied liability. Like the IC, the ISO should seek input on the worst-case scenario from the ASO-HM and third-party experts who have no stake in the incident.

When evaluating the pace of the hazmat incident, the ISO should adopt the time-tested phrase, "If you don't know, don't go, 'cause it might blow." A slow, methodical, and intellectual approach is the best pace for hazmat incidents **FIGURE 14-3**.

FIGURE 14-3 A slow, methodical, and intellectual approach is the best pace for hazmat incidents.

At structure fires, the ISO needs to project on-scene time. Doing so may not be practical at the hazmat incident due to its size, complexity, and slow pace. Once a hazmat incident becomes a technician-level operation, time seems to stand still. Technicians must follow a prescribed approach that includes lots of research, planning, equipment/PPE preparation, and slow, deliberate tactical application. Add to these tasks chemical identification/testing FIGURE 14-4 , decontamination, and rehab and you can easily see that the response effort will take lots of time. The ISO should anticipate this and manage the impacts of time passage. Attention to human needs, pace, and the distraction of boredom (especially for non-technician responders) is essential. Reflex time for any unplanned event is delayed—and should be—so that ramifications can be studied. Assigned ASOs and the ASO-HM should watch for crews or individual responders who are out of synch with the incident pace; they are taking an inappropriate risk.

Safety System Effectiveness at Hazmat Incidents

Accountability Systems

Two systems can be at play: one for the hazmat team members and one for support responders. Cross-communication between the two needs to be encouraged. Realistically, the ISO should deal with strategic accountability (support personnel) and ASOs with tactical accountability (technicians). The planned, deliberate, and slow approach to hazmat calls tends to minimize issues associated with accountability systems—although the ISO should make sure that is the case.

Control Zones

Two issues apply to control zones at the hazmat incident: language and appropriate PPE. In other portions of this text we have outlined a zoning identification and marking scheme that includes:

- No-entry zone: Red-and-white diagonally striped or chevron tape
- Hot zone: Solid red tape
- Warm zone: Solid yellow tape
- Cold zone: Solid green tape

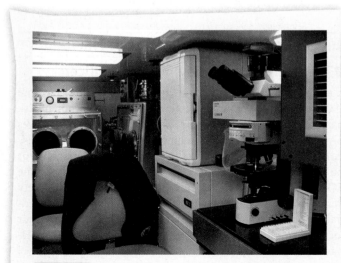

FIGURE 14-4 Mobile chemical labs like this one may be necessary to help identify hazards—adding to incident duration.

For the hazmat incident, control zones need to be expanded to include:

- Escape area for hot zone
- Contaminated safe refuge area
- Decontamination corridor
- Contamination reduction zone
- Zone gateways

The escape area is a hazmat zone used for technicians who are operating in a hot zone. The dedicated technician rapid intervention team (HAZRIT) often stages in the escape area. A contaminated safe refuge area is a zone established for obvious or potentially contaminated victims who are awaiting mass decon efforts and assistance. The decontamination corridor refers to an area where progressive mass decon is taking place for victims. A contamination reduction zone is an area where progressive decon is accomplished for hazmat teams (and support personnel). Zone gateways are simply the entry and exit points for each established zone.

Initial zoning determinations are usually made by first responders and then adjusted by the ISO. The development of formal zones for the actual hazmat technician operation should be made with input from the ASO-HM. There are several factors that can affect the establishment and adjustment of control zones:

- The principal hazard (chemical properties) of the event
- Containment challenges, including form (solid, liquid, gas) and environmental factors (e.g., terrain, drainage, weather)
- Ongoing input or discoveries made by the TECHREF
- Number and location of victims contaminated
- Travel distances and transportation modalities available between zones
- The complexity of decontamination procedures required

The effective formal zoning plan should include travel paths and gateways between zones. Drawing a simple diagram can help responders and entry teams visualize them FIGURE 14-5 . The drawing can help support personnel deploy the appropriate cones (gateways), color-coded zoning tape, and directional signs that help everyone understand what has been established.

Additionally, the ASO-HM should take the lead in dictating the level of PPE required for responders working in each zone and should verify that it is used. Persons who transition from one zone to another should follow a prescribed pathway. Prior to leaving the contamination reduction zone, persons may need to be checked with instrumentation in an attempt to verify that the contamination has been decreased.

Communications

Technological developments have minimized some communication challenges presented by technician-level operations. However, several issues still exist, and some have even been created by the technology. Hazmat protective ensembles may not allow for effective radio communications, or they may introduce multiple radio types and frequencies that need to be monitored. In addition to general radio monitoring, the ISO and ASO-HM must see that backup communication systems are in place. Hand signals, message boards, and tag-line signals can be used as backup systems. Although these communication systems are best preplanned and practiced, the ISO/ASO-HM

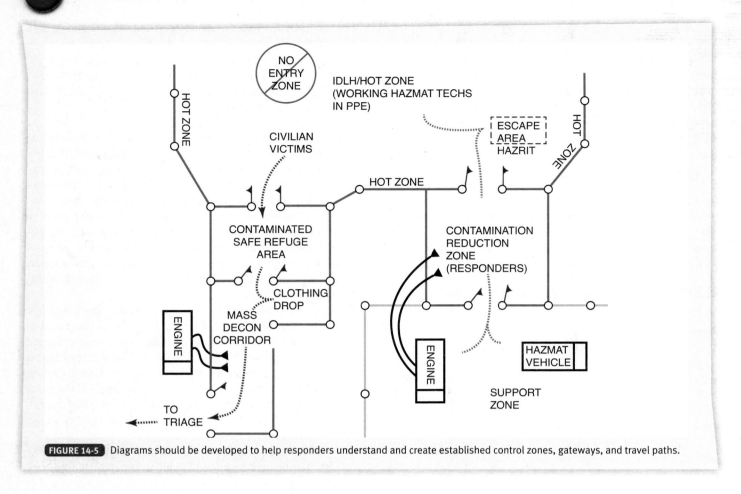

FIGURE 14-5 Diagrams should be developed to help responders understand and create established control zones, gateways, and travel paths.

team may suggest developing a spontaneous system on scene to address unique communication issues. When spontaneous communications systems are developed, they need to be explained to affected incident personnel prior to use.

Rehabilitation

The potential duration of a hazmat operation may require rehab components to span hours or even days. While all the rehab components are applicable, the hazmat incident requires that the ISO/ASO-HM pay particular attention to the evaluation of certain areas:

- *Medical monitoring.* In the structure fire arena, medical monitoring takes place *after* a given assignment has been completed. For hazmat incidents, it is necessary to establish baseline medical monitoring *before* technician stabilization efforts. The stresses created while operating in a totally encapsulating suit can be significant. Having before-and-after responder vital signs can help decision makers make judgments and adjustments regarding working times, rest periods, and active cooling strategies.
- *Sanitation needs.* Extended operations can create sanitation concerns, which are presumably addressed by personnel assigned to the logistics function. These concerns may include the disposal of human waste, the removal of garbage and/or recyclable materials, and the disposition of contaminated clothing or equipment. The best hazard mitigation approach in dealing with these issues is *separation.*

- *Food service.* As wrong as it sounds, feeding firefighters at a structure fire rarely addresses issues of distance and cleanliness. At the hazmat incident, these two issues *must* be addressed. Food storage, preparation (if any), and distribution should be well clear of working areas. The food service area should include areas for additional hand and face washing (further decon!). The ISO is responsible for evaluating and adjusting the rehab needs to meet these requirements.

Unique Considerations for Hazmat Incidents

■ The Need for an Assistant Safety Officer

In addition to the assignment of an ASO-HM and one or more ASOs, the IC may request technical specialists, corporate risk managers, process experts, and public health representatives to respond and assist with specific planning functions. These representatives can also provide safety-specific information for the ISO. If responders have been exposed to chemicals, the ISO should consider calling the department health and safety officer (HSO) or infection control officer to help with procedural issues and documentation for the exposure. The need for documentation and the development of written safety briefings and site safety plans is mandatory for technician-level incidents. The ISO should consider not only having ASOs to help with safety functions while briefings/plans are developed but also having a trained personal assistant to help with documentation. Some departments train their civilian administrative staff to assist with incident record keeping so that firefighters can remain available for support functions that require more "hands-on" tactics and the use of PPE.

The wise ISO should provide a separate safety briefing for civilians who have been requested to help with the incident (personal documentation assistants, public health representatives, risk managers, etc.). The briefing should emphasize the need to stay in a defined cold zone, any evacuation directives, and basic safety expectations (rehab, weather concerns, and the like).

Documentation

The need for documentation is paramount for hazmat incidents that involve mass decon of victims and those that are technician level. A technician-level stabilization effort requires the *formal* development and delivery of a site safety plan and safety briefings. In this context, "formal" means written documentation. The IC is tasked with developing a written IAP and site safety plan and typically delegates the site safety plan to the ISO. Federal law specifically requires the written site safety plan to include the following elements.[3] Listed items with an asterisk (*) should also be included in a written safety briefing.

- Safety, health, and hazard risk analysis, including entry objectives*
- Site organization, including the training and qualifications of responders
- Identification of the exact type of PPE required for the tasks performed by responders*
- Medical monitoring procedures*
- Environmental monitoring and sampling procedures
- Site control measures, including exact control zone locations and gateway marking*
- Decontamination procedures*
- Predefined responder emergency plans (for fires, medical emergencies, and rapid intervention)*
- Confined space entry procedures, including intervention and escape plans*
- Spill containment procedures, including container-handling measures

Documenting that incident responders have received appropriate safety briefings is included in the site safety plan requirement. Additionally, the ISO or ASO-HM may have to sign off on other developed plans (see the *Getting the Job Done* box).

The ISO is likely to be required to fill out some sort of unit log to document his or her efforts and actions during the incident. Be sure to set aside time to document all requirements as the incident evolves and finally concludes. Again, the use of a trained personal assistant can help with real-time documentation (as opposed to follow-up documentation from memory). Hazmat documentation is not subject to any statute of limitations and may be used for litigation purposes years—or even decades—after the incident. Be thorough!

An additional reporting responsibility exists at incidents where first responders engaged in rapid-rescue activities or where responders were unknowingly exposed to products prior to the hazmat team arrival. It is best to treat these cases as if all the first responders had an exposure. The ISO may suggest to the IC that the department HSO or infection control officer

---

Getting the Job Done

Hazmat Incident Plans That May Require ISO Sign-Off
In addition to the site safety plan, the ISO and/or ASO-HM may have to sign off on numerous plans that have been developed for the hazmat incident:

- IAP
- Communications plan
- Exposure protection plan
- Evacuation plan
- Mass-victim decontamination plan
- Responder decontamination plan
- Traffic plan
- Medical plan
- Demobilization plan
- Other site-specific plans

---

become involved to help document the exposure. Although it may seem like a lot of effort, especially if no symptoms exist, the documentation may prove essential if health issues crop up days, weeks, or years into the future.

Clandestine Drug Labs

The response of a fire department to a fire, explosion, or EMS call at a suspected or confirmed clandestine drug lab and/or waste disposal site taxes even the most experienced responder. Drug labs and waste sites can be found in homes, businesses, hotel rooms, and vehicles of all shapes and sizes. A law enforcement raid on a suspected drug lab may call for the standby of fire personnel and hazmat team responders. Firefighters may discover the remains of an active or abandoned drug lab in the course of other incident responses or activities. Large-scale drug labs are often located in rural areas, where the risk of detection is lower.

These potential locations will create challenges for fire department first responders who may not be trained or experienced in dealing with the associated hazards. In either case, the responding ISO/ASO-HM has to include additional considerations in evaluating clandestine drug lab safety hazards.

The response to a clandestine drug lab or waste site typically involves a multiagency effort that includes law enforcement, EMS, fire, public health, environmental protection regulators, and even social services. At any time, the primary management authority of the incident may change. These events must be preplanned, and the ISO must understand who has primary control authority for any phase of the incident. Other than fire control and victim treatment, the law enforcement agency usually has primary control responsibilities for criminal investigative purposes.

The suspected clandestine lab presents alarming hazards to fire department personnel who are actively involved with fire control or stabilization activities. These hazards include:

- Poor ventilation
- Flammable/toxic atmospheres
- Incompatible chemicals
- Chemical reactions in progress
- Unidentified chemicals and/or containers
- Unstable and/or leaking containers

- Booby traps (improvised firearms, incendiary/explosive devices, and other such devices)

When a suspected drug lab is discovered as part of a fire, odor investigation, EMS, or other non-hazmat response, the fire department should immediately notify law enforcement and technician-level hazmat teams. The ISO should meet with the IC and offer solutions to initiate a careful (but immediate) withdrawal of responders. Chemical exposure, decontamination, isolation procedures, and evidence protection become priorities. Routine firefighting tasks, such as utility control, overhaul, and debris removal, should not be attempted unless directed by law enforcement or forensic chemists familiar with clandestine drug lab intricacies.

■ Weapons of Mass Destruction

Once an incident has been classified as a suspected terrorist event involving a WMD, the Federal Bureau of Investigation (FBI) takes the lead, as provided by Presidential Decision Directive Thirty-Nine (PDD 39). When a suspected WMD event occurs, it will likely trigger a response of predesignated and staged teams and equipment. Large, municipal fire departments may already have knowledge of, or be part of, these teams. Smaller suburban and rural fire departments that are unaware of these teams may be surprised when crews of people in dark-painted, unmarked, high-tech vehicles arrive and begin setting up **FIGURE 14-6**. The local ISO is usually replaced by a safety officer who responds as part of an incident management team (IMT) following a terrorist attack. But the transition time between the onset of the suspected WMD event and the pass-off to an IMT can be considerable, lasting from a few hours to several days depending on the magnitude, geographical location, and competing demands of other simultaneous events.

Developing a local WMD plan that addresses ISO functions assists in implementing an organized approach to the event. When a WMD event (or suspected one) does occur, the ISO must remain *strategic* and focus on the basic safety issues associated with the *Office of Domestic Preparedness Emergency Responder Guidelines*:

- Recognition
- Detection
- Self-protection
- Crime scene preservation
- Scene security
- Notifications

In brief, the ISO should support the IC and use a *quick-in/quick-out* approach for immediate rescues, then adopt a *back-off* posture. Victims and exposed firefighters should be isolated until *clean* or *contaminated* determinations can be made. Staged equipment should be out of the sight of gathered spectators because the terrorist may be among them and waiting for an opportunity to compound the event by attacking the responders. If it is not possible to stage out of sight, then security measures are warranted (up to and including armed guards!).

Once the initial overall strategic concerns are addressed, the ISO should move to create an integrated approach to incident-wide safety. The ISO should recognize that the WMD incident will become a multiagency event and will likely require ASOs to deal with the high pressure and time constraints of the

FIGURE 14-6 An actual or suspected WMD incident will trigger the response of national assets that are staged and trained to deal with the event.

postdisaster environment. The strategic goals of the ISO and ASOs should be:

- *Gather recon and threat information.* Use ASOs and representatives from law enforcement and other emergency response agencies to ascertain the threat potential, resource status, chemical agent type, and possible acute or chronic health concerns. Some of this information may be guarded (confidential) for security purposes; the point is that the ISO and ASOs should try to obtain what they can.
- *Analyze options.* Draw from the technical expertise of multiple responding organizations and lean toward the worst case. Remember that worst-case analysis is for planning and not for public information. Protect this analysis to prevent panic.
- *Develop a safety action plan.* The action plan should focus on sustainability measures to protect the health of responders across organizational boundaries. Ancillary agencies may not be familiar with the safety officer function; communication in a simple, yet compelling manner helps non-fire service responders understand and comply with safety plans. Accountability, PPE, rehab, and zoning issues are a great starting place for the safety action plan. Once the plan has been delivered to responders, the ISO should focus on expanding his or her role into manageable parts and begin addressing safety and health issues prior to the arrival of an IMT.

Chief Concepts

- Hazmat incidents have become the most regulated of all the incidents to which fire departments might respond. Never has it been more important for a duty ISO or potential ASOs to do their homework. For most hazmat incidents, the assignment of an ISO is mandated by federal law (OSHA CFR 1910.120).
- The ISO at a hazmat incident must meet the competencies listed in NFPA 472. If the ISO doesn't have that level of training, an ASO-HM (assistant safety officer–hazmat) who meets the competencies must be appointed.
- The ASO-HM actually works with three people in an IMS structure: the overhead ISO, hazmat group supervisor/branch director, and any assigned technical specialists.
- Upon arrival and assignment at a hazmat incident, the ISO must:
  1. Ascertain the level of control actions from the IC (mass decontamination, zone and isolate, initial control actions)
  2. Assist with developing/adjusting zones
  3. Discuss the IAP and appropriate risk management levels with the IC
  4. Request ISO additional assistance (through the IC) to include an ASO-HM, third-party technical expert, and trained personal assistant to help with documentation
- The ISO does need to make a recon trip around the incident perimeter, but most of the useful recon information will come from a hazmat team technical reference specialist (TECHREF).
- Risk-taking judgments are found in the evaluation of chemical properties, environmental integrity, other hazards, and resource effectiveness. The chemical properties information will come from the TECHREF, integrity reports from the ASO-HM and recon, other hazard-related information from recon, and resource effectiveness updates from the ASO-HM.
- Resource allocation may take time, as regional teams need to gather and assemble tools, address training needs, practice/prep for entry, and prepare for decontamination. The needs for mass decontamination for victims are significant, including a decontamination corridor, towels, sheets, blankets, and modesty needs. The ISO is usually responsible for identifying these needs.
- The ASO-HM is responsible for ensuring the correct level of PPE and decontamination efforts required for responders working in support zones. Additionally, the ASO-HM is responsible for ensuring that a hazmat rapid intervention team (HAZRIT) is dressed and staged to back up entry teams.
- The ISO and ASO-HM must agree on the risk management criteria at play for a given incident and communicate this determination to the IC for inclusion in the IAP. They are also responsible for developing a control zone plan (diagram) for the IAP and for support personnel to start marking the zones (cones, flags, tape, etc.).
- The ISO must ensure that a slow, methodical, and intellectual approach is being practiced by everyone on the hazmat scene.
- In addition to no-entry, hot, warm, and cold zones, the hazmat incident will require the establishment of escape areas, contaminated safe refuge areas, a decontamination corridor, a contamination reduction zone, and zone gateways.
- Factors that influence the development or adjustment to control zones include chemical properties, containment challenges, revisions directed by the TECHREF, number and location of victims, travel distances, and complexity of decontamination efforts.
- The ISO's responsibility to evaluate rehab efforts at hazmat incidents needs to have more focus on medical monitoring before and after entry, sanitation needs (separation), and food services (greater attention to decon).
- The tremendous documentation demands for hazmat incidents require the ISO to request additional ASOs and technical experts. Some consideration should be given to preplanning and training non-firefighter department personnel who can assist with documentation for those departments with limited firefighter resources. At a minimum, the ISO needs to document his or her actions, formal safety briefings, and a 10-point site safety plan. The ISO may have to review and sign off on other documents, including the IAP, exposure protection plan, decontamination plans, medical plan, and demobilization plan, among others.
- Initial responders may have been unknowingly exposed to chemicals. The wise ISO should request (through the IC) an HSO or infection control officer to help document exposure reports. There is no statute of limitations for hazmat incident documents and exposure reports.
- When a clandestine drug lab is found or suspected, the ISO should meet with the IC and offer solutions to initiate a careful (but immediate) withdrawal of responders. Chemical exposure, decontamination, isolation procedures, and evidence protection become priorities.
- Weapons of mass destruction (actual or suspected) will trigger the response of staged national assets that will likely take over the incident. Until that happens, the ISO and IC should treat the event using a quick-in/quick-out approach for immediate rescues, then adopt a back-off posture. Victims and exposed firefighters should be isolated until clean or contaminated determinations can be made.

Key Terms

assistant safety officer–hazmat (ASO-HM) A person who meets or exceeds the NFPA 472 requirements for a hazardous materials technician and is trained in the responsibilities of the ISO position as it relates to hazmat response.

contaminated safe refuge area A hazmat zone established for obvious or potentially contaminated victims who are awaiting mass decontamination efforts and assistance.

contamination reduction zone An area where progressive decontamination is accomplished for hazmat teams (and support personnel).

decontamination corridor An area where progressive mass decontamination is taking place for victims.

escape area A hazmat zone used for technicians who are operating in a hot zone. The dedicated technician rapid intervention team often stages there.

Review Questions

1. List the federal regulations that may have an impact on ISO functions.

2. To whom does the ASO-HM report at a hazmat technician-level incident?

3. With whom does the ASO-HM likely work with at a typical hazmat technician-level incident?

4. What are the IAP components that the ISO and ASO-HM are responsible for developing?

5. Persons not trained for their hazmat incident assignments create two risks. What are they?

6. In addition to the classic hot, warm, and cold zones, what other zones and areas need to be created for a hazmat incident?

7. List the three hazmat rehab components that require close evaluation.

8. What are the 10 federally required components of a hazmat response site safety plan?

9. List five hazmat ancillary plans that may require ISO sign-off.

10. List five or more alarming hazards at a clandestine drug lab incident.

11. List and describe the three strategic goals for the ISO at a WMD/terrorist incident.

References and Additional Resources

Lake, William, Stephen Divarco, Peter Schulze, and Robert Gougelet. *Guidelines for Mass Casualty Decontamination During an HazMat/Weapons of Mass Destruction Incident: Volumes I and II*. Aberdeen Proving Grounds, MD: Edgewood Chemical Biological Center, U.S. Army Research, Development, and Engineering Command, August 2013.

National Institute for Occupational Safety and Health. *Protecting Emergency Responders, Volume 3*. NIOSH Publication No. 2004-144. http://www.cdc.gov/niosh/docs/2004-144/. Accessed May 24, 2015.

Nedder, J. *Fire Service Rapid Intervention Crews, Principles and Practice*, Burlington, MA: Jones & Bartlett Learning, 2015.

NFPA 472, *Standard for Professional Competence for Responders to Hazardous Materials Incidents*. Quincy, MA: National Fire Protection Association, 2013.

NFPA 472, *Standard for Competence of Responders to Hazardous Materials/Weapons of Mass Destruction Incidents*, Quincy, MA: National Fire Protection Association, 2018.

Noll, G. and M. Hildebrand. *Hazardous Materials Managing the Incident*. 4th ed. Burlington, MA: Jones & Bartlett Learning, 2014.

Schnepp, R. *Hazardous Materials Awareness and Operations*. 2nd ed. Burlington, MA: Jones & Bartlett Learning, 2016.

U.S. Department of Transportation. *2012 Emergency Response Guidebook: A Guidebook for First Responders During the Initial Phase of a Dangerous Goods/Hazardous Materials Transportation Incident*. Washington, DC: DOT, Pipeline and Hazardous Materials Safety Administration, 2012. Available through http://www.phmsa.dot.gov/hazmat/library/erg as a PDF file or smartphone app.

Ward, M. *Fire Officer Principles and Practice*. Enhanced 3rd ed. Burlington, MA: Jones & Bartlett Learning, 2016.

Endnotes

1 OSHA 29 CFR 1910.120(q)(3)(G) requires the assignment and use of a safety officer at hazmat incidents..

2 U.S. Department of Transportation. *2012 Emergency Response Guidebook: A Guidebook for First Responders During the Initial Phase of a Dangerous Goods/Hazardous Materials Transportation Incident*. Washington, DC: DOT, Pipeline and Hazardous Materials Safety Administration, 2012.

3 Paraphrased from OSHA 29 CFR 1910.120.

Smoke: © Greg Henry/ShutterStock, Inc.

The fire department was dispatched to a report of a vehicle fire with structures threatened. Upon arrival, the first-due engine found an old school bus with fire coming from its rear interior portion. The bus appeared to have been converted to a recreational vehicle with a liquid propane (LP) tank strapped to the outside. Five feet from the bus was a small detached garage that was beginning to ignite. Two lines were deployed to attack the fire and keep the LP tank cool. Once the fire was knocked down from the outside, one attack team moved to the front bus access door to extinguish the remaining fire. Upon entry, they found an unresponsive smoke inhalation victim. The victim was removed and passed to ambulance personnel.

During the attack and rescue, the IC and ISO began their respective duties. At some point, a later-arriving sheriff deputy approached the IC and said that the bus was a suspected clandestine drug lab, according to an undercover drug agent who was in the area. The sheriff went on to explain that one of the bystanders didn't want to break "cover" and was actually on surveillance at a nearby home when the fire was discovered.

The IC immediately relayed this info to the ISO and the two agreed to order a hazmat team response. Although the fire was still burning, crews were withdrawn, and the department's clandestine drug lab protocol was implemented.

1. Generally speaking, what issues would you expect to arise when the incident transitions from a fire attack to a hazmat incident?

2. As the IC, would you let the crews finish extinguishing the fire? If so, what restrictions would you communicate to the crews?

3. Whom would you consider to be "contaminated" during the initial attack and rescue at this incident?
   A. Only the smoke inhalation victim
   B. The victim and the ambulance personnel
   C. The victim, the fire attack/rescue team, and the ambulance personnel
   D. Any person within a 2000' radius of the incident

4. What should the ISO at this incident accomplish next while waiting for the hazmat team to arrive?
   A. Request a personal assistant (through the IC) to help start documentation
   B. Define initial control zones and communicate them to the IC and responders
   C. Assist with rapid gross decontamination efforts for the responders
   D. Ascertain the drug type and suspected chemicals from the deputy sheriff and undercover drug agent

Knowledge Objectives

Upon completion of this chapter, you should be able to:

- Identify imminent threats to firefighter safety.
  (NFPA 5.2.4 , pp 218–227)
- Describe the incident scene conditions that are monitored as part of an ongoing incident. (NFPA 5.2.5 , pp 220–227)
- Describe motor vehicle hazards present at an incident scene. (NFPA 5.2.8 , pp 226–227)
- Identify methods of temporary traffic control used to mitigate hazards. (NFPA 5.2.8 , pp 226–227)
- Describe the types of incidents at which additional ASOs or technical specialists are required due to corresponding hazards of the incident type. (NFPA 5.2.10 , pp 217–218)
- Define the process for making recommendations to an incident commander to request ASOs or a technical specialist. (NFPA 5.2.10 , p 218)
- Identify the hazards presented at a designated helicopter landing zone. (NFPA 5.2.11 , p 220)
- Describe the helicopter landing zone requirements at an emergency incident. (NFPA 5.2.11 , pp 220–221)
- Identify the types of technical rescue incidents that would require the assignment of an ISO or ASO trained as a rescue technician. (NFPA 5.4.1 , pp 218–222)
- Identify the standards and regulations that would require the assignment of an ISO or ASO trained for a technical rescue incident. (NFPA 5.4.1 , p 217)
- Describe the qualifications of individuals assigned as the ISO at a technical rescue incident. (NFPA 5.4.1 , pp 217–218)
- Describe the elements of an incident action plan (IAP) related to safety plan development and safety briefings at a technical rescue incident. (NFPA 5.4.2 ; NFPA 5.4.3 , pp 218, 223–224)
- Describe risk management principles used in the safety component of an IAP to identify corrective or preventive actions at a technical rescue incident. (NFPA 5.4.2 , pp 218–219)
- Define the elements of a technical rescue IAP that the ISO would be responsible for reviewing. (NFPA 5.4.2 , p 218)
- List the three rehab issues that require special attention at technical rescue incidents, and describe the "on-deck" system for crew rotation. (pp 219–220)

- Name the four ways to classify a building collapse. (p 221)
- List five hazards associated with industrial entrapments. (p 223)
- Define *LCES* and how it can be used at a cave-in incident. (p 223)
- List six hazards associated with water rescues. (p 224)
- List five hazards associated with high-angle rescues. (p 225)
- Name five circumstances in which a duty ISO should implement a discretionary response to motor vehicle accidents and diagram a strategic approach to protect rescuers at roadway incidents. (pp 226–227)

Skills Objectives

Upon completion of this chapter, you should be able to:

- Demonstrate the ability to recognize hazards at a helicopter landing zone. (NFPA 5.2.11 , pp 220–221)
- Demonstrate the ability to select a potential landing zone with all potential hazards identified. (NFPA 5.2.11 , p 220)
- Perform a review of a technical rescue incident action plan. (NFPA 5.4.2 , pp 218–220)
- Conduct a technical rescue safety briefing. (NFPA 5.4.3 , pp 223–224)
- Demonstrate the ability to evaluate the placement and use of apparatus or other temporary traffic control devices at an emergency incident scene. (NFPA 5.2.8 , pp 226–227)

After the recent hazmat incident, your chief has asked you to coordinate efforts with the technical rescue team leaders to develop a set of standard action plans and safety responsibilities for both technician-level safety officers and the ISO (who is you). You have identified several NFPA and local standards that identify what types of responses would require and benefit from both ISO levels being present. You plan to meet later in the month with the team leaders to discuss this program.

1. What types of technical rescue incidents does your department respond to? At what level do the members operate according to NFPA 1006?
2. What common duties and incident-specific duties for the ISO should be identified in your safety action plans?

Introduction: The Technical Rescue Trap

Addressing the safety-related aspects of the various types of technical rescue incidents is like trying to explain the variations in musical genres: There are seemingly infinite styles and subsets of styles! However, one thing is certain: If someone becomes trapped (and injured), the fire department is called. Regardless of how a victim became trapped or what is causing the entrapment, the fire department is called upon to find a positive solution. Herein lies the trap. Firefighters' "can-do" attitude compels them to action, whether they are trained for the specific situation or not. Resourcefulness, inventiveness, willingness, and compassion drive firefighters to a solution. What a great problem to have! These attributes are what endear firefighters to their communities. From the incident safety officer (ISO) perspective, the same attributes lead directly to firefighter injuries and deaths, compounding the incident. Firefighting history is full of examples of willing firefighters who got caught up in a situation for which they were ill prepared and soon overwhelmed, resulting in tragedy. Repeated tragedies have spawned regulations for especially risky rescue incidents. Additionally, fire service members have pushed for specific training standards to prevent reoccurrences.

The concept of technical rescue is open to some broad interpretation. The official NFPA definition of a technical rescue is:

> The application of special knowledge, skills, and equipment to resolve unique and/or complex search and rescue situations.[1]

Firefighters routinely perform rescues that could fall into the category of technical rescue, although they may laugh at the idea of calling it a technical rescue just because a porta-power was used for a unique call. NFPA goes on to define a technical rescue *incident* as:

> Complex search and/or rescue incidents requiring specialized training of personnel and special equipment to complete the mission.

For the sake of this chapter, we will use the words "complex" and "specialized" to address technical rescue incidents that should include an ISO as part of the response. To that end, technical rescues could include those involving:

- Ropes, including high-angle, low-angle, and hauling
- Confined space, including trench and cave-ins
- Structural collapse
- Machinery, including industrial, farm, and transportation modalities
- Water, including dive, swift, surface, and ice
- Wilderness
- Mine, tunnel, and cave

This chapter presents tangible, broad-based issues for the ISO to consider when performing at the technical rescue incident. As with hazardous materials (hazmat) incidents, a technical rescue incident may fall under federal regulations. Being familiar with these regulations is essential. The following federal regulations—defined in the Occupational Safety and Health Administration's (OSHA) Code of Federal Regulations (CFR)—may have an effect on the functions of the ISO at a technical rescue:

- 29 CFR 1910.95, *Occupational Noise Exposure Limits*
- 29 CFR 1910.120, *Hazardous Waste Operations and Emergency Response Solutions*
- 29 CFR 1910.134, *Respiratory Protection*
- 29 CFR 1910.146, *Permit-Required Confined Spaces*
- 29 CFR 1910.147, *The Control of Hazardous Energy (Lockout/Tagout)*
- 29 CFR 1910.1030, *Blood-Borne Pathogens*
- 29 CFR 1910.1200, *Hazard Communication*
- 29 CFR 1910.1926, *Excavations, Trenching Operations*

When there is regulation, legal liability exists. These two challenges have led fire departments to develop technical rescue response systems that address procedures, training, equipment, and command elements. The assignment of an incident safety officer is *mandatory* for confined space, trench (including cave-ins), and hazmat incidents. For these three rescue environments, the ISO should have the professional competencies for the level of incident involved. The competencies for confined space, trench, and cave-ins are defined in various chapters of NFPA 1006, *Standard for Technical Rescue Personnel Professional Qualifications* (2017). The 1006 standard covers many more incident types, and each type is divided into job performance requirements (JPRs) for awareness-, operations-, and technician-level training. The standard has no other specific requirements for persons serving as a safety officer, like the hazmat standard does. As a result, and to satisfy OSHA regulations, the ISO needs to match the technician-level requirements of NFPA 1006 for confined space, trench, and cave-in rescues.

Some confusion regarding training levels occurs when NFPA 1670, *Standard on Operations and Training for Technical Search & Rescue Incidents* (2017), is introduced into the conversation. The 1670 standard defines the awareness-, operations-, and technician-level capabilities for myriad tech-rescue incident types. The 1006 defines the JPRs that an individual must meet to provide the capabilities in 1670. In simple speak, 1670 defines the capability an organization offers, and 1006 defines the specific knowledge and skills that an individual must have for any given level and tech rescue incident type.

When the ISO does not have technician-level training and/or certification, an assistant safety officer–rescue technician (ASO-RT) should be appointed to help with technician safety functions. An ASO-RT is a person who meets or exceeds rescue technician-level competencies. When an ASO-RT is appointed, the ISO retains overall safety function responsibility, while the ASO-RT works with the technician-level group or branch.

As with the hazmat ASO (ASO-HM), the ASO-RT may be titled differently based on how the incident management system has been scaled. Some of these titles are:

- Rescue safety officer
- Rescue group safety officer
- USAR safety officer (for organized urban search and rescue teams)

In this text, ASO-RT is used to identify the person fulfilling safety functions for the technician-level components at an incident. Organizationally, the ASO-RT should report to and work with the overhead ISO. In reality, the ASO-RT works with three or more people: the ISO, the rescue branch director (or rescue group supervisor), and any technical specialists involved **FIGURE 15-1**.

The purpose of this chapter is to address specific ISO/ASO-RT challenges that need to be considered in performing the assigned safety functions at the technical rescue incident. This chapter is *not* designed to replace technical rescue professional competencies. We depart from the format used in other incident-specific chapters of this text because each type of technical rescue has unique characteristics and one model does not fit all. However, we start with the general ISO

functions applicable to all technical rescues and assume that the ISO is in a command staff role and an ASO-RT is assigned for technician-level components.

General ISO Functions at Technical Rescue Incidents

Because of the vast array of technical rescue incidents, the ISO must gain a strong sense of the situation status (sit-stat), which includes victim location, predicament, rescue likelihood, and the integrity of the surrounding environment. Additionally, the ISO must gain an understanding of committed resources (restat) and what is being planned. The best place to get this information and awareness is at the command post. Once in possession of a sitstat and restat, the ISO should inquire about established control zones, identified hazards, and other personal safety systems that are already in place. Armed with this information, the ISO should request an ASO-RT from the incident commander (IC) if the ISO does not have the qualifications/training for the given incident. The ISO is then likely to do a recon if pressing incident action plan (IAP) or control zone issues don't take precedence. The elements of a technical rescue IAP that the ISO would be responsible for reviewing/adjusting include:

- Overall risk-taking profile (rescue mode, recovery mode, incident stabilization mode)
- Control zones and required level of PPE for each
- Personnel safety systems (accountability, communications, rehab)
- Traffic issues and interface with medical evacuation (medevac) helicopters

■ Risk Evaluation at Technical Rescue Incidents

As at a structure fire, the ISO must evaluate the rescue profile of victims who are entrapped or thought to be entrapped. A similar approach is used for technical rescues, although the risk level language is tied to operational modes that reflect the obvious and perceived conditions present:

- Rescue mode: Victims obviously need rescue and/or the conditions present indicate that a search and rescue operation can lead to savable victims.
- Recovery mode: Victims have been determined to be deceased or the conditions present are incompatible with survivability.
- Incident stabilization mode: Victims have been rescued (or recovered when appropriate), yet the conditions are such that a potential for further harm exists and must be addressed.

The nature of building collapses is such that the rescue profile is an unknown; for example, a void space may have trapped savable victims. In other cases, the rescue profile of a victim is obvious. When a victim is obviously deceased (or has been determined to be deceased), a *recovery mentality* should be adopted (reduce risk-taking). The declaration of *recovery mode* should be made to all responders in a discreet manner to help keep from triggering further bystander/media

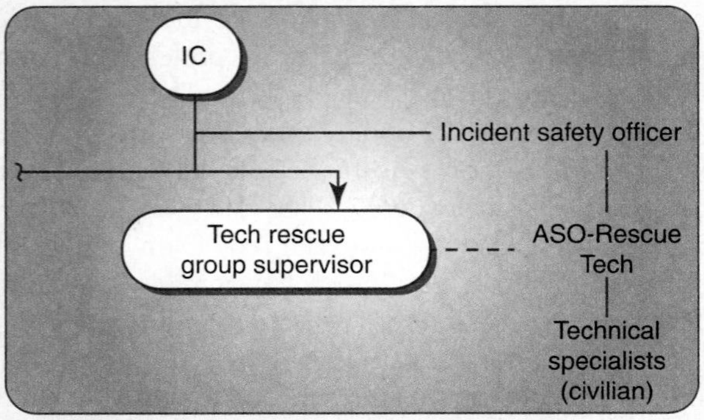

FIGURE 15-1 The ASO-RT actually works for three or more persons.

emotional reactions. The IC or appointed public information officer (PIO) is responsible for keeping others informed. When recovery mode is declared, the ISO needs to close the loop with the IC and make sure the IAP reflects a change in operational risk-taking.

When the fire department effort switches to a reduced-risk recovery mode, distraught onlookers may attempt to jump in, presenting a difficult situation that can lead to emotional outbursts and expectation stress on responders. When this situation develops, the IC or PIO should notify police and ask them for assistance. Additionally the IC or PIO should speak with the coworkers, relatives, and Good Samaritans, explaining that their involvement will only complicate the situation and divert responder attention away from victim recovery efforts.

■ Control Zones

The differences between no-entry, hot, warm, and cold zones may be measured in inches at some technical rescue calls. At times, the marking of zones (with tape) is not practical due to the closeness of them, especially the delineation between no-entry and hot zones. Communication is essential. The ASO-RT is usually the person responsible for delineating the various zone boundaries, and the ISO relays the ASO-RT directives to the IC (for IAP inclusion) and to responders, who help mark and enforce the zones when needed. Everyone shares the responsibility of monitoring the established zones and offering reminders to others when marking is impractical.

Establishing the required levels of personal protective equipment (PPE) for support personnel at the technical rescue often falls on the ISO, who uses information from the ASO-RT to make this judgment. The ISO should establish PPE requirements based on a realistic worst-case scenario—which may not be a popular decision **FIGURE 15-2**. The PPE requirements may be different for each of the established zones, and everyone needs to help monitor and enforce compliance.

■ Safety System Effectiveness at Technical

FIGURE 15-2 The ISO needs to establish the level of PPE required for various technical rescue control zones. Where assigned, the ASO-RT can assist the ISO in establishing the PPE and zone requirements.

Rescue Incidents
Accountability Systems

Tracking assigned resources at the technical rescue should follow the established procedures applicable to other types of incidents. The real accountability issue at technical rescues is the potential for freelancing and self-deployment, as firefighters rush against the clock to save a victim. The ISO and any assigned ASOs need to understand that it is in firefighters' nature to do what they think is best, yet freelancing is inherently unsafe. This is an issue that may have to be addressed when risks outweigh benefits. Firm interventions are often warranted and need to be coached with a keen sense of risk-taking appropriateness.

Communications

The nature of some technical rescue events requires technicians to communicate constant instructions via radio: "A little to the left—hold," "A little to the right—hold," "More slack," etc. These transmissions can tie up radio frequencies. Some technical rescue teams use small talk-around radios to accomplish their communications needs. These small commercially available radios can free up tactical channels, provided their users can also monitor the tactical channel (note: some of these radios are not intrinsically safe). The ASO-RT should have the ability to monitor talk-arounds. Hand signals, message boards, and tag-line signals may also be used. Make sure support personnel are clear on the use of these devices; it may have been several years since they used them. To help with clarity, a cheat sheet should be created so that support responders do not miss an important signal.

Rehabilitation

The duration of a technical rescue operation may require rehab components to span hours or even days. While all the rehab components are applicable, the technical rescue incident requires particular attention in the evaluation of certain areas:

- *Perceived comfort.* When a rescue is under way, responders may not admit that they are in need of rehab. Careful monitoring is essential. Do not allow teams to make rehab decisions based on their perceived comfort. Creating a defined work period helps. Like hockey players, responders can be assembled in line changes and rotated on and off the bench. This approach can be further enhanced if the line changes practice an "on-deck" system. An <u>on-deck system</u> is an organized system in which a working team is replaced with another working team that is already dialed in and ready to replace them. Once replaced, the rotated workers go to rehab, while a crew from rehab rotates into the vacated on-deck position. Those in rehab should have an opportunity to disengage mentally. Only when they rotate to the on-deck position should they reengage. The on-deck system may not be doable for rope incidents but should be effective for most other situations.

- *Energy replacement.* Most technical rescue incidents challenge firefighters' concentration and physical stamina. The *Reading Firefighters* chapter outlines efficient fueling strategies that can ensure that muscle and brain cells work at optimal levels.

- *Medical observation.* When a rescuer is physically tired, concentration is also lost. Empowering personnel responsible for medical observation to make judgments regarding a rescuer's physical and *mental* readiness to reengage in a technical environment is essential—and often forgotten.

■ Traffic and Medevac Helicopter Issues

A dramatic or unusual rescue attempt undoubtedly attracts media coverage, which increases the numbers of onlookers. Congestion is a real issue that requires attention. While traffic sounds like a law enforcement duty, the ISO needs to consider the impact of the congestion and media coverage. Firefighters who have never "performed" in front of the media and dozens (if not hundreds) of onlookers may become stressed or distracted. The ISO/ASOs should watch for signs that a firefighter is becoming distracted and offer friendly or humorous concentration reminders.

In addition to roadway congestion, the ISO must be alert for safety hazards associated with railways, air traffic, and waterways. Make sure a travel corridor is maintained for additional resources and equipment shuttling. In some cases, the IC may ask the ISO to develop a traffic/staging plan for IAP inclusion. We will address roadway incidents more specifically later and offer some suggestions that can help the ISO develop these plans.

When rescue helicopters and/or air ambulances are ordered, additional hazards are introduced and certain requirements need to be assured prior to their arrival **FIGURE 15-3**. Ideally, firefighters have received training from their local air medevac service and departments have developed standard operation procedures (SOPs) to help ensure that hazards and requirements are addressed. Often, the air medevac service will provide the training and recommended SOPs at no cost. While locally developed SOPs take precedence, some general helicopter interface hazards and procedures can be addressed here (also see the *Safety Tip* box).

FIGURE 15-3 Medevac helicopters introduce additional hazards, and certain requirements have to be addressed prior to their arrival.

Helicopter Hazards

The general hazards associated with interfacing with medevac helicopters deal with the rotary wing and tail rotor and the potential for a crash. Specific hazards include, but are not limited to:

- *Rotor wash.* The rotor stirs up dust and debris and creates turbulent winds.
- *Main (overhead) rotor blade.* This large rotating blade is unprotected and may be within the overhead reach of responders (strike hazard). It may become "invisible" due to its rapid rotation.
- *Tail rotor.* The tail rotor is an extreme strike hazard. Stay away from it, and never approach the rear of a helicopter.
- *Mishap.* If a rotor strikes an object during take-off or landing, debris can be scattered at tremendous velocity and can kill or maim ground personnel. A fuel fire may ensue if the craft breaks apart.

Safety Tip

Helicopter Safety Procedures
While medevac crews are lifesaving and their pilots are well trained, the chance for a mishap is always present. Interfacing with medevac helicopters requires responders to follow some strict procedures.

1. Assign a landing zone (LZ) officer to prepare the LZ and brief responders.
2. Choose an area that has a clear zone diameter of 100 feet (30 m) (nothing over 12 inches [30 cm] in height) and is smooth, level, and relatively free of loose material.
3. Park standby apparatus outside the 100-foot (30-m) clear zone.
4. At night or in low-light conditions, use spotlights to illuminate power poles/lines or other overhead obstructions that are within 300 feet (90 m) of the LZ. Place a small, but weighted flashing light in the center of the LZ. Apparatus headlights can help illuminate the LZ, but turn off most strobe warning lights to prevent pilot night blindness.
5. Keep spectators back 300 feet (30 m).
6. Standby responders in the LZ must have full PPE, including eye and hearing protection.
7. Establish radio contact with the incoming pilot and advise of ground wind speed, nearby obstructions, and the LZ location.
8. As the helicopter approaches, have standby crews take shelter in the apparatus or be in a position shielded by it in case a mishap occurs.
9. Never approach a helicopter unless the pilot signals to do so.
10. When signaled, approach the helicopter from the front or side, in the pilot's view. Always approach and depart a helicopter with a slight crouch—a reminder to NEVER lift your arms over your head.
11. NEVER approach the rear of a helicopter. The tail rotor is at body level and may be unseen due to rotation.

ISO/ASO Helicopter Issues

The ISO or ASO should evaluate the landing zone (LZ) prior to helicopter arrival and ensure that an LZ officer is in charge and that safety issues have been addressed. Pay particular attention to overhead obstructions (two sets of eyes are better than one), crowd separation, and protective measures for standby personnel. Some SOPs require the standby crew to have full PPE with self-contained breathing apparatus (SCBA), a charged hand line, and forcible entry tools at the ready. Remind crews to seek a protected (shielded) position behind apparatus as the helicopter lands or departs: A rotor strike, rough landing, or crash can send debris flying in all directions and can actually cut a person in half if struck.

Considerations at Specific Technical Rescue Incidents

Technical rescue incidents, while diverse, can be classified into several categories that, in some cases, are guided by regulated and response/training standards **TABLE 15-1**. Some may argue that knowing all the information contained in the guiding documents is impossible. The intent is that you are aware of them and have perused their content as part of your professional front-loading. Having the ability to retrieve the critical information in the documents could be useful during the incident, and evolving technology (laptops, tablets, smartphones, and other devices) can help you with information retrieval.

We will now discuss the unique hazards and considerations for the ISO/ASO-RT in each of the technical rescue categories.

■ Building Collapse

Classifying building collapse incidents is a good starting place for understanding associated hazards **FIGURE 15-4**. While different classification methods exist—and what is suggested here may be overly simple for the rescue technician—the ISO can gain some hazard forecasting ability by classifying the incident (see the *Awareness Tip* box).

Moderate and heavy collapses should trigger the ISO to request an ASO-RT from the IC if one has not already been

Awareness Tip

Building Collapse Classifications

Basic/surface collapse: In this collapse, victims are easily accessible and trapped by surface debris. Loads are minimal and easily moved by rescuers. The threat of secondary collapse is minimal.

Light collapse: This is usually a light-frame (wood) partition collapse, and common fire department equipment (from engine and truck companies) can access or shore areas for search and extrication. The threats of a secondary collapse can be mitigated easily.

Moderate collapse: This is an ordinary construction collapse that involves masonry materials and heavier wood. Lightweight construction with unstable, large, open spans should also be classified as moderate. Significant void space concerns are present. Victims may be trapped by load-bearing members requiring heavy lifting equipment. Serious attention should be given to secondary collapse.

Heavy collapse: In this collapse, stressed concrete, reinforced concrete, and steel girders are impeding access. Heavy collapses require the response of urban search and rescue teams and specialized heavy equipment. Collapses that threaten other structures or significant secondary collapse also fall into this category.

TABLE 15-1	Sample of Technical Rescue Categories and Associated Documents
Incident Type	**Guiding Documents**
Building Collapse	• 29 CFR 1901.146, Confined Spaces • 29 CFR 1910.147, Hazardous Energy • 29 CFR 1910.132, PPE • 29 CFR 1910.1030, Bloodborne Pathogens • NFPA 1670, Technical Rescue Incidents
Industrial Entrapment	• 29 CFR 1910.95, Noise • 29 CFR 1910.120, Hazmat • 29 CFR 1910.147, Hazardous Energy (lockout/tagout) • 29 CFR 1910.132, PPE • 29 CFR 1910.1030, Bloodborne Pathogens
Trench/Earthen/Material Cave-In	• 29 CFR 1926.650, Trench/Collapse • NFPA 1670, Technical Rescue Incidents
Water	• Professional Association of Diving Instructors (PADI), Rescue Diver Training
High Angle	• NFPA 1985, Ropes and Harnesses
Confined Space	• 29 CFR 1910.146, Confined Spaces • 29 CFR 1910.147, Hazardous Energy • 29 CFR 1910.132, PPE • NFPA 1670, Technical Rescue Incidents
Roadway/Transportation Modes	• 29 CFR 1910.120, Hazmat

FIGURE 15-4 Classifying building collapses is a good start for identifying hazards that might be present.

TABLE 15-2 Structural Types Used by USAR

Structural Identifier	General Description
W	Wood buildings of all types
S1	Steel moment-resisting frames
S2	Braced steel frames
S3	Light metal buildings
S4	Steel frames with cast-in-place concrete shear walls
C1	Concrete moment-resisting frames
C2	Concrete shear wall buildings
C3/C5	Concrete or steel frame buildings with unreinforced masonry in-fill walls
TU	Tilt-up buildings
PC2	Precast concrete frame buildings
RM	Reinforced masonry
URM	Unreinforced masonry

appointed. Additional ASOs may also be required, based on the size and complexity of the incident, to address collapse hazards (see the *Safety Tip* box).

Safety Tip

Building Collapse Hazards
- Falling/loose debris
- Instability
- Secondary collapse
- Poor air quality/dust
- Unsecured hazardous energy
- Weather exposure
- Blood-borne pathogens
- Difficult access/escape options
- Sharp or rugged debris
- Poor footing

The urban search and rescue (USAR) community has developed some building construction classification language that is a bit different than the *type/era/use/size* method covered in the chapter *Reading Buildings* **TABLE 15-2**. Likewise, the ISO/ASO-RT should be familiar with the Federal Emergency Management Agency's USAR Task Force building marking system that is used for structural assessments **FIGURE 15-5**.

In addition to general duties, the ISO/ASO-RT should consider specific evaluations and actions that can improve responder safety, including technical assistance, air monitoring, and improvisation monitoring at collapses.

Technical Assistance

Experience has repeatedly shown that consultation with a structural engineer pays dividends. Some departments have retained structural engineers for emergency call-in. Large cities may have structural engineer employees in the building department. Regardless of their origin, structural engineers provide invaluable assistance.

2'×2' Box, spray-painted on Building near most accessible access. Orange paint

☐ = Relatively safe for search

▨ = Significantly damaged. May require shoring, bracing, & hazard removal for search

⊠ = Not safe, sudden collapse potential, rescue efforts will be high risk

← = Arrow to indicate direction of safest entry

HM = Hazmat present

31 Jul 16/1430 hr. = Date/time of assessment

Co2 = Task Group Designator (Colorado USAR task group performed the assessment)

FIGURE 15-5 USAR Task Force structural assessment marking system.

Drywall/concrete dust, asbestos, and mold exposure can lead to acute and chronic health concerns. Involving a public health official or respiratory specialist can help the ISO develop mitigation strategies to prevent associated ailments. The fire department health and safety officer (HSO) can assist with exposure reports and documentation of the inevitable barrage of "nuisance" injuries (cuts, blisters, and contusions) that always seem to accompany collapse incidents.

Air Monitoring

Rescue technicians (and the ASO-RT) usually begin air monitoring early in their efforts. During basic and light collapse incidents, this essential task may be forgotten. Simple four-gas monitors can help responders become aware of oxygen levels and the presence of sewer gas. Natural and propane gas detectors are also useful.

Improvisation Monitoring

"Adapt and overcome" is a phrase embraced by firefighters. How talented firefighters improvise their way through a situation is truly amazing, and the building collapse incident often showcases these abilities. From the ISO perspective, improvisation should be continually evaluated to identify when responders are pushing the envelope. Use your building construction knowledge (imposition of loads, material characteristics, etc.).

■ Industrial Entrapment

The hazards and considerations associated with industrial accidents are as numerous as the types of processes and plants. Unusual machines, conveyors, bizarre chemicals, technologically advanced processes, and supersized equipment can present challenges and injury potential for responders. Hazards at industrial entrapments include:

- Heavy machinery
- Complicated access
- Unsecured hazardous energy
- Hazardous materials
- Noise
- Interfaced and/or automated systems
- Security system impediment
- Mega-sized equipment
- Pinch hazards
- Equipment congestion
- Exotic materials
- Material stockpiling

In most industrial rescue cases, on-site employees shut down operating equipment that may further affect the victim. The ISO should double-check that lock-out/tag-out measures have been implemented. Reliance on on-site expertise at the industrial plant presents a double-edged sword. Usually, nobody knows the equipment better than those who operate it, yet their familiarity with the equipment may undervalue the hazards that the responder faces.

When evaluating rescue efforts, remember the basic law of motion: For every action, there is an equal and opposite reaction. Engineered components may be load stressed and spring out when cut. Rescue tools and equipment may be pushed beyond their designed limits; watch (and listen) for power equipment that is bogging down under load. As in all technical rescues, anyone performing high-concentration tasks needs an opportunity to take a mental (and physical) break.

■ Cave-Ins and Trench Rescue

The generic term "cave-in" can be applied to trench collapses, earthen slides (mud and rock), avalanches, and material entrapments (e.g., grain, sand, logs). Like other technical rescues, cave-ins come with their own set of hazards:

- Shifting/unstable material
- Hidden infrastructure
- Oxygen deficiency
- Weather exposure
- Difficult slope or grade
- Poor footing
- Sink potential

- Secondary collapse
- Crush potential

Trench rescues require specific regulated procedures (29 CFR 1926.650). The ISO must develop a site safety plan, emergency procedures, and safety briefings (see the following Confined Spaces section).

Using the LCES approach (lookouts, communications, escape routes, and safety zones) for developing safety briefings is useful for all types of cave-ins and can remind the ISO and responders of essential safety elements:

- *Lookouts.* ASOs, soil engineers, and briefed support personnel can serve as lookouts. Binoculars and signal devices (like a whistle) can help them in this function.
- *Communications.* Visual and voice communications are often a viable communications tool. Do not forget to communicate the IAP to all responders as part of the safety briefing. Everyone should know the "all-evac" signal that lookouts will use should the unexpected occur.
- *Escape routes.* Escape ladders and boarded footpaths should be used to create reliable escape routes. Tethered rescuers may require mechanical or powered assistance for rapid escape. In these cases, technician-level rescue personnel should be used as the rescuer and escape assistant.
- *Safe zones.* Developing suitable safe zones can prove challenging. A separate shore or refuge area may need to be erected prior to victim extrication efforts. Natural and structural barriers should be exploited.

Other unique hazards may require ISO attention. Exhaust fume accumulation, ground vibration, and specialized hydro-vac equipment require awareness and attention **FIGURE 15-6**. Finally, do not forget that one undeniable force: gravity.

■ Confined Spaces

As with hazmat and trench operations, confined space rescues are governed by OSHA regulations (29 CFR 1910.146). The use of an ISO/ASO-RT, the development of a site safety/emergency plan, and safety briefings are mandatory. Much like the hazmat incident, the IC is tasked with developing a written IAP and site safety plan and typically delegates the site safety plan to the ISO. Federal law specifically requires the written site

FIGURE 15-6 A trench rescue requires compliance with OSHA safety precautions prior to entry.

safety plan to include the following elements. Listed items with an asterisk should also be included in a written safety briefing.

- Safety, health, and hazard risk analysis, including entry objectives*
- Site organization, including the training and qualifications of responders
- Identification of the exact type of PPE required for the tasks performed by responders*
- Medical monitoring procedures*
- Environmental (air) monitoring
- Site control measures, including exact control zone locations*
- Decontamination procedures (if necessary)*
- Predefined responder emergency plans (for fires, medical emergencies, and rapid intervention)*
- Confined space entry procedures, including intervention and escape plans*

Documenting that incident responders have received appropriate safety briefings is included in the site safety plan requirement. Additionally, the ISO or ASO-RT may have to sign off on other developed plans.

Hazards that may require intervention at the confined space incident include:

- Limited access/escape options
- Toxic/flammable atmospheres
- Oxygen deficiency
- Hazardous energy
- Communication difficulties
- Collapse
- Cramped quarters, limited mobility
- Distance that exceeds air lines, ropes, etc.
- Rust and mold, residues

■ Water Rescues

One would think that firefighters' intimate relationship with water would prepare them to understand its daunting force. Yet perplexingly, the very ingredient that firefighters regularly use has caused numerous firefighter deaths. Water incidents can include swift water, lake, oceanic, flood, and ice situations. Each can present hazards for the rescuer (see the *Awareness Tip* box).

Awareness Tip

Hazards at Water Incidents
- Swift/hidden currents
- Low-head dams
- Submerged entrapment hazards
- Floating debris
- Electrocution
- Hypothermia
- Reduced visibility (murky water and water/salt spray)
- Fragile and/or shifting ice
- Marine life
- Frightened animals
- Distance to solid ground
- Crushing wave forces and undertows or riptides

Issues relating to protection from elements and appropriate PPE and personal floatation device (PFD) commonly lead the list of ISO concerns at the water incident. Spontaneous or initial-arrival water rescues often find responders without PFDs and a rope throw bag. Minimum PPE and equipment for a first responder to flooding situations should include a PFD, helmet, and rope throw bag. Planning for rapid rescuer intervention should be weighted heavily. Timekeeping can help the IC and ISO make judgments about rescue profiles and risk/benefit decisions. Responders certified in dive rescues can offer judgment regarding rescue profiles. Once a decision is made that a dive operation has changed from a rescue to a recovery, risk-taking should be reduced **FIGURE 15-7**.

Flood incidents can present multiple rescue situations —and very difficult access issues to make those rescues **FIGURE 15-8**. The power and damaging effects of riverine flooding cannot be overemphasized. Flood events can also overtax resources, as evidenced during hurricane Katrina in 2005 and the subsequent levy breaks in New Orleans. Responders to Katrina were faced with overwhelming difficulties in all phases of the search and rescue effort. From the numerous accounts of the Katrina response, you may glean some useful vicarious lessons that may help you perform ISO functions at a flood. Of particular interest are the numerous health issues that plagued the responders, who reported working in "cesspool"-like conditions as the water receded. Soliciting input from public health and environmental professionals can help the ISO address these concerns. It is advisable to include the department HSO to help document exposures.

FIGURE 15-7 Responders certified in dive rescues can offer judgment regarding the rescue-versus-recovery risk decision.
© Keith Muratori. Used with permission.

■ Rope Rescues

The popular trend of extreme sports has heightened fire department awareness of their ability (or inability) to rescue victims from radio/cell towers, elevated water storage tanks and bridge spans, high-rises, and precarious cliffs. Likewise, trapped or injured maintenance workers can find themselves in need of rescue from some amazingly challenging locations **FIGURE 15-9**. High-angle rescues can present hazards that can test even the most capable firefighter (see the *Awareness Tip* box).

FIGURE 15-8 Riverine flooding can present difficult access issues to make victim rescues. Do not underestimate the power of riverine flooding.

Courtesy of Justin King, Loveland Fire Rescue Authority, CO

Awareness Tip

Rope Rescue Hazards
- Limited access
- Dizzying heights
- Limited escape routes
- Slip/fall hazards
- Lightning/wind
- Limited anchor options
- Electrocution
- Heights beyond equipment capabilities
- Use of helicopters
- Equipment failure
- Falling debris
- Dropped equipment

Risk/benefit evaluations are paramount when performing ISO functions at the high-angle incident. Secondarily, the ISO should ascertain whether the rescuers are trained and *willing* to engage in the operation. (While the fear of public speaking is the number one fear of most adults, the fear of falling from heights is number two.) Some firefighters freely admit that they are uncomfortable with dizzying heights; others may not be so forthcoming. While it sounds unconscionable that a firefighter would not be willing to work from heights, it is a reality. Thankfully, most firefighters can work through and overcome their fear—*most*. The ISO should watch for signs that a firefighter is losing concentration or becoming stressed by fear (rapid breathing, wide eyes, and/or uncontrollable shaking).

As a matter of course, rescuers trained to hang from ropes typically double- and triple-check anchors and rigging. Using an ASO-RT can add redundancy to these checks. When it looks like rescuers will be committed to a rope (or climb) for an extended period, prehydration and intake of energy-nourishing foods are warranted. If possible, suggest to the rescuers that they take a water bottle and an extra energy snack as they climb. While the ASO-RT monitors the rescue technicians, the ISO should monitor the actions and focus of support personnel and the surrounding environment. Understandably, support personnel (and onlookers) spend an inordinate amount of time looking up (or down, as the case may be) during the rescue. Other hazards may catch individuals off-guard as they train their attention on the rescuers and victim; be their "wingman" and continually scan the surrounding environment.

Obtaining weather forecasts and watching the sky (see the chapter *Reading Hazardous Energy*) can help the ISO be proactive in preparing rescuers for wind gusts, lightning, and precipitation. Nighttime operations present additional concerns. The introduction of artificial light, while logical, may actually cause further problems. As an example, rescuers climbing on a radio tower may experience night blindness as spotlights are trained on them as they ascend or descend; it is better to "lead" them with the light. The use of glow sticks and soft headgear lighting by the climbers may be preferable on the ascent and descent. Spotlights can illuminate ascent and descent hazards in the path of the rescuers and/or illuminate the victim for reassurance and give

FIGURE 15-9 Thrill seekers and maintenance personnel injured or trapped in elevated places can present challenging rescues for fire departments.

rescuers a target point. The progress of a scaling rescuer can be monitored using a thermal imaging camera (TIC) in some cases.

The ISO also has to deal with the concerns resulting from the gathering of crowds and the media (see general ISO functions in the preceding section).

■ Roadway/Transportation Incidents

The risk of being struck by other vehicle traffic at roadway incidents appears to be increasing as communities struggle with the ever-increasing number of vehicles that are overtaxing the traffic infrastructure. Although exact data are difficult to assemble, most fire departments will agree that the number of responses to traffic-related incidents is increasing. To battle the increasing risks of being struck, fire departments must be vigilant to improve response and safety control measures at roadway incidents. Rarely does an ISO respond (or get appointed) to a typical motor vehicle accident (MVA) unless the situation becomes a "working" incident with the need for more resources. Duty ISOs should consider a discretionary response when dispatch information paints a picture of a particularly risky response. Details that should raise the ISO's alarm level include:

- Multiple vehicles, transport trucks or buses
- Long response time

- Involvement of hazardous energy, including power utilities, hydrants, bridges, and so on (The *Reading Hazardous Energy* chapter includes threat information on vehicle and other transportation modalities.)
- Extreme weather
- Hazardous materials, high-angle situations, and the like

In addition to roadway incidents, these considerations apply to railway, subway, and aircraft incidents. Each can present hazards that need to be evaluated by the ISO (see the *Awareness Tip* box).

> **Awareness Tip**
>
> **Hazards at Roadway/Transportation Incidents**
> - Other traffic or congestion
> - Threat of nearby secondary crash
> - Limited access or escape options
> - Hazardous materials/munitions
> - Ignition of fuels and/or presence of alternative fuels
> - Damaged infrastructure
> - Hazardous energy
> - Heavy entanglement
> - Weather exposure
> - Instability
> - Vehicle hazards
> - Blood-borne pathogens

Roadway Incidents

Saving lives is what firefighters are trained to do, and most of them would risk their lives to save lives. However, the roadway incident should be an exception. Given the trends in traffic-related firefighter deaths, responders must save their own lives first. The number one safety consideration at roadway incidents is the threat of being struck by other traffic **FIGURE 15-10**. Creating barriers, work zones, and traffic-calming processes are essential and should be the first priority for roadway incidents. The reality is stunning: We are at tremendous risk just investigating motor vehicle accidents.

FIGURE 15-10 The number one safety consideration at roadway incidents is the threat of being struck by other traffic.

Courtesy of Justin King, Loveland Fire Rescue Authority, CO

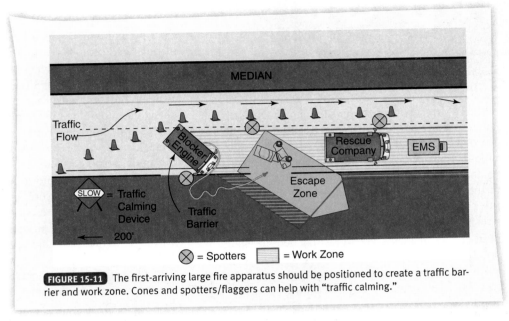

FIGURE 15-11 The first-arriving large fire apparatus should be positioned to create a traffic barrier and work zone. Cones and spotters/flaggers can help with "traffic calming."

All other things being equal, the ISO at an MVA should focus more on surrounding elements (especially other traffic) than on the rescue itself. Certainly rescuers working near damaged pad transformers, downed wires, and rushing water are at risk and may need ISO attention, but if they are hit in a secondary crash, that is sure to complicate things. There are still cases in which law enforcement personnel are reluctant to shut down traffic in support of MVA rescues. In some cases, fire officers argued with law enforcement officers that traffic needed to be shut down and were arrested and went to jail! With the increase in secondary crashes that have injured police and fire personnel, the reluctance seems to be ebbing away, although it may still be present in a few jurisdictions. Still, the ISO may want traffic stopped, or diverted from the accident scene, and must use well-articulated facts to make that happen.

As already mentioned, the basic strategy for improving rescuer safety at roadway incidents is to establish a system of traffic barriers, traffic-calming strategies, work areas, and escape zones **FIGURE 15-11**. A traffic barrier can be defined as an object (like a large fire apparatus) that can absorb the impact of a secondary crash to protect rescuers. When positioning the barrier (usually the first-arriving large apparatus), make sure that, if hit, it is driven away from the rescuers and other traffic. If the blocking apparatus is already committed, consider using a secondary large apparatus (unused) as a blocker. The original traffic barrier should create a work zone, shielded from moving traffic, for rescuers. Traffic-calming strategies include efforts to slow down approaching traffic: traffic cones, spotters or flaggers, arrow sticks, flashing lights, and warning signs. Using a traffic barrier without traffic-calming strategies is *not* advisable. Worse, using traffic-calming strategies without a traffic barrier is downright dangerous.

Another consideration for the ISO: At night, have apparatus operators minimize the use of white lights and strobes that can cause other drivers to experience night blindness, which can increase the risk of a secondary crash. Additional roadway safety initiative resources are listed at the end of this chapter.

Railway/Subway Incidents

The disaster drills that start off simple—only to have the facilitator throw in a hazmat incident, then a collapse, then an equipment malfunction, and then an explosion—may seem unrealistic. But it can all get real at the train derailment. The ISO must expect the worst to happen at a railway or subway incident. Imagine a subway train that has derailed between stops underground. In essence, you have a confined space, hazardous material, industrial entrapment, and structural collapse—all rolled into one. Throw in a fire, and you have the ingredients of the most challenging incidents. As the ISO, you must become strategic and rely on ASOs to monitor rescuers. Use the considerations in relevant preceding sections to address the ISO functions at the subway (or railway) incident.

Aircraft Incidents

Like the railway incident, an aircraft crash can present a variety of challenges. Using building collapse classifications (basic, light, moderate, heavy) can help the ISO judge the magnitude and potential hazards associated with the crash. The classification of the aircraft incident is influenced by the size of the aircraft (passenger or cargo load) and the size/type/use of the building that was hit. Regardless of the incident magnitude, a rescue profile needs to be established.

When the situation is viewed to be recovery in nature, the ISO should implement risk-reduction strategies. In the United States, representatives from the National Transportation Safety Board or Department of Defense (for military aircraft) respond to the crash site and ask that the incident be treated like a crime scene. If this mind-set is taken and communicated before representatives arrive, risk reduction can be achieved as working responders try to minimize the disruption and destruction of potential evidence.

Catastrophic crashes introduce the need for greater attention to the hazards of blood-borne pathogens. Gearing up for this type of incident should be suggested by the ISO. Unburned jet fuel residue can be very damaging to protective equipment, and decontamination efforts may have to be implemented. Likewise, jet fuel vapors, burnt plastics, and composite metal dust are respiratory irritants and/or toxins; do not be quick to allow responders to doff SCBA during operations at aircraft incidents.

Wrap-Up

Chief Concepts

- The technical rescue incident creates a trap in that firefighters' can-do attitude compels them to action, whether they are trained for the specific situation or not. From the ISO perspective, the can-do attribute can lead directly to firefighter injuries and deaths that only compound the incident. The ISO needs to take immediate corrective action where personnel are freelancing at a technical rescue incident.

- Technical rescue incidents are those that require specially trained personnel and equipment to complete the mission and may include incidents involving ropes, confined spaces, collapse, machinery, water, wilderness, and tunnels or caves.

- Confined space and trench (including cave-ins) incidents require the adherence to the OSHA CFR, including the assignment of an ISO for safety functions. The ISO must have the competencies of a technician-level responder for the incident type (NFPA 1006). When the ISO does not have the technician-level qualification, an ASO needs to be assigned.

- Initial activities of the ISO at a technical rescue include meeting with the IC to ascertain sitstat/restat and to ensure that any control zones, safety systems, and the IAP are in place. From there, the ISO is likely to do a recon if no pressing IAP or control zone issues need to be addressed immediately.

- IAP elements that need to be evaluated/adjusted by the ISO include overall risk-taking profile, control zones (and required PPE), personnel safety systems (accountability, communications, and rehab), and issues with traffic and medevac helicopters.

- Risk-taking is usually tied to the mode of operation: active rescue, recovery, or incident stabilization. Risk-taking should be reduced for recovery and stabilization modes.

- The general hazards associated with interfacing with medevac helicopters deal with the rotary wing and tail rotor and the potential for a crash.

- The ISO or ASO needs to recon the medevac helicopter LZ to ensure that an LZ officer is overseeing the interface, specific procedures are being followed, and crews have been briefed on procedures and hazards.

- Building collapses can be classified as surface, light, moderate, or heavy based on the type of construction and severity of damage. The ISO/ASOs should be familiar with USAR building classification and structural assessment marking systems.

- Of particular concern at collapses are the needs for technical assistance from structural engineers, air monitoring, and improvisation monitoring.

- Cave-ins and trench rescues are governed by OSHA regulations and require the ISO to develop formal safety briefings, a situational safety plan, and other documentation. The acronym LCES (lookouts, communications, escape routes, and safety zones) can be used to help the safety officer develop safety briefings.

- Like the trench rescue, confined space incidents require compliance with OSHA regulations.

- The initial response to water rescues typically requires that rescuers use PPE and a personal floatation device (PFD) and involves issues associated with environmental exposure. Dive rescue incidents require accurate time keeping to help make rescue/recovery decisions (when to reduce risk-taking). Flooding rescues can challenge access options and overtax resources. Flood incident operations may include the need to decontaminate responders and document exposures for acute and chronic health concerns.

- Rope rescues should be evaluated by an ASO-RT to make sure rigging has been triple-checked. The ISO should focus on hazard issues that surround the incident (onlookers, lighting, physiological reaction of responders).

- The primary threat that requires immediate attention at roadway/transportation incidents is the likely chance of responders being struck by other traffic. The use of traffic barriers (large apparatus) and traffic-calming devices (cones, flaggers, flashing lights) is mandatory.

Key Terms

assistant safety officer–rescue technician (ASO-RT) A person who meets or exceeds rescue technician-level competencies.

basic/surface collapse A collapse in which victims are easily accessible and trapped by surface debris. Loads are minimal and easily moved by rescuers. The threat of secondary collapse is minimal.

heavy collapse A collapse in which stressed concrete, reinforced concrete, and steel girders are impeding access. Included are collapses that require the response of urban search and rescue teams and specialized equipment, and collapses that threaten other structures or that involve the possibility of significant secondary collapse.

incident stabilization mode An approach to incident response used when victims have been rescued (or recovered when appropriate) yet the conditions are such that a potential for further harm exists and must be addressed.

light collapse A collapse in which usually a light-frame (wood) partition collapses and common fire department equipment (from engine and truck companies) can access or shore areas for search and extrication. The threats of secondary collapse can be mitigated easily.

moderate collapse A collapse of ordinary construction that involves masonry materials and heavy wood. Lightweight construction with unstable large open spans should also be classified as moderate. Significant void space concerns are present.

on-deck system An organized system in which a working team is replaced with another working team that is already dialed in and ready to replace them. Once replaced, the rotated workers go to rehab, while a crew from rehab rotates into the vacated on-deck position.

recovery mode An approach to incident response used when victims have been determined to be deceased or the conditions present are incompatible with survivability.

rescue mode An approach to incident response used when victims obviously need rescue and/or the conditions present indicate that a search and rescue operation can lead to savable victims.

traffic barrier An object (like a large fire apparatus) that can absorb the impact of a secondary crash to protect rescuers (often referred to as a "blocker"). These objects should be used to create a work zone, shielded from moving traffic, for rescuers.

traffic-calming strategy An effort to slow down approaching traffic: traffic cones, spotters or flaggers, arrow sticks, flashing lights, and warning signs.

Review Questions

1. List several regulations that outline response requirements for technical rescue incidents.

2. When should an ASO-RT be assigned at a technical rescue incident?

3. For whom does the ASO-RT work at a technical rescue incident?

4. What is meant by the "on-deck" system for crew rotation? What types of incidents should it be used for?

5. List four specific hazards that are present when interfacing with medevac helicopters.

6. A collapsed building made of cinder block walls and small timber beams should be classified as which type of collapse?

7. List the issues that the ISO should address when developing a formal safety briefing for trench and confined space rescues.

8. Describe the lighting tactics that should be used for a nighttime rope rescue event that requires responders to climb.

9. Name five circumstances in which a duty ISO should implement a discretionary response to a reported motor vehicle accident.

10. What is the difference between traffic barriers and traffic-calming strategies?

11. Diagram a strategic approach to protect rescuers at roadway incidents.

12. What do jet fuel, burnt plastics, and composite metals have in common at an aircraft incident?

References and Additional Resources

Department of Homeland Security, Federal Emergency Management Agency. *National Urban Search and Rescue System.* 2013. https://www.fema.gov/urban-search-rescue. Accessed May 26, 2015.

International Association of Fire Fighters. *Best Practices for Emergency Vehicle and Roadway Operations Safety in the Emergency Services.* Washington, DC: IAFF, 2010. http://iaff.org/hs/EVSP/guides.html. Accessed June 12, 2015.

Linton, Steven, and Damon Rust. *Ice Rescue.* Fort Collins, CO: International Association of Dive Rescue Specialists, 1982.

Martinette, C.V. *Trench Rescue Levels I & II Principles and Practice.* 3rd ed. Burlington, MA: Jones & Bartlett Learning, 2016.

National Institute for Occupational Safety and Health. Safety Management in Disaster and Terrorism Response. *Protecting Emergency Responders, Volume 3.* NIOSH Publication No. 2004-144. http://www.cdc.gov/niosh/docs/2004-144/. Accessed May 26, 2015.

NFPA 1006, *Standard for Technical Rescue Personnel Professional Qualifications.* Quincy, MA: National Fire Protection Association, 2017.

NFPA 1670, *Standard on Operations and Training for Technical Search and Rescue Incidents.* Quincy, MA: National Fire Protection Association, 2017.

Norman, John. *Fire Department Special Operations.* Tulsa, OK: Fire Engineering Books and Video, PennWell, 2009.

U.S. Fire Administration. *Emergency Vehicle Safety Initiative.* FA-336. Emmitsburg, MD: USFA, 2014.

Vines, T., and S. Hudson. *High Angle Rope Rescue Techniques Levels I & II.* 4th ed. Burlington, MA: Jones & Bartlett Learning, 2016.

Endnotes

1   NFPA 1670, *Standard on Operations and Training for Technical Search and Rescue Incidents.* 3.3.146, 3.3.147. Quincy, MA: National Fire Protection Association, 2017.

Smoke: © Greg Henry/ShutterStock, Inc.

A fire department was notified of several cars stranded due to heavy amounts of rain and subsequent flooding. A crew was dispatched to the scene at approximately 1700 hours to assist motorists stranded by the floodwaters. Upon arrival, the captain sent two firefighters to check for any motorists in need of assistance. The two waded through approximately knee- to waist-high water for two blocks, checking several cars floating in the water. After approximately 15 minutes on the scene, the crew radioed their captain that all of the civilian motorists had exited their cars. After the crew determined that there were no civilians in the cars, they directed traffic away from the pooled water and waited until the police arrived to take over scene control. While the two firefighters were waiting for the police to arrive, they were verbally summoned by a civilian bystander to help a female civilian stranded in the water. The civilian was observed holding onto a traffic sign pole in a large pool of water. The crew believed the water was only 3 feet deep and that the civilian was standing on the ground. The civilian was actually standing on the top edge of a culvert on a large slope into the pool of water, which was approximately 10 feet deep.

Both of the firefighters responded to the location of the female civilian and attempted a rescue. The firefighters were wearing bunker pants, coats, boots, gloves, and helmets. Neither had received water rescue training. The first firefighter to enter the water was quickly pulled under by the undertow. The second firefighter then entered the water to aid his fellow firefighter. A grab was made and they both struggled to the edge of the water. The first firefighter, with his back to the water, climbed onto the bank, coughing from water he had swallowed. When the second firefighter reentered the water to assist the civilian, he told his partner to radio for help. As the first firefighter turned around, his partner was gone and his helmet was circling on the surface of the water. Seeing this, he removed his bunker coat and told a witness to use the radio and call for help. He reentered the water and assisted the stranded civilian to safety. Witnesses found a welding cable and tied it around the firefighter's waist. He reentered the water and began to frantically search for his partner under the surface of the water. At 1744 hours, a female civilian witness used the radio to call for help. The captain, several hundred feet to the south of the scene, was confused about who was on the radio. Since his crew did not have any females on duty, he first thought it was a firefighter on another scene.

After an unsuccessful search for his partner, the firefighter exited the water and called on the radio for help, stating that a firefighter was down. The captain and another firefighter then ran to the location.

For several hours, dive crews and firefighters made numerous attempts to locate and rescue the victim. At approximately 2245 hours, the victim was found several blocks from the original location of the attempted rescue. He was pronounced dead at the scene.

The NIOSH investigation outlined several recommendations to prevent a reoccurrence. Some of the recommendations follow:

- Ensure that a proper scene size-up is conducted before performing any rescue operation and that applicable information is relayed to the officer in charge.
- Ensure that all rescue personnel are provided and wear appropriate personal protective equipment when operating at a water incident.
- Ensure that firefighters who could potentially perform a water rescue are trained and utilize the "reach, throw, row, and go" technique.
- Develop site surveys for existing water hazards.
- Ensure that standard operating procedures (SOPs) are developed and used when water rescues are performed.

1. What would you consider to be the prevailing factor contributing to this tragedy?

2. What personal safety system elements do you think were missing (or incomplete)?

3. One recommendation is the provision of appropriate PPE for water rescues. For the first-due engine, what should the minimal PPE include?
   A. Structural PPE clothing, helmet, and rescue gloves
   B. Water-resistive uniform, breathing device, and rescue harness
   C. Rescue harness, no-slip gloves, and helmet
   D. Personal floatation device (PFD), rope throw bag, and helmet

4. Risk-taking at a technical rescue incident is tied to an operational mode. An operational mode where victims have been rescued, yet the conditions are such that further harm may be incurred, would be classified as:
   A. rescue mode.
   B. recovery mode.
   C. incident stabilization mode.
   D. zero risk mode.

Note: This case study was developed using NIOSH firefighter fatality investigative report 2001-02. Available at: www.cdc.gov/niosh/fire/reports/face200102.html.

Postincident Responsibilities and Mishap Investigations

Flames: © Ken LaBelle NRIFirePhotos.com

Knowledge Objectives

Upon completion of this chapter, you should be able to:

- Describe the elements of a postincident analysis (PIA) related to the ISO's perspective of an incident. (NFPA 5.7.1 , pp 234–236)
- Identify the standards that require a PIA. (NFPA 5.7.1 , p 233)
- Define the goals of a PIA. (NFPA 5.7.1 , p 233)
- Describe how concerns and recommendations are identified and communicated as a result of a PIA. (NFPA 5.7.2 , p 234)
- Describe the elements of a safety investigation process. (NFPA 5.6.1 , pp 238–240)
- Identify the types of incidents that should be investigated. (NFPA 5.6.1 , p 238)
- Describe situations that would create a conflict of interest during an investigation. (NFPA 5.1.2 , p 237)
- Identify the documentation process used during a safety investigation. (NFPA 5.6.1 , pp 238–240)
- Describe the data needed to conduct a safety investigation. (NFPA 5.6.1 , p 238)
- Identify the information needed from incident witnesses. (NFPA 5.6.1 , p 239)
- Describe the witness interview techniques used to gather information regarding an incident. (NFPA 5.6.1 , p 239)
- Determine what equipment or information would be considered evidence used to determine cause of injury, death, or property damage. (NFPA 5.6.1 , p 239)
- Identify the goal of a safety investigation. (NFPA 5.6.1 , pp 236, 240)
- Describe the relationship between the ISO and the health and safety officer (HSO) during the investigative process. (NFPA 5.6.1 , p 236)
- Describe the types of incidents that may create atypical stress exposure to emergency responders. (NFPA 5.2.12 , p 241)
- Identify the signs and symptoms of exposure to occupational stress. (NFPA 5.2.12 , p 241)
- Describe the characteristics of an incident debriefing. (NFPA 5.2.12 , p 242)
- Describe the characteristics of an incident defusing. (NFPA 5.2.12 , p 242)
- Identify the types of resources available to responders who are exposed to occupational stress. (NFPA 5.2.12 , p 242)

Skills Objectives

Upon completion of this chapter, you should be able to:

- Demonstrate the ability to communicate the need for support teams or other resources for members who are exposed to stressful occurrences. (NFPA 5.2.12 , pp 241–242)
- Complete a written post-incident analysis from an ISO perspective. (NFPA 5.7.1 , pp 234–236)
- Communicate findings of a post-incident analysis. (NFPA 5.7.2 , pp 234–236)
- Demonstrate the ability to conduct a safety investigation of an incident close-call or injury. (NFPA 5.6.1 , pp 236–240)

One of the final pieces remaining to be placed in the puzzle that is your ISO program will become a major change in department operations. Your department has never routinely conducted any form of postincident analysis or investigation of close calls or injuries. You are concerned that implementing this component will create a lack of trust in the ISO program, as members will be worried about documentation and presentations of mistakes or errors in judgment they may have made. You are planning a brainstorming session with the training division and command staff to discuss your concerns.

1. Review your department's policy and/or forms process used for postincident analysis or investigations. Do you believe the process is aimed primarily at improving safety and operations or at finding fault and blaming members?
2. What steps are involved in conducting an initial accident investigation in your department?

Introduction: Paying It Forward

For every 1 serious firefighter injury, over 600 near misses or close calls could easily have been serious.[1] Firefighters sometimes wear a close call as a badge of courage and use exaggerated tales to fuel firehouse coffee talk. Firefighters involved with a close call may trivialize or minimize the brush with injury or death. Often, the closer firefighters come to serious injury, the more they minimize the storytelling, perhaps indicating that the event really got their attention. At what point does the incident safety officer (ISO) need to follow up on a near miss and work toward the prevention of a similar event that may not have such a "lucky" outcome?

Ideally, the lessons learned from *any* close call should be folded into training and used for ongoing efforts to avoid similar situations in the future. In other words, *pay it forward*. Often, the war stories that arise from close calls are invaluable tools in the teaching of new firefighters in academies nationwide. The key to making the lessons productive is an accurate portrayal of the facts and actions. Collecting information quickly and from multiple sources will help to ensure that information is captured accurately. The reconnaissance responsibility places the ISO in the best position to capture and document incident activities for a postincident analysis (PIA), critique, or after-action review. Further, the ISO can use the information to begin an investigative process if an injury or fatality has occurred. Paying it forward also helps those who are dealing with any psychological imprint that may have occurred as the result of an unusual, emotionally draining, or especially gruesome incident—those atypical events.

This chapter explores the responsibilities and duties of the ISO for postincident activities, PIAs, atypical events, and accident investigations.

Postincident Activities

Accurate data are spotty, but many injuries seem to occur while crews are packing up to leave an incident. Common postincident injuries include strains, sprains, and being struck by objects. Postincident injuries seem ironic in a profession whose hallmark is aggressive and calculated risk-taking. For each cause of postincident injuries, the ISO can take preventive steps to reduce their likelihood of happening. Typical causes of postincident injuries are usually tied to postincident thought patterns and chemical imbalance.

■ Postincident Thought Patterns

One cause of postincident injuries has to do with the little-studied concept of postincident thought patterns. Postincident thought patterns are the reflective or introspective mental preoccupations that firefighters experience just after incident control. These thought patterns can lead to inattentiveness and, consequently, injuries. In cases of especially difficult, unusually spectacular, or particularly challenging incidents, firefighters tend to reflect on their actions. The replay of the incident starts almost instantly when the order is given to "pick up." Introspection is normal **FIGURE 16-1**. The switch from activities requiring mental sharpness and physical effort to an activity that is so rudimentary as to be dull presents a letdown trap. Herein lies the problem.

Some incident commanders (ICs) may release the ISO from the event after the "serious stuff" is done. This is an error. After a working incident, the ISO should circulate among those involved in the pick-up and keep an eye out for inattentiveness. Signs may include faraway stares or robot-like actions. Firefighters might stop and look about as if they have forgotten

FIGURE 16-1 Postincident introspection is normal but can lead to inattentiveness and injury.

their task. Simple reminders or jocularity can help them regain focus and reduce their injury potential. One method to reduce the effect of postincident thought patterns is to take a brief time-out and have everyone gather for a quick incident summary and safety reminder. These huddles can be effective for all personnel on scene or even small groups **FIGURE 16-2**. Even a casual coachlike approach that emphasizes the need to stay alert and not fall into an injury trap can be useful.

■ Chemical Imbalance

A successful rehabilitation program prevents firefighters from experiencing fatigue and mental drain. Nevertheless, some may not have benefitted from rehab, or the incident was simply draining due to the duration or demand of the incident; the human form can only do so much. Even with effective rehab, the end of an incident, especially one requiring significant physical output, can cause chemical imbalance. With the end of an incident comes the relaxation of the firefighters' minds and the shutoff of protective chemicals that stimulate performance. The adrenaline rush is over and the firefighters' metabolisms return to a "repair" state; this reaction also causes a mental slowdown that can lead to unclear thinking and injuries.

Another way to see the chemical (and mind) imbalance is to look at a firefighter's tools from a layperson's point of view.

FIGURE 16-2 Calling a huddle before incident pick-up creates an opportunity to remind firefighters of lingering injury threats.

Laypersons have to concentrate on carrying an axe, pike-pole, chainsaw, roof ladder, or hose length in order to not cause harm to themselves or anyone around them. Put them in bulky, restrictive, sweat-soaked clothing and heavy boots, and you can see that they would probably get hurt doing almost any task. The firefighter performs these tasks after incredible energy bursts under frightening conditions. Familiarity may help ensure some degree of safety, but concentration is still required. The combination of fatigue and the signal to relax (because the incident is "over") creates the imbalance. Yet the concentration requirement remains the same, and the potential for injury rises.

Whether the issue is chemical imbalance or postincident thought patterns, the ISO must stay alert, pick up signs of potential injury, and take steps to remind crews of that potential. The ISO may also consider asking the IC for a fresh crew to respond and assist with pick-up when the ISO feels that crews are at risk of injury due to their exhaustion.

Postincident Analysis

"Tailgate talk," "after-action review," "critique," "slam session," "incident review," and "Monday morning quarterbacking" are all labels the fire service has applied to the PIA. The PIA is a formal and/or informal reflective discussion that fire departments use to summarize the successes and the areas requiring improvement discovered from a given incident. Successful fire officers learn something from every working incident they are involved in. Each and every firefighter involved in an incident has a viewpoint or an opinion regarding specific circumstances or the general outcome of an operation, and these are important. The IC and the ISO bring perspectives to the incident overview. In fact, the ISO should contribute—officially—to the PIA.

NFPA 1500 requires that the ISO be involved with the PIA. NFPA 1521 lists several job performance requirements (JPRs) that the ISO needs to meet in order to maximize his or her PIA involvement (listed in the Knowledge/Skills Objectives at the beginning of this chapter). To maximize the effect of safety-related input on a PIA, the ISO must understand the essential philosophy of PIA as well as ISO issues surrounding PIAs. This section contains a simple process to ensure that the ISO covers the appropriate information for the PIA.

■ PIA Philosophy

The ISO should approach any formal or informal PIA with an attitude of positive reinforcement for safe habits and an honest, open desire to prevent future injuries. In most cases, the PIA is nothing more than a discussion of what went right and what should be different next time. This may sound simple, but it is often hard to achieve, especially in light of a close call or a significant operational mistake that could have easily led to an injury.

When an operational mistake has been made, the ISO should first consider the outcome: Did an injury occur or was there reportable property damage? If so, the PIA becomes secondary to accident investigation (covered later in this chapter). If no injury or damage occurred, the stage is set for lesson-learning. To capture the lesson, the ISO should start by employing a philosophy of *discovery*. The ISO can discover the general feeling of crews by making some inquiries regarding the operational environment, ascertaining the perceptions of crews regarding the seriousness of the mistake, and performing some general fact-finding. The discovery philosophy can also give crews a voice for their concerns and may even assist in getting someone to acknowledge an error. When making inquiries, stay away from questions that can be answered as yes or no. Framing open-ended inquiries is the best approach. For example:

- "Explain the location in regard to . . . ."
- "Describe the conditions you faced."
- "What was your thought process for . . . ?"
- "How do you feel regarding . . . ?"
- "How can we proceed/prevent/improve . . . ?"

Occasionally, this approach does not work. Crews may spend great amounts of energy "explaining" their actions in an attempt to justify them. In these cases, the ISO needs to act as a sounding board and not offer judgments. If the discovery process leads to individuals or crews getting overly defensive or emotional with each other (the blame game), it may be best to separate the parties and let them cool off.

While the general approach of the PIA is to look *back* at an incident, the overriding goal is to look forward to the future. Through this process, departments seek to discover what went well (reinforcement) and what didn't (prevention) so that the next incident will have a better outcome.

■ The PIA Process

Fire departments should have a standard operating procedure/guideline (SOP/SOG) that suggests the *level* of PIA that should be used for various incidents. Where none exists, the ISO (through the IC) can use the following guide:

- *Informal, on-scene PIA.* Used for most simple incidents with no major issues and no reportable injuries or property damage.
- *Informal incident debriefing PIA.* Used for working incidents that require some information gathering (e.g., dispatch log, scene sketches) and a thoughtful, organized approach to capture lessons. A debriefing typically occurs at a nearby fire station shortly after the incident (after equipment reservicing and cleanup).
- *Formal PIA.* Used for significant incidents that require an effort to collect information from multiple sources in order to prepare an official written report.

From the ISO's perspective, some types of incidents should trigger a formal (written) PIA. These include the following:

- Firefighter injury (requiring medical transport) or line-of-duty death (LODD)
- Apparatus or equipment failure that changed the incident outcome
- Incident at which a firefighter declared a Mayday or at which a rapid intervention crew (RIC) was activated
- High-profile incident, such as a large property- or life-loss event
- Large-scale incident requiring many resources (e.g., multiple alarms)
- Unusual or uncommon incident
- Specialized or technical-level incident (e.g., technical rescue, hazardous materials [hazmat] incident)

This list is not meant to be all-inclusive. Some fire departments require a formal PIA for all structure fires or working incidents—especially those departments that have only a few structure fires a year! The scope of a formal PIA can range from a few days of preparation to one requiring months of documentation and presentation development (e.g., reports, graphics, facilitation).

As stated, there are various levels of PIAs. Regardless, the ISO needs to capture observations and be prepared to contribute in a positive manner. First, the ISO should ascertain from the IC whether he or she wishes to host an informal, debriefing, or formal PIA. If the IC does not wish to do a PIA of any type, the ISO should still prepare some documentation (covered later). The ISO should then employ a few simple steps to maximize the opportunity to increase everyone's ability to make solid risk decisions and prevent future injuries. Let's take a look at these simple steps.

On-Scene Preparation for a PIA

Regardless of the type of PIA that will be used, as the ISO, you should collect some information while still on scene to better prepare your input. Before crews pick up to leave, make it a point to check in with them and hear what they are saying and how they are saying it (these are emotional clues; we'll address incident stress later in the chapter). Also take the opportunity to ask a few questions. A successful technique is to ask crews if they noticed any hazards that you, the ISO, may have missed. You should also ask if everyone on the crew is feeling OK or has received any minor injuries that you should know about. A parting positive comment regarding the crew's effort should always be included.

At a minimum, the ISO should document a quick summary of his or her actions and the hazards addressed (see the *Getting the Job Done* box and **FIGURE 16-3**. Also include a summary of building construction features, any unique features of the operational environment, and an incident timeline. A quick call to the dispatch center can help you record times. The chronological succession of events is often questioned during PIAs; these questions can be clarified if the ISO has a documented timeline.

FIGURE 16-3 Documentation is essential! It doesn't have to be fancy; some simple memory joggers can help capture the ISO's perspectives for the PIA and archive purposes.

Getting the Job Done

ISO Documentation Issues

The ISO should get in the habit of documenting his or her actions for *all* incidents. In some cases, the documentation is mandatory (e.g., hazmat incident, confined space rescue, departmental SOP requirements). The ISO's documentation efforts are beneficial in numerous ways, but they are especially helpful in developing a PIA (formal or informal) and in resolving follow-up inquiries that may not be evident at the time of the incident.

In addition to basic incident information (e.g., assigned incident number, location, timeline), the ISO should document the following:

1. Overview of the incident conditions found on arrival and notable changes that occurred
2. Suggested changes to the IAP: What alternatives were proposed, and were they adopted? (If the suggested change is not adopted and the incident goes "sideways" on the IC, how is the ISO to defend his or her suggestions?)
3. Soft interventions, stern advisories, and firm interventions
4. Summary of meetings with the IC
5. Substance of safety briefings delivered to responders
6. Follow-up items that may need to be addressed within the framework of departmental improvement (reoccurring problematic areas like PPE, rehab, or training issues)
7. Unusual or strange events or conditions
8. Other forms or required reporting documents

If a formal PIA is scheduled, the ISO should spend more time with documentation and confer with the IC prior to the session. This preparation helps the IC and ISO avoid displaying differing points of view in front of crews, which can have the effect of devaluing safety. Remember, the focus should be "paying it forward"; you are capturing lessons in a way that can prevent similar occurrences in the future.

It is not at all uncommon for the ISO to take the lead in preparing the entire final PIA report for significant incidents. This may seem like a daunting task, especially for ISOs who have never prepared such a report. To help, the ISO assigned the preparation task can do a web search of PIAs that have been posted by hundreds of fire departments. These PIAs can provide a general template for the ISO to follow and, ideally, reduce the stress of having to format a report from scratch.

Trend Spotting

The ISO who finds a recurring problem or concern should take the time to jot down some thoughts, then share them with a supervisor, training officer, or the department's health and safety officer (HSO). Remember: Many people can spot and articulate problems; only a few can present a problem along with some reasonable solutions.

■ ISO PIA Contributions

By monitoring an incident for potentially unsafe situations, the ISO brings many valuable observations to the PIA. The functional areas addressed in the *hazard MEDIC* approach to ISO duties (monitor, evaluate, develop measures, intervene, and communicate) are also of worth to the PIA effort. Nevertheless, the PIA is a time for crews to share and reflect and take home a message. A long dissertation from the ISO can easily negate any such message. However, the ISO should comment on some key issues, including the following:

- *General risk profile of the incident.* The ISO can share the overall picture from a risk management point of view. Items such as risk/benefit, pace, and impressions about appropriateness of the risks taken can be discussed. If a situation developed that placed a crew at risk, the ISO may find it valuable to call on the crew to relay his or her thoughts or perceptions. These observations may have to be built on so that everyone takes away some value.
- *Effectiveness of crew accountability.* The ISO can yield to an accountability system manager for some of this type of information. Observations about crew freelancing (working in conflict with the action plan), individual freelancing (working without a partner), and reinforcement about successful tracking and communication should be shared.
- *Rehab effectiveness.* Even though they should be, ISOs are seldom "processed" through rehab. How then can the ISO comment on the effectiveness of rehab? The ISO can share observations of the pace, energy, and focus trends throughout the incident, as well as the duration or rotation of work efforts. If injuries resulted during the incident, an investigation is likely. But for postincident purposes, some exploration of rehab as a contributing factor may be discussed.
- *Personal protective equipment (PPE) use.* Although the ISO normally defers individual PPE concerns to the company officer or crew leader, ongoing PPE issues can be addressed by the ISO. As an example, the choice to do overhaul at an incident without self-contained breathing apparatus (SCBA) may have been premature. Likewise, a four-gas monitor may have been used to make the decision to go "packs off" and use simple dust masks. These decisions can be discussed and reinforced as appropriate.
- *Close calls.* The circumstances surrounding a near injury should be detailed from all participants' points of view. The ISO should minimize his or her contributions to close-call events and reserve judgment because an actual investigation may be warranted. In some cases, a close-call event may be an invaluable lesson to share with others. Several fire service organizations have joined to create a near-miss reporting system to help share experiences with other firefighters (see the *Fire Marks* box).
- *Injury status.* If no injuries have been reported, this positive outcome should be reinforced. Most likely,

good practice and procedure led to no injuries. If this belief is not shared by all, then the role of "luck" should be discussed; this can very well raise safety awareness for the next incident. If a firefighter injury required transporting someone to a medical facility, firefighters will want an update FIGURE 16-4. The ISO should be cautious. Issues of medical confidentiality, investigative needs, and other ramifications may limit the amount of information that can be shared. In these cases, the ISO should keep the discussion centered on the efforts under way to care for the injured.

If a firefighter injury was significant, or if a firefighter fatality occurred, the ISO should use any initial PIA session as a tool to listen to firefighters (there will likely be a more formal PIA in the future, once the investigation and written report are finished). While listening, try to keep an open mind. Often, firefighters may appear to be blaming when, in reality, they are venting, displacing stress, or even grieving. These reactions are normal.

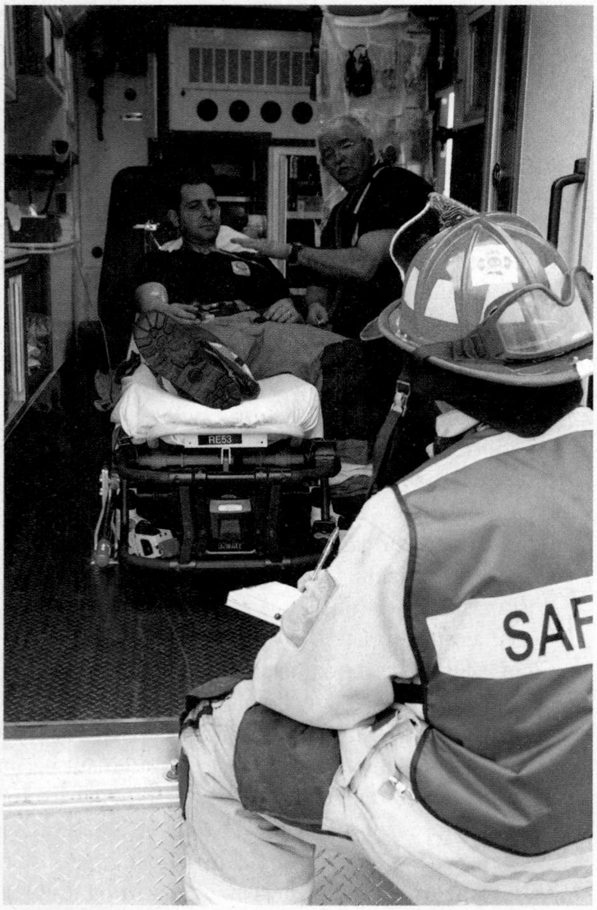

FIGURE 16-4 Firefighters will want an update on the status of an injured firefighter. The ISO should limit the update to the efforts under way to care for the injured.

Fire Marks

The National Firefighter Near-Miss System
The National Firefighter Near-Miss System is a voluntary, confidential, nonpunitive, and secure reporting system with the goal of improving firefighter safety. The system is funded by the Assistance to Firefighters Grant Program of the U.S. Department of Homeland Security and is hosted in affiliation with the International Association of Fire Fighters (IAFF) and the International Association of Fire Chiefs (IAFC).

This web-based system allows individuals to submit a report. An easy-to-follow template helps guide inputs that are used to categorize the incident. Filed reports are reviewed by fire service professionals, and lessons learned are reposted on the website in an anonymous fashion.

The system is also designed to generate a database that can be used to spot trends, develop training programs, and offer suggestions. Anyone can access the system to review reports or sign up for a news service that automatically emails the report of the week to subscribers. The system can be accessed at http://www.firefighternearmiss.com.

Accident Investigation

An accident investigation is a critical step in avoiding future injuries and deaths. Often, the results of the investigation can lead to changes in SOPs, unsafe situations, habits, or equipment not only for the originating department but for departments nationwide. This vicarious learning is essential. The Sofa Super Store tragedy (June 2008) in Charleston, South Carolina, is a perfect example.

In Charleston, the fire department responded to a fire in the rear loading dock area of a large furniture store. Initial control efforts were not successful and the fire spread into a warehouse as well as the showroom. The conditions in the showroom deteriorated quickly, leading to firefighter Mayday situations and the eventual ignition of all the smoke in the showroom (approximately 33,000 square feet [3,065 m²] of floor space). Nine firefighters were unable to make it out and perished. The incident led to several investigative and technical reports that are now being used to help change fire department procedures, training, and tactical operations. Further, the reports have become an impetus for sprinkler ordinances, building inspection programs, and community fire protection planning around the country.

Close calls or near misses should also be investigated. Technically speaking, the phrases "close call" and "near miss" are interchangeable. Anecdotally, the phrase "near miss" can be regarded as a near hit. Regardless, we must learn from these events to prevent future injuries. Although the notion of a close call or near miss is subjective, it can be loosely defined as an unintentional, unsafe occurrence that could have produced an injury, fatality, or property damage; only a fortunate break in the chain of events prevented the undesirable outcome. An open, nonjudgmental attitude toward close calls can help a fire department realize the many warning signs, situational occurrences, and contributing factors that precede an injury.

Often, the person to begin an accident investigation following a close call, firefighter injury, fatality, or equipment mishap is the ISO, given the nature of his or her assignment as a command staff member. NFPA 1561 specifically requires the ISO to "investigate accidents that have occurred within the incident area" as a major responsibility.[2] For most accident investigative purposes (property or bodily harm), the ISO is well positioned for this function as a result of 360-degree scene monitoring. To prepare for accident investigations, the ISO needs to understand the components of the accident chain, issues that create a conflict of interest, types of mishaps that require investigation, and the general steps to conducting an investigation.

■ The Accident Chain

Accidents are the result of a series of conditions and events that lead to an unsafe situation that result in injury and/or property damage. Many call this series of conditions and events the accident chain. The accident chain is a "sequence of events" model derived from *The Domino Theory of Accident Causation* developed by H.W. Heinrich (1931). The investigation of an accident is actually the discovery and evaluation of the accident chain, which has five components **FIGURE 16-5**:

- *Environment.* The physical surroundings, such as weather, surface conditions, access, lighting, and barriers
- *Human factors.* The components of human (or social) behavior—training, the use of or failure to use recognized practices and procedures, fatigue, fitness, and attitudes
- *Equipment (including PPE).* Limitations and restrictions of equipment, its maintenance and serviceability, the appropriateness of its application, and, some may argue, its misuse (a human factor)
- *Event.* A scenario that brings the first three accident chain components together in such a way as to create an unsafe or unfavorable condition
- *Injury.* The injury or property damage associated with the accident (Because a near miss or close call is an accident without physical injury, for the sake of the accident chain, the injury can be supposed.)

Ideally, the ISO should be able to stop a potential accident by eliminating one or more of the elements in the chain *during* the incident or by creating barriers between the elements (recall hazard MEDIC).

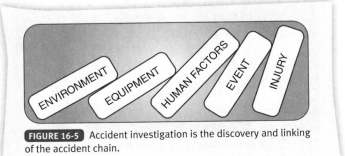

FIGURE 16-5 Accident investigation is the discovery and linking of the accident chain.

■ Investigation Issues

The ISO should be aware of issues and concerns that arise regarding involvement in accident investigation, one of the biggest of which is liability. Picture this: A firefighter is seriously injured while working a commercial structure fire where an ISO is functioning. In the ensuing investigation, the question arises, if a safety officer was present, should an injury have occurred? The conclusion of some may be that the ISO obviously did not do his or her job; it may lead to doubts about the objectivity of the ISO in investigating the incident. Did the ISO cover up anything in an attempt to deflect blame? This scenario may seem far-fetched, but it is a reality in our litigious society. For this reason, there are circumstances where ISOs should recuse themselves from the investigation process:

- The ISO is the one injured.
- The ISO was directly involved in the property damage.
- An LODD occurred.
- Multiple, and significant, injuries to firefighters occurred.

For significant (or multiple) injuries and death(s), the ISO should request HSO assistance for a simple reason: The role of the ISO is part of the equation that needs to be investigated. More specifically, some may reason that an assigned ISO should have prevented the outcome, and therefore his or her actions should be questioned. In these cases, a conflict of interest exists for the ISO when assigned the function of accident investigation. Outlining fire department procedures that need to be implemented following a fatality (or an injury that may result in a fatality) is beyond the scope of this text. Such information is available through the U.S. Fire Administration, IAFF, and IAFC.

How does the ISO perform safety tasks on scene *and* conduct an honest, meaningful investigation following an accident? The answer is simple: Do both with due diligence. Due diligence is a legal phrase for the effort to act in a reasonable or prudent way, given the circumstances, with due regard to laws, standards, and accepted professional conduct.

The ISO who acts in a prudent manner—uses the hazard MEDIC action model, takes steps to eliminate or communicate hazards, and works within established standards (NFPA) and laws (Occupational Safety and Health Administration's [OSHA] Code of Federal Regulations [CFR])—has taken significant steps in reducing liability. Added to this is a long-standing legal principle of *discretionary function*, which recognizes that certain activities require a value judgment among competing goals and priorities. In these cases, nonliability exists (*Nearing v. Weaver*).[3]

Another issue the ISO must be aware of in accident investigation is the involvement of outside agencies with an interest in the accident. State and/or federal OSHA and/or National Institute for Occupational Safety and Health (NIOSH) officials, labor group investigators, insurance investigators, and law enforcement officials are often involved in a significant injury or death investigation. In many cases, these agencies can help the ISO; most likely, however, an investigation that has reached this magnitude signals the end of the ISO's need to lead or even participate in the investigation (the ISO becomes a witness).

Events Requiring an Investigation

The types of mishaps that should be investigated can be influenced by federal, state, and local laws and by individual department SOPs and insurance carrier requirements. As a guideline, however, events that should be investigated from a prevention (pay-it-forward) perspective include the following:

- *Firefighter injury.* The local fire department should have a policy in place regarding the types of injuries that require the ISO to investigate. Minor injuries that don't require medical attention are often documented on simple forms, filled out by the injured person's supervisor, for archive and workers' compensation purposes (most workers' compensation programs require an investigation and documentation before claims can be processed). More serious injuries that require medical attention and have the potential for lost time are likely candidates for an ISO investigation **FIGURE 16-6**.

- *Firefighter LODD.* As stated earlier, a firefighter LODD event will require a comprehensive investigation involving liaison with several agencies. The ISO is likely to *start* the LODD investigation process (evidence protection, initial documentation) but will likely be recused. The department HSO or other uninvolved officer (preferably a chief officer) should take the investigative lead.

- *Close call.* Incidents involving near-injury or near-death circumstances can benefit from an investigation. Events such as flashover, building collapse, firefighter Mayday (no injury), and RIC activation are strong candidates for investigation.

- *Fire department equipment failure.* Stuff breaks all the time. The point here is to investigate equipment/apparatus failures that negatively altered the incident outcome. Issues that impact product liability, recall potential, maintenance, or serviceability can benefit from an investigation also.

- *Apparatus mishaps and property damage.* Apparatus involved in mishaps are typically required to be investigated by insurance carriers, and the ISO may have to take the lead. Unintentional damage to property, caused by fire department actions, may warrant an investigation. The key word here is unintentional. Firefighter actions often cause damage (cutting a hole in the roof), but that is intentional. Examples of unintentional damage can include water run-off flooding of exposures, collapse of streets/culverts due to apparatus weight, and damage to parked cars when deploying equipment.

The Investigative Process

Where does the ISO begin to investigate an injury or mishap? Of the many investigative models to choose from, the most common is a simple three-step approach.[4]

Step 1: Information Collection

Numerous sources of information should be collected following an incident. These can be divided into six categories:

1. *Incident data.* Included is factual information, such as incident number, chronological time of events, weather conditions, apparatus assigned, personnel assigned (by name and assignment), and documented benchmarks (primary search complete, incident under control).

2. *Witness statements.* Statements may be difficult to gather, and assistance may be required (law enforcement officials can usually help). An attempt is made to gain as many perspectives as possible **FIGURE 16-7**. While keeping the witness speaking in facts is important, so is gaining a sense of the witness's perspective (see the *Getting the Job Done* box). Remember that much of what a firefighter does requires rapid judgment and execution.

3. *Scene sketches/diagrams.* Accuracy is critical: Be as precise as possible. Quick hand sketches work well for apparatus, hose, and crew placement, as long as measured distances are included so that a more precise drawing can be rendered later.

4. *Photographs/video.* If you noticed video footage being taken during an incident, attempt to gain it from the videographer. Media sources may be helpful as may

FIGURE 16-6 Firefighter injuries requiring medical attention should be investigated by the ISO. A Kansas firefighter suffered a leg injury in this building fire collapse.

Courtesy of Corey Kneedler

FIGURE 16-7 The ISO should support an accident investigation with many witness reports.

posts on social networks. Follow-up video or still photography can capture the results of the mishap.

5. *Physical evidence.* Protective equipment, damaged equipment, or other physical forms of evidence should be retained **FIGURE 16-8**. Once again, law enforcement officials and fire origin/cause investigators are a good source of expertise in the collection and documentation of physical evidence. For serious firefighter injuries and/or LODD, the PPE used by the victim (including SCBA, personal alert safety system [PASS], radio, thermal imaging camera [TIC], and escape devices) should be collected and treated as evidence. EMS personnel and those treating the victim at a medical facility may need a reminder that the PPE and clothing worn by the patient must be retained as evidence and not discarded.

6. *Existing records.* Equipment maintenance records, policy and procedure manuals, training records, and other documents are useful when it comes time to analyze the factors leading to the accident. At times, the extensive search may require going back many years to make a discovery that may have set the stage for the mishap. Likewise, the research may reveal that the proper maintenance, training, and other conditions were in place.

Step 2: Analysis and Reconstruction

The ISO reads through the accumulated data and separates facts, perceptions, and unknowns and determines the need for more information. At times, irrelevant data can be discarded. Once the information is analyzed, the ISO can reconstruct the accident. Utilizing the accident chain concept as inputs, the ISO can chart the accident sequence as part of the reconstruction. The reconstruction is designed to find causal factors, which are the events or conditions in the accident sequence that contributed to the unwanted outcome. Causal factors include:

- *Direct cause.* The immediate event/condition that caused the accident.
- *Root cause.* The factor(s) that, if corrected, would prevent reoccurrence. Root causes are usually high-order, fundamental factors such as training, equipment, and procedures.

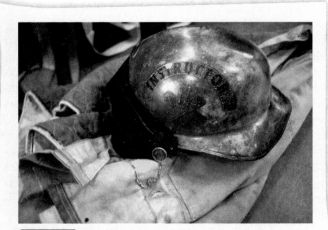

FIGURE 16-8 Protective equipment and other physical evidence must be retained, marked, tagged, and identified as evidence.

Getting the Job Done

Witness Interview Techniques

Gathering witness statements from firefighters involved in an incident at which another was injured can be challenging. Likewise, a situation that warrants an ISO investigation (equipment failure, property damage) can be perceived as a fault-finding mission by those being interviewed. In both cases, the ISO can help collect information in a meaningful way if some basic interviewing techniques are utilized.

First, it is best to interview witnesses individually and to choose a location that is relatively free of distractions. Next, the ISO should follow an interview sequence and use effective communication techniques.

Interview Sequence

1. Establish the need for the interview and attempt to minimize anxiety: Explain the investigative process, emphasize the discovery philosophy, and reinforce the point that everyone has a viewpoint and/or perception that is appreciated.
2. Collect basic information from the witnesses, including name, position/rank, supervisor, and contact numbers.
3. Ask witnesses to describe their relationship to the event: Where were they assigned? Who was with them? What were the conditions like?
4. Ask witnesses to mentally recreate the circumstances of the incident: What did they do? What did they see? What did they feel, and how do they feel now?
5. If witnesses seem to be having difficulty finding the right words, encourage them to use nonverbal communication, such as gestures, to demonstrate actions, or ask them to draw a diagram.
6. Near the end of the interview, ask witnesses, "Is there anything else I should have asked or that you'd like to add?"
7. Conclude the interview by asking witnesses to contact the investigator if they recall any additional information.

Interview Communication Techniques

- Ask open-ended questions: Lead-in words like what, how, describe, and tell will help discourage yes/no answers.
- Encourage witnesses to volunteer information without prompting, and remind them that all details—even seemingly trivial ones—can be important.
- Avoid leading questions.
- Avoid interrupting witnesses, and don't be too quick to ask the next question. A pause allows witnesses to collect their thoughts, which may result in them providing more information.

Obviously, the interviewer needs to document the witness's statements. During the interview, jotting down notes is expected as long as it doesn't overly interfere with the dialogue. Using a recording device is helpful but may present some legal issues. If it hasn't been done, it is advisable to have the witness write a narrative also. Following the interview, the investigator should complete his or her note taking.

- *Contributing causes.* The factors that collectively increased the likelihood of an accident but individually did not cause the accident.

Once the causal factors are discovered through deductive reasoning, some attention must be given to *barrier* analysis. As we know, firefighting is a risky business accomplished in very hazardous environments. We train and equip firefighters to face those risks and hazards using many administrative (SOPs and training) and physical (PPE and equipment) barriers to separate the hazards from the targets (firefighters). The performance and application of these barriers must be analyzed as part of the reconstruction effort.

Often, the reconstruction analysis points to an unintended result that occurred during human performance (an error). Human performance errors can be viewed as a mismatch between the human condition and environmental factors at a given moment or during a course of actions. When a human performance error is discovered, the analysis effort should continue to help uncover certain precursors that may have been present. The use of a *Task, Work Environment, Individual Capabilities, and Human Nature* (TWIN) model can assist as a diagnostic tool for analyzing human error precursors **TABLE 16-1** .[5]

Step 3: Recommendations

Charting the accident chain and analyzing the causal factors, barriers, and error precursors set the stage for answering two really important questions: What happened, and why did it happen? Recommendations are then developed to keep the *what* and the *why* from happening again. Most often, the

TABLE 16-1	Error Precursors
Task Demands	**Individual Capabilities**
Time pressure (in a hurry)	Unfamiliarity with task / First time
High workload (large memory)	Lack of knowledge (faulty mental model)
Simultaneous, multiple actions	New techniques not used before
Repetitive actions / Monotony	Imprecise communication habits
Irreversible actions	Lack of proficiency / Inexperience
Interpretation requirements	Indistinct problem-solving skills
Unclear goals, roles, or responsibilities	Unsafe attitudes
Lack of or unclear standards	Illness or fatigue; general poor health or injury
Work Environment	**Human Nature**
Distractions / Interruptions	Stress
Changes / Departure from routine	Habit patterns
Confusing displays or controls	Assumptions
Work-arounds	Complacency / Overconfidence
Hidden system / Equipment response	Mind-set (intentions)
Unexpected equipment conditions	Inaccurate risk perception
Lack of alternative indication	Mental shortcuts or biases
Personality conflict	Limited short-term memory

Courtesy of US Department of Energy

recommendations fall into the areas of equipment, policy and procedure, or personnel (training, attitude, fitness). There may be a tendency to focus on one solution. Be inventive and force yourself to develop more than one solution. After multiple solutions are developed, evaluate each and focus on the approach that you believe would best prevent a reoccurrence.

Nowhere in the accident investigative process are the words "blame" or "discipline" used. This is important. Placing blame or recommending discipline has a tendency to close minds and erect acceptance barriers. If the ISO is to remain effective, it is best to state recommendations in the form of future accident prevention. Upon discovering a case of complete disregard for safe practice (a form of negligence), the ISO should meet with a supervisor or chief officer and allow the department to handle the issue administratively. Although there may be some backlash and tension regarding the ISO's investigative effort, most safety-conscious firefighters and officers will applaud the actions of the department and the ISO.

Accident investigation is not a fun task, but it is vital to the reduction of future injuries. If the ISO can demonstrate good intent, the investigation serves as an investment in making a difference and paying it forward.

Stressful Events and Trauma Support

Stress happens—to everyone. For our purposes, stress can be defined as the physical, mental, or emotional tension and strain resulting from a situation in which a person feels pressured or threatened. The year 2019 started off with the job-posting website CareerCast listing firefighting as the second most stressful job in America. The scoring system measures 11 stress factors, including:

- Physical demands
- Environmental conditions
- Hazards encountered
- Risk to own life
- Risk to others
- Time constraints

Although some may view the ranking as dubious, it is probably safe to say that most working adults view firefighting as a relatively stressful endeavor. Those same adults also expect firefighters to tolerate the job stress—and for the most part, firefighters do.

Accepting the inherent stresses of the job is one thing, managing that stress is another. To manage stress, people find relief in recreation, religion, socializing, family time, and physical fitness regimens, to name but a few outlets. Still others resort to undesirable coping mechanisms such as drug and alcohol abuse. Some say they "have had enough" and choose to leave the profession. At its worst, the short- and long-term effects of stress can lead to behavioral issues, including acute stress disorder (leading to post-traumatic stress disorder) and suicide. The rate of suicides among emergency responders is alarming: The Firefighter Behavioral Health Alliance reports that there were 103 validated responder suicides in 2017.[6]

The point here is to recognize that the job is stressful and that people react differently to those stresses. Even though we still have much to learn about the behavioral sciences, we do know that stress-related issues can evolve from an unusually traumatic incident (atypical event) and/or repeated exposure (burnout) from more common incidents. Fire departments are encouraged to address both by accessing behavioral health programs and professionals (some available at no cost through county health services). With that said, the ISO does have a role in the recognition of occupational stress. Namely, the ISO must be able to identify the types of incidents that might be atypical, describe some signs and symptoms of occupational stress, and initiate some first aid measures for addressing potential occupational stress issues.

■ Atypical Stressful Events

An atypical stressful event (ASE) is an incident that presents mental or emotional pressure or strain circumstances that are outside the ordinary experience of the responders. Some refer to an ASE as a potentially traumatic event (PTE). Examples of an ASE/PTE might include:

- Mass-casualty incident
- Firefighter LODD
- Witnessing of a coworker's suicide
- An unusually difficult or long-duration rescue or recovery effort with gruesome circumstances
- A death caused by the responders

Some may regard *ASE* and *PTE* as fancy terms to describe what used to be called a critical incident that triggered the use of critical incident stress management (CISM) teams and debriefings (CISDs). Although CISM teams and CISDs have been used for decades, research has shown that they don't necessarily prevent behavioral issues and, in some cases, inhibit natural recovery. Using *ASE* or *PTE* recognizes that responders have different resiliency levels toward stress and marks a new approach and model to help with firefighter behavioral health. We will use *ASE* in our discussion.

■ Signs and Symptoms of Incident-Related Stress

The way responders react to an ASE is diverse and subjective. Reactions to trauma can range from a simple pause to a complete behavioral shutdown. Regardless, the ISO is perhaps the first person to see that an ASE is challenging the emotions of one or more of the responders. Listing signs and symptoms of incident-related stress can be tricky in that some of them are shared with the signs and symptoms that accompany overexertion, postincident thought patterns, and chemical imbalance, including:

- Far-away stares
- Task forgetfulness
- Lethargy
- Reserved verbal responses to routine conversations

On-scene signs and symptoms that are more aligned to an ASE (as opposed to other causes) can include:

- Outward crying
- Emotional outbursts of anger and intolerance toward coworkers
- Total shutdown of interaction with others

The signs and symptoms included in these lists may be found at the incident and warrant attention by the ISO. Some responders may not show any indication of incident-related stress but are nonetheless impacted. As days and weeks pass, the internalized stress from the incident may become apparent.

Acute stress disorder (ASD) refers to behavioral issues that arise in the days following an ASE. Flashbacks, bad dreams, depression, and worry are common symptoms of ASD. Some recover quickly from ASD with little or no intervention. If symptoms persist for longer than a week or two, or go untreated, the ASD may proceed to post-traumatic stress disorder (PTSD). PTSD is a mental health disorder that can develop in individuals who have experienced a terrifying ordeal that involved physical harm or the threat of harm. PTSD can cause individuals to experience intense, potentially debilitating, emotional distress that can keep an otherwise healthy person from leading a normal life. According to the National Institute of Mental Health, signs and symptoms of PTSD include thought perseveration (reexperiencing of thoughts), avoidance behaviors, and hyperarousal issues.

■ Initiating Stress Exposure Support

As mentioned, everyone reacts to incident-related stress differently, and the methods used to address the stress must also be varied. One size does not fit all. Establishing a department protocol for addressing ASE issues is encouraged and should guide the ISO's actions. A sample protocol is available for fire departments through the National Fallen Firefighters Foundation's Everyone Goes Home website. Relevant details are filed under Firefighter Life Safety Initiative 13, "Psychological Support."

The new model for addressing ASE exposure is tiered and relies on a team approach that includes access to trained behavioral health professionals. The ISO's role in supporting an ASE exposure program has to do with initiating stress exposure protocols, facilitating a "time-out" defusing, and encouraging responders to use stress first aid measures.

1. *Initiating stress exposure protocols.* At an incident, the ISO considers the incident circumstances and responder signs and symptoms to make a judgment regarding the need to initiate the exposure protocol. When in doubt, lean toward initiation. Confer with the IC to begin the process.
2. *Facilitate a time-out defusing.* The time-out concept is borrowed from the military as part of the "hot-wash" after-action review model. Basically, the responders take a time-out to gather and briefly review what actually happened, what went well, and what could be improved. Participants are encouraged to share what they felt and what they are currently feeling. This process is similar to the informal PIA but also allows an opportunity to put the event into perspective and to relieve anxiety and uncertainty from a peer-to-peer perspective. The intent is to provide a safe opportunity for responders to discuss any emotional impacts of the event. The savvy ISO uses the time-out defusing not only to gauge the signs and symptoms of stress but to start formulating input for any upcoming PIA (or accident investigation).
3. *First aid for stress.* Once the ASE is over and crews are released, a method of ongoing awareness and action is needed. In the *peer-to-peer model* of stress first aid, peers are encouraged to initiate further interventions when they observe a change in functioning, hear statements of internal stress, and perceive a need to help

with an individual's sense of confidence or competence following an ASE. Stress first aid can include many measures, including further peer defusing, debriefing, and access to organizational support such as a trauma screening questionnaire that may indicate the need for further assessment and treatment.

The terms defusing and <u>debriefing</u> are used often when discussing incident stress programs. The two are different, and the ISO should be aware of those differences. Defusing (as mentioned in the time-out defusing above) refers to a gathering during or immediately after the incident and is characterized by an informal, peer-to-peer discussion format designed to share observations, actions, and feelings. For some, the defusing is all that is needed and they are mentally prepared to move on. A debriefing is typically a scheduled event and is characterized by a more formal agenda designed to promote healing or closure and to outline the process or options for accessing more assistance.

To this point, we have looked at the ISO's approach to dealing with potential stress-related issues associated with incident handling. Although it is past the scope of this chapter, some mention needs to be made of the fire department's organizational approach to member assistance and the resources available for those in need. Ideally, the fire department has created a behavioral health assistance program, as required by NFPA 1500 (2018 edition, Chapter 12). The program should include the capability to provide clinical assessments, basic counseling, and stress-crisis intervention as well as processes to address drug and alcohol abuse, depression, and other personal problems that can adversely affect fire department work performance.

Where no program is defined, individual responders (and ISOs) should be aware of the many resources available to assist with issues associated with occupational stress. Many of these resources are no-cost. They include, but are not limited to:

- County health services
- Department chaplain or other clergy
- The National Institute of Mental Health (http://www.nimh.nih.gov/index.shtml)
- The Firefighter Behavioral Health Alliance (http://www.ffbha.org/)
- PTSD Support Services (http://www.ptsdsupport.net/)

Wrap-Up

Chief Concepts

- The reconnaissance effort of ISOs places them in a desirable position to help capture and document incident activities for a postincident analysis (PIA). The same information can be used for an accident investigation should one be required.
- Postincident thought patterns are the reflective or introspective mental preoccupations that firefighters experience just after incident control. These thought patterns can lead to inattentiveness and, consequently, injuries. Chemical imbalance can also impact postincident "pick-up" activities and lead to accidents. For these reasons, an ISO should not be released prematurely from an incident; the ISO should remain and help monitor actions and remind responders to be safe.
- A PIA is a formal and/or informal reflective discussion that fire departments use to summarize the successes and areas requiring improvement discovered from a given incident. NFPA 1500 requires that the ISO participate in, and contribute to, the PIA.

- The overriding goal of a PIA is to prevent future injuries or mishaps. The ISO facilitates this outcome by adopting a positive, discovery-oriented philosophy as crews discuss what happened, what went well, and what could be improved.
- PIAs can be informal, on scene; informal debriefing style; or formal (written). Formal PIAs should be used for incidents involving an LODD, significant apparatus or equipment failure, a Mayday call or RIC activation, large life or property loss, large-scale resources, and specialty teams (technician level).
- The ISO's involvement with PIAs starts with information collection on scene by visiting with crews and asking simple questions regarding the incident. That is followed up with documentation, which includes observation notes, scene sketches, and a timeline. The ISO is also encouraged to spot trends so that information can be shared with training officers and supervisors in the spirit of improving or correcting developing issues.
- The ISO should get in the habit of documenting his or her actions and observations of conditions for all incidents, not just those involving a PIA. The documentation can help archive the incident for issues that might arise later.
- During the actual PIA, the ISO's primary function is to listen to others and encourage their reflection. The ISO

should, however, be prepared to comment on issues pertaining to risk-taking, crew accountability, rehab effectiveness, PPE use, and any close calls. For incidents at which a responder injury occurred, the ISO should be cautious and limit input to efforts to help the injured.

- NFPA 1561 requires the ISO to investigate accidents that have occurred within the incident area. The purpose of the investigation is to prevent future occurrences and is accomplished through a discovery and evaluation of the accident chain. The accident chain has five components: the environment, human factors, equipment, an event or series of events, and the actual injury or mishap.

- There are times when it is not advisable for the ISO to conduct the accident investigation because a conflict of interest might exist. Those incidents include:
  - The ISO is the one injured.
  - The ISO was directly involved in the property damage.
  - An LODD occurred.
  - Multiple or significant injuries to a firefighter occurred.
  In these cases, ISOs should recuse themselves from the investigation and pass the investigation on to the department HSO or other chief officer who was uninvolved with the incident.

- Mishaps that should be investigated include firefighter injuries, LODDs, close calls, significant equipment failure, and apparatus mishaps or unintentional property damage caused by the fire department.

- There are many processes that can be used to begin an investigation. A simple three-step process is most common:
  1. Collection of information
  2. Analysis and reconstruction
  3. Recommendations

- The information that needs be collected includes incident data, witness statements, sketches/diagrams, photographs/video, physical evidence (especially PPE/SCBA), and existing departmental records (like training documentation and SOPs).

- Witness statements should be obtained using a recognized sequence and effective communication techniques. Law enforcement and fire origin/cause investigators can help obtain witness statements.

- The analysis and reconstruction step attempts to chart the accident chain and discover causal factors such as the direct cause, root cause, and contributing causes. Once discovered, the ISO needs to also do a *barrier analysis*, which is an examination of the performance or application of administrative controls (SOPs/training) and physical controls (PPE/equipment) that worked or didn't work. Human performance errors that have been discovered should be further analyzed to help uncover error precursors. Using the *Task, Work Environment, Individual Capabilities, and Human Nature* (TWIN) matrix can help identify those precursors.

- The recommendation step should include several solutions designed to prevent an accident reoccurrence. Blame and discipline discussions should be omitted

from recommendations. In cases where a complete disregard for safety has been uncovered, the ISO should yield to the department chief or other supervisory officer to handle disciplinary actions.

- The firefighting job is plenty stressful, and people react differently to those stresses. Even though we still have much to learn about the behavioral sciences, we do know that stress-related issues can evolve from an unusually traumatic incident (atypical event) and/or repeated exposure (burnout) from more common incidents. An atypical stressful event (ASE), sometimes referred to as a potentially traumatic event (PTE), is an incident that presents mental or emotional pressure or strain circumstances that are outside the ordinary experience of the responders. Examples might include mass-casualty incidents, LODDs, witnessing of a coworker's suicide, especially gruesome and prolonged rescue/recovery events, and deaths caused by the actions of responders.

- Signs and symptoms of incident stress can include task forgetfulness, reserved demeanor, and lethargy. More pronounced signs can include outward crying, emotional outbursts, and total shutdown of interactions with others.

- The ISO's role in supporting an ASE exposure program has to do with initiating stress exposure protocols, facilitating a "time-out" defusing, and encouraging responders to use stress first aid measures.

- The terms defusing and debriefing are used often when discussing incident stress programs. Defusing refers to a gathering during or immediately after the incident and is characterized by an informal, peer-to-peer discussion format designed to share observations, actions, and feelings. A debriefing is typically a scheduled event and is characterized by a more formal agenda designed to promote healing or closure and to outline the process or options for accessing more assistance.

- NFPA 1500 requires fire departments to have a behavioral health assistance program in place to address issues of clinical assessments, basic counseling, and stress-crisis intervention. Individuals are encouraged to access these programs but may also find assistance from county health services, clergy, and several networks, including the Firefighter Behavioral Health Alliance.

Key Terms

accident chain A series of conditions and events that lead to an unsafe situation that results in injury and/or property damage. It typically has five components: environment, human factors, equipment, an event, and the injury.

acute stress disorder (ASD) The behavioral issues that arise in the days following an atypical stress event (ASE).

atypical stressful event (ASE) An incident that presents mental or emotional pressure or strain circumstances that are outside the ordinary experiences of the responders.

close call An unintentional, unsafe occurrence that could have produced an injury, fatality, or property damage; only a fortunate break in the chain of events prevented the undesirable outcome.

debriefing A scheduled event that is characterized by a formal agenda designed to promote healing or closure and to outline the process or options for accessing more assistance.

defusing A gathering during or immediately after the incident that is characterized by an informal, peer-to-peer discussion format designed to share observations, actions, and feelings.

due diligence A legal phrase for the effort to act in a reasonable or prudent way, given the circumstances, with due regard to laws, standards, and accepted professional conduct.

postincident analysis (PIA) A formal and/or informal reflective discussion that fire departments use to summarize the successes and areas requiring improvement discovered from an incident.

postincident thought patterns Reflective or introspective mental preoccupations that firefighters experience just after incident control.

post-traumatic stress disorder (PTSD) A mental health disorder that can develop in individuals who have experienced a terrifying ordeal that involved physical harm or the threat of harm.

Review Questions

1. What is meant by postincident thought patterns?

2. Compare and contrast the role of the ISO in informal and formal PIA.

3. Explain the role of the ISO in accident investigation according to NFPA standards.

4. What types of mishaps should be investigated by the ISO, and when should the ISO recuse himself or herself from an investigation?

5. What are the five parts of the accident chain?

6. What are the three steps of accident investigation?

7. What are the six pieces of information that should be collected for an accident investigation?

8. What basic information should be collected from a witness during an investigation?

9. How should a complete disregard for safety be handled during the recommendation phase of an accident investigation?

10. What is meant by an atypical stressful event? What types of incidents could it refer to?

11. List several signs and symptoms that might suggest a responder is experiencing occupational stress.

12. What is a "time-out" defusing, and how is it different than a debriefing?

13. List several resources that can be used to assist those who are exposed to occupational stress.

References and Additional Resources

The American Institute of Stress website offers many stress-related features, resources, and links. It is available at http://www.stress.org/.

U.S. Department of Energy. *Accident and Operational Safety Analysis, Volume I: Accident Analysis Techniques.* Handbook 1208. Washington, DC: U.S. Department of Energy, 2012.

The Fire Near Miss website hosted by the International Association of Fire Chiefs describes near-miss reports and reporting procedures. It is available at http://www.everyonegoeshome.com/.

The Firefighter Close Calls website hosted by FireCompanies.com offers fire ground stories, near-miss reports, and lessons learned at structure fires. It is available at http://www.firefighterclosecalls.com/.

International Association of Fire Fighters. *Firefighter Line-of-Duty Death or Injury Investigation Manual.* Revised as a Cooperative Effort Between the IAFF and IAFC. Washington, DC: IAFF, 2010.

The National Fallen Firefighters Foundation's Everyone Goes Home website provides links to information relating to LODD procedures, support, and sample protocols. It is available at http://www.everyonegoeshome.com/.

National Institute of Mental Health. Post-Traumatic Stress Disorder (PTSD). Washington, DC: National Institutes of Health. http://www.nimh.nih.gov/health/topics/post-traumatic-stress-disorder-ptsd/index.shtml. Accessed June 1, 2015.

NFPA 1500, *Standard on Fire Department Occupational Safety and Health Program.* Quincy, MA: National Fire Protection Association, 2018.

Endnotes

1   Jones, Dennis G. Accident Investigation Analysis. *Health & Safety for Fire and Emergency Service Personnel*, 8, no. 9, September 1997.

2   NFPA 1561, *Standard on Emergency Service Incident Management System and Command Safety.* 5.9.7.3(5). Quincy, MA: National Fire Protection Association, 2020.

3   *Nearing v. Weaver*, 295 Ore. 702, 670 P.2d 137, 143 (Ore 1983). Citation from Callahan, Timothy, and Charles W. Bahme. *Fire Service and the Law.* Quincy, MA: National Fire Protection Association, 1987.

4   Hopkins, Ronald H. Accident Investigation. *Health & Safety for Fire and Emergency Service Personnel*, 5, no. 9, September 1994.

5   U.S. Department of Energy. *Accident and Operational Safety Analysis, Volume I: Accident Analysis Techniques.* Handbook 1208. Washington, DC: U.S. Department of Energy, 2012.

6   Firefighter Behavioral Health Alliance. http:www.ffbha.org. Accessed September 3, 2019.

Firefighters Thompson and Leon were assigned to secure utilities at a single-family residential fire. After shutting off the natural gas feed, they proceeded to the electrical service. The breaker box was locked; after some discussion, they decided they could access the meter and shut off power that way. While pulling the meter, Firefighter Thompson felt a slight sting, accompanied by a brief electrical arc flash. Firefighter Thompson shook it off and finished his assignment.

As was customary for their fire department, all the firefighters involved in the house fire met for a brief informal postincident analysis. As each firefighter summarized their actions, it became apparent that Firefighters Thompson and Leon experienced a close call. The incident commander, Battalion Chief Sears, commented that pulling an electrical meter is quite dangerous and is not usually practiced in this fire department. The ensuing discussion revealed that the regional recruit academy had taught firefighters how to pull a meter and that individual local fire department policy should dictate whether this practice is allowed. Unfortunately, no written policy existed for Battalion Chief Sears' fire department. The assigned incident safety officer, Captain Drowns, was busy taking notes during the discussion and decided to reserve judgment.

1. What type of follow-up do you think the safety officer should pursue in this case?

2. Do you think a formal investigation is warranted for this close call? Why or why not?

3. During a postincident analysis, the assigned ISO should be prepared to comment on which of the following?
   A. Risk-taking issues
   B. Human error precursors
   C. Firefighter injury details
   D. Close-call investigative needs

4. Identifying causal factors (direct, root, and contributing causes) is an important part of which step in the accident investigation process?
   A. Information collection
   B. Witness interviews
   C. Analysis and reconstruction
   D. Recommendations

Knowledge Objectives

Upon completion of this chapter, you should be able to:

- List four safety-minded values that should be present for training drills. (p 248)
- List four nonincident events that can benefit from the assignment of a separate, dedicated safety officer. (p 248)
- Describe the preevent planning issues that a safety officer should cover with an instructor-in-charge. (p 249)
- Define *shadowing* and describe how it is part of a three-step approach to new ISO training. (pp 249–250)
- List the three training/exercise steps that should take place prior to a full-scale drill for non-fire-agency participants. (p 250)
- List at least seven items that should be included in a safety officer–prepared, predrill written safety briefing. (p 250)
- Describe the pros and cons of a no-notice drill, and list several guidelines that can be established to make a no-notice drill safer. (pp 253–254)
- Identify the steps that can be taken to reduce liabilities associated with live-fire firefighter training events. (pp 254–255)
- Define the instructor ratios and assignments that are mandated for live-fire training events. (pp 255, 259)
- Describe the makeup of a live-fire training safety team. (pp 259–260)

Skills Objectives

There are no ISO skills objectives for this chapter.

A large-scale house burn is being planned by the training division. Many mutual-aid departments have been invited, and planning is almost complete for the session. The training officer asks you to visit the site with him to review the water flow calculations and apparatus positioning planned for the training. The house is in an isolated area without any exposures or other hazardous conditions in the area. You walk up the narrow driveway and see that the house is in disrepair and appears to be in a fairly weakened state. The training officer is planning two interior evolutions on each floor of the building.

1. What NFPA standards exist to guide decision making and recommended practices for live-fire training? What is the most current edition of each of these standards?

2. How should the training officer proceed in the planning of this evolution? Would your department proceed with the training burn under these conditions? Can a reduced-scale evolution still have training benefit?

Introduction: Practice Makes Perfect?

The popular expression "practice makes perfect" is perhaps overused and may even set the stage for failure. The legendary professional football coach Vince Lombardi understood this when he said:

Practice doesn't make perfect. Only perfect practice makes perfect.[1]

And while we are mentioning famous quotations, there is another one applicable to this chapter:

We are what we repeatedly do.

This from the Greek philosopher, understudy of Plato, and advisor to Alexander the Great: Aristotle.[2] Fire department training drills (and other nonemergency planned events) are opportunities to set a positive, safety-minded approach that we use for actual incident response. The planning nature of each type of event creates an arena for a perfect "practice incident." The opportunity can extend to participating firefighters, officers, and safety officers. In fact, training drills and planned events can provide newly assigned or potential ISOs an opportunity to gain experience under the tutelage of an experienced one.

Some may argue that training drills should not bear the pressure of achieving perfection; they are for discovering and improving weaknesses in a safe environment. That point is well taken and brings up two more points.

First, the type of training practice we're talking about here has to do with *in-context* drills or evolutions as opposed to classroom lessons and skills introduction. Lombardi would introduce a new play on a chalkboard, let the players walk through it a few times, then have the players practice it in real time. The practice would continue until the play was repeated perfectly (practice 'til you can't get it wrong). Every drill opportunity is a practice incident, and we should do it until we can't get it wrong.

Second, the "safe environment" for discovery and improving weaknesses continues to be misapplied, as evidenced by the history of firefighter line-of-duty deaths (LODDs) associated with training activities. Research reveals an average of 8 training-related LODDs per year in the past 10 years.[3] The environment may not be as safe as we believe it to be.

This chapter is new to this text and is presented in response to the many comments and suggestions shared with the author since the book was first published. In it we will discuss the responsibilities of the incident safety officer (ISO) at fire department training activities, such as drills and live-fire events. Some of the information in this chapter is applicable to smaller nonemergency events like a planned local festival or small crowd event. Events that have the potential for large crowds, limited emergency access, or civil disturbances are beyond the scope of this chapter and typically involve a significant planning effort by fire department chief policy makers and emergency managers.

The Need for a Safety Officer at a Training Drill

There seems to be a resurgence of priority given to ongoing training as a primary responsibility of firefighters. Those of us who value training (and education) are jubilant! After all, "We are what we repeatedly do." Unfortunately, the increase in training drills also brings the potential for more injuries (and deaths) as well-meaning training officers and fire officers attempt to make that training more realistic. The potential has become reality in too many cases, making the use of an ISO at reality-based training drills seem prudent.

While training should be an organized, preplanned event designed to teach and reinforce *safe* accomplishment of skills, realism in training does make it more practical. Thus, some believe that the training and reinforcement of skills must be accomplished in an environment similar to that which firefighters will likely encounter (heat, smoke, changing conditions, high stress). Therein lies the foundation for injuries and fatalities. Given this inherent danger in training, it falls to the assigned training officer (TO), instructor-in-charge (IIC), and/or ISO at training events to reinforce the concept of *risk-taking*. In real operations, there are times when firefighters must take great risks, such as when making rescues. In the context of a planned training drill, there are no lives to save; therefore, risk-taking should be minimized. The TO, IIC, and ISO (or any officer, for that matter) need to approach training with a safety-minded value system that includes some key overhead commitments:

- The safety of participants trumps performance expectations.
- Train in accordance with established standard operating procedures/guidelines (SOPs/SOGs), national standards, and best practices.
- Train only in environments that can be quickly controlled through preplanned and communicated protective measures.
- All participants are empowered to stop unsafe acts and alter any activity (for themselves or others) that presents an imminent threat of injury.

The responsibility for planning and safely conducting training activities lies with the IIC. Likewise, the company officer is the default IIC when a crew or company decides to conduct a spontaneous skills practice. In either case, the IIC is the *safety officer* for the training activity unless he or she delegates the safety function to a separate individual **FIGURE 17-1**. Certain training activities can benefit from the assignment of a separate individual, independent of the IIC, to serve as a safety officer. (For training drills, we will now call the ISO an SO.) The activities that can benefit from a separate SO should include:

- Live-fire training
- Full-scale drills involving multiple companies
- Multiagency drills
- Non-fire training evolutions involving some degree of inherent risk (rope rescue, dive/water rescue, collapse tunneling/shoring)
- Events that include multiple hands-on skill stations such as a hosted hands-on training seminar
- Nonemergency planned event like a small community festival.

In the case of live-fire training (e.g., dedicated burn building, exterior prop, acquired structure), the assignment of a separate, dedicated SO is mandatory according to NFPA 1403, *Standard on Live Fire Training Evolutions*. Similarly, some state and local laws require that a dedicated SO (separate from the IIC) be assigned for firefighter live-fire training drills **FIGURE 17-2**.

FIGURE 17-1 The assigned instructor at a training drill is the default SO for the activity unless the role has been delegated.

FIGURE 17-2 NFPA 1403 requires a dedicated SO, separate from the IIC, for firefighter live-fire training drills.
Courtesy of Clallam County Fire District 2

Let's look at the general responsibilities of the SO assigned to any type of training activity, and then we'll cover drills and live-fire training more specifically.

General Safety Officer Functions at Training Events

The functions of the SO at a training event can be similar to those at an actual incident, with a few notable adjustments. Using the *hazard MEDIC* mnemonic to remind the SO of his or her functions is, for the most part, applicable:

M = Monitor
E = Evaluate
D = Develop preventive measures
I = Intervene
C = Communicate

However, applying the MEDIC functions at a training event differs from actual incidents in several ways. Because training events typically are planned, the evaluation (*E*) of environmental conditions and the development of preventive measures (*D*) can be addressed prior to the actual training. For this reason, the SO should meet with the assigned IIC and do some pretraining planning. Likewise, the assigned SO has an opportunity to develop a safety briefing that should be shared with those participating. Ideally, the IIC and the SO should co-deliver the safety briefing to show unity and emphasize safety. They should agree on content, divide it up, and deliver it to the participants in a unified manner.

■ Safety Officer Pretraining Planning

The SO assigned to a training activity should invest in some preplanning to help ensure a safe, accident-free event. Acquired structure live-fire, full-scale, and multiagency drills will typically require planning meetings and several site visits in preparation for the event (covered later). For most other training activities, the SO can meet with the IIC a few hours prior to the training. At a minimum, the SO and IIC should discuss and agree on the following:

- Training description
- Objectives
- Minimum personal protective equipment (PPE) required by the participants *and* instructors/assistants
- Rehabilitation needs, including EMS standby
- Rapid intervention crew (RIC) requirements
- Conditions that would terminate the training (e.g., weather, interruption by an actual incident)
- Communication methods and the signal used by anyone to stop the training

The SO should also tour the drill site and evaluate the hazards that are present. The site evaluation should be relatively simple for known department training facilities, although the evaluation should still be accomplished to check for serviceability, maintenance issues, varmint infestation, or other hazards. Drills conducted in places other than a known fire department facility will require more time to survey, evaluate, and prepare. Items to evaluate and address include:

- Permission/notification of person(s) responsible for the site
- Hazardous energy
- Lighting issues and trip/fall/sharp-edge hazards
- Traffic and parking issues
- Potential to draw onlookers
- Potential to do unintended damage (liabilities)
- Off-limits areas
- Water resources
- Asbestos/lead presence
- Number and location of fires
- Fire load
- Hazards such as holes in the floor, walls, or roof

Many issues can be addressed by simply flagging or zoning the hazard using marking tape. Red-and-white diagonally striped tape can be used to mark areas that are off limits to firefighters participating in the drill or to flag for hazardous energy. Solid yellow tape should be used to warn potential onlookers of hazards and to prevent their entry into dangerous areas. "Training in progress" or "exercise ahead" signs are a wise investment. They can advise potential onlookers of the activity and help reduce the number of inquiry calls to the dispatch center or department administration **FIGURE 17-3**.

Information and actions derived from the instructor meeting and site visit are then used to develop a safety briefing, which is delivered to the drill participants by the IIC and SO. As mentioned previously, the risk-taking level at a training incident should be minimized. The SO is wise to use the initial safety briefing to inform the participants that risk-taking should be reduced and that interventions will take place as warranted. Likewise, the SO should inform participants that there will be more emphasis placed on rehabilitation efforts, as strains/sprains and overexertion are the leading types of injuries at training events. The

FIGURE 17-3 "Training in progress" signs can help calm the public and reduce the number of inquiry calls to dispatch.

balance of the SO MEDIC functions can proceed as if the training activity were an incident.

Training New Safety Officers

Departmental drills and evolutions provide a perfect environment to help prepare "rookie" ISOs for their functions. The methods of shadowing and coaching are particularly useful. Training and preparing new ISOs involve a three-step approach:

1. *Step 1: Introductory training.* New or prospective ISOs should be trained in a classroom environment that includes information front-loading and learning assessments (group projects and scenarios).
2. *Step 2: Shadowing.* Shadowing is a training approach in which a trainee closely observes experienced mentors as they perform their assigned duties. By allowing the ISO trainee to observe and interact with a mentor, the process allows the trainee to vicariously (and thus safely) experience live, evolving situations. It is important that the mentor focus more on the event taking place than on training the new ISO. Mentors are wise to not only set a good example (discipline, responsibility) but also verbalize some of their thoughts as they address the issues that they encounter. Shadowing is an often-missed step when training new people for a role.

3. *Step 3: Coaching.* Coaching and shadowing are different. Coaching places the trainee in the hot seat to perform ISO functions. The mentor becomes a watchful supervisor who encourages the trainee and advises him or her when a hazard issue goes unnoticed.

Ideally, the new ISO should experience shadowing and coaching at training drills and planned events first, then move to shadowing and coaching at actual incidents. At some point, the mentor should "sign off" that the ISO trainee can be assigned as a solo ISO at incidents and training events.

Specific Safety Officer Functions at Drills

The general functions covered so far apply to all training activities at which an SO is assigned. Certain training activities, such as multiagency drills and the so-called "no-notice" drills, present the potential for more safety issues and require the SO to do more preparation.

■ Multiagency Drills

The historical lessons of natural and human-caused disasters have compelled fire departments to work with multiple agencies in addressing the challenges of such events. The development of disaster plans (all-hazard emergency management plans), initial training, and the practice of exercises (drills) are ongoing, and now commonplace, processes in most communities. In many cases, the fire department (typically the department's emergency manager) serves as the lead coordinator for the many stakeholders involved in a disaster response/recovery. Some of the stakeholder agencies may not be accustomed to working within the framework of an incident management system (IMS) and may not fully understand the role of the SO within a supporting incident command system (ICS). Scheduling a full-scale disaster drill that involves these agencies is sure to be a disaster without careful planning, initial training, and skill-building exercises.

Predrill Preparation

The SO should be aware of the general training and stakeholder preparation steps needed to prepare for a full-scale, multiagency drill. Briefly, they include:

- *Initial indoctrination and classroom training.*
- *Table-top drill.* A representative from each agency attends and talks through a scenario.
- *Functional drill.* This no-pressure walk-through of a scenario includes the establishment of a command post, use of radio/communication links, and the response of involved agencies that stage, deploy, and simulate the accomplishment of their tasks. Most functional drills are done in street clothes or a duty uniform with no risk-taking or need for PPE.
- *Full-scale drill.* This takes the functional drill scenario to the next level and can include victim actors and actual performance of skill sets by responders using apparatus, PPE, props, and other resources to help achieve a level of realism.

As mentioned, the fire department's emergency manager will likely take the lead in facilitating the events listed above, often with the help of the department's training officer. The wise emergency manager should select an ISO-trained individual to serve as the SO for the upcoming drills. The selected SO is encouraged to be involved in every step of the preparation process. Doing so can help the SO anticipate and prepare preventive measures for hazards and safety issues that may eventually present themselves at the full-scale drill.

Once the logistical planning process begins for a full-scale drill, the SO should be in attendance at each meeting. During these meetings, the SO should be processing the information and decisions being made and mentally ask, "What could possibly go wrong if they do that?" Answering that question for various aspects of the drill becomes the substance of a drill safety plan **FIGURE 17-4**.

As the date of the full-scale drill approaches, the designated SO may have to evaluate props, deliver written predrill safety briefings, sign off on developed scenario and response plans, and recruit assistant safety officers (ASOs) to help ensure a safe and successful event. Of particular importance is the predrill safety briefing. In essence, the prebriefing, delivered before the drill date, is a tool to help the involved agencies and responders understand certain safety expectations. It has the added benefit of helping to create a positive safety attitude or culture. At a minimum, the safety prebriefing should include:

- A safety overview (expectations, risk-reduction philosophy)
- Chain of command and contact numbers
- A traffic plan (parking, staging, and apparatus movement areas)
- The level of PPE required for various functions
- Procedures for a true emergency or injury
- Expected hazards and the marking scheme used for zoning or flagging
- Responsibilities for each participant (e.g., be rested, hydrated, nourished)
- The rehabilitation resources that will be available
- The identification method used for SOs and drill monitors (vest or armband color) so that rapid recognition is possible in case assistance is needed

Full-scale, multiagency drills will likely exceed the abilities of a single SO to address the hazard MEDIC functions. Assembling and preparing a "safety team" is prudent. The safety team can include ASOs (trained and competent ISOs) as well as drill monitors (non-fire-trained assistants). The role of the drill monitors is to provide an easy-to-spot assistant who can answer questions or provide immediate communication to the drill coordinators and SO or ASOs. The monitors can also serve to help document observations. Clearly, the primary SO needs to brief monitors on reporting procedures and provide them with identification and communication tools.

Drill-Day Issues

Good planning, training, and predrill safety briefings should combine to make a safe and successful drill. Still, issues can

Training Safety Plan

Drill Date:		Time:		Shift:	

Drill Location:

Type of Training: *(Check all that apply)*

	Fire Suppression		EMS		Technical Rescue
	Live Fire Training		Vehicle/Machinery Extrication		Hazardous Materials/WMD
	Driver Training		Other Acquired Structure Training		Water/Dive Rescue
	Apparatus Operation		Preplan Survey or Simulation		Physical Fitness Activity
	Other:				

Drill Risk Assessment: ☐ High   ☐ Medium   ☐ Low

Maximum Student/Instructor Ratio:		to		Safety Officer Needed: (Required on High Risk Drills)	

Instructor PPE Requirements:

☐ SCBA	☐ Full PPE	☐ Helmet	☐ Eye Protection	☐ Filter Mask	☐ Hearing
☐ Gloves	☐ Radio	☐ Lights	☐ High Visibility Vest		

Drill Objective(s): *(Brief explanation of objectives)*

Description of Training: *(e.g., extricate victim from vehicle, victim search in limited visibility)*

PPE/Equipment Required for Each Participant: *(Check all that apply)*

	Helmet		Personal Floatation Device
	Eye Protection		Buoyancy Compensator
	Hearing Protection		Mask/Snorkel/Fins
	Gloves (*Type*):		SCBA
	Bunker Coat		SCUBA
	Hood		Other Respiratory Protection (*Type*):
	Bunker Pants		Hazardous Materials CPC (*Type*):
	Safety Boots		Radio
	Other (*Specify*):		

Department Related SOPs or Technical References: *(List number and name)*

FIGURE 17-4 A sample drill safety plan.   (continues)

Courtesy of Forest Reeder

Hazards and Control Measures: *(Check all hazards AND write in control measure)*

Atmospheric (e.g., smoke, dust, low oxygen):	Sewage/Septic:
Combustible/Flammable Environment:	Sharp Edges/Objects:
Confined Space:	Structural:
Electrical:	Terrain:
Elevation:	Traffic:
Hazardous Substances (e.g., asbestos, chemicals):	Water:
Nighttime Conditions:	Weather:
Other:	

Accountability: *(Check all that apply)*

Buddy System
Visual
Passport
Dive Master Control Sheet
Other:

Communications:

Radio—Primary Frequency:
Radio—Secondary Frequency:
Hand Signals
Rope Line
Lights
Other:

In Case of Emergency: *(Check all that apply)*

Code or Signal Used:
RIT Assigned:
ALS Standby:

Resources Assigned: *(Check all that apply AND fill in designated unit)*

Battalion Chief(s):
Rehab Officer/Area:
Rescue Unit(s):
Safety Officer:
Specialty Unit(s):
Suppression Unit(s):
Other Resources/Equipment:

Rehabilitation Plan: *(Describe rehabilitation plan and guidelines)*

Job Safety Analysis

1. Identify level of required PPE for each participant.

2. List basic steps required to safely complete evolution.

3. Identify potential accidents or hazards that may occur.

4. Determine recommended safe procedures.

Safety Planning Notes: *(e.g., site plan, drawings)*

Lead Instructor:	
Signature:	Date:
Reviewed by *(print)*:	
Signature:	Date:

FIGURE 17-4 A sample drill safety plan.

Courtesy of Forest Reeder

crop up and the ever-present Murphy's law can lead to unforeseen problems.

The SO will likely have to attend a drill-day meeting to get final instructions and to deliver a final safety message for the overhead team. He or she will need to plan time to have a separate final meeting with the safety team so that responsibilities, communications, and last-minute inspections of the drill site can be assigned and confirmed.

The SO may choose to keep a presence at the command post during the drill, depending on the number of ASOs assigned to fill the monitoring, evaluation, and intervention functions. History tells us that no drill is perfect and issues will arise. The SO and ASOs should not hesitate to make interventions! Remember, it is a training event, and injury prevention takes precedence over task completion or performance. Don't be afraid to terminate the drill (or a portion of the drill) if an unacceptable hazard or responder injury is presented. Common drill issues that require SO/ASO focus and interventions include:

- Weather-related problems
- Traffic congestion and apparatus movement
- Failure to rehabilitate (overexertion and strain/sprain prevention)
- Overzealous actors and/or responders
- Control zone discipline problems

As with incidents, the SO should address postincident responsibilities (covered in the *Postincident Responsibilities and Mishap Investigations* chapter) as the drill terminates. That includes circulating among crews and inquiring about their well-being. Likewise, the SO should have a postdrill huddle with ASOs and any monitors to collect their thoughts, being sure to take notes. Lastly, the SO must document actions and observations to help prepare for the formal, postdrill debriefing and/or critique.

Bonus time: If you replace the word "drill" with "planned event" in the preceding discussion, you have a perfect planning and delivery tool for SOs assigned to common community activities such as festivals, parades, concerts in the park, and small- to medium-sized gatherings (less than 500 people), all of which have the potential to challenge normal fire department incident responses.

■ No-Notice Drills

The "no-notice" drill has been, and may continue to be, a controversial doctrine among fire service officers. Although some professionals (chiefs and some TOs) support the basic evaluation-tool concept of no-notice drills, others (SOs and some TOs) question the spontaneity of the event.

There are two schools of thought at play here. The first school of thought can be seen in those TOs who practice no-notice drills with all the top-secret security of a *DEFCON 1* nuclear missile launch. They argue that the only way to truly evaluate whether firefighters know their stuff is to observe them acting spontaneously, with absolutely no notice or hint of the upcoming drills; after all, that's what they have to do at an actual incident. One corollary to this school is the chief administrator or officer who conducts a no-notice drill without even informing the TO or SO.

Members of the other school contend that no-notice drills undermine the learning process and, more often than not, emphasize "sneaky testing" versus improving. They go on to argue that a no-notice drill can be dangerous and that responders may take unnecessary risks or experience unnecessary stress. These safety-doctrine adherents would rather give full notice of upcoming drills and even schedule the exact time and place of the drill. Further, they may outline the expected evolution and the desired outcome and coach the participants to use a slow pace for drill accomplishment.

For our purposes, the two schools have been displayed at their extremes. The idea is to show that no-notice drills are controversial. Not surprisingly, they are criticized by those required to participate (line responders). However, perhaps a middle ground (or third school of thought) can help people accept the use of no-notice drills. In the words of Alan Brunacini, "Standard outcomes are a result of firefighters doing standard things in a standard, predictable way."[4] The quote is applicable to incident actions as well as the no-notice drill; the drill is merely a way of observing firefighters doing what they do. Couple this attitude with some simple preestablished guidelines, and the no-notice drill can become a *successful*, *safe*, and *minimal-stress* training tool. Preestablished guidelines for no-notice drills include the following:

1. *Make sure the drill has been previously presented as a training evolution that has been practiced by all potential responders.* Without a doubt, this is a key to eliminating the fear and apprehension of no-notice drills. Firefighters and fire officers who have had the opportunity to practice a given evolution tend to gain confidence in the proper learning environment. TOs should take advantage of successful training drill sessions to announce that the crew may be asked to perform the evolution in a no-notice drill in the future. Let the crew know how the drill will be announced and what to expect in the way of evaluation and critique.

2. *Don't wait for information "half-life."* Within a given department, firefighters will retain, gain, and lose skill, knowledge, and abilities over time. If a given evolution is practiced in March, it may not be appropriate to conduct a no-notice drill the following November if no opportunity to practice or perform the evolution has happened in between. This information half-life varies based on the complexity, necessity, and repetition of the evolution.

3. *Have a safety plan.* Writing out a department-recognized guideline or SOP regarding no-notice drills can establish safe practices and expected behavior before the drill is ever announced. Main components of that SOP/SOG include the following:

   - Preface initial radio dispatching using the phrase, "This is a drill." Follow-up radio transmissions should be prefaced with "drill message." If a radio message needs to be made that is outside the drill evolution, consider prefacing it with "real-world message."

- Apparatus response to the drill will be non-emergent, following all traffic laws (no lights or siren).
- Play as you've been trained: Do predictable things in a predictable way.
- Have a signal or phrase that *anyone* can use to terminate the drill because of a safety issue.

4. *Prenotify key personnel.* Department administrators, dispatchers, and allied agencies have real-world responsibilities and may not appreciate the interruption of a no-notice drill that might impact them.

5. *Use an SO.* A predesignated SO is essential for multicompany, complex, or multiagency no-notice drills. The SO should be part of the drill facilitation team and work the scene as he or she would an actual incident.

6. *If the drill involves live fire, scrap the no-notice approach.* Stick to safe guidelines (like NFPA 1403), careful safety planning, and participant prebriefings when dealing with live-fire events (or other drills that are inherently a higher risk).

7. *Be practical.* A relevant axiom would read "Leave your imagination at home." Most veteran fire officers have experienced the following scenario: They're told to respond to a no-notice drill involving the collision of a school bus and truck. The responders begin size-up and triage only to have a note handed to the incident commander (IC) explaining that the truck has released some unpronounceable chemical. After adjusting and dealing with that, the IC receives another note saying that a military jet just crashed into the accident scene. To compound things, mutual-aid companies are not available due to a multiple-alarm commercial fire. Spare us! Make the drill realistic with a chance for success. Keep the "surprises" minimal and practical. Save the doomsday imagination for table-top exercises.

8. **Congratulate** *before you* **evaluate.** This should be obvious. Nothing reinforces performance better than praise. If you believe that the drill performance was so bad that you want to give up TO/SO responsibilities and return to the backseat, maybe the crew wasn't ready (see number 1 above).

If practiced right, no-notice drills can be an effective reinforcement and evaluation tool for trainers and a practice tool for SOs. Perhaps these guidelines can help you reconsider the use of no-notice drills or help you be more successful with your current use of this evaluation and learning tool.

Safety Officer Functions for Live-Fire Training

Most firefighters will never forget their first working structure fire. Likewise, most fire academy graduates will never forget their first live-fire event in a burn house or acquired structure. The training can be that powerful. In many ways, live-fire training is the pinnacle of all fire training events. It can also be the most hazardous.

NFPA 1403, *Standard on Live Fire Evolutions*, has been developed to address the inherent dangers and risks associated with live-fire training. The standard should be mandatory reading for any SO or instructor involved with any form of firefighter live-fire training. Major topics covered within the standard include:

- General requirements
- Acquired structures
- Gas-fired live-fire training structures
- Non-gas-fired live-fire training structures
- Exterior live-fire training props

Of note within the general section are requirements for student prerequisites, a mandate for a designated SO (and responsibilities), IIC responsibilities, and planning/preparation procedures.

In some jurisdictions, instructors and SOs must be *certified* to perform those functions and must follow, by law, the requirements set forth in the standard when conducting live-fire training events. SOs *must* research and adhere to local and state laws, permit processes, and other requirements that might be applicable prior to any live-fire training activity. Where no requirements exist, the SO should, at a minimum, follow the 1403 standard. The reasons for this strict directive are founded in a history of deaths associated with live-fire training and the liabilities that may be incurred.

■ Safety Officer Liabilities at Live-Fire Training Events

If court case history is any indication, the liability for injury or loss resulting from a live-fire training incident will fall directly on the IIC *and* the designated SO.[5] An argument can be made that a designated SO is ultimately responsible for the safety of everyone participating at live-fire training. That argument is found by comparing the NFPA 1403 safety responsibilities for the IIC to those for the SO:

Instructor-in-Charge:
4.7.4 *It shall be the responsibility of the instructor-in-charge to coordinate overall fireground activities to ensure correct levels of safety.*

Safety Officer:
4.5.1 *A safety officer shall be appointed for all live fire training evolutions.*
4.5.3 *The safety officer shall have the authority, regardless of rank, to intervene and control any aspect of operations when, in his or her judgment, a potential or actual danger, potential for accident, or unsafe condition exists.*
4.5.4 *The responsibilities of the safety officer shall include, but not be limited to, the following:*
(1) *Prevention of unsafe acts*
(2) *Elimination of unsafe acts*
4.5.5 *The safety officer shall provide for the safety of all persons on the scene, including students, instructors, visitors, and spectators.*

This language differs significantly from ISO requirements and responsibilities for actual incidents (NFPA 1561). For incidents, the IC is ultimately responsible for the safety of all responders.

A finding of SO liability is more probable for live-fire training (versus IC liability for an incident) simply because training events are *planned* and actual incidents are not. NFPA 1403 can be viewed as a "nationally recognized standard of care" for live-fire training and includes many planning and preparation requirements to help minimize the chance of injuries or deaths. Failure to follow a recognized standard for a planned event is likely to be viewed as a form of negligence in the eyes of a judge or jury.

Fire Marks

ISFSI Live Fire Instructor Credential Program

The International Society of Fire Service Instructors (ISFSI) has developed a Live Fire Instructor Credential Program to address the responsibilities of instructors who facilitate and deliver live-fire training events. The program is designed to satisfy the live-fire training requirements found in NFPA standards 1231, 1402, 1403, and 1500. Its content is based on NFPA 1403, *Standard on Live Fire Training Evolutions*, and it references the textbook *Live Fire Training: Principles and Practice* by Jones and Bartlett Learning. After successful completion of the program, students are eligible for an area-specific ISFSI Live Fire Instructor Credential. The area-specific credentials are as follows:

1. Live Fire Instructor—Fixed Facility (online course plus an ISFSI-instructed hands-on academy program)
   - Gas Fire Facilities
   - Class A Facilities
2. Live Fire Instructor—Acquired Structure (online program only)
3. Live Fire Instructor—Instructor-in-Charge (online program only)

To enroll in the credentialing programs, candidates must already be certified as an Instructor I or higher and be a certified Firefighter II or higher. Credentialing lasts for three years but requires a minimum of 4 hours of live-fire or NFPA 1403 continuing education credits from ISFSI and a minimum of one (1) documented burn in each year (to maintain credentialed status).

The program can benefit anyone who is interested in making a difference in the creation and execution of a safe and controlled live-fire experience, including:
- Training officers
- Instructors
- Company officers
- Safety officers

More information can be found at the ISFSI website (http://isfsi.org/).

The point here is not to discourage the use of live-fire training or to scare an individual away from filling the role of the SO at those events. The path to minimizing SO injury and death liability for live-fire training can be found by following some simple guidelines:

- Know the content of NFPA 1403 and ensure that the IIC is adhering to its planning, preparation, and delivery requirements. (It contains many useful checklists!)
- Participate in the planning and preparatory activities.
- Perform the SO live-fire training functions with due diligence and good intent.
- Don't hesitate to intervene if you judge something to be unsafe or a potential for injury exists.
- Consider becoming *credentialed* as a live-fire training instructor (see the *Fire Marks* box).

■ Safety Officer Focus Areas for Live-Fire Training Events

Historical experience lessons and some suggestions can be added to the content and training information in NFPA 1403 to help the SO prevent injuries at the live-fire training event. These additions include building preparation, instructor issues, and the establishment of a safety team.

Building Preparation

Buildings that are designed and built for the purpose of live-fire training typically have minimal safety hazards that require SO evaluation and intervention. That is not to say the SO shouldn't inspect the structure prior to drills. NFPA 1403 includes a burn facility inspection sheet that can assist the SO.

Acquired structures require much more SO evaluation. There are two template forms in 1403 that are used for acquired structures: a site inspection worksheet that has "fill-in" space to document items such as building materials and wastes, and a live-fire evolution checklist **FIGURE 17-5** that includes a section for training structure preparation. The IIC may have filled in the required information and "checked all the boxes," but many of the items are subjective. For example, one checklist item states, "Unnecessary inside and outside debris removed." Another directs that "All extraordinary exterior and interior hazards are remedied." These items are open to interpretation. The SO should consider the spirit of the checklist and ensure that items are satisfied that lead to prevention of mishaps and injuries.

Instructor Issues

The IIC is responsible for assigning additional instructors as necessary to ensure a ratio of one instructor to a maximum five students (1:5 ratio); a ratio of 1:3 is better **FIGURE 17-6**. All instructors must be familiar with NFPA 1403 and be approved by the department (or other regulatory jurisdiction) to serve as a live-fire instructor. Additionally, the instructors need to be

LIVE FIRE EVOLUTION SAMPLE CHECKLIST

PERMITS, DOCUMENTS, NOTIFICATIONS, INSURANCE.

_____ 1. Written documentation received from owner:
- ❏ Permission to burn structure
- ❏ Proof of clear title
- ❏ Certificate of insurance cancellation
- ❏ Acknowledgment of post-burn property condition

_____ 2. Local burn permit received

_____ 3. Permission obtained to utilize fire hydrants

_____ 4. Notification made to appropriate dispatch office of date, time, and location of burn

_____ 5. Notification made to all affected police agencies:
- ❏ Received authority to block off roads
- ❏ Received assistance in traffic control

_____ 6. Notification made to owners and users of adjacent property of date, time, and location of burn

_____ 7. Liability insurance obtained covering damage to other property

_____ 8. Written evidence of prerequisite training obtained from participating students from outside agencies

PREBURN PLANNING.

_____ 1. Preburn plans made, showing the following:
- ❏ Site plan drawing, including all exposures
- ❏ Floor plan detailing all rooms, hallways, and exterior openings
- ❏ Location of command post
- ❏ Position of all apparatus
- ❏ Position of all hoses, including backup lines
- ❏ Location of emergency escape routes
- ❏ Location of emergency evacuation assembly area
- ❏ Location of ingress and egress routes for emergency vehicles

_____ 2. Available water supply determined

_____ 3. Required fire flow determined for the acquired structure/live fire training structure/burn prop and exposure buildings

_____ 4. Required reserve flow determined (50 percent of fire flow)

_____ 5. Apparatus pumps obtained that meet or exceed the required fire flow for the building and exposures

_____ 6. Separate water sources established for attack and backup hose lines

_____ 7. Periodic weather reports obtained

_____ 8. Parking areas designated and marked:
- ❏ Apparatus staging
- ❏ Ambulances
- ❏ Police vehicles
- ❏ Press vehicles
- ❏ Private vehicles

_____ 9. Operations area established and perimeter marked

_____ 10. Communications frequencies established, equipment obtained

TRAINING STRUCTURE PREPARATION.

_____ 1. Training structure inspected to determine structural integrity

_____ 2. All utilities disconnected (acquired structures only)

_____ 3. Highly combustible interior wall and ceiling coverings removed

_____ 4. All holes in walls and ceilings patched

_____ 5. Materials of exceptional weight removed from above training area (or area sealed from activity)

_____ 6. Ventilation openings of adequate size precut for each separate roof area

_____ 7. Windows checked and operated, openings closed

_____ 8. Doors checked and operated, opened or closed, as needed

_____ 9. Training structure components checked and operated:
- ❏ Roof scuttles
- ❏ Automatic ventilators
- ❏ Mechanical equipment
- ❏ Lighting equipment
- ❏ Manual or automatic sprinklers
- ❏ Standpipes

_____ 10. Stairways made safe with railings in place

_____ 11. Chimney checked for stability

_____ 12. Fuel tanks and closed vessels removed or adequately vented

_____ 13. Unnecessary inside and outside debris removed

_____ 14. Porches and outside steps made safe

_____ 15. Cisterns, wells, cesspools, and other ground openings fenced or filled

FIGURE 17-5 Live Fire Evolution Sample Checklist from NFPA 1403 (2018).   (continues)

LIVE FIRE EVOLUTION SAMPLE CHECKLIST *(continued)*

_____ 16. Hazards from toxic weeds, hives, and vermin eliminated

_____ 17. Hazardous trees, brush, and surrounding vegetation removed

_____ 18. Exposures such as buildings, trees, and utilities removed or protected

_____ 19. All extraordinary exterior and interior hazards remedied

_____ 20. Fire "sets" prepared:
- ❏ Class A materials only
- ❏ No flammable or combustible liquids
- ❏ No contaminated materials

PREBURN PROCEDURES.

_____ 1. All participants briefed:
- ❏ Training structure layout
- ❏ Crew and instructor assignments
- ❏ Safety rules
- ❏ Training structure evacuation procedure
- ❏ Evacuation signal (demonstrate)

_____ 2. All hose lines checked:
- ❏ Sufficient size for the area of fire involvement
- ❏ Charged and test flowed
- ❏ Supervised by qualified instructors
- ❏ Adequate number of personnel

_____ 3. Necessary tools and equipment positioned

_____ 4. Participants checked:
- ❏ Approved full protective clothing
- ❏ Self-contained breathing apparatus (SCBA)
- ❏ Adequate SCBA air volume
- ❏ All equipment properly donned

POSTBURN PROCEDURES.

_____ 1. All personnel accounted for

_____ 2. Remaining fires overhauled, as needed

_____ 3. Training structure inspected for stability and hazards where more training is to follow (see Training Structure Preparation)

_____ 4. Training critique conducted

_____ 5. Records and reports prepared, as required:
- ❏ Account of activities conducted
- ❏ List of instructors and assignments
- ❏ List of other participants
- ❏ Documentation of unusual conditions or events
- ❏ Documentation of injuries incurred and treatment rendered
- ❏ Documentation of changes or deterioration of live fire training structure
- ❏ Acquired structure release
- ❏ Student training records
- ❏ Certificates of completion

_____ 6. Building and property released to owner release document signed

- -

RELEASE FORM

Having agreed with the Building Official, City of _____, that a structure owned by me and located at _____ is unfit for human habitation and is beyond rehabilitation, I further agree that the structure should be demolished. In order that demolition may be accomplished, I give my consent to the City of _____ to demolish, by burning or other means, the said structure.

I further release the City of _____ from any claim for loss resulting from such demolition.

Fire Department _____

Address _____

City, State _____

Date _____

Owner/Agent _____

Owner/Agent _____

Witness _____

(continues)

FIGURE 17-5 Live Fire Evolution Sample Checklist from NFPA 1403 (2018).

RESPONSIBILITIES OF PERSONNEL

INSTRUCTOR-IN-CHARGE

____ 1. Plan and coordinate all training activities

____ 2. Monitor activities to ensure safe practices

____ 3. Inspect training structure integrity prior to each fire

____ 4. Assign instructors:
   - ❏ Attack hose lines
   - ❏ Backup hose lines
   - ❏ Functional assignments
   - ❏ Teaching assignments

____ 5. Brief instructors on responsibilities:
   - ❏ Accounting for assigned students
   - ❏ Assessing student performance
   - ❏ Clothing and equipment inspection
   - ❏ Monitoring safety
   - ❏ Achieving tactical and training objectives

____ 6. Assign coordinating personnel, as needed:
   - ❏ Emergency Medical Services
   - ❏ Communications
   - ❏ Water supply
   - ❏ Apparatus staging
   - ❏ Equipment staging
   - ❏ Breathing apparatus
   - ❏ Personnel welfare
   - ❏ Public relations

____ 7. Ensure adherence to this standard by all persons within the training area

INSTRUCTOR

____ 1. Monitor and supervise assigned students (no more than five per instructor)

____ 2. Inspect students' protective clothing and equipment

____ 3. Account for assigned students, both before and after evolutions

SAFETY OFFICER

____ 1. Prevent unsafe acts

____ 2. Eliminate unsafe conditions

____ 3. Intervene and terminate unsafe acts

____ 4. Supervise additional safety personnel, as needed

____ 5. Coordinate lighting of fires with instructor-in-charge

____ 6. Ensure compliance of participants' personal equipment with applicable standards:
   - ❏ Protective clothing
   - ❏ Self-contained breathing apparatus (SCBA)
   - ❏ Personal alarm devices, where used

____ 7. Ensure that all participants are accounted for, both before and after each evolution

STUDENT

____ 1. Acquire prerequisite training

____ 2. Become familiar with building layout

____ 3. Wear approved full protective clothing

____ 4. Wear approved SCBA

____ 5. Obey all instructions and safety rules

____ 6. Provide documentation of prerequisite training, where from an outside agency

FIGURE 17-5 Live Fire Evolution Sample Checklist from NFPA 1403 (2018).

FIGURE 17-6 Live-fire training should not exceed a 1:5 instructor-to-student ratio. A ratio of 1:3 is better!

© Kim Fitzsimmons. Used with permission.

assigned such that no less than one instructor is in a position to directly supervise the following:

- Each functional crew (e.g., attack, ventilation, search)
- Each backup line
- Outside students (uninvolved or staged students awaiting rotation)

Additionally, a <u>fire control team</u> must be established that includes a minimum of two qualified firefighters (not a student or the SO). One of the members of the fire control team is designated as the *ignition officer* and is responsible for starting the fire. The other member serves as a ready observer (in full PPE) to rescue the ignition officer if something goes wrong. In most cases, the fire control team supervises the fuel delivery and setup process for repeated or rotational fire sets. The SO must serve as a "quality control" officer to make sure these instructor/fire control team requirements are maintained as part of the monitoring/evaluation function. The SO is also responsible for giving the ignition officer and IIC a final go/no-go signal for each fire set.

An area that bears SO attention—and is not specifically addressed in the 1403 standard—deals with instructor rehabilitation. Making rehabilitation provisions available for all participants (students and instructors) is the responsibility of the IIC, and assistant instructors are usually good at making sure their students rotate to rehabilitation. Too often, though, the instructors themselves don't rotate to rehabilitation. There are many historical, anecdotal, and research cases where live-fire instructors were found to have dangerously high core temperatures, overexertion symptoms, and dehydration (see the *Safety Tip* box). The SO needs to intervene as necessary to make sure instructors are also rotating to rehabilitation.

Safety Team Establishment

While not required by NFPA 1403, some live-fire training specialists[6] establish a <u>safety team</u> to help manage the training evolution. The safety team typically comprises:

Safety Tip

ALERT! Instructor Rehab

Providing rehabilitation resources for students participating at live-fire events is essential, and rehab is likely enforced and reinforced by event instructors. An irony exists when the live-fire instructors themselves don't adequately partake in rehab. In one example, a veteran live-fire instructor of thousands of live-fire drills (over a 10-year span) developed a heart condition that was directly attributed to repeated exposure to elevated core temperatures and dehydration. This instructor is known to be in outstanding physical condition with a calm, self-deprecating nature and possessing instructor skills that are above reproach. Still, the heart condition and subsequent surgery almost ended the instructor's fire service career as a suppression company officer and regional fire school instructor.

The need for live-fire instructor rehab, including active cooling, hydration, electrolyte replacement, and nourishment, cannot be overemphasized. In fact, an argument can be made that instructors who cycle multiple students through repeated fires in a burn facility are in need of more rehab than the students themselves.

When assigned as the SO at live-fire drills, you should monitor rehab efforts for all participants—including the instructors. A friendly (and discrete) reminder may be all that is needed. Oh, and don't forget to be the *example* and rehab yourself!

- The designated SO
- An experienced interior standby team that includes an officer, nozzle person, and door control person (no students)
- An outside building monitor who is dedicated to evaluating structural issues, smoke flow paths, and fire behavior
- A fuel shutoff officer (for facilities with gas-fired devices) or a fuel-set supervisor (for repeated class A fuel fires)
- An ignition officer (the ignition officer and fuel shut-off officer serve as the NFPA 1403-required fire control team)

The benefits of a safety team are numerous—the most obvious being the interior standby team (independent of instructors or students) that is ready to make an immediate fire suppression or rescue intervention should something go wrong. The SO is wise to establish a safety team (especially for acquired structures) to help maximize safety efforts. An ICS-inspired organization chart that shows the reporting relationship of the IIC, SO, safety team, and other training participants is shown in **FIGURE 17-7**.

The SO functions, liabilities, and focus areas for live-fire training evolutions presented in this chapter are not meant to replace the requirements and responsibilities contained in NFPA 1403. Rather, the intent is to highlight certain areas and make recommendations to enhance the standard and help improve an SO's ability to prevent training-related mishaps, injuries, and deaths.

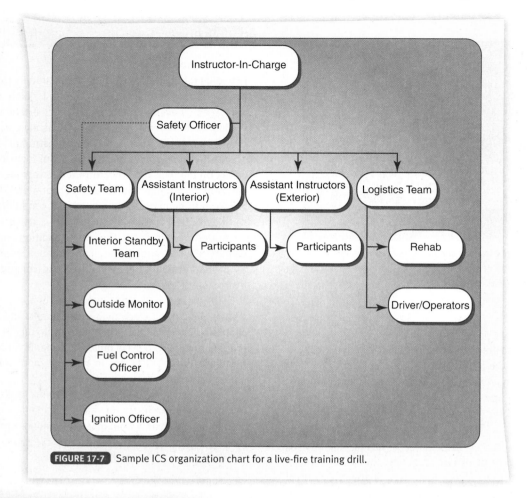

FIGURE 17-7 Sample ICS organization chart for a live-fire training drill.

Wrap-Up

Chief Concepts

- The preplanned nature of training activities provides an opportunity for "practice incidents" to help develop firefighters, officers, and SOs.
- All fire officers should be committed to four safety-minded values during training events: (1) Safety trumps performance expectations, (2) train in accordance with established SOPs/SOGs and national standards, (3) train only in environments that can be quickly controlled through preplanned measures, and (4) empower everyone with the responsibility to stop unsafe acts.
- Nonincident activities that can benefit from the assignment of a separate, dedicated SO include live-fire training (SO required), full-scale multiagency drills, inherently risky training activities (such as rope rescue drills), and events that include multiple skill stations that participants rotate through.
- The SO and instructor-in-charge (IIC) of a training event should preplan training events together and agree on the training description and objectives, minimum PPE required by instructors/participants, rehabilitation needs, RIC requirements, termination triggers (weather, etc.), and communication methods used by anyone to stop the training.
- Shadowing is a training approach in which a trainee closely observes experienced mentors performing their assigned duties. The concept of shadowing is part of a three-step approach when training and preparing new ISOs: introductory classroom training, shadowing, and coaching (where the ISO trainee is supervised while filling the role).

- Multiagency drills can include participants who are not familiar with ICS or the role of the SO. Therefore it is best to prepare them for a full-scale drill by first exposing them to a three-step training/exercise process that includes initial classroom indoctrination followed by a table-top drill (talk-through in a classroom) and then a functional drill (walk-through using a command post, use of radio/communication links, and the response of involved agencies who stage, deploy, and simulate the accomplishment of their task).

- Multiagency drills benefit from a predrill, written safety briefing (developed by the SO) that includes items such as a traffic plan, level of PPE required for various roles, expected hazards and defined control zones, rehabilitation measures, and participant responsibilities.

- The SO assigned to a multiagency drill should not hesitate to intervene when any situation is deemed an unacceptable hazard. Common drill-related intervention issues include weather-related problems, traffic, failure to rehabilitate, and overzealous actors or participants.

- The SO processes used to plan and perform duties for a multiagency drill can be equally applied to planned, nonemergency events like festivals, parades, park concerts, and small-crowd gatherings that have the potential to challenge normal fire department incident responses.

- No-notice drills typically induce participant stress and may be viewed as unsafe. When performed, simple guidelines that are contained in written SOPs/SOGs can help lead to a successful, safe, and minimally stressful event.

- NFPA 1403, *Standard on Live Fire Evolutions*, addresses the inherent dangers and risks associated with live-fire training. Any SO or instructor involved with any form of firefighter live-fire training must adhere to this standard.

- The SO must provide for the safety of all persons on the scene of a live-fire training event, including students, instructors, visitors, and spectators (NFPA 1403). This requirement leads to the potential of SO liability if a mishap, injury, or death occurs. The following steps will reduce SO liability:

  1. Know the content of NFPA 1403 and ensure that the IIC is adhering to its planning, preparation, and delivery requirements.
  2. Participate in planning and preparations activities that are managed by the IIC.
  3. Perform the SO live-fire training functions with due diligence and good intent.
  4. Don't hesitate to intervene if you judge something to be unsafe or a potential for injury exists.
  5. Consider becoming credentialed as a live-fire training instructor through the ISFSI live-fire training credential program.

- Focus areas for the designated SO at live-fire training events include building preparation (especially for acquired structures), instructor requirements, and the creation of a safety team.

- The IIC is responsible for maintaining an instructor-to-student ratio of 1:5 (maximum); 1:3 is better. A separate instructor is required for each functional crew (e.g., attack, ventilation, search), for each backup line, and for outside crews awaiting rotation. The IIC must also designate a two-person *fire control team*. The SO serves as a quality control officer to make sure these requirements are met.

- Live-fire instructors exposed to multiple fire evolutions (rotating student groups through repeated drills) are candidates for dangerously high core temperatures and dehydration. The SO should pay particular attention to instructor rehab and provide discreet rehab reminders to help prevent acute and chronic health issues.

- The wise SO will establish a live-fire *safety team*, comprising an interior standby team (experienced officer and two firefighters), an outside monitor for building and smoke conditions, a fuels-control officer, and an ignition officer.

Key Terms

fire control team A minimum of two qualified firefighters (not a student or the safety officer) used as an ignition officer and a ready observer (in full PPE) to rescue the ignition officer should something go wrong at live-fire trainings.

safety team When applied to live-fire training, a team comprising the designated safety officer; an experienced interior standby team made up of an officer, nozzle person, and door control person; an outside building monitor; a fuel shutoff officer or a fuel-set supervisor; and an ignition officer.

shadowing A training approach in which a trainee closely observes experienced mentors performing their assigned duties.

Review Questions

1. List four planned, nonincident events that can benefit from the assignment of a separate, dedicated SO.
2. What are the planning issues that the SO should cover with an instructor-in-charge?
3. What is meant by *shadowing*, and how does it fit into a process to train new ISOs?
4. List the three training/exercise steps that should take place prior to a full-scale drill for non-fire-agency participants.

5. What steps can be taken to minimize stress and improve the safety of no-notice drills?

6. Identify the steps that can be taken to reduce liabilities associated with live-fire firefighter training events.

7. Define the instructor ratios and assignments that are mandated for live-fire training events.

8. What is a live-fire training safety team and who should it include?

References and Additional Resources

Karter, Michael J., National Fire Protection Association. *Patterns of Firefighter Fireground Injuries.* 2013. http://www.nfpa.org/research/reports-and-statistics/the-fire-service/fatalities-and-injuries/patterns-of-firefighter-fireground-injuries. Accessed June 2, 2015.

Live Fire Training Principles and Practice. 1st ed. revised. Burlington, MA: Jones & Bartlett Learning, 2016.

National Institute for Occupational Safety and Health. *NIOSH Alert: Preventing Deaths and Injuries to Firefighters During Live-Fire Training in Acquired Structures.* Publication No. 2005-102, January 2005. http://www.cdc.gov/niosh/docs/wp-solutions/2005-102/. Accessed June 2, 2015.

NFPA 1403, *Standard on Live Fire Training Evolutions.* Quincy, MA: National Fire Protection Association, 2018.

Reeder, Forest, and Alan Jobs. *Fire Service Instructor: Principles and Practice.* 2nd ed. Burlington, MA: Jones & Bartlett Learning, 2014.

Endnotes

1   BrainyQuote.com. Vince Lombardi. http://www.brainyquote.com/quotes/quotes/v/vincelomba138158.html. Accessed January 27, 2015.

2   Davis, Wynn. *The Best of Success.* Lombard, IL: Great Quotations Publishing Company, 1988.

3   The author conducted an online research of data posted on the U.S. Fire Administration website: http://www.usfa.fema.gov/fireservice/firefighter_health_safety/firefighter-fatalities. The 10-year average is collected from 2009 through 2018.

4   Brunacini, Alan. Quote from an International Fire Department Instructors Conference 2013 workshop. Verified via personal contact January 2015.

5   *People v. Baird*, 2 A.D.3d 1433, 768 N.Y.S.2d 88, 2003 N.Y. Slip Op. 20227. The IIC of a live-fire training event was found guilty of criminally negligent homicide following the death of firefighter participant.

6   The title *live-fire training specialist* refers to veteran fire instructors who are well versed in NFPA 1403 and provide live-fire training services to fire departments or regional training academies on a contract or reciprocal basis. Many are certified and/or credentialed.

In 2007, a career probationary firefighter died while participating in a live-fire drill in a condemned, three-story townhome. The victim was part of an initial attack team (victim plus three recruit students), led by an adjunct instructor, and was tasked with advancing a hose line to the third floor to find and extinguish any fire found on the third floor. A second attack team was assigned to fight any fire found on the second floor. An academy instructor and three students served as the rapid intervention crew. Additionally, three truck companies (each with an instructor and three or four students) participated in the drill. A battalion chief, engine company, and truck company (fully staffed, no students) were on scene but did not participate in the drill. The instructor in charge was an academy instructor lieutenant. A medic unit with two paramedics served in a standby role on scene.

A total of nine fires were set (one on the first floor, six on the second floor, and two on the third floor) using wood pallets and excelsior bales. The victim's team was instructed to bypass any fire they found on the second floor. As the drill proceeded, the initial attack team encountered heavy fire conditions on the second floor and the stairwell to the third floor. The adjunct instructor knocked down some of the second floor fire so they could proceed to the third floor. As they reached the third floor, they found moderate heat and gray smoke, but no visible fire. The other initial attack team members were assisting with hose advancement in the stairway as the victim and instructor searched for the fire.

The second attack team became delayed as they needed to add hose sections to reach the second floor. The second attack team also encountered fire conditions on the first floor, which furthered delayed their advancement to the second floor. Subsequently, a team assigned to truck functions placed an extension ladder to a second floor window for ventilation. Once opened, fire immediately blew from the window.

Conditions in the stairwell between the second and third floors deteriorated rapidly causing the initial attack team back-up members to seek shelter on the third floor (one member receiving burns). As the third floor became untenable, the instructor and one backup team member escaped out a third floor window and onto the roof of the second floor. The victim became stuck in the same window, which was 41 inches above the landing floor. The instructor tried several times to free the victim who was now screaming as flames vented out the same window. The second attack team controlled the fire in the stairwell, and their adjunct instructor raced up the stairs to the third and assisted in freeing the victim who had become unresponsive. He then went downstairs and summoned the second attack crew to come fight the growing fire on the third floor. The victim was immediately transported to a trauma center and then pronounced dead. The initial attack team adjunct instructor and one back-up team member received first, second, and third degree burns to various body parts.

The investigation into this incident revealed numerous NFPA 1403 noncompliance issues including:

- Inadequate preparation of the acquired structure
- Too many fire sets in multiple locations
- Absence of pre-burn briefings and walk-through
- Extensive fuel loads
- Adjunct instructors who did not meet minimum instructor training requirements for live-fire evolutions

1. What additional NFPA 1403 noncompliance issues seem to have contributed to this case study?

2. How could the resources present at this drill be better assigned to help prevent injuries and deaths?

3. What does NFPA 1403 mandate as the minimum instructor-to-student ratio for live-fire evolutions?
   A. 1:3
   B. 1:5
   C. 1:7
   D. 1:10

4. A live-fire safety officer should establish a safety team that includes which of the following?
   A. Instructor rehab crew
   B. Exterior rapid intervention crew (minimum of an instructor and three students)
   C. Interior standby team (no students)
   D. Instructor in charge

Note: This case study was developed from NIOSH Firefighter Fatality report #2007-09 available at www.cdc.gov/niosh/fire.

APPENDIX

A

Sample SOPs

© Photos.com

Standard Operating Procedure		*SOP#* 2.10
Division: Fire Suppression		*Revision Number:* V3.1
Topic: Incident Safety Officer—Utilization Requirements		
Approval:		*Date:* 06 June 2015

Purpose

The department recognizes that certain incidents present a significant or increased risk to firefighters. With these incidents come an increased responsibility to monitor firefighting actions and environmental conditions. The appointment of an incident safety officer (ISO) can increase an incident commander's (IC's) effectiveness in protecting firefighters. Therefore, the department has developed a system that provides a dedicated, qualified fire officer to respond to predesignated and significant incidents in order to serve the IC as the ISO. This SOP outlines general and specific procedures that support the use of an ISO at incidents and those required to help support a duty safety officer program.

Responsibilities

1. Command officers: Responsible for the support and utilization of duty safety officers at incidents outlined in this SOP. Where no duty safety officer is available, command officers shall appoint a qualified fire officer to serve the role of ISO.
2. Duty safety officers: Responsible for maintaining skills and pursuing professional development that supports the duties and functions of an effective and efficient ISO. Additionally, duty safety officers shall respond to predefined incidents to serve as an ISO. Duty safety officers shall perform other duties as assigned by the chief of safety when not engaged in incident activities.
3. Company officers/firefighters: Responsible for working with any appointed ISO to recognize and minimize risks associated with incident environments and operations.

Procedure

■ General Requirements

1. The IC shall appoint an ISO early during a working incident to maximize the effectiveness of the IC–ISO team. Company officers and firefighters shall report hazards to the ISO in the course of incident operations.
2. This procedure in no way diminishes the responsibility of each and every responder to work in a safe and predictable manner at all times. All responders are expected to communicate unsafe conditions and actions to help prevent injuries and deaths.
3. The department shall support and maintain a 24/7 duty safety officer position. The department shall select, train, and appoint qualified fire officers to fill the position as outlined in this SOP. The duty safety officer's primary function is to respond to predesignated incident types and serve the IC as a command staff member (ISO). In cases where the duty safety officer is not available for incident response (e.g., committed to another incident, delayed response time), the IC shall appoint a qualified fire officer to serve as the ISO for those incidents outlined below or as a discretionary IC function.

4. Predesignated incident types that require a proactive duty safety officer response (and appointment of an ISO by the IC) include:
   - Report and/or confirmation of a structure fire
   - Wildland–urban interface fire
   - Specialty (technician) team incident (i.e., hazardous materials, confined space, trench, cave-in, and water/ice rescue)
   - Any type of incident at a "target" hazard facility as preplanned in the computer-aided dispatch system
   - Aircraft incident
   - Railroad train derailment
   - Vehicle incident reported as severe or involving multiple trapped victims, and those involving a school bus, tractor/trailer rollover, or structural damage (e.g., vehicle into a building, bridge)

5. The duty safety officer has the discretion to monitor incident call-outs and self-initiate a response based on dispatch information, weather extremes, or other compelling situations that might benefit from an ISO.

6. Certain incident situations require the IC to delegate the safety function to a qualified ISO where none has been appointed. They include:
   - Working incidents (all responders committed to control efforts and more resources are needed)
   - Incidents requiring the expansion of the ICS to include an operations, plans, and/or logistics section
   - Incidents requiring mutual-aid response
   - Incidents that include a Mayday declaration or those with a report of missing or down firefighters (rapid intervention crew [RIC] activation)
   - Incidents at which the size, duration, or circumstances require the IC to establish a formal rehab component

7. The IC may delegate the safety function to a qualified fire officer, or request the duty safety officer to respond, for any incident at which he or she feels that responder safety can be enhanced.

8. Any fire officer who has been assigned by the IC as an ISO or assistant safety officer (ASO) shall have the authority of the IC to stop, alter, suspend, or redirect any activity that, in the judgment of the ISO/ASO, presents an imminent threat to responders. An imminent threat is an act or condition that is judged to present a danger to persons or property that is so urgent and severe that it requires immediate corrective or preventive action.

9. The ISO/ASO shall notify the IC when an intervention has been made to stop, alter, suspend, or redirect an incident activity. The notification shall be accomplished forthwith—as soon as possible given the safety concerns of the intervention.

■ Incident Safety Officer Qualifications

The department recognizes that the ISO function requires knowledge, skills, and experience that exceed that of a company officer. A minimum level of training, specific to the ISO function, shall be provided by the department for duty safety officers. ISO-specific training shall also be provided to all chief officers. Company officers who are preparing for promotional opportunities are encouraged to pursue ISO training and may be granted department sponsorship based on space availability and budgeting priorities. Minimum qualifications for the duty safety officers and those delegated to the ISO function by the IC include:

1. Rank of captain or chief officer within the department
2. Fire Officer I certification
3. NIMS ICS 400 completion
4. Completion of the Fire Department Safety Officers Association (FDSOA) ISO Academy
5. Certification as an ISO through the FDSOA
6. Completion of the worker injury and risk reduction workshop provided by the city's risk manager
7. Completion of the department ISO task book (shadowing/supervised ISO performance) and sign-off by the chief of safety

To maintain status as a duty safety officer or qualified ISO, members shall:

1. Attend and document 4 hours of ISO-related proficiency training each month
2. Perform the function of a safety officer at two or more incidents each year (training drills and nonemergency community events may satisfy this requirement, subject to the chief of safety's approval)
3. Maintain Fire Officer I and FDSOA certifications

Standard Operating Procedure	SOP# 2.11
Division: Fire Suppression	**Revision Number:** V1.3
Topic: Incident Safety Officer—Duties and Functions	
Approval:	**Date:** 06 June 2015

Purpose

The appointment of an incident safety officer (ISO) can make a difference in the prevention of firefighter injuries and deaths and can assist the incident commander (IC) in recognizing changing incident conditions that might impact the IC's incident action plan (IAP). Likewise, the ISO can offer judgment to the IC regarding the effectiveness of crews performing tactical assignments. To achieve these desirable outcomes, an assigned ISO needs to understand certain expectations and perform position functions with due diligence. This SOP addresses the duties and responsibilities for any fire officer filling the ISO role at an incident as well as those specific to assigned duty safety officers.

Responsibilities

1. Command officers: Ultimately responsible for the safety of all responders at an incident. Additionally, the IC is responsible for communicating the IAP, as well as any known or suspected hazards or concerns, to the assigned ISO. During the course of an incident, the IC shall maintain a communication link with the ISO to ensure that provisions of this SOP are accountable.
2. Incident safety officers: Responsible for performing the duties contained in this SOP and accountable to the IC for all ISO functions.
3. Duty safety officers: Responsible for the ISO preceding responsibilities and other required readiness and incident follow-up functions that are included in this SOP.
4. Company officers/firefighters: Responsible for understanding the intent of this SOP and for working with any assigned ISO or the duty safety officer to help reduce injury and death potential.

Procedure

■ ISO Functions

1. When appointed as an ISO at an incident, obtain a briefing from the IC that includes the IAP, recognized hazards, and the resource commitment. Prepare for the ISO function by establishing communication links with the IC (radio frequency, face-to-face schedule), donning appropriate personal protective equipment (PPE) and "safety" vest, collecting needed equipment (e.g., box light, zoning tape), and checking into the accountability system.
2. The essential functions of the ISO can be remembered using the MEDIC mnemonic:

   M = Monitor the incident (recon to achieve situational awareness)
   E = Evaluate hazards and unsafe conditions (to judge responder risk-taking)
   D = Develop preventive measures (adjustments to reduce harm—zoning)
   I = Intervene (as necessary to prevent harm from immediate threats)
   C = Communicate (e.g., face-to-face, zone marking, safety briefings)

   The MEDIC essential ISO functions are purposefully broad in scope. The ISO must remain diligent in the cyclic application of these functions and must communicate findings to the IC on a regular basis. The remaining line items in this section are specific ISO functions that must be addressed while the ISO cycles through the MEDIC functions.
3. The ISO shall evaluate all incident actions and make a judgment regarding the appropriateness of risk-taking. Risk-taking should be judged as unacceptable when it falls outside of the department's risk management policy. When risk-taking is judged to be unacceptable, the ISO shall make an intervention (see next item).

4. When the ISO discovers a condition or action that presents an imminent threat, an intervention (direct order) shall be made to stop, alter, suspend, or redirect responders. The ISO shall advise the IC of the intervention as soon as possible given the circumstances. When a condition or action is not considered imminent (a potential threat), the ISO shall advise the responders facing the potential to alter or correct their actions to avoid the threat. When a situation is developing that could potentially create a threat (no immediate intervention required), the ISO should develop a preventive measure and present it to the IC for inclusion in the IAP.

5. The ISO shall evaluate fire extension (read smoke) and building collapse potentials (read building) and advise the IC of forecasted events that could pose a threat to operating crews.

6. The ISO shall offer judgment to the IC on establishing control zones and ensure that those zones are communicated to all responders. The department recognizes a zoning scheme that includes the following:
   - No-entry zone: Marked with red-and-white striped tape
   - Hot zone: Marked with red tape
   - Support zone: Marked with yellow tape

7. The ISO shall assess rehab efforts under way (self-rehab or formal) and make suggestions to the IC for improvement when warranted.

8. The ISO shall meet with any established rapid intervention crew (RIC) and share observations and/or concerns with the RIC officer.

9. The ISO shall communicate to the IC the need for assistant safety officers (ASOs) (or technical assistance) due to the size, duration, or complexity of the incident.

10. The ISO typically monitors the incident just outside of the hot zone (immediately dangerous to life and health [IDLH]). In situations where the ISO is needed within the hot zone, the ISO shall utilize appropriate PPE, request a partner, and report status to the IC on a regular basis.

11. Situations involving a declared Mayday; report of missing, trapped, or injured firefighters; and activation of the RIC require the ISO to report the command post to assist the IC. An ASO shall be assigned to help monitor rescue efforts.

12. The ISO shall be responsible for initiating the documentation and investigative process for firefighter injuries requiring transportation to a medical facility. Minor injuries that are easily handled by on-scene assistance should be noted on the ISO's activity log; the injured person's supervisor is responsible for filling out the form "First Report of an Employee Injury" for the city's risk manager.

13. The ISO shall document observations, interventions, preventive measures, and other pertinent ISO activities or issues on the ISO activity log prior to release by the IC. The ISO may be tasked by the IC to initiate the procedures for an after-action review. If not, the ISO is encouraged to check in with the IC and ascertain the IC's direction regarding the need for, and level of, an after-action review.

■ Duty Safety Officer Functions

The preceding procedures reflect the functions of any fire officer who is delegated the safety function by the IC. Fire officers who are assigned the readiness position of *duty safety officer* have additional responsibilities:

1. Duty safety officers must remain available for incident response during their assigned work period. Being available involves the same expectations as all suppression forces:
   - Maintain PPE ensembles in serviceable condition and ready for donning
   - Maintain assigned equipment and keep ready for use
   - Remain physically and mentally prepared for incidents: hydrated, nourished, and rested
   - Maintain a readiness profile while engaged in daily routines, training, and other nonincident activities
   - Remain within the department's jurisdictional boundaries
   - Advise dispatch and the shift battalion chief when "out of service"

2. If the duty safety officer must go out of service, he or she is responsible for finding another ISO-qualified person to fill in. Options include the chief of safety, other chief officer, or an off-duty duty safety officer (subject to overtime approval).

3. The duty safety officer is considered a suppression-force resource and is accountable to the on-duty command officer (battalion chief) for readiness and incident response issues. For administrative and performance review issues, the duty safety officer reports to the chief of safety.

4. The duty safety officer shall maintain his or her response vehicle (and all equipment/supplies within) in accordance with department policies and SOPs. The duty safety officer is authorized to respond emergent to incidents and is required to park the duty safety officer vehicle near the command post in such a way as to not impede other arriving apparatus.

5. If the duty safety officer arrives first to a dispatched incident (or discovers an incident in the course of the work period), the duty safety officer shall conduct a size-up, communicate conditions to other dispatched responders, and assume command. Likewise, if the duty safety officer arrives at an incident before the responding command officer but after the first-due apparatus, the duty safety officer shall obtain a quick briefing from the first-due officer and then establish a stationary command. After arrival, the dispatched command officer will determine whether to assume command or serve as the ISO (with the duty safety officer retaining command).

6. Nonincident functions of the duty safety officer include, but are not limited to:
   - Documenting incident safety and, when directed, developing an after-action review (as determined by the IC)
   - Conducting follow-up investigations for any incident close calls, firefighter injuries, and/or mishaps
   - Undertaking ongoing professional development
   - Serving as the safety officer for training drills and other nonemergency events
   - Providing instruction to other department members as directed by the chief of safety
   - Undertaking area familiarization and building hazard surveys (preincident intelligence as opposed to code-compliance inspections)
   - Developing and reviewing SOPs as assigned by the chief of safety

Checklists

Incident Safety Officer Checklist (All-Hazard)

Incident Name:_____ IC:_____ ISO:_____ Arrival Time:_____

Initial Actions

IC	Risk Profile	Tools/Actions

☐ Confirm ISO assignment
☐ IAP—Initial strategy
☐ SitStat/ReStat
☐ Radio freq:_____/_____
☐ RIC established
☐ Hazard issues (add to activity log)
☐ Check into accountability
☐ Announce "safety" on radio
☐ Start incident time-keeping
	☐ High: Life rescue/search
	☐ Medium: Stabilize/control
	☐ Low: Lost/property only
	
	Benchmarks
	
	☐ 360 complete
	☐ All clear/primary complete
	☐ Under control
	☐ Loss stopped
		☐ Don PPE and "safety" vest
		☐ Box light, zone tape, ISO kit
		☐ Request ASOs as necessary
		☐ Prioritize actions:
		  • Recon/MEDIC functions
		  • Develop briefings
		  • Develop site safety plan
		  • Establish control zones
		  • Meet with plans section if established

ISO Hazard "MEDIC" Functions

Monitor: 360 Recon of Environment and Activities (cyclic throughout incident)

☐ Read smoke: volume, velocity, density, and color ☐ Read the building (size, type, era, use): _____ ☐ Access/egress ☐ Weather/wind issues	☐ Hazardous energy issues (utilities notified?) ☐ Traffic and apparatus issues ☐ Firefighter actions/standard outcomes/freelancing ☐ Firefighter support systems: REHAB, PPE, RIC, PARs ☐ Risk-taking ☐ Radio traffic

Evaluate: Risk and Hazard Potentials

☐ Define principal hazard (collapse, hostile event, etc.):
_____
☐ Environment integrity:
  ☐ Stable, no change   ☐ Stable, may change
  ☐ Unstable, slow change   ☐ Unstable, rapidly changing
☐ Contributing hazards: terrain, barriers, energy
	☐ Resource effectiveness:
	  Staffing matches IAP—tools, tasks, trip/fall/struck hazards
	☐ Operational effectiveness—outcomes meet IAP?
	☐ Are safety systems appropriate and sufficient?
	☐ Rehab effectiveness:
	  ☐ Rest ☐ Energy nutrition ☐ Hydration
	  ☐ Accommodate weather ☐ BLS/ALS standby
	☐ **Is risk-taking appropriate?**

Develop: Preventive Measures

☐ Establish control zones (no-entry, hot, support, cold) ☐ Post sentries, flag hazards ☐ Develop safety messages/briefings ☐ Improve rehab ☐ Improve lighting	☐ **Improve cancer prevention through exposure- and contamination-reduction processes** ☐ **Document developed measures on activity log**

Interventions

☐ Imminent threats: *firm* order to stop/alter/redirect ☐ Communicate *firm* interventions to IC ☐ Offer *soft* intervention suggestions for potential threats ☐ **For Mayday/RIC, report to IC and assign ASO**	☐ **Document interventions on activity log**

Communicate

☐ Control zones (no-entry, hot, support, cold)
☐ Urgent safety messages
☐ Face-to-face with RIC officer
	☐ **Face-to-face with IC:**
	  ☐ IAP/risk adjustments ☐ Building hazards/collapse potential
	  ☐ Flashover potential ☐ Hazards/zones ☐ Rehab needs

Postincident Actions

☐ Monitor crews as they demobilize (overexertion issues) ☐ Confirm final PARs with IC ☐ Confirm no injuries/mishaps ☐ Confirm desired after-action review approach with IC ☐ Document actions on ISO activity log	As needed: ☐ Document issues for formal after-action reviews ☐ Begin injury/mishap documentation/investigation ☐ Start atypical stress event (ASE) protocol

© Photos.com

Safety Officer Position Checklist

The following checklist should be considered as the minimum requirements for this position. Note that some of the tasks are one-time actions; others are ongoing or repetitive for the duration of the incident.

☑	**Task**
☐	1. Obtain briefing from Incident Commander and/or from initial on-scene Safety Officer.
☐	2. Identify hazardous situations associated with the incident. Ensure adequate levels of protective equipment are available, and being used.
☐	3. Staff and organize function, as appropriate: • In multi-discipline incidents, consider the use of an Assistant Safety Officer from each discipline. • Multiple high-risk operations may require an Assistant Safety Officer at each site. • Request additional staff through incident chain of command.
☐	4. Identify potentially unsafe acts.
☐	5. Identify corrective actions and ensure implementation. Coordinate corrective action with Command and Operations.
☐	6. Ensure adequate sanitation and safety in food preparation.
☐	7. Debrief Assistant Safety Officers prior to Planning Meetings.
☐	8. Prepare Incident Action Plan Safety and Risk Analysis (USDA ICS Form 215A).
☐	9. Participate in Planning and Tactics Meetings: • Listen to tactical options being considered. If potentially unsafe, assist in identifying options, protective actions, or alternate tactics. • Discuss accidents/injuries to date. Make recommendations on preventative or corrective actions.
☐	10. Attend Planning meetings:

Sample Planning Meeting Agenda		
	Agenda Item	**Responsible Party**
1	Briefing on situation/resource status.	Planning/Operations Section Chiefs
2	Discuss safety issues.	Safety Officer
3	Set/confirm incident objectives.	Incident Commander
4	Plot control lines & Division boundaries.	Operations Section Chief
5	Specify tactics for each Division/Group.	Operations Section Chief
6	Specify resources needed for each Division/Group.	Operations/Planning Section Chiefs
7	Specify facilities and reporting locations.	Operations/Planning/Logistics Section Chiefs
8	Develop resource order.	Logistics Section Chief
9	Consider communications/medical/transportation plans.	Logistics/Planning Section Chiefs
10	Provide financial update.	Finance/Administration Section Chief
11	Discuss interagency liaison issues.	Liaison Officer
12	Discuss information issues.	Public Information Officer
13	Finalize/approve/implement plan.	Incident Commander/All

☐ 11. Participate in the development of Incident Action Plan (IAP):
- Review and approve Medical Plan (ICS Form 206).
- Provide Safety Message (ICS Form 202) and/or approved document.
- Assist in the development of the "Special Instructions" block of ICS Form 204, as requested by the Planning Section.

☐ 12. Investigate accidents that have occurred within incident areas:
- Ensure accident scene is preserved for investigation.
- Ensure accident is properly documented.
- Coordinate with incident Compensation and Claims Unit Leader, agency Risk Manager, and Occupational Safety and Health Administration (OSHA).
- Prepare accident report as per agency policy, procedures, and direction.
- Recommend corrective actions to Incident Commander and agency.

☐ 13. Coordinate critical incident stress, hazardous materials, and other debriefings, as necessary.

☐ 14. Document all activity on Unit Log (ICS Form 214).

INCIDENT OBJECTIVES (ICS 202)

1. Incident Name:	2. Operational Period:	Date From: Time From:	Date To: Time To:

3. Objective(s):

4. Operational Period Command Emphasis:

General Situational Awareness

5. Site Safety Plan Required? Yes ☐ No ☐

Approved Site Safety Plan(s) Located at: _____

6. Incident Action Plan (the items checked below are included in this Incident Action Plan):

☐ ICS 203   ☐ ICS 207   Other Attachments:
☐ ICS 204   ☐ ICS 208   ☐ _____
☐ ICS 205   ☐ Map/Chart   ☐ _____
☐ ICS 205A   ☐ Weather Forecast/Tides/Currents   ☐ _____
☐ ICS 206   ☐ _____

7. Prepared by:	Name:	Position/Title:	Signature: _____

8. Approved by Incident Commander:	Name:		Signature: _____

ICS 202	IAP Page	Date/Time:

ICS 202

■ Incident Objectives

Purpose. The Incident Objectives (ICS 202) describes the basic incident strategy, incident objectives, command emphasis/priorities, and safety considerations for use during the next operational period.

Preparation. The ICS 202 is completed by the Planning Section following each Command and General Staff meeting conducted to prepare the Incident Action Plan (IAP). In case of a Unified Command, one Incident Commander (IC) may approve the ICS 202. If additional IC signatures are used, attach a blank page.

Distribution. The ICS 202 may be reproduced with the IAP and may be part of the IAP and given to all supervisory personnel at the Section, Branch, Division/Group, and Unit levels. All completed original forms must be given to the Documentation Unit.

Notes:
- The ICS 202 is part of the IAP and can be used as the opening or cover page.
- If additional pages are needed, use a blank ICS 202 and repaginate as needed.

Block Number	Block Title	Instructions
1	**Incident Name**	Enter the name assigned to the incident. If needed, an incident number can be added.
2	**Operational Period** ☐ Date and Time From ☐ Date and Time To	Enter the start date (month/day/year) and time (using the 24-hour clock) and end date and time for the operational period to which the form applies.
3	**Objective(s)**	Enter clear, concise statements of the objectives for managing the response. Ideally, these objectives will be listed in priority order. These objectives are for the incident response for this operational period as well as for the duration of the incident. Include alternative and/or specific tactical objectives as applicable. Objectives should follow the SMART model or a similar approach: **S**pecific – Is the wording precise and unambiguous? **M**easurable – How will achievements be measured? **A**ction-oriented – Is an action verb used to describe expected accomplishments? **R**ealistic – Is the outcome achievable with given available resources? **T**ime-sensitive – What is the timeframe?
4	**Operational Period Command Emphasis**	Enter command emphasis for the operational period, which may include tactical priorities or a general weather forecast for the operational period. It may be a sequence of events or order of events to address. This is not a narrative on the objectives, but a discussion about where to place emphasis if there are needs to prioritize based on the Incident Commander's or Unified Command's direction. Examples: Be aware of falling debris, secondary explosions, etc.
	General Situational Awareness	General situational awareness may include a weather forecast, incident conditions, and/or a general safety message. If a safety message is included here, it should be reviewed by the Safety Officer to ensure it is in alignment with the Safety Message/Plan (ICS 208).
5	**Site Safety Plan Required?** Yes ☐ No ☐	Safety Officer should check whether or not a site safety plan is required for this incident.
	Approved Site Safety Plan(s) Located At	Enter the location of the approved Site Safety Plan(s).
6	**Incident Action Plan** (the items checked below are included in this Incident Action Plan): ☐ ICS 203 ☐ ICS 204 ☐ ICS 205 ☐ ICS 205A ☐ ICS 206 ☐ ICS 207 ☐ ICS 208 ☐ Map/Chart ☐ Weather Forecast/Tides/Currents Other Attachments:	Check appropriate forms and list other relevant documents that are included in the IAP. ☐ ICS 203 – Organization Assignment List ☐ ICS 204 – Assignment List ☐ ICS 205 – Incident Radio Communications Plan ☐ ICS 205A – Communications List ☐ ICS 206 – Medical Plan ☐ ICS 207 – Incident Organization Chart ☐ ICS 208 – Safety Message/Plan
7	**Prepared by** • Name • Position/Title • Signature	Enter the name, ICS position, and signature of the person preparing the form. Enter date (month/day/year) and time prepared (24-hour clock).
8	**Approved by Incident Commander** • Name • Signature • Date/Time	In the case of a Unified Command, one IC may approve the ICS 202. If additional IC signatures are used, attach a blank page.

ASSIGNMENT LIST (ICS 204)

1. Incident Name:	2. Operational Period: Date From:		Date To:	3.
	Time From:		Time To:	Branch:
4. Operations Personnel:	Name		Contact Number(s)	Division:
Operations Section Chief:				Group:
Branch Director:				Staging Area:
Division/Group Supervisor:				

5. Resources Assigned:		# of Persons	Contact (e.g., phone, pager, radio frequency, etc.)	Reporting Location, Special Equipment and Supplies, Remarks, Notes, Information
Resource Identifier	Leader			

6. Work Assignments:

7. Special Instructions:

8. Communications (radio and/or phone contact numbers needed for this assignment):

Name	/Function		Primary Contact: indicate cell, pager, or radio (frequency/system/channel)
	/		
	/		
	/		
	/		

9. Prepared by:	Name:		Position/Title:	
ICS 204	IAP Page		Date/Time:	

ICS 204

■ Assignment List

Purpose. The Assignment List(s) (ICS 204) informs Division and Group supervisors of incident assignments. Once the Command and General Staffs agree to the assignments, the assignment information is given to the appropriate Divisions and Groups.

Preparation. The ICS 204 is normally prepared by the Resources Unit, using guidance from the Incident Objectives (ICS 202), Operational Planning Worksheet (ICS 215), and the Operations Section Chief. It must be approved by the Incident Commander, but may be reviewed and initialed by the Planning Section Chief and Operations Section Chief as well.

Distribution. The ICS 204 is duplicated and attached to the ICS 202 and given to all recipients as part of the Incident Action Plan (IAP). In some cases, assignments may be communicated via radio/telephone/fax. All completed original forms must be given to the Documentation Unit.

Notes:
- The ICS 204 details assignments at Division and Group levels and is part of the IAP.
- Multiple pages/copies can be used if needed.
- If additional pages are needed, use a blank ICS 204 and repaginate as needed.

Block Number	Block Title	Instructions
1	**Incident Name**	Enter the name assigned to the incident.
2	**Operational Period** • Date and Time From • Date and Time To	Enter the start date (month/day/year) and time (using the 24-hour clock) and end date and time for the operational period to which the form applies.
3	**Branch** **Division** **Group** **Staging Area**	This block is for use in a large IAP for reference only. Write the alphanumeric abbreviation for the Branch, Division, Group, and Staging Area (e.g., "Branch 1," "Division D," "Group 1A") in large letters for easy referencing.
4	**Operations Personnel** • Name, Contact Number(s) – Operations Section Chief – Branch Director – Division/Group Supervisor	Enter the name and contact numbers of the Operations Section Chief, applicable Branch Director(s), and Division/Group Supervisor(s).
5	**Resources Assigned**	Enter the following information about the resources assigned to the Division or Group for this period:
	• Resource Identifier	The identifier is a unique way to identify a resource (e.g., ENG-13, IA-SCC-413). If the resource has been ordered but no identification has been received, use TBD (to be determined).
	• Leader	Enter resource leader's name.
	• # of Persons	Enter total number of persons for the resource assigned, including the leader.
	• Contact (e.g., phone, pager, radio frequency, etc.)	Enter primary means of contacting the leader or contact person (e.g., radio, phone, pager, etc.). Be sure to include the area code when listing a phone number.
	• Reporting Location, Special Equipment and Supplies, Remarks, Notes, Information	Provide special notes or directions specific to this resource. If required, add notes to indicate: (1) specific location/time where the resource should report or be dropped off/picked up; (2) special equipment and supplies that will be used or needed; (3) whether or not the resource received briefings; (4) transportation needs; or (5) other information.
6	**Work Assignments**	Provide a statement of the tactical objectives to be achieved within the operational period by personnel assigned to this Division or Group.
7	**Special Instructions**	Enter a statement noting any safety problems, specific precautions to be exercised, dropoff or pickup points, or other important information.
8	**Communications** (radio and/or phone contact numbers needed for this assignment) • Name/Function • Primary Contact: indicate cell, pager, or radio (frequency/system/channel)	Enter specific communications information (including emergency numbers) for this Branch/Division/Group. If radios are being used, enter function (command, tactical, support, etc.), frequency, system, and channel from the Incident Radio Communications Plan (ICS 205). Phone and pager numbers should include the area code and any satellite phone specifics. In light of potential IAP distribution, use sensitivity when including cell phone number. Add a secondary contact (phone number or radio) if needed.
9	**Prepared by** • Name • Position/Title • Signature • Date/Time	Enter the name, ICS position, and signature of the person preparing the form. Enter date (month/day/year) and time prepared (24-hour clock).

MEDICAL PLAN (ICS 206)

1. Incident Name:		2. Operational Period:	Date From:	Date To:
			Time From:	Time To:

3. Medical Aid Stations:

Name	Location	Contact Number(s)/Frequency	Paramedics on Site?
			☐ Yes ☐ No
			☐ Yes ☐ No
			☐ Yes ☐ No
			☐ Yes ☐ No
			☐ Yes ☐ No
			☐ Yes ☐ No

4. Transportation (indicate air or ground):

Ambulance Service	Location	Contact Number(s)/Frequency	Level of Service
			☐ ALS ☐ BLS
			☐ ALS ☐ BLS
			☐ ALS ☐ BLS
			☐ ALS ☐ BLS

5. Hospitals:

Hospital Name	Address, Latitude & Longitude if Helipad	Contact Number(s)/ Frequency	Travel Time		Trauma Center	Burn Center	Helipad
			Air	Ground			
					☐ Yes Level: ____	☐ Yes ☐ No	☐ Yes ☐ No
					☐ Yes Level: ____	☐ Yes ☐ No	☐ Yes ☐ No
					☐ Yes Level: ____	☐ Yes ☐ No	☐ Yes ☐ No
					☐ Yes Level: ____	☐ Yes ☐ No	☐ Yes ☐ No
					☐ Yes Level: ____	☐ Yes ☐ No	☐ Yes ☐ No

6. Special Medical Emergency Procedures:

☐ Check box if aviation assets are utilized for rescue. If assets are used, coordinate with Air Operations.

7. Prepared by (Medical Unit Leader):	Name:	Signature: _____
8. Approved by (Safety Officer):	Name:	Signature: _____
ICS 206	IAP Page	Date/Time:

ICS 206

■ Medical Plan

Purpose. The Medical Plan (ICS 206) provides information on incident medical aid stations, transportation services, hospitals, and medical emergency procedures.

Preparation. The ICS 206 is prepared by the Medical Unit Leader and reviewed by the Safety Officer to ensure ICS coordination. If aviation assets are utilized for rescue, coordinate with Air Operations.

Distribution. The ICS 206 is duplicated and attached to the Incident Objectives (ICS 202) and given to all recipients as part of the Incident Action Plan (IAP). Information from the plan pertaining to incident medical aid stations and medical emergency procedures may be noted on the Assignment List (ICS 204). All completed original forms must be given to the Documentation Unit.

Notes:
- The ICS 206 serves as part of the IAP.
- This form can include multiple pages.

Block Number	Block Title	Instructions
1	**Incident Name**	Enter the name assigned to the incident.
2	**Operational Period** • Date and Time From • Date and Time To	Enter the start date (month/day/year) and time (using the 24-hour clock) and end date and time for the operational period to which the form applies.
3	**Medical Aid Stations**	Enter the following information on the incident medical aid station(s):
	• Name	Enter name of the medical aid station.
	• Location	Enter the location of the medical aid station (e.g., Staging Area, Camp Ground).
	• Contact Number(s)/Frequency	Enter the contact number(s) and frequency for the medical aid station(s).
	• Paramedics on Site? ☐ Yes ☐ No	Indicate (yes or no) if paramedics are at the site indicated.
4	**Transportation** (indicate air or ground)	Enter the following information for ambulance services available to the incident:
	• Ambulance Service	Enter name of ambulance service.
	• Location	Enter the location of the ambulance service.
	• Contact Number(s)/Frequency	Enter the contact number(s) and frequency for the ambulance service.
	• Level of Service ☐ ALS ☐ BLS	Indicate the level of service available for each ambulance, either ALS (Advanced Life Support) or BLS (Basic Life Support).
5	**Hospitals**	Enter the following information for hospital(s) that could serve this incident:
	• Hospital Name	Enter hospital name and identify any predesignated medivac aircraft by name a frequency.
	• Address, Latitude & Longitude if Helipad	Enter the physical address of the hospital and the latitude and longitude if the hospital has a helipad.
	• Contact Number(s)/Frequency	Enter the contact number(s) and/or communications frequency(s) for the hospital.
	• Travel Time • Air • Ground	Enter the travel time by air and ground from the incident to the hospital.
	• Trauma Center ☐ Yes Level:_____	Indicate yes and the trauma level if the hospital has a trauma center.
	• Burn Center ☐ Yes ☐ No	Indicate (yes or no) if the hospital has a burn center.
	• Helipad ☐ Yes ☐ No	Indicate (yes or no) if the hospital has a helipad. Latitude and Longitude data format need to compliment Medical Evacuation Helicopters and Medical Air Resources
6	**Special Medical Emergency Procedures**	Note any special emergency instructions for use by incident personnel, including (1) who should be contacted, (2) how should they be contacted; and (3) who manages an incident within an incident due to a rescue, accident, etc. Include procedures for how to report medical emergencies.
	☐ Check box if aviation assets are utilized for rescue. If assets are used, coordinate with Air Operations.	Self explanatory. Incident assigned aviation assets should be included in ICS 220.
7	**Prepared by** (Medical Unit Leader) • Name • Signature	Enter the name and signature of the person preparing the form, typically the Medical Unit Leader. Enter date (month/day/year) and time prepared (24-hour clock).
8	**Approved by** (Safety Officer) • Name • Signature • Date/Time	Enter the name of the person who approved the plan, typically the Safety Officer. Enter date (month/day/year) and time reviewed (24-hour clock).

SAFETY MESSAGE/PLAN (ICS 208)

1. Incident Name:	2. Operational Period:	Date From: Time From:	Date To: Time To:

3. Safety Message/Expanded Safety Message, Safety Plan, Site Safety Plan:

4. Site Safety Plan Required? Yes ☐ No ☐
Approved Site Safety Plan(s) Located At:

5. Prepared by:	Name:	Position/Title:	Signature: _____
ICS 208	**IAP Page**	Date/Time:	

ICS 208

■ Safety Message/Plan

Purpose. The Safety Message/Plan (ICS 208) expands on the Safety Message and Site Safety Plan.

Preparation. The ICS 208 is an optional form that may be included and completed by the Safety Officer for the Incident Action Plan (IAP).

Distribution. The ICS 208, if developed, will be reproduced with the IAP and given to all recipients as part of the IAP. All completed original forms must be given to the Documentation Unit.

Notes:

- The ICS 208 may serve (optionally) as part of the IAP.
- Use additional copies for continuation sheets as needed, and indicate pagination as used.

Block Number	Block Title	Instructions
1	**Incident Name**	Enter the name assigned to the incident.
2	**Operational Period** • Date and Time From • Date and Time To	Enter the start date (month/day/year) and time (using the 24-hour clock) and end date and time for the operational period to which the form applies.
3	**Safety Message/Expanded Safety Message, Safety Plan, Site Safety Plan**	Enter clear, concise statements for safety message(s), priorities, and key command emphasis/decisions/directions. Enter information such as known safety hazards and specific precautions to be observed during this operational period. If needed, additional safety message(s) should be referenced and attached.
4	**Site Safety Plan Required?** Yes ☐ No ☐	Check whether or not a site safety plan is required for this incident.
	Approved Site Safety Plan(s) Located At	Enter where the approved Site Safety Plan(s) is located.
5	**Prepared by** • Name • Position/Title • Signature • Date/Time	Enter the name, ICS position, and signature of the person preparing the form. Enter date (month/day/year) and time prepared (24-hour clock).

ACTIVITY LOG (ICS 214)

1. Incident Name:	2. Operational Period:	Date From: Time From:	Date To: Time To:
3. Name:	4. ICS Position:	5. Home Agency (and Unit):	

6. Resources Assigned:

Name	ICS Position	Home Agency (and Unit)

7. Activity Log:

Date/Time	Notable Activities

8. Prepared by:	Name:	Position/Title:	Signature: _____
ICS 214, Page 1		Date/Time:	

ACTIVITY LOG (ICS 214)

1. Incident Name:	2. Operational Period:	Date From: Time From:	Date To: Time To:

7. Activity Log (continuation):

Date/Time	Notable Activities

8. Prepared by:	Name:	Position/Title:	Signature: _____
ICS 214, Page 2		Date/Time:	

ICS 214

■ Activity Log

Purpose. The Activity Log (ICS 214) records details of notable activities at any ICS level, including single resources, equipment, Task Forces, etc. These logs provide basic incident activity documentation, and a reference for any after-action report.

Preparation. An ICS 214 can be initiated and maintained by personnel in various ICS positions as it is needed or appropriate. Personnel should document how relevant incident activities are occurring and progressing, or any notable events or communications.

Distribution. Completed ICS 214s are submitted to supervisors, who forward them to the Documentation Unit. All completed original forms must be given to the Documentation Unit, which maintains a file of all ICS 214s. It is recommended that individuals retain a copy for their own records.

Notes:
- The ICS 214 can be printed as a two-sided form.
- Use additional copies as continuation sheets as needed, and indicate pagination as used.

Block Number	Block Title	Instructions
1	**Incident Name**	Enter the name assigned to the incident.
2	**Operational Period** • Date and Time From • Date and Time To	Enter the start date (month/day/year) and time (using the 24-hour clock) and end date and time for the operational period to which the form applies.
3	**Name**	Enter the title of the organizational unit or resource designator (e.g., Facilities Unit, Safety Officer, Strike Team).
4	**ICS Position**	Enter the name and ICS position of the individual in charge of the Unit.
5	**Home Agency** (and Unit)	Enter the home agency of the individual completing the ICS 214. Enter a unit designator if utilized by the jurisdiction or discipline.
6	**Resources Assigned**	Enter the following information for resources assigned:
	• Name	Use this section to enter the resource's name. For all individuals, use at least the first initial and last name. Cell phone number for the individual can be added as an option.
	• ICS Position	Use this section to enter the resource's ICS position (e.g., Finance Section Chief).
	• Home Agency (and Unit)	Use this section to enter the resource's home agency and/or unit (e.g., Des Moines Public Works Department, Water Management Unit).
7	**Activity Log** • Date/Time • Notable Activities	• Enter the time (24-hour clock) and briefly describe individual notable activities. Note the date as well if the operational period covers more than one day. • Activities described may include notable occurrences or events such as task assignments, task completions, injuries, difficulties encountered, etc. • This block can also be used to track personal work habits by adding columns such as "Action Required," "Delegated To," "Status," etc.
8	**Prepared by** • Name • Position/Title • Signature • Date/Time	Enter the name, ICS position/title, and signature of the person preparing the form. Enter date (month/day/year) and time prepared (24-hour clock).

INCIDENT ACTION PLAN SAFETY ANALYSIS (ICS 215A)

1. Incident Name:		2. Incident Number:			
3. Date/Time Prepared:		4. Operational Period:		Date From:	Date To:
Date:	Time:			Time From:	Time To:
5. Incident Area	6. Hazards/Risks			7. Mitigations	
8. Prepared by (Safety Officer):		Name:		Signature: _____	
Prepared by (Operations Section Chief):		Name:		Signature: _____	
ICS 215A				Date/Time:	

ICS 215A

■ Incident Action Plan Safety Analysis

Purpose. The purpose of the Incident Action Plan Safety Analysis (ICS 215A) is to aid the Safety Officer in completing an operational risk assessment to prioritize hazards, safety, and health issues, and to develop appropriate controls. This worksheet addresses communications challenges between planning and operations, and is best utilized in the planning phase and for Operations Section briefings.

Preparation. The ICS 215A is typically prepared by the Safety Officer during the incident action planning cycle. When the Operations Section Chief is preparing for the tactics meeting, the Safety Officer collaborates with the Operations Section Chief to complete the Incident Action Plan Safety Analysis. This worksheet is closely linked to the Operational Planning Worksheet (ICS 215). Incident areas or regions are listed along with associated hazards and risks. For those assignments involving risks and hazards, mitigations or controls should be developed to safeguard responders, and appropriate incident personnel should be briefed on the hazards, mitigations, and related measures. Use additional sheets as needed.

Distribution. When the safety analysis is completed, the form is distributed to the Resources Unit to help prepare the Operations Section briefing. All completed original forms must be given to the Documentation Unit.

Notes:

- This worksheet can be made into a wall mount, and can be part of the IAP.
- If additional pages are needed, use a blank ICS 215A and repaginate as needed.

Block Number	Block Title	Instructions
1	**Incident Name**	Enter the name assigned to the incident.
2	**Incident Number**	Enter the number assigned to the incident.
3	**Date/Time Prepared**	Enter date (month/day/year) and time (using the 24-hour clock) prepared.
4	**Operational Period** • Date and Time From • Date and Time To	Enter the start date (month/day/year) and time (24-hour clock) and end date and time for the operational period to which the form applies.
5	**Incident Area**	Enter the incident areas where personnel or resources are likely to encounter risks. This may be specified as a Branch, Division, or Group.
6	**Hazards/Risks**	List the types of hazards and/or risks likely to be encountered by personnel or resources at the incident area relevant to the work assignment.
7	**Mitigations**	List actions taken to reduce risk for each hazard indicated (e.g., specify personal protective equipment or use of a buddy system or escape routes).
8	**Prepared by** (Safety Officer and Operations Section Chief) • Name • Signature • Date/Time	Enter the name of both the Safety Officer and the Operations Section Chief, who should collaborate on form preparation. Enter date (month/day/year) and time (24-hour clock) reviewed.

SITE SAFETY AND CONTROL PLAN (ICS 208 HM)

1. Incident Name:	2. Date Prepared:	3. Operational Period: Time:

Section I. Site Information

4. Incident Location:

Section II. Organization

5. Incident Commander:	6. HM Group Supervisor:	7. Tech. Specialist - HM Reference:
8. Safety Officer:	9. Entry Leader:	10. Site Access Control Leader:
11. Asst. Safety Officer - HM:	12. Decontamination Leader:	13. Safe Refuge Area Mgr:
14. Environmental Health:	15.	16.

17. Entry Team: (Buddy System)			18. Decontamination Element:		
	Name:	PPE Level		Name:	PPE Level
Entry 1			Decon 1		
Entry 2			Decon 2		
Entry 3			Decon 3		
Entry 4			Decon 4		

Section III. Hazard/Risk Analysis

19. Material:	Container type	Qty.	Phys. State	pH	IDLH	F.P.	I.T.	V.P.	V.D.	S.G.	LEL	UEL

Comment:

Section IV. Hazard Monitoring

20. LEL Instrument(s):	21. O_2 Instrument(s):
22. Toxicity/PPM Instrument(s):	23. Radiological Instrument(s):

Comment:

Section V. Decontamination Procedures

24. Standard Decontamination Procedures:	YES:	NO:

Comment:

Section VI. Site Communications

25. Command Frequency:	26. Tactical Frequency:	27. Entry Frequency:

Section VII. Medical Assistance

28. Medical Monitoring:	YES:	NO:	29. Medical Treatment and Transport In-place:	YES:	NO:

Comment:

Section VIII. Site Map

30. Site Map:

Weather ☐	Command Post ☐	Zones ☐	Assembly Areas ☐	Escape Routes ☐	Other ☐

Section IX. Entry Objectives					
31. Entry Objectives:					

Section X. SOP S and Safe Work Practices

32. Modifications to Documented SOP s or Work Practices:	YES:	NO:
Comment:		

Section XI. Emergency Procedures

33. Emergency Procedures:

Section XII. Safety Briefing

34. Asst. Safety Officer - HM Signature: _____	Safety Briefing Completed (Time):
35. HM Group Supervisor Signature: _____	36. Incident Commander Signature: _____

INSTRUCTIONS FOR COMPLETING THE SITE SAFETY AND CONTROL PLAN

■ ICS 208 HM

A Site Safety and Control Plan must be completed by the Hazardous Materials Group Supervisor and reviewed by all within the Hazardous Materials Group prior to operations commencing within the Exclusion Zone.

Item Number	Item Title	Instructions
1.	Incident Name/Number	Print name and/or incident number.
2.	Date and Time	Enter date and time prepared.
3.	Operational Period	Enter the time interval for which the form applies.
4.	Incident Location	Enter the address and or map coordinates of the incident.
5–16.	Organization	Enter names of all individuals assigned to ICS positions. (Entries 5 & 8 mandatory). Use Boxes 15 and 16 for other functions: i.e. Medical Monitoring.
17–18.	Entry Team/Decon Element	Enter names and level of PPE of Entry & Decon personnel. (Entries 1 - 4 mandatory buddy system and back-up.)
19.	Material	Enter names and pertinent information of all known chemical products. Enter UNK if material is not known. Include any which apply to chemical properties. (Definitions: ph = Potential for Hydrogen (Corrosivity), IDLH = Immediately Dangerous to Life and Health, F.P. = Flash Point, I.T. = Ignition Temperature, V.P. = Vapor Pressure, V.D. = Vapor Density, S.G. = Specific Gravity, LEL = Lower Explosive Limit, UEL = Upper Explosive Limit)
20–23.	Hazard Monitoring	List the instruments which will be used to monitor for chemical.
24.	Decontamination Procedures	Check NO if modifications are made to standard decontamination procedures and make appropriate Comments including type of solutions.
25–27.	Site Communications	Enter the radio frequency(ies) which apply.
28–29.	Medical Assistance	Enter comments if NO is checked.
30.	Site Map	Sketch or attach a site map which defines all locations and layouts of operational zones. (Check boxes are mandatory to be identified.)
31.	Entry Objectives	List all objectives to be performed by the Entry Team in the Exclusion Zone and any parameters which will alter or stop entry operations.
32–33.	SOP s, Safe Work Practices, and Emergency Procedures	List in Comments if any modifications to SOP s and any emergency procedures which will be affected if an emergency occurs while personnel are within the Exclusion Zone.
34–36.	Safety Briefing	Have the appropriate individual place their signature in the box once the Site Safety and Control Plan is reviewed. Note the time in box 34 when the safety briefing has been completed.

APPENDIX

D

© Photos.com

NFPA 1521 - Incident Safety Officer - 2020 Edition						
IMPORTANT: The language from the standard on the AMMs is truncated. When completing the AMMs, the agency must refer to the NFPA standards for the complete text and a comprehensive statement of each Job Performance Requirement (JPR), Requisite Knowledge (RK), Requisite Skill (RS), and any applicable annex or explanatory information.						
INSTRUCTIONS: Please review the instructions for filling out this Assessment Methodology Matrix at http://theproboard.org/AMM.htm. The submission of this form by an agency is affirmation that it is filled out in accordance with the instructions listed above.						
AGENCY NAME:					**DATE COMPILED:**	
	OBJECTIVE / JPR, RK, RS		**COGNITIVE**	**MANIPULATIVE**			
SECTION	**ABBREVIATED TEXT**	**WRITTEN TEST**	**SKILLS STATION**	**PORTFOLIO**	**PROJECTS**	**OTHER**
5.1.1	The fire department incident safety officer shall meet the requirements of Fire Officer Level I specified in NFPA 1021 and the JPR's defined in section 5.2 through 5.7	**Prerequisite requirements should be met according to the approved agency policy**				
5.1.2	A fire department ISO shall recuse himself/herself from any investigatory process where a conflict of interest exists					Chapter 16 (p 237)
5.2.1	Perform the role of ISO within an ICS at an incident or planned event					Chapter 1 (p 12), Chapter 2 (pp 17-26), Chapter 3 (pp 33-34), Chapter 10 (pp 151, 154-155), Chapter 11 (pp 165-167), Chapter 12 (pp 176-177, 181-183, 187)
5.2.1(A)	RK: Understand accepted safety and health principles					Chapter 1 (p 12), Chapter 2 (pp 17-26), Chapter 3 (pp 33-34), Chapter 10 (pp 151, 154-155), Chapter 11 (pp 165-167)
5.2.1(B)	RS: Prioritizing tasks, making decisions in an environment with a large number of unknowns					Chapter 12 (pp 176-177, 181-183, 187)
5.2.2	Monitor the IAP, conditions, activities, and operations					Chapter 3 (pp 33-34), Chapter 7 (pp 97-100), Chapter 10 (pp 147-148, 151-155), Chapter 13 (pp 197-198)
5.2.2(A)	RK: Comprehensive knowledge of incident hazards					Chapter 3 (pp 33-34), Chapter 7 (pp 97-100), Chapter 10 (pp 147-148, 151-152, 154-155), Chapter 12 (pp 178-181), Chapter 13 (pp 197-198)

NFPA 1521 - Incident Safety Officer - 2020 Edition

IMPORTANT: The language from the standard on the AMMs is truncated. When completing the AMMs, the agency must refer to the NFPA standards for the complete text and a comprehensive statement of each Job Performance Requirement (JPR), Requisite Knowledge (RK), Requisite Skill (RS), and any applicable annex or explanatory information.

INSTRUCTIONS: Please review the instructions for filling out this Assessment Methodology Matrix at http://theproboard.org/AMM.htm. The submission of this form by an agency is affirmation that it is filled out in accordance with the instructions listed above.

AGENCY NAME:					DATE COMPILED:	
OBJECTIVE / JPR, RK, RS		**COGNITIVE**	**MANIPULATIVE**			
SECTION	**ABBREVIATED TEXT**	**WRITTEN TEST**	**SKILLS STATION**	**PORTFOLIO**	**PROJECTS**	**OTHER**
5.2.2(B)	RS: Ability to apply knowledge of fire behavior and fire dynamics					Chapter 10 (pp 152-155), Chapter 12 (pp 178-181)
5.2.3	Mange the transfer of ISO duties					Chapter 10 (p 156)
5.2.3(A)	RK: AHJ's procedures for transfer of duty					Chapter 10 (p 156)
5.2.3(B)	RS: Conducting a transfer briefing meeting					Chapter 10 (p 156)
5.2.4	Stop, alter, or suspend operations based on imminent threats posed to fire fighter safety					Chapters 7 (pp 96-100), Chapter 10 (pp 151-152), Chapter 11 (pp 166-168), Chapter 12 (pp 177-178, 181-182), Chapter 13 (pp 194-196), Chapter 14 (pp 206, 208, 212), Chapter 15 (pp 218-227)
5.2.4(A)	RK: Knowledge of what constitutes imminent hazards					Chapter 7 (p 96), Chapter 10 (pp 148, 151-152), Chapter 11 (p 168), Chapter 12 (pp 177-178, 181-182) Chapter 13 (pp 194-196), Chapter 14 (pp 206, 208, 212), Chapter 15 (pp 218-227)
5.2.4(B)	RS: Ability to evaluate hazards					Chapter 7 (pp 97-100), Chapter 10 (152-155), Chapter 11 (pp 166-167)
5.2.5	Monitor and determine the incident scene conditions					Chapter 11 (p 166), Chapter 12 (pp 181-182), Chapter 13 (pp 194-198), Chapter 14 (pp 206-208), Chapter 15 (pp 220-227)
5.2.5(A)	RK: Knowledge of what constitutes hazards at an emergency incident					Chapter 11 (p 166), Chapter 13 (pp 194-198), Chapter 14 (pp 206-208), Chapter 15 (pp 220-227)
5.2.5(B)	RS: Ability to evaluate hazards					Chapter 12 (pp 181-182), Chapter 13 (pp 194)
5.2.6	Monitor the accountability system					Chapter 11 (pp 169-170), Chapter 12 (pp 181-183)
5.2.6(A)	RK: Knowledge of incident management system					Chapter 11 (pp 169-170), Chapter 12 (p 183)
5.2.6(B)	RS: Ability to recognize inadequacies in the use of accountability systems					Chapter 11 (pp 169-170), Chapter 12 (pp 181-183)
5.2.7	Determine hazardous incident conditions					Chapter 8 (pp 107, 118-119), Chapter 11 (p 171)
5.2.7(A)	RK: Comprehensive knowledge of hazardous conditions					Chapter 8 (pp 107, 118-119), Chapter 11 (p 171)
5.2.7(B)	RS: Ability to evaluate the effect of proximity for incident hazards					Chapter 11 (p 171)

(Continued)

		NFPA 1521 - Incident Safety Officer - 2020 Edition					
	**IMPORTANT: The language from the standard on the AMMs is truncated. When completing the AMMs, the agency must refer to the NFPA standards for the complete text and a comprehensive statement of each Job Performance Requirement (JPR), Requisite Knowledge (RK), Requisite Skill (RS), and any applicable annex or explanatory information.**

| | **INSTRUCTIONS: Please review the instructions for filling out this Assessment Methodology Matrix at http://theproboard.org/AMM.htm. The submission of this form by an agency is affirmation that it is filled out in accordance with the instructions listed above.** |

AGENCY NAME:					**DATE COMPILED:**		
	OBJECTIVE / JPR, RK, RS		**COGNITIVE**	**MANIPULATIVE**			
SECTION	**ABBREVIATED TEXT**		**WRITTEN TEST**	**SKILLS STATION**	**PORTFOLIO**	**PROJECTS**	**OTHER**
5.2.8	Identify motor vehicle incident scene hazards						Chapter 15 (pp 226-227)
5.2.8(A)	RK: Knowledge of hazards associated with vehicle incidents						Chapter 15 (pp 226-227)
5.2.8(B)	RS: Ability to apply knowledge of hazards and regulations to an incident						Chapter 15 (pp 226-227)
5.2.9	Monitor radio transmissions						Chapter 10 (pp 152-153)
5.2.9(A)	RK: Knowledge of radio protocols and transmission procedures						Chapter 10 (pp 152-153)
5.2.9(B)	RS: Ability to recognize missed, unclear, or incomplete communications						Chapter 10 (pp 152-153)
5.2.10	Identify the incident strategic requirements						Chapter 10 (pp 155-158), Chapter 12 (p 185), Chapter 14 (pp 209-211), Chapter 15 (pp 217-218)
5.2.10(A)	RK: Comprehensive knowledge of incident hazards						Chapter 10 (pp 155-158), Chapter 12 (p 185), Chapter 14 (pp 209-211), Chapter 15 (pp 217-218)
5.2.10(B)	RS: Ability to recognize the types of hazards that might require additional ISO's						Chapter 14 (pp 209-211)
5.2.11	Determine the hazards associated with the designation of a landing zone and interface with helicopters						Chapter 15 (pp 220-221)
5.2.11(A)	RK: Helicopter and landing zone requirements						Chapter 15 (pp 220-221)
5.2.11(B)	RS: Ability to recognize landing zone locations and hazards						Chapter 15 (pp 220-221)
5.2.12	Notify the IC of the need for intervention resulting from an occupational exposure to atypical stressful events						Chapter 16 (pp 241-242)
5.2.12(A)	RK: Knowledge of incidents that can lead to occupational exposure to atypical stress						Chapter 16 (pp 241-242)
5.2.12(B)	RS: Ability to recognize signs and symptoms of occupational exposure to atypical stress						Chapter 16 (pp 241-242)
5.2.13	Determine hazardous energy sources that can affect responder health and safety						Chapter 8 (pp 105-119)
5.2.13(A)	RK: Common component assemblies for hazardous energy sources						Chapter 8 (pp 105-119)

NFPA 1521 - Incident Safety Officer - 2020 Edition						
IMPORTANT: The language from the standard on the AMMs is truncated. When completing the AMMs, the agency must refer to the NFPA standards for the complete text and a comprehensive statement of each Job Performance Requirement (JPR), Requisite Knowledge (RK), Requisite Skill (RS), and any applicable annex or explanatory information.						
INSTRUCTIONS: Please review the instructions for filling out this Assessment Methodology Matrix at http://theproboard.org/AMM.htm. The submission of this form by an agency is affirmation that it is filled out in accordance with the instructions listed above.						
AGENCY NAME:				**DATE COMPILED:**		
OBJECTIVE / JPR, RK, RS		**COGNITIVE**	**MANIPULATIVE**			
SECTION	**ABBREVIATED TEXT**	**WRITTEN TEST**	**SKILLS STATION**	**PORTFOLIO**	**PROJECTS**	**OTHER**
5.2.13(B)	RS: Critical identification, analysis, and judgment abilities					Chapter 8 (pp 105-117)
5.2.14	Monitor conditions, including weather, firefighter activities, and work cycle durations					Chapter 9 (pp 128-131, 135-138)
5.2.14(A)	RK: Comprehensive knowledge of heat and cold assessment criteria					Chapter 9 (pp 128-131, 135-138)
5.2.14(B)	RS: Ability to recognize signs of cardiac, heat, and cold stress					Chapter 9 (pp 135-137)
5.2.15	Identify contaminates and the need for contamination reduction procedures					Chapter 6 (pp 80-82, p 28)
5.2.15(A)	RK: Common byproducts, NFPA 1851, and methods of control					Chapter 6 (pp 80-82, p 87), Chapter 12 (pp 187)
5.2.15(B)	RS: Ability to read conditions, recognize exposure, and judge exposure reduction efforts					Chapter 6 (pp 80-82, p 87) Chapter 12 (p 187-188)
5.3.1	Determine incident environmental and operational factors					Chapter 12 (pp 184-185)
5.3.1(A)	RK: RIC criteria for NFPA 1500,1561,1710,1720					Chapter 12 (pp 184-185)
5.3.1(B)	RS: Interpret applicable regulations, guidelines, procedures					Chapter 12 (pp 184-185)
5.3.2	Communicate fire behavior					Chapter 5 (p 74), Chapter 12 (p 185)
5.3.2(A)	RK: Structural/compartmentalized fire behavior					Chapter 5 (p 74), Chapter 12 (p 185)
5.3.2(B)	RS: Ability to interpret fire suppression hazards and operations					Chapter 12 (p 185)
5.3.3	Identify and estimate building/structural collapse hazards					Chapter 5 (pp 58-61, 65-69, 71-74)
5.3.3(A)	RK: Building construction classifications and associated hazards					Chapter 5 (pp 58-61, 65-69, 70-74)
5.3.3(B)	RS: Critical identification, analysis, and judgment abilities					Chapter 5 (pp 64-67, 70-74)
5.3.4	Determine flashover and hostile fire event potential at building fires					Chapter 6 (pp 80-90), Chapter 12 (pp 180-181)
5.3.4(A)	RK: Compartmentalized fire behavior theory					Chapter 6 (pp 80-90), Chapter 12 (pp 180-181)
5.3.4(B)	RS: Critical identification, analysis, and judgment abilities					Chapter 6 (pp 86-90)

(Continued)

NFPA 1521 - Incident Safety Officer - 2020 Edition						
IMPORTANT: The language from the standard on the AMMs is truncated. When completing the AMMs, the agency must refer to the NFPA standards for the complete text and a comprehensive statement of each Job Performance Requirement (JPR), Requisite Knowledge (RK), Requisite Skill (RS), and any applicable annex or explanatory information.						
INSTRUCTIONS: Please review the instructions for filling out this Assessment Methodology Matrix at http://theproboard.org/AMM.htm. The submission of this form by an agency is affirmation that it is filled out in accordance with the instructions listed above.						
AGENCY NAME:				DATE COMPILED:		
OBJECTIVE / JPR, RK, RS		COGNITIVE	MANIPULATIVE			
SECTION	ABBREVIATED TEXT	WRITTEN TEST	SKILLS STATION	PORTFOLIO	PROJECTS	OTHER
5.3.5	Determine fire growth and blow up, given wildland and cultivated vegetation fires					Chapter 13 (pp 194-195)
5.3.5(A)	RK: Wildland and vegetation fire behavior					Chapter 13 (pp 194-195)
5.3.5(B)	RS: Critical identification, analysis, and judgment abilities					Chapter 13 (p 194)
5.3.6	Determine the suitability of building entry and egress options at building fires					Chapter 12 (pp 178-180, 185)
5.3.6(A)	RK: Building construction access and egress challenges					Chapter 12 (pp 178-180)
5.3.6(B)	RS: Critical identification, analysis, and judgment abilities					Chapter 12 (p 185)
5.4.1	Determine the need for a rescue technician-trained ISO or assistant ISO					Chapter 15 (pp 217-222)
5.4.1(A)	RK: Technical rescue incident types as defined in NFPA 1006					Chapter 15 (pp 217-222)
5.4.1(B)	RS: Identifying technical rescue incident resource needs					Chapter 15 (pp 217-222)
5.4.2	Prepare a safety plan that identifies corrective or preventive actions, given a technical rescue incident					Chapter 15 (pp 218-224)
5.4.2(A)	RK: Risk management principles; technical rescue operations strategies					Chapter 15 (pp 218-224)
5.4.2(B)	RS: Critical identification, analysis, and judgment abilities					Chapter 15 (pp 218-220)
5.4.3	Deliver a safety briefing for technical rescue incident response members					Chapter 15 (pp 218, 223-224)
5.4.3(A)	RK: OSHA 29 CFR 1910.146 requirements for a site safety and health plan					Chapter 15 (pp 218, 223-224)
5.4.3(B)	RS: Ability to communicate critical messages in written and oral formats					Chapter 15 (pp 223-224)
5.5.1	Determine the need for a hazardous materials technician trained ISO or assistant ISO					Chapter 14 (pp 205-206)
5.5.1(A)	RK: Hazardous materials incident types as defined in NFPA 472					Chapter 14 (pp 205-206)
5.5.1(B)	RS: Identifying hazardous materials incident resource needed					Chapter 14 (pp 204-205)

NFPA 1521 - Incident Safety Officer - 2020 Edition

IMPORTANT: The language from the standard on the AMMs is truncated. When completing the AMMs, the agency must refer to the NFPA standards for the complete text and a comprehensive statement of each Job Performance Requirement (JPR), Requisite Knowledge (RK), Requisite Skill (RS), and any applicable annex or explanatory information.

INSTRUCTIONS: Please review the instructions for filling out this Assessment Methodology Matrix at http://theproboard.org/AMM.htm. The submission of this form by an agency is affirmation that it is filled out in accordance with the instructions listed above.

AGENGY NAME:					DATE COMPILED:	

SECTION	OBJECTIVE / JPR, RK, RS — ABBREVIATED TEXT	COGNITIVE WRITTEN TEST	MANIPULATIVE SKILLS STATION	PORTFOLIO	PROJECTS	OTHER
5.5.2	Prepare a safety plan that identifies corrective or preventive actions, given a hazmat incident					Chapter 14 (pp 208-209, 211)
5.5.2(A)	RK: Risk management principles; hazmat operations strategies					Chapter 14 (pp 208-209, 211)
5.5.2(B)	RS: Critical identification, analysis, and judgment abilities					Chapter 14 (p 211)
5.5.3	Deliver a safety briefing for hazmat incident response members					Chapter 14 (p 211)
5.5.3(A)	RK: OSHA 29 CFR 1910.120 requirements for a site safety and health plan					Chapter 14 (p 211)
5.5.3(B)	RS: Ability to communicate critical messages in written and oral formats					Chapter 14 (p 211)
5.5.4	Identify that hazard materials incident control zones have been established and communicated to personnel on the scene					Chapter 14 (pp 209-210)
5.5.4(A)	RK: Common zoning strategies for hazmat operations					Chapter 14 (pp 209-210)
5.5.4(B)	RS: Ability to adapt zoning strategies to individual incident challenges					Chapter 14 (pp 209-210)
5.6.1	Conduct a safety and health investigative process					Chapter 16 (pp 236-240)
5.6.1(A)	RK: Procedures for conducting, documenting, recording, and reporting a safety investigation					Chapter 16 (pp 236, 238-240)
5.6.1(B)	RS: Analyzing information from different data sources					Chapter 16 (pp 236-240)
5.7.1	Prepare a written post incident analysis from the ISO perspective					Chapter 16 (pp 233-236)
5.7.1(A)	RK: NFPA 1500, PIA reporting criteria, and AHJ SOP/G's for PIA's					Chapter 16 (pp 233-236)
5.7.1(B)	RS: Transferring incident observations into field notes					Chapter 16 (pp 234-236)
5.7.2	Report observations, concerns, and recommendations given a witnessed incident					Chapter 16 (pp 234-236)
5.7.2(A)	RK: Group dynamics in problem solving					Chapter 16 (p 234)
5.7.2(B)	RS: Active listening skills					Chapter 16 (pp 234-236)

Glossary

accident chain A series of conditions and events that lead to an unsafe situation that results in injury and/or property damage. It typically has five components: environment, human factors, equipment, an event, and the injury.

acclimation An individual's gradual process of becoming accustomed to an environment.

accommodation The efforts to alter or adjust the environment, worker relationship, or task to reduce injury potential.

action model A template that outlines a mental or physical process that considers inputs that lead to an output or outcomes.

active cooling The process of using external methods or devices (e.g., hand and forearm cold water/ice immersion, covering the head and neck with cold wet towels, and gel cooling vests) to reduce elevated body core temperatures.

acute stress disorder (ASD) The behavioral issues that arise in the days following an atypical stress event (ASE).

assistant safety officer (ASO) A member of the fire department appointed by the incident commander to assist the ISO in the performance of the ISO functions at an incident scene.

assistant safety officer–hazmat (ASO-HM) A person who meets or exceeds the NFPA 472 requirements for a hazardous materials technician and is trained in the responsibilities of the ISO position as it relates to hazmat response.

assistant safety officer–rescue technician (ASO-RT) A person who meets or exceeds rescue technician-level competencies.

atypical stressful event (ASE) An incident that presents mental or emotional pressure or strain circumstances that are outside the ordinary experiences of the responders.

atypically stressful incident An incident that involves an unusually gruesome situation, serious firefighter injury, firefighter death, or other potential psychological stress.

axial load A load that is imposed through the centroid of another object.

backdraft An explosive event that occurs when air is suddenly reintroduced into a closed space that is filled with pressurized, ignition-temperature, and oxygen-deprived products of combustion and pyrolysis.

balloon framing A construction method in which continuous wood studs run from the foundation to the roof, and floors are placed on a shelf—called a ribbon board—that hangs on the interior surface of the studs.

basic/surface collapse A collapse in which victims are easily accessible and trapped by surface debris. Loads are minimal and easily moved by rescuers. The threat of secondary collapse is minimal.

beam A structural element that transfers loads perpendicularly to the imposed load.

black fire A slang term for smoke that is high-volume, has turbulent velocity, is ultra-dense, and is deep black; a sign of impending autoignition and flashover.

blowup A wildland fire term used to describe the sudden advancement and increase in fire intensity due to wind, prewarmed fuels, or a topographical feature such as a narrow canyon or "chimney."

brittle Description for a material that will fracture or fail as it is deformed or stressed past its design limits.

building geometry The constellation of elements that must be taken into consideration at a structure fire, including issues associated with the building's layout, size, number of floors, access options, and other features.

cantilever beam A beam supported at only one end, or a beam that extends well past a support in such a way that the unsupported overhang places the top of the beam in tension and the bottom in compression.

case law A precedent established over time through the judicial process.

circadian rhythm A person's physiological response to the 24-hour clock, which includes sleep, energy peaks, and necessary body functions.

close call An unintentional, unsafe occurrence that could have produced an injury, fatality, or property damage; only a fortunate break in the chain of events prevented the undesirable outcome.

code A work of law established or adopted by a rule-making authority. It serves to regulate an approach, system, or topic for which it is written.

Code of Federal Regulations (CFR) OSHA regulations that often outline the equipment required to accomplish a given process.

cold zone The area surrounding a warm zone that is used for an incident command post, support-agency interfacing, and staging of personnel and equipment.

collapse zone The area that is exposed to trauma, debris, and/or thrust should a building or part of a building collapse. It is a more specific form of a no-entry zone.

column A structural element that transmits a compressive force axially through its center.

compression A force that causes a material to be crushed or flattened axially through the material.

connection A structural element used to attach other structural elements to one another.

contaminated safe refuse area A hazmat zone established for obvious or potentially contaminated victims who are awaiting mass decontamination efforts and assistance.

contamination reduction zone An area where progressive decontamination is accomplished for hazmat teams (and support personnel).

continuous beam A beam that is supported in three or more places

control line All constructed or natural barriers and treated fire edges used to control a wildland fire.

countermeasure An action used to effect hazard mitigation.

curtain wall A non-load-bearing wall that supports only itself and is used only to keep weather out.

dead load The weight of the building itself and anything permanently attached to it.

debriefing A scheduled event that is characterized by a formal agenda designed to promote healing or closure and to outline the process or options for accessing more assistance.

decontamination corridor An area where progressive mass decontamination is taking place for victims.

defusing A gathering during or immediately after the incident that is characterized by an informal, peer-to-peer discussion format designed to share observations, actions, and feelings.

ductile Description for a material that will bend, deflect, or stretch as a force is resisted, yet retain some strength.

due diligence A legal phrase for the effort to act in a reasonable or prudent way, given the circumstances, with due regard to laws, standards, and accepted professional conduct.

eccentric load A load that is imposed off-center to another object.

education The process of developing one's analytical ability using principles, concepts, and values.

emergency abandonment A strict order for all crews to *immediately escape* from a building interior or roof, leaving hose lines and tools that can impede rapid retreat behind.

engineered wood A host of products that consist of many pieces of native wood (chips, veneers, and sawdust) glued together to make a sheet, a long beam, or a strong column.

ergonomics The science of adapting work or working conditions to a worker and the study of problems associated with people adjusting to their work environment.

escape area A hazmat zone used for technicians who are operating in a hot zone. The dedicated technician rapid intervention team often stages there.

explosive growth phase A rapid fire growth phenomenon that occurs when combustion air is reintroduced into a ventilation-controlled fire, leading to smoke flame-over and room flashovers.

false work Temporary shoring, bracing, or formwork used to support incomplete structural elements during building construction.

fire control team A minimum of two qualified firefighters (not a student or the safety officer) used as an ignition officer and a ready observer (in full PPE) to rescue the ignition officer should something go wrong at live-fire trainings.

fire line The part of a control line that is scraped or dug to mineral soil.

fire storm A violent convection caused by a large continuous area of intense fire. They are characterized by violent surface in-drafts near the fire perimeter and occasional tornado-like whirls.

firm intervention A direct order to immediately stop, alter, suspend, or withdraw personnel, activities, and operations due to an imminent threat.

flame-over A hostile fire event that includes the ignition and sustained burning of the overhead smoke layer within a room and/or hallway.

flare-up A sudden, but short-lived, rise in wildland fire intensity that is usually attributed to wind, fuel, or topographical changes.

flashover A sudden hostile fire event that occurs when all the surfaces and contents of a space reach their ignition temperature nearly simultaneously, resulting in full-room fire involvement.

flow path An avenue that heat, smoke, flames, and combustion air follow.

formal process A process defined in writing. It can take on many forms: standard operating procedures, standard operating guidelines, departmental directives, temporary memorandums, and the like.

freelancing (1) The act of working outside the parameters of an IAP, or (2) an individual performing incident functions out of the sight or voice range of others.

general collapse The complete failure of a building to resist gravity.

ghosting A hostile fire event warning sign that is characterized as the intermittent ignition of small pockets of smoke; usually seen as fingers of flame that dance through the upper smoke layer.

girder A beam that carries other beams.

ground gradient Electrical energy that has established a path to ground through the earth and that is now energizing the ground.

guide A publication that offers procedures, directions, or standards of care as a reasonable means to address a condition or situation.

guideline An adaptable template that offers wide flexibility in application.

hazardous energy The unintended and often sudden release of stored, residual, or potential energy that will cause harm if contacted.

health and safety officer (HSO) The individual assigned and authorized by the fire chief as the manager of the health and safety program.

heavy collapse A collapse in which stressed concrete, reinforced concrete, and steel girders are impeding access. Included are collapses that require the response of urban search and rescue teams and specialized equipment, and collapses that threaten other structures or that involve the possibility of significant secondary collapse.

high rescue profile A classification indicating a high likelihood that a person located within a fire-involved structure can be successfully rescued.

hostile fire event A fire behavior phenomenon that can suddenly harm firefighters; events include explosive growth phase, flashover, backdraft, smoke explosion, and flame-over.

hot zone The area immediately surrounding a hazardous area that can be considered an immediately dangerous to life and health (IDLH) environment requiring appropriate PPE, accountability procedures, and a standby rapid intervention crew.

hybrid building A building that is a mix of multiple NFPA 220 types or that does not fit into any of the five types.

imminent threat An act or condition that is judged to present a danger to persons or property that is so urgent and severe that it requires immediate corrective or preventive action.

incident benchmark A key phrase that signals the completion of significant tactical objectives. Common examples include *360 complete, all clear, under control,* and *loss stopped.*

incident safety officer (ISO) A member of the command staff responsible for monitoring and assessing safety hazards or unsafe situations and for developing measures to ensure personnel safety.

incident stabilization mode An approach to incident response used when victims have been rescued (or recovered when appropriate) yet the conditions are such that a potential for further harm exists and must be addressed.

informal process A process or operation that is part of a department's routine but that is not written. Because such processes are not written, they are typically learned through new member training, on-the-job training, and day-to-day routine.

law An enforceable rule of conduct that helps protect a society. From a legal perspective, laws are divided into *statutory laws* and *case laws.*

light collapse A collapse in which usually a light-frame (wood) partition collapses and common fire department equipment (from engine and truck companies) can access or shore areas for search and extrication. The threats of secondary collapse can be mitigated easily.

lintel A beam that spans an opening in a load-bearing masonry wall, such as over a garage door opening (often called a "header" in street slang).

live load Any force or weight, other than the building itself, that a building must carry or absorb.

marginal rescue profile A classification indicating a moderate likelihood that a person located within a fire-involved structure can be successfully rescued; in this scenario, there is clear danger to occupants and rescuers but no compelling evidence to indicate a high or zero rescue profile.

mitigation The overall strategy of hazard control.

mitigation hierarchy A preferred order of hazard control strategies: elimination, reduction, adaptation, transfer, and avoidance.

moderate collapse A collapse of ordinary construction that involves masonry materials and heavy wood. Lightweight construction with unstable large open spans should also be classified as moderate. Significant void space concerns are present.

National Incident Management System (NIMS) An incident response system developed by the Department of Homeland Security.

National Interagency Incident Management System (NIIMS) An incident response system developed by the National Wildfire Coordinating Group.

no-entry zone The area in which no responders are allowed to enter, regardless of PPE, due to dangerous conditions.

on-deck system An organized system in which a working team is replaced with another working team that is already dialed in and ready to replace them. Once replaced, the rotated workers go to rehab, while a crew from rehab rotates into the vacated on-deck position.

partial collapse An event in which the building can accept the failure of a single component and still retain some strength (such as a curtain wall collapse).

passive cooling The use of shade, air movement, and rest to bring down core temperatures.

personnel accountability report (PAR) An organized reporting activity (usually via radio communication) designed to account for all personnel working an incident. To be truly effective, it should include radio transmissions that include confirmation of the assignment, location, and number of people in each assignment.

platform framing A construction method in which a single-story wall is built and the next floor is built on the tops of the wall studs, creating vertical fire-stopping to help minimize fire spread.

post-traumatic stress disorder (PTSD) A mental health disorder that can develop in individuals who have experienced a terrifying ordeal that involved physical harm or the threat of harm.

postincident analysis (PIA) A formal and/or informal reflective discussion that fire departments use to summarize the successes and areas requiring improvement discovered from an incident.

postincident thought patterns Reflective or introspective mental preoccupations that firefighters experience just after incident control.

potential threat An activity, condition, or inaction that is judged to have the capability, but not immediacy, to cause harm to persons or property and thus warrants monitoring and/or operational modification.

precautionary withdrawal A directive for crews to exit a building interior or roof in an orderly manner, bringing hoses and tools along.

procedure A strict directive that must be followed with little or no flexibility.

pyrolysis Also referred to as *pyrolitic decomposition*, the chemical breakdown of compounds into other substances by heat alone.

raker A diagonal brace that serves primarily as a column but must absorb some beam forces as well.

recognition-primed decision making (RPD) A mental model that suggests that many quick decisions are made using mental templates from previous experiences that fit the images that you are currently witnessing.

reconnaissance An exploratory examination of the incident scene conditions and activities.

recovery mode An approach to incident response used when victims have been determined to be deceased or the conditions present are incompatible with survivability.

regulation A rule, issued by an executive government authority, that outlines details and procedures that have the force of law.

rescue mode An approach to incident response used when victims obviously need rescue and/or the conditions present indicate that a search and rescue operation can lead to savable victims.

rescue profile A classification of the probability that a victim will survive a given environment within a given building space or compartment. Classifications include *high rescue profile, marginal rescue profile*, and *zero rescue profile*.

risk The chance of damage, injury, or loss.

risk management The process of minimizing the chance, degree, or probability of damage, loss, or injury.

safety team When applied to live-fire training, a team comprising the designated safety officer; an experienced interior standby team made up of an officer, nozzle person, and door control person; an outside building monitor; a fuel shutoff officer or a fuel-set supervisor; and an ignition officer.

safety trigger An approach that an ISO uses to help remind firefighters to be situationally aware and to perform in a safe and predictable manner.

scratch line A preliminary control line hastily constructed as an emergency measure to check fire spread.

shadowing A training approach in which a trainee closely observes experienced mentors performing their assigned duties.

shear A force that causes a material to be torn in opposite directions perpendicular or diagonal to the material.

situational awareness (SA) The degree of accuracy to which one's perception of his or her current environment mirrors reality.

smoke The products of incomplete combustion and pyrolysis; it includes an aggregate of particles, aerosols, and fire gases that are toxic, flammable, and volatile.

smoke explosion A hostile fire event that occurs when a spark or flame is introduced into a pocket of smoke that is below ignition temperature but above some aggregate flashpoint. The result is a split-second ignition (and rapid expansion) of that pocket with no sustained burning.

soft intervention Awareness or suggestive communication made to crews or command staff that causes them to modify their observations and activities to prevent injury from a potential threat.

spalling The crumbling and loss of concrete material when exposed to heat.

spreader A seemingly decorative star or other metal plate used to distribute force over more bricks or blocks as part of an unseen corrective measure that exists inside a building.

standard A rule, procedure, or professional measurement established by an authority. To have the effect of law, it must be adopted by an authority with the legal responsibility to enact the standard as law (that is, promulgate it).

statutory law A rule of conduct in civil and criminal matters.

structural element The primary load-bearing column, beam, or connection used to erect a building.

tension A force that causes a material to be stretched or pulled apart in line with the material.

thermal protective performance (TPP) A value given to the protective (insulative) quality of structural firefighting PPE and equipment.

torching The burning of the foliage of a single tree, or small bunch of trees, from the bottom up (sometimes called "candling").

torsional load A load that is imposed in a manner that causes another object to twist.

traffic barrier An object (like a large fire apparatus) that can absorb the impact of a secondary crash to protect rescuers (often referred to as a "blocker"). These objects should be used to create a work zone, shielded from moving traffic, for rescuers.

traffic-calming strategy An effort to slow down approaching traffic: traffic cones, spotters or flaggers, arrow sticks, flashing lights, and warning signs.

training The process of learning and applying knowledge and skills.

transient heat fatigue (THF) An early warning sign that the core temperature is elevated, as characterized by the onset of physical exhaustion that is remedied by rest and hydration only to return quickly and more profoundly upon engagement with the hot environment.

truss A series of triangles used to form an open-web structural element to act as a beam (in many ways, a "fake" beam because it uses geometric shapes, lightweight materials, and assembly components to transfer loads just like a beam).

valued property Physical property whose loss will cause harm to the community.

veneer wall A decorative wall finish that supports only its own weight.

vicarious learning The process of observing others to develop knowledge, skill, or experience base.

warm zone The area surrounding the hot zone where personnel, equipment, and apparatus are operating in support of an operation.

wet line Water or a water agent sprayed on the ground as a temporary control line for a low-intensity fire or to ignite a burn-out.

wildland–urban interface (WUI) Also referred to as the I-zone; areas where homes and businesses have minimal separation from, or are interspersed with, natural-growing wildland areas.

work hardening Efforts to improve an individual's strength, flexibility, and aerobics to help prevent overexertion at incidents.

working incident An incident at which the initial response assignment will be 100% committed and more resources are needed.

zero rescue profile A classification indicating no likelihood that a person located within a fire-involved structure can be successfully rescued; in this scenario, conditions are simply not survivable.

Abbreviations

AES	alternative energy source
ALS	advanced life support
ARFF	aircraft rescue and firefighting
ASD	acute stress disorder
ASE	atypical stressful event
ASO	assistant safety officer
ASO-HM	assistant safety officer—hazmat
ASO-RT	assistant safety officer—rescue technician
BLEVE	boiling liquid expanding vapor explosion
CAFS	compressed-air foam system
CAN report	a crew report of conditions, actions, and needs
CFR	Code of Federal Regulations (as in OSHA regulations)
CMU	concrete masonry unit (also known as a concrete block or cinder block)
CNG	compressed natural gas
CO	carbon monoxide
DC	direct current
DHS	Department of Homeland Security
ED	energy drinks
EPA	Environmental Protection Agency
EPS	expanded polystyrene
FDSOA	Fire Department Safety Officers Association
FEMA	Federal Emergency Management Agency
FIRESCOPE	Firefighting Resources of Southern California Organized for Potential Emergencies
GACC	Geographic Area Coordination Center
Hazmat	hazardous material
HCN	hydrogen cyanide
HSO	health and safety officer
IAFF	International Association of Fire Fighters
IAP	incident action plan
IC	incident command *or* incident commander
ICF	insulated concrete forming
IDLH	immediately dangerous to life and health
IIC	instructor-in-charge
IMT	incident management team
ISO	incident safety officer
I-Zone	wildland–urban interface area
JPR	job performance requirement
K&T	knob and tube wiring
LCES	lookouts, communication methods, escape routes, and safety zones
LNG	liquefied natural gas
LODD	line-of-duty death
LOTO	lockout–tagout
LPG	liquefied petroleum gas
LVL	laminated veneer lumber
LZ	landing zone
MEDIC	monitor, evaluate, develop preventive measures, interventions, and communicate (general ISO functions)
MVA	motor vehicle accident
NFA	National Fire Academy
NFPA	National Fire Protection Association
NFR	national response framework
NIC	NIMS Integration Center
NICC	National Interagency Coordination Center
NIFC	National Interagency Fire Center
NIIMS	National Interagency Incident Management System (used by the National Wildfire Coordinating Group)
NIMO	National Incident Management Organization (Team)
NIMS	National Incident Management System (used by the Department of Homeland Security)
NIOSH	National Institute for Occupational Safety and Health
NOAA	National Oceanic and Atmospheric Administration
NWCG	National Wildfire Coordinating Group
OSB	oriented strand board
OSHA	Occupational Safety and Health Administration
PAR	personnel accountability report
PASS	personal alerting safety system
PCB	polychlorinated biphenyl
PFD	personal flotation device
PIA	postincident analysis
PPA	positive-pressure attack
PPE	personal protective equipment
PTE	potentially traumatic event
PTSD	posttraumatic stress disorder
PUVCE	percussive unconfined vapor cloud explosion (pronounced "poov-see")
PV	photovoltaic (solar panel)
REHAB	rest, energy nutrition, hydration, accommodation of weather conditions, and BLS medical observation and care (essential components of a firefighter rehabilitation effort)
RIC	rapid intervention crew (or company)
RPD	recognition-primed decision making
SA	situational awareness

© Photos.com

SCBA	self-contained breathing apparatus	THF	transient heat fatigue
SD	sports drinks	TIC	thermal imaging camera
SIP	structural insulated panel	TO	training officer
SO	safety officer	TPP	thermal protective performance
SOF1	Type 1 safety officer	UL	Underwriters Laboratories
SOF2	Type 2 safety officer	USAR	urban search and rescue
SOFR	NIMS compliant acronym for the safety officer assigned to incidents. Also used for an ASO working for a SOF1 or SOF2	USCG	U.S. Coast Guard
		USFA	U.S. Fire Administration
		UPS	uninterrupted power supply
SOG	standard operating guideline	URM	unreinforced masonry
SOP	standard operating procedure	VVDC	(*referring to smoke*) volume, velocity, density, and color
SPE	severity, probability, exposure (U.S. Coast Guard risk-taking model)	WMD	weapon of mass destruction
TECHREF	technical reference specialist (a member of the hazmat team)	WUI	wildland–urban interface

Index

Note: Page numbers followed by *f*, or *t* indicate materials in figures or tables respectively.